Bulk Chemicals from Petroleum

Dr. B.K. Bhaskara Rao
Professor (Retd.)
Deptt. of Chemical Engineering
Indian Institute of Technology, Kharagpur, (W.B.)

Presently Visiting Professor
M.V.G.R. College of Engineering
Vizianagaram (A.P.)

KHANNA BOOK PUBLISHING CO. (P) LTD.

Publisher of Engineering and Computer Books

4C/4344, Ansari Road, Darya Ganj, New Delhi-110002
Phone : 011-23244447-48 **Mobile:** +91-99109 09320
E-mail : contact@khannabooks.com
Website : www.khannabooks.com

Dear Readers,

To prevent the piracy, this book is secured with HIGH SECURITY HOLOGRAM on the front title cover. In case you don't find the hologram on the front cover title, please write us to at contact@khannabooks.com or whatsapp us at +91-99109 09320 and avail special gift voucher for yourself.

Specimen of Hologram on front Cover title:

Moreover, there is a SPECIAL DISCOUNT COUPON for you with EVERY HOLOGRAM.

How to avail this SPECIAL DISCOUNT:

Step 1: Scratch the hologram
Step 2: Under the scratch area, your "coupon code" is available
Step 3: Logon to www.khannabooks.com
Step 4: Use your "coupon code" in the shopping cart and get your copy at a special discount
Step 5: Enjoy your reading!

Price: ₹ 450.00

ISBN: 978-93-86173-99-7

Bulk Chemicals from Petroleum
by Dr. Bhaskar Rao

First Edition: 2019
Reprint: 2022, **2025**

Published by:
Khanna Book Publishing Co. (P) Ltd.
CIN: U22110DL1998PTC095547

Visit us at: **www.khannabooks.com**
Write us at: **contact@khannabooks.com**

Printed in Bharat.

Copyright © Publishers

No part of this publication may be reproduced, stored in a retrieval system or transmitted, in any form or by any means, electronic, mechanical, photocopying, recording or otherwise without prior permission of the publisher.

This book is sold subject to the condition that it shall not, by way of trade, be lent, re-sold, hired out or otherwise disposed of without the publisher's consent, in any form of binding or cover other than that in which it is published.

Disclaimer: The website links provided by the author in this book are placed for informational, educational & reference purpose only. The Publisher do not endorse these website links or the views of the speaker/ content of the said weblinks. In case of any dispute, all legal matters to be settled under Delhi Jurisdiction only.

Preface

Petroleum is a source of host of chemicals termed as petrochemicals. Chemical industry includes basic chemicals and its product, petrochemicals, fertilizers, paints & varnishes, gases, soap and detergent, perfumes, pharmaceuticals and covers thousands of products, which are finding use in our daily life from industrial to household goods. Various products are finding use in various fields packaging to agriculture, automobiles to telecommunication, construction to home appliances, health care to personal care, explosive, pesticides to fertilizer, textile to tire cord, chemicals to pharmaceuticals .

This book contains 13 chapters covering petrochemical industry feedstocks, chemicals derived from C1, C2, C3, C4 & higher hydrocarbon atoms, synthesis gas & chemicals and petroleum aromatics. Besides, it contains comprehensive informations pertaining to polymers which include plastics, synthetic fibers & elastomers and synthetic detergents.

This book will serve as as reference material for the students , teachers and practicing engineers in the field of chemical, petroleum and petrochemical engineering.

Suggestions for further improvement are welcome.

B.K.B.Rao

Contents

CHAPTER 1

Petrochemical Industry—Feed Stocks

1.0 OVERVIEW

Table 1.1 (a)

IPCL PRODUCTS

AROMATICS

A	**Xylenes**	
	Process Licensor : KRUPP KOPPERS, FRG.	
	Products	Licensed capacity (MTA)
	Ortho and mixed xylenes	23,500
	Paraxylene	17,000
	Solvent CIX	10,000
	Hcpton	20,000
B.	**DMT**	
	Process Licensor : DYNAMIT NOBEL, FRG	
	Products	Licensed capacity (MTA)
	Dimethyl terephthalate	40,000
	Methyl benzoate	300
	Dimethyl isophthalate (impure)	200
	OLEFINS	
A.	**NAPHTHA CRACKER**	
	Process Licenser : LUMMUS CO., UK	
	Products	Licensed capacity (MTA) Present capacity 3 lakh ton
	Ethylene	1,30,000
	Propylene (polymer grade)	38,000
	Propylene (chemical grade)	43,300
	Carbon black feed stock	17,900
B.	**BUTADIENE EXTRACTION**	
	Process Licensor : UNIVERSAL OIL PRODUCTS, U.S.A.	
	Products	Licensed capacity (MTA)

	Butadiene	22,000
	C_4 Raffinate	8,200
C.	**AROMATICS EXTRACTION**	
	Process Licensor : UNIVERSAL OIL PRODUCTS, U.S.A.	
	Products	Licensed capacity (MTA)
	Benzene	23,600
	Cixon	12,000
D.	**PYROLYSIS GASOLINE HYDROGENATION**	
	Process Licensor : INDIAN INSTITUTE OF PETROLEUM, DEHRADUN, INDIA; IFP, FRANCE,	
	Products	Licensed capacity (MTA)
	Light pyrolysis gasoline	29,300
	Heavy pyrolysis gasoline	34,600
	ACRYLONITRILE	
	Process Licensor : SOHIO, U.S.A.	
	Products	Licensed capacity (MTA)
	Acrylonitrile	24,000
	Acetonitrile	900
	Hydrocyanic acid	3,600
	POLYBUTADIENE RUBBER	
	Process Licensor : POLYAR INTERNATIONAL S.A., SWITZERLAND	
	Product	Licensed capacity (MTA)
	Polybutadiene rubber (Cisamer)	20,000
	LINEAR ALKYL BENZENE	
	Process Licensor : UNIVERSAL OIL PRODUCTS, U.S.A.	
	Products	Licensed capacity (MTA)
	Linear alkyl benzene	43,500
	Poly alkyl benzene	3,500
	Heavy n-paraffins	25,000
	Ceton	16,000
	L.R. 2030	15,000
	POLYPROPYLENE HOMOPOLYMER	
	Process Licensor : MONTEDISON, ITALY	
	Products	Licensed capacity (MTA)
	Polypropylene homopolymer	
	(Koylene)	30,000
	Atactic Polymer	3,000

	ETHYLENE GLYCOL	
	Process Licensor : SCIENTIFIC DESIGN CO., U.S.A.	
	Products Ethylene glycol	Licensed capacity (MTA) 20,000
	Polyethylene glycol	2,400
	Ethylene oxide	5,000
	Diethylene glycol	2,500
	Triethylene glycol	250
	LOW DENSITY POLYETHYLENE	
	Process Licensor : AQUITAINE TOTAL ORGANIC CHIMIE	
	(ATO CHIMIE), FRANCE	
	Product	Licensed capacity (MTA)
	Low density polyethylene (Indothene)	80,000
	MONOCOMPONENT ACRYLIC FIBRE ACN	30,000
	Process Licensor : ASAHI CHEMICAL CO., JAPAN	
	Product	Licensed capacity (MTA)
	Monocomponent acrylic fiber	
	(Indacrylon)	12,000
	VINYL CHLORIDE / POLYVINYL CHLORIDE	
A.	**EDC/VCM**	
	Process Licensor : STAUFFER CHEMICAL CO. U.S.A.	
	Products	Licensed capacity (MTA)
	Ethylene dichloride	1,00,000
	Vinyl chloride	57,300
B.	**PVC**	
	Process Licensor : B.P.GOODRICH, U.S.A. Product	Licensed capacity (MTA)
	Polyvinyl chloride (Indovin)	55,000
	ACRYLATES	
	Process Licensor : NATIONAL CHEMICAL LABORATORY, PUNE, INDIA	
	Products	Licensed capacity (MTA)
	Methyl acrylate	2,000
	Ethyl acrylate	3,000
	Butyl acrylate	4,000
	2-Ethyhexyl acrylate	1,000
	Ammonium sulphate	12,000

	PETROLEUM RESIN	
	Process Licensor : CdF CHIMIE, FRANCE	
	Product	Licensed capacity (MTA)
	Aromatic hydrocarbon resin (Petrez)	5,000

1.1 PLAY OF TRANS NATIONAL CORPORATIONS (TNC)

Biggest boost to the petrochemical industry given by IPCL is equally supplemented by the Trans National Corporations with active collaborations and new ventures. India's march in the last two decades in different fields has been justified due to the permeation of petrochemicals into all walks of life. The large necessities of imports to face the race in petrochemical field has generated an urgent wave of establishing various petrochemical industries.

The Sixth Plan period has barged to establish a firm footing in oil and petrochemical industries. The Seventh Plan witnessed a huge leap in petrochemicals displacing the oil industry to second place. The most important industry in this plan is the gas industry. The biggest gas cracker complex in Maharashtra (11,670 million rupees) (see Fig. 1.1 (*b*)) is a right beginning in this direction, under the surveillance of IPCL. Ten major gas based fertilizer units along with the self supporting energy schemes and transportation of Bombay High-Bas-sein gas to Jagdishpur (HBJ-gas line) are some of the major ventures in this Plan.

HBJ Gas-Grid (Fig. 1.1.c)

Associated gas comes out along with oil. The quantity of gas dissolved depends upon the saturated pressure of the reservoir. Commonly about 500 SCF dissolves in a barrel of oil under a pressure of 2500 psi. when the production of oil has to be stepped up naturally the associated gas quantum also increases. During the initial stages of Bombay High development there were no infrastructure facilities to consume the gas hence it was flared. During 80-90 the quantum of gas flared was enormous. About two decades back gas was ranked second to oil, but today the thinking has changed. Right from a boiler fuel to petrochemical feed stock and a clean fuel for automobiles; the utility and serviceability of gas has brought increase its creditability to No. 1 fuel. Today GAS Refinery is a reality. The enhanced gas consumption during last decade and ever increasing demand is a real meter to show how it is being regarded as most important material in daily life. GOI has started looking for Hydrocarbon deposits throughout the country. The results of which were significant. Rajasthan, Gujarat, Tripura, Palk straits etc. are some of the examples. KG basin discovery in the late 80s brought significant change in the hydrocarbon activity in this region.

To successfully utilise the enormous amount of associated gas from Bombay High, a gas processing plant at Hazira was commissioned and this sweetened gas is transported through a pipe line known as Hazira-Bijapur-Jagdishpur Pipe line (HBJ). The length of this pipe line initially was 1700km, (presently it is 2300km long) and in the second phase was able to transport 20MMCMD gas. The gas is supplied to ten fertiliser plants, and power house for generation of 4000MW besides to a gas cracker plant at Gandhar and to IPCL. With branch line drawn up to Kanpur, Faridabad and Salimpur Petrochemical complex (originally planned) if works in full capacity; the line can transport 33 MMMD of gas.

KG basin is the next biggest Off-Shore activity in India, approximate reserves are 7 trillion CM, With the entrance of Reliance Industries with the partnership of Niko Resources of Canada into this basin remarkable achievements were attained. With the success in Gas discovery in M-l and H-l wells it has plans to produce 45MMCMD of gas from its structure D6 (The reserves of which are 1.45 trillion cubic meters gas). Presently 60,000 to one lakh cu.meters of gas is being supplied to Power house in Kovvur 80Km away from KG basin and other gas consumers. The future of gas is secure in KG basin and will provide gas to Nagarjuna Fertilisers and other gas based power houses when the production reaches one million cu. meters.

In a remarkable turnaround, Hazia Gas Processing Complex of Oil and Natural Gas Corporation Ltd. (ONGC) has reached a throughput level of 28 million metric standard cubic metres a day (mmscmd) of Gas to the HVJ Pipeline, 70 per cent of its pre-shut-down level. Domestic gas supplies may be available from the KG basin from D-6 blocks of RIL in 2008-09. On the other hand the supply projected by ONGC in the Xlth plan period is likely to witness fall from 47.28 MMSCMD in 2007-08 to 41.08 MMSCMD in 2011-12., which is a concern for fertilizer units. However the supply from private players/JVs is expected to increase from 23.26 MMSCMD to about 57.22 MMSCMD in 2011-12. DGH has projected additional supply of 20, 30 MMSCMD and 40 MMSCMD from RIL fields in 2009-10, 2010-11 and 2011-12, respectively, and 54 MMSCMD from GSPC in each of the above years. This shows that the gas supply will be 70.54 MMSCMD in 2007-08, and is expected to grow to 192.30 MMSCMD in 2011-12. The demand-supply gap projected in the Xlth Five-Year Plan period clearly indicates the need for imports. —LNG supply will reach a level of 23.75 MMTPA by the year 2011-12. Though there is a possibility of availability of gas from Iran through Pakistan and Myanmar-India pipeline another 30-40 MMSCMD in 2010-11 may be available.

HBJ pipeline project was conceived in year '82. This pipeline of 1758 km long at a cost of 1760 crores of rupees is capable of transporting 30 million Cu. meters of gas at peak demands. Presently, the gas production of Bombay High South Bassein and satellite platforms amount to 26 M^2 cu meters a day, the present processing capacity of Hazira terminal; which in two or three years will touch a new high mark of 50 M^2 CMD.

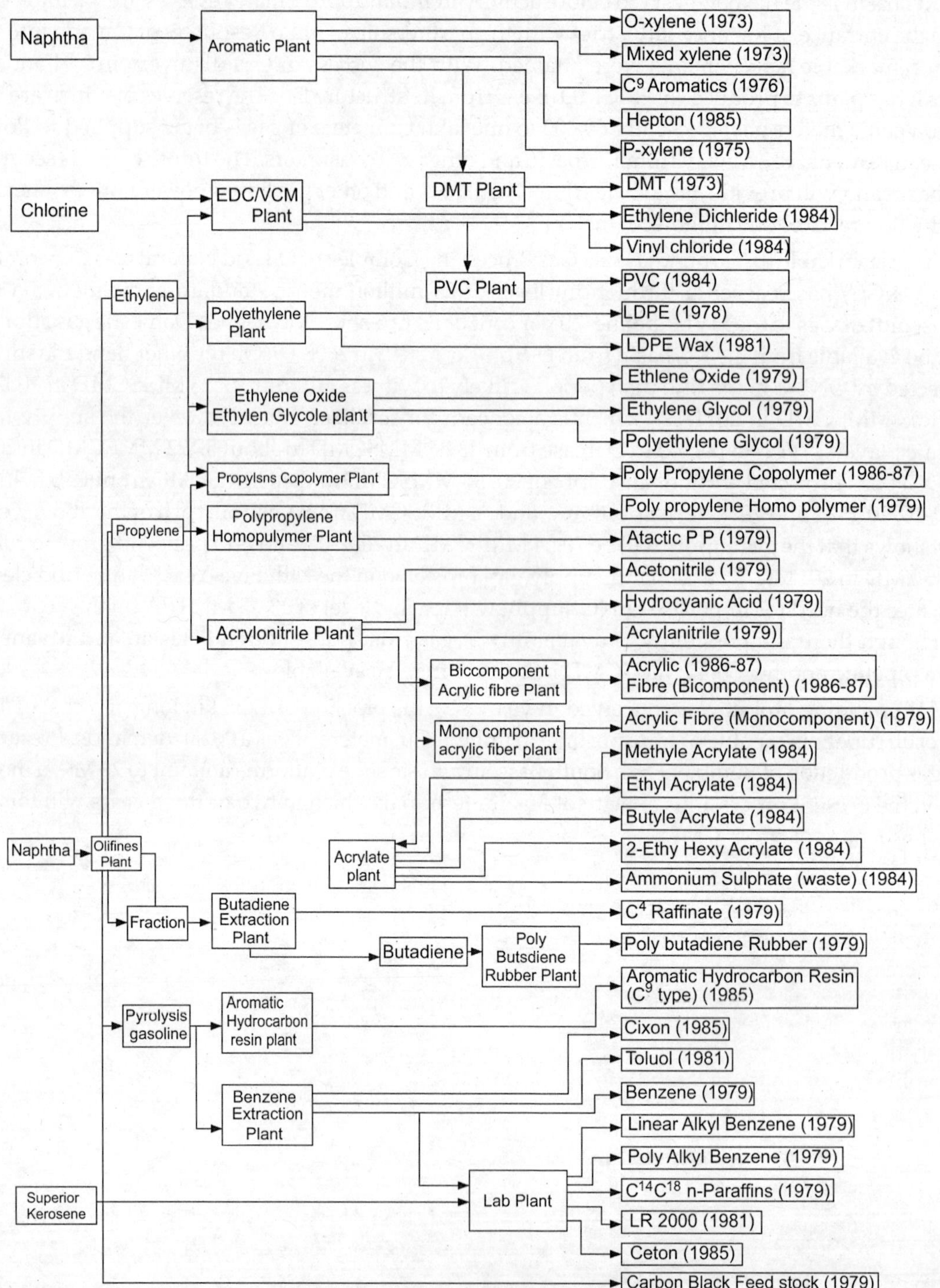

Fig. 1.1 (a): Flow Sheet of IPCL-Products

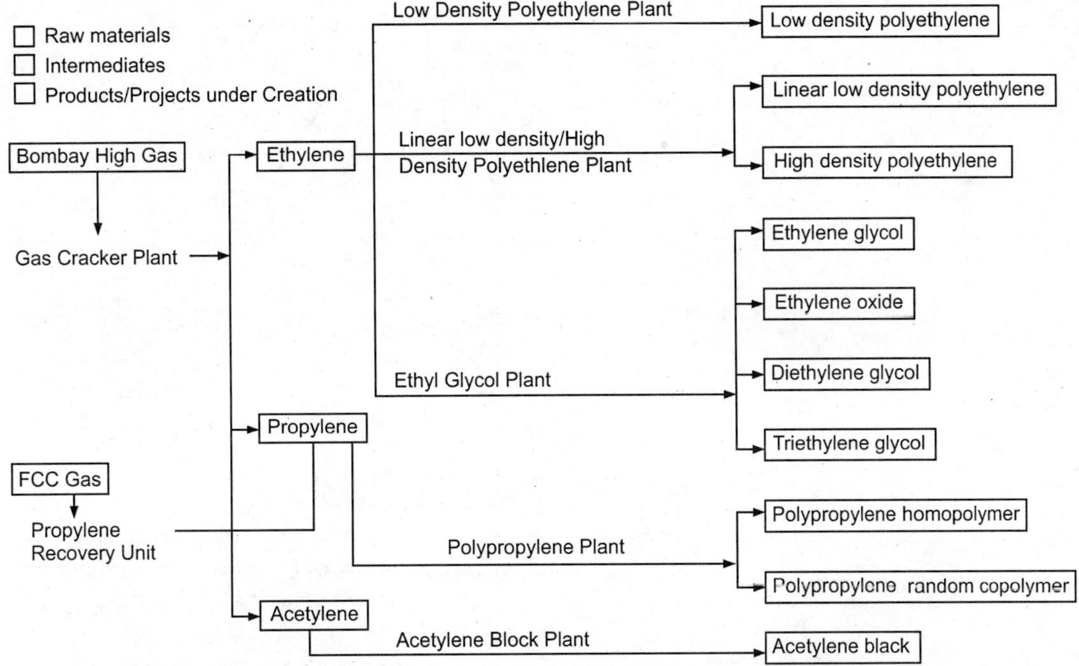

Fig. 1.1 (b): Maharashtra gas cracker Plant-Operational Schemes.

The consumption of gas by 6 fertilizer plants, petrochemical units, power and other industries will roughly need a 30-36 M^2 CMD of gas which may thus permit about 20 M^2 CMD gas to south zone, if pipeline connections are laid. Thus the total needs of south India amounting to 20 M^2 CMD for fertilizer, thermal power plants and other units can be safely met. K.G. and Caveri basins are likely to yield about 6.5 and 3.5 M^2 CuM gas respectively per day by 1994-95. Fig. 1.1(*c*) shows HBJ pipeline.

This is in congruence with new alignment policy of the government with respect to increasing gas and oil strikes. Presently, the production of gas on all the fronts is around 60 M^2. Cu. M per day (1995) and 30% of the gas is still flared. Government had set up a working group on petrochemicals for the Seventh Plan to work out demand projections for various petrochemicals and their indigenous production (shown in Table 1.1*b*). Consumption of petrochemicals were 3.28 lakh tonnes in 1982-83 and increased at an average annual rate of 5% till 1990, it is likely to increase further. The anticipated demands for olefins would be met by Maharashtra Gas Cracker Complex and another one in Gujarat. Three aromatic recovery units one each in Kerala, Maharashtra and U.P. during Sixth Plan would supply the BTX streams. Salempur Petrochemical Complex has been already initiated and it is based on naphtha from Mathura Refinery.

Salempur project likely to produce (Investment '625 × 10^7).

Benzene	1.6×10^5 tonnes
O-xylene	53×10^3 tonnes
p-xylene	1.53×10^5 tonnes
Alkyl-benzene	17.5×10^3 tonnes
Dimethyl tere phthalate	1.5×10^3 tonnes
Cyclohexane	90×10^3 tonnes

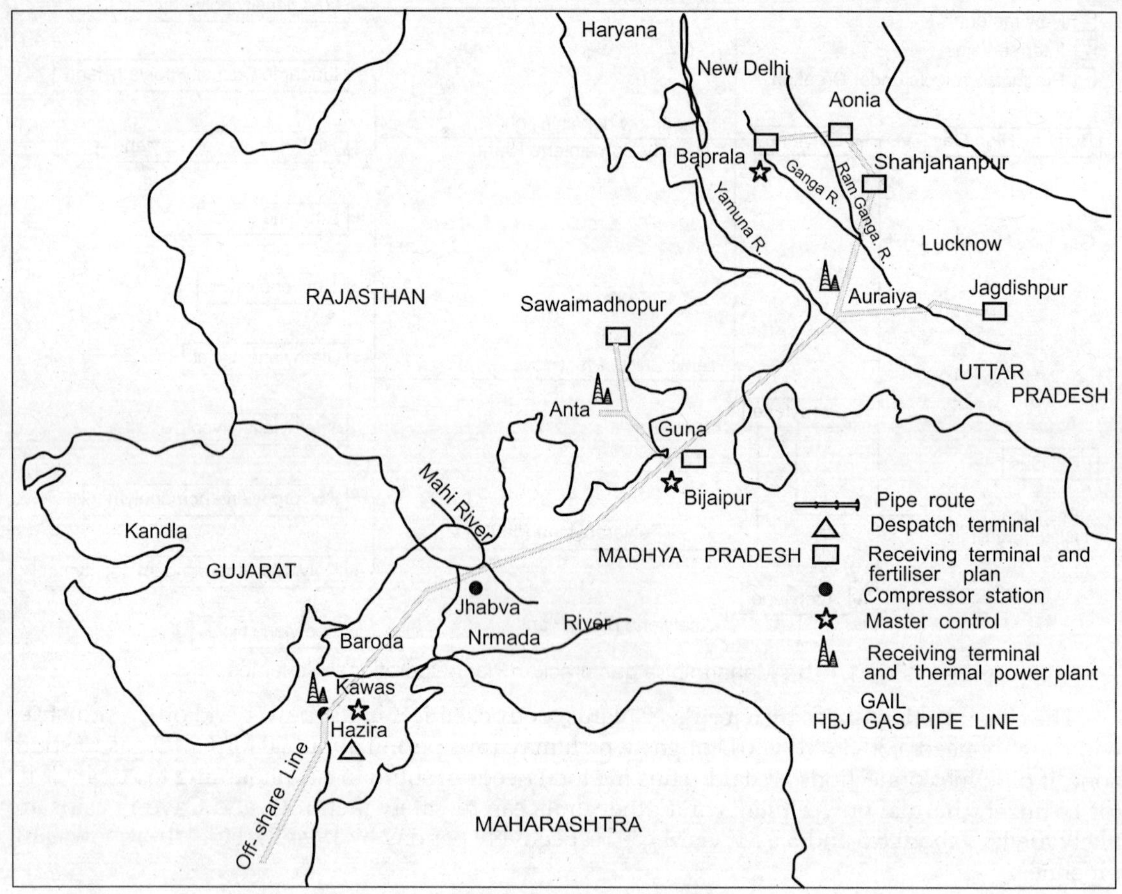

Fig. 1.1 (c)

The picture is not the same always; for instance the utilisation of full capacity in some cases for PVC, LDPE is remote. Hence the production, demand and consumption has to be linked always.

Fast advances elsewhere would cause dismay at the technology and the uneconomical methods adopted in our country. These developments indicate formidable odds, which India has to reckon with while continuing the production on the basis of imported technology. Thus, keeping track of advances on the subject and concerned technology is essential to relegantly face the march. Revamping, retrofitting of the existing complexes by jettisoning old technology and or by adopting assured technology on the basis of feed-stocks availability are some of the prime essentialities of the pollution free industrial future.

The major industries have been surveyed in depth and can be divided into four broad groups such (2) as :

(*a*) Basic bulk chemicals

(*b*) First generation intermediates

(*c*) Second generation intermediates, and

(*d*) End products.

Table 1.1 (b): Physical Targets in Seventh Plan

Plastics (thousand tons)	1984-85	1989-90
LDPE	110	211
HDPE	39	113
Poly-propylene	26	103
PTFE	0.5	—
PVC	80	197
Ethylene Oxide	30000	50000
Acetone	40000	125000
Caprolactum	60000	130000
Butadiene	50000	65000
LAB	80000	120000
Poly Styrene	19.6	331.6
SBR	41	71.5
PBR		
NBR	5	—
ABS	10	—
Synthetic Fibres		
Nylon filament yarn	30	70
Polyester staple and filament	95	191
Moulding powders		
PF	4.6	
MF	3.5	
UF	3.5	
CA	0.55	

Petrochemical industry is dominated by two dozen TNCS some of them being petroleum companies. The design and construction are strict monopoly of these TNCS and developing countries are at the mercy of such companies for know-how.

1.1.1 Plastics and Synthetic Fibre Industry

Plastics as well as synthetic fibres development is a post war activity. In case of plastics, the real starting commenced with establishing a synthetic rubber plant at Bareilley (1962, product SBR 30,000 tonnes). The raw material used was alcohol and the same old technology is continued till now. Soon ICI started manufacturing polyethylene; again based on alcohol; a polystyrene plant was also commissioned by Hindustan Polymers now known as LG Polymers at Vizag using alcohol as feed material. This shows the early plastic industry was based upon the then available alcohol raw material. Even today, India is one of the largest producers of alcohol (1300 million litres). The new industries like NOCIL, Hindustan Organics, Poly Olefins, etc. were based on the petroleum feed stocks. Some of the earlier plastics, like Urea-phenol/formaldehyde combinations, are even today operating on the same technology. However, the new industries are clearly shifting towards petroleum feed stocks. Establishment of other petrochemical industry has been a great fillip to the growth of plastic industry. Initially, two petrochemical industries, Union Carbide and NOCIL were set up with a limited capacity based upon naphtha. IPCL was the real institution to bring about a transformation in the country by familiarising the plastics, like LDPE, Polypropylene, PVC, Methacrylates. Though engineering plastics and speciality polymers have not been developed in par with the progress, the diffusion of plastics into all walks of life, has made them essential ingredients of the modern industry. Presently, consumption of plastic raw materials is a meagre 4.5 lakh tonnes virgin and 75,000 tonnes reprocessed; the target for 1990 is fixed at one million tonnes. Poly Urethane-foams (present production 5000 tpa) a leading plastic has not been fully given importance till now. Production of isocynates has not been given any importance at all. In 1987-88, about 66,000 tonnes of SBR was produced for which 210 million litres of alcohol was required.

1.1.2 Synthetic Fibres

Except cellulose fibres, all the synthetic fibres are based upon the petroleum feed stocks and these industries are relatively new. In fact the activity of these would be touching all time high at the end of Seventh Plan. Even today, the consumption of cellulosic fibres tops as shown here.

Consumption of fibres	*1983*	*1984*	*1985*
Total (10^3) (tonnes)	1,518	1,547	1,672
Cotton	1,341	1,371	1,500
Viscose Rayons	151	116	105
Polyester	39	40	45
Acrylic	12	15	17
Miscellaneous	5	5	5

Synthetic Fibres (Production) 1989-1990

Cotton	2.8 lakh tons
PSF (Filament)	1.6 lakh tons
Fibre	1.1 lakh tons
Acrylic	32,000 tons
Nylon	37,726 tons
Polypropylene	911 tons
Viscose (filament yarn)	46,000 tons
Cellulosic fibre	1,92,435 tons
Total man made fibres/yarns	4,40,000 tons

Production of Fibres (In TMT)

Group	*2002-03*	*2003-04*	*2004-05*	*2005-06*	*2006-07*
Synthetic Fibres	1669	1755	1868	1875	1906
Fibre intermediates	2417	2648	2691	2851	2963

Thus, India has a distinction of being one of the countries with the lowest per capita consumption of fabrics (14.8 metres/head).

The growth of polyester staple fibre in India has not been in a significant manner, because of the policy of government *i.e.* classifying these fibres meant for affluents only and raising the cost. However, the folly has gradually disappeared and it is ultimately realised that synthetic fibre industry has to go along with natural fibres and cannot be dispensed with. Presently, there are 7 nylon plants, and 5 polyester plants. One acrylic fibre plant would soon be joining the national stream. The capacity in 1985, was 54,200 tonnes of polyester. In 1996, installed capacity polyester fibre is 4.95 lakh tonnes.

Existing Units	*Capacity production (Thousand tonnes)*		
	1985	*1987*	*1990*
IEL (CAFI Division)	10	16	30
Indian Organic	12	18	30
Swadeshi Polytex	12.62	12.2	15
Calico	7.6	15	15
J.K. Synthetics	12.2	18	30
New Units			
Bongaigon		30	30
India Polyfibres		15	25
Orissa Synthetics		15	25
Madhya Pradesh Fibres		15	25
Punjab Polyfibres		15	25
Reliance Industries		45	45

1.1.3 Rubbers

Though Indian Natural Rubber Industry is century old, synthetic rubbers industry has not yet come out of ages. Synthetic rubbers based on butadiene, which is mostly derived from alcohol till now are the only rubbers produced. However, the butyl rubber, though the potentiality of the stock is about 89,000 tpa as shown below; is yet only 25,000 tpa produced.

	i-Butylene available
IPCL	9,000 tonnes
Koyali	11,000
HP-Bombay	9,000
Cochin Refinery	12,000
Mathura Refinery	11,000
Madras Refinery	7,400
BPCL	12,000
Panipat	10,000
NOCIL	4,000
WBIDC, Haldia (not yet begun)	7,000
Vizag Refinery	7,400
Total	89,800 tpa

Natural rubber production in the country is about 1.86 lakh tonnes in 1985 and there is no hope of further increase. However, the expansion of refineries specially by instituting FCC-units, the production of butylenes has increased, that would solve the production of butyl rubber.

1.1.4 Nitrogen Industries

With reference to the present context, nitrogen industries are mainly regarded as ammonia and urea industries i.e. fertiliser, being large tonnage product. The first fertiliser unit (ammonia was built by Maharaja of Mysore, as an essential step of industrialisation of Mysore), soon Maharaja of Travancore followed by establishing a unit based on wood gasification. This continued till the post independence period, and the industrial activity started from 1950 and continued with rapid strides. Started at 60,000 tonnes in 1951-52 the production of nitrogenous fertilisers rose to 5.235 x 10^6 tonnes in 1984-85 and expected to cross ten million tonnes shortly. The first coke based fertiliser plant at Sindri became the center of activity of R and D plans under Fertiliser Corporation of India Ltd. Initial capacity of the ammonia and urea plants were of 600 tonnes each. These gradually expanded to 900 tonnes and 1350 tonnes per day of ammonia and urea, but today 1350 tonnes ammonia and 1500 tpa urea plants are under construction. Presently, there are 41 units producing 5.235 million tonnes of nitrogen (NPK) fertiliser. R and D wing of FCI and Projects and Development (India)Ltd., (which was a part of FCI till 1978) are actually involved in the creation of fertiliser producing capacity ; other R and D wing of FCI being FACT. The requirement of food grains by 2000 AD when the population reaches around a billion is about 236 M^2 tpa (food grains production in 1997-98 was about 199.3 M^2T ; in 1998-99 it was about 200 M^2T, 231 M^2T in 2007-08.

The development of the industry systematically scanned for various raw materials and still continuing. Coal gas, coal-coke, lignite, are the chosen raw materials in early sixties, afterwards naphtha reforming and fuel oil partial oxidation technologies have been introduced. Presently, natural gas is envisaged for the production of fertilisers. Dearth of oil shortage and abundant low grade coals have forced to go for coal gasification units, the results of which took the shape of Ramagundam and Talcher fertiliser units. The success of gas strikes in different parts of the country, not to speak of existing Bombay High-Bassein and Cambay sites, have prompted to go for gas based fertiliser units, the result of which is Hazira-Bijapur-Jagdispur (HBJ) pipe line to transport the natural gas from Bassein to Thal Vaishet plants in Maharashtra and Hazira in South Gujarat. This pipe line would link 10 gas based fertiliser units with the supporting electric power units. Gujarat Narmada Valley Fertiliser (GNFC)[5.8] unit at Baruch is the biggest (1800 tpd) single train urea plant in India based on gasification of fuel oil. Similarly, Hazira plant is the biggest urea producing plant in the world. The plans, if successful would utilise only 80% of the gas and probably 20% of the gas would be flared.

The planning of these fertiliser units is based upon the available raw materials and the prominence has been shifted from one feed stock to the other in due course of time, Naphtha gas as well as LSHS are equitable but from 1990 onwards gas was major feed-stock for nitrogenous as well as petrochemical industries. Upgrading the older plants and integrating with the modern technology and advanced instrumentation to save energy and increase efficiency and production have been also given highest priority. Presently, several fertiliser plants are producing about 80% of the capacity, though a few plants warrant substantial improvement in working.

In 1998-99, 78 million cubic meters of gas per day has been produced. The demand for nitrogen fertilizer will be crossing 10 million ton by 2000 AD. The capacity installed till 1997 was 9.8 million tons and the production was about 9.6 million tons. The fertiliser units working on various feed stocks and the trend in share of feed stocks can be seen from Table 1.1(*d*). By the end of the Eighth Plan, plants working are shown in the Table 1.1(*e*). The utilisation of installed capacity in public sector and other sectors is shown in the following Table 1.1.

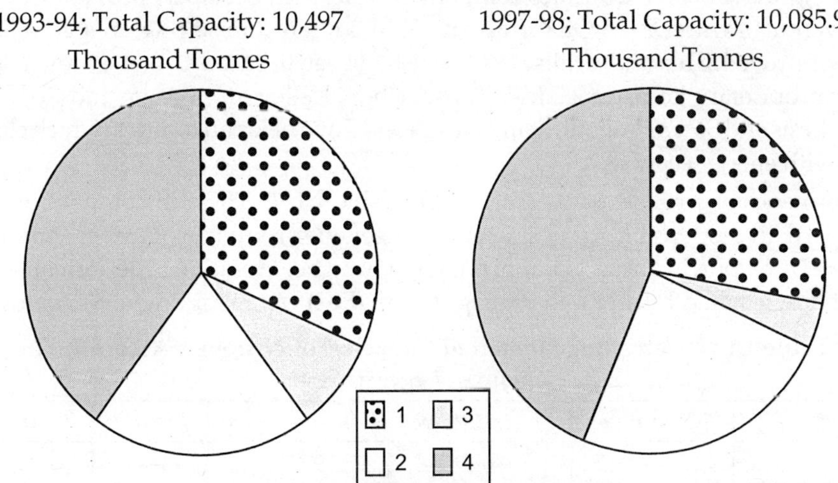

1 : Vlable units 2 : Loss making units 3 : Co-operatives 4 : Private sector

Table 1.1 (c): Installed Capacity of Nitrogen Fertilisers in thousand tonnes

	1993-94	1997-98
	Public Sector	
Viable	3306	3041.6
Loss making	1011	299.3
Co-operative	2268	2310.7
Private	3910	4434.3
Total	**10,497**	**10,085.90**

Similarly the Ministry of Fertilisers and Chemicals has outlined the necessary targets of production and consumption of fertilisers during Ninth Plan and projected outlay at 11,448 crores of rupees. This would enhance the nitrogen fertiliser capacity installed from the 9.77 MMTPA to 14 million tons. This all depends upon the availability of gas in, Western and KG Basin and regular supply to fertiliser units. Thus in Western sector Aonla, Bijapur and Jagdishpur could not run to full capacity. With the KG basin now able to produce 3 MMCUMPD the two fertiliser units at Kakinada may touch full capacity. The main objective of Ninth Plan will be to ensure regular feed stock availability at steady prices and locating the producing units in deficit areas. Thus new projects to be implemented are :

FACT's replacement of ammonia plant at Udyogamandal with a capacity of 2.97 lakh tons, GSFC's ammonia plant at Baroda with a capacity of 4.45 lakh tons, IFFCO's Kandla expansion project 3.7 lakh tons of NPK and 2.27 lakh tons of di-ammonium phosphate, Chambal Fertiliser and Chemicals unit at Gadepan with an expansion to 7.75 lakh tons of urea, Oswal Chemicals, and Fertilisers Paradeep project of 15 lakh ton of diammonium phosphate are almost ready and are likely to work by the end of 1999. Again KRIBHCO's proposal for third ammonium plant to produce 7.26 lakh tons of ammonia based fertilisers and another new plant at Gorakhpur with identical capacity Rashtriya Chemicals and Fertiliser expansion project at Thal Vaishet for another 7.26 lakh tons urea. SPIC and Jordan Phosphatic Mines company have already set-up a phosphatic unit in Jordan. Similar joint ventures with Oman Oil Company; SPIC's interest in setting up a gas based plant in Iran has been approved. Thus the fertiliser industry had been licensed with the proven technologies from the four proprietary licensors Halder Topse, Uhde, Kellog and Brown and Roots of USA for ammonia synthesis and Snam Progetti, Stamicarbon and Toyo of Japan for urea; which claim superior technologies with energy economy.

The performance of fertiliser units is shown in Table 1.1(*d*). The stiff competition from liberal imports was partially responsible to cut short the indigenous production. Further some of the public sector units like Durgapur, Namrup, Barauni (HFC) showed deterioration in performance. Gorakhpur unit was shut down, while Talcher and Ramagundum (FCI) were working under below capacity.

Table 1.1 (*d*) Percentage Installed Capacity[8] of Nitrogen According to Source of Feedstock

Period	*Naphtha*	*Natural gas*	*Electric Power*	*Coke-oven gas*	*Lignite*	*Coal*	*Fuel oil*
1960	-	-	0.8	99.2	—	—	—
1970	65.3	10.2	6.0	13.3	5.2	—	—
1980	51.7	13.0	1.7	1.7	—	9.9	19.6
1984	48.0	14.2	1.5	1.3	—	8.8	22.6
1990	27.4	45.8	0.9	0.8	—	4.9	14.2

Table 1.1 (e) Capacity of N and P Plants Factory-wise[8] and the Expected Date of Commissioning (7th Plan Period)

(Thousand tons)

	Plant		*Capacity N*	P_2O_5	*Date of Commissioning*
1.	RCF, Thal, Vaishet (2 Plants) (Maharashtra)	Plant I Plant II	341.5 341.5	— —	April '85 July '85
2.	MCFL, Mangalore Expn. Karnataka		25.0	63.0	April '87
3.	KRIBHCO, Hazira, Gujarat	Plant I	334.0	—	Dec. '85
		Plant II	334.0	—	April '86
4.	Hindustan Lever, Haldia (West Bengal)		29.0	73.0	Dec. '85
5.	HFCL, Namrup III, Assam		152.0	—	Oct. '86
6.	HFCL, Haldia, West Bengal		152.0	75.0	April '86
7.	Paradeep Phosphates, Paradeep (Orissa)	Phase I Phase II	58.5 58.5	150.0 150.0	April '86 Jan. '98
8.	Godavari Fertilisers and Chemicals, Kakinada (A.P.)		54.0	138	Jan. '88

9.	GSFC, Sikka (Gujarat)		59.0	150.0	Oct. '87
10.	SPIC, Tuticorin (Retrofitting)		19.0	48.0	Jan. '87
11.	NFL, Vijaypur (Dist. Guna), MP		334.0	—	Dec. '87
12.	Indo-Gulf Fertilisers and Chemicals, Jagdishpur (Dist. Sultanpur), UP		334.0	—	June '88
13.	IFFCO, Anola (Dist. Bareilly), UP		334.0	—	April '88
14.	Aravali Fertilisers, (Dist. Sawai Mad-hopur), Rajasthan		334.0	—	Jan. '89
15.	Tata, Babrala (Dist. Bedaun), UP		334.0	—	July '89
16.	Caparo Group (Vlth gas based Plant), Shahjahanpur, UP		334.0	—	Oct. '89
17.	Nagarjuna Fertilisers, Kakinada (A.P.)	Ammonia Urea	1900 tpd 3000 tpd	—	1998 1998

1.1.4.1 Technology-TNCS

There are at present eight major technologies available on gas based fertiliser units. They are Halder Topsoe, ICI, Kinetic Technology International, Humphreys and Glasgow, Kellog Lummus, C.F. Braun, Snam Progetti. All these technologies are directed at maximising[9] energy saving and elimination of pollution. Energy saving may be an outcome of choosing active catalysts enabling the operation at low temperatures and less pressures, or the reactor modification etc. and successful utilisation of energy saved. In nitrogen fertilisers urea will account for 84% and the share of production in different sectors is as shown below:

	%	N_2 *share in total capacity*
	1984-85	*1989-90*
Public Sector	57	49
Private Sector	34	35
Cooperative Sector	9	16

The consumption of fertilisers and the feed stock distribution and the addition of plants, technology, that took place during these years are shown in Table 1.2. Some of the major fertiliser units capacities feed stocks etc. are presented.

1.1.4.2 Fertilisers

The consumption of fertilisers (NPK) has been stagnant around 12 million tonnes between 1990-91 and 1993-94. Subsequently, it increased to reach the level of 14.3 million tonnes in 1996-97. There were changes in the policy on fertilisers in the 1990s. Phosphatic and potassic fertilisers were decontrolled in August 1992. Only urea (nitrogenous fertiliser) continued to be under the price control system and involves a heavy subsidy for keeping the farm gate prices low. Consequent to decontrol in 1992, prices of phosphatic and potassic fertilisers rose sharply. The gap between the controlled price of urea and the decontrolled prices of phosphatic and potassic fertilizers increased leading to an imbalance in their use. Against the optimum N, P and K ratio of 4 : 2 : 1 and an actual ratio of 5.9 : 2.4 : 1 just before the decontrol, the ratio during the end of the Eighth Plan worked out to 10 : 2.9 : 1. It may be, however, noted that this ratio differs from area to area.

The consumption of fertilizers (NPK) during 1996-97 was 14.31 million tonnes. The total nutrient requirement in the terminal year of the Ninth Plan is projected at 20.0 million tonnes. During the Ninth Plan greater use of biofertilisers and bio-technological research in this direction will be encouraged. To support the programme for achievement of a target of 234 million tonnes of foodgrains, in the terminal year of the Ninth Plan, a very strong network of all associated activities would be required to be created in the country. This would include, inter-alia, the Integrated Plant Nutrient Management system, use of organic sources and bio-fertilisers, use of legumes by farmers for generating and sustaining the inherent nutrient potentiality of soils, followed by application of chemical fertilisers. Programmes of campaigning for making the farming community aware of micro-nutrient deficiencies and their use would be launched.

1.1.4.3 Some Important Fertiliser Units

As of February 2000, the installed capacity is around 11.07 mtpa of nitrogen and 3.6 mtpa of phosphate nutrients, making India the third largest fertilizer producer in the world. The country has 64 large size fertilizer plants besides 79 medium and small single superphosphate units. The country is self-sufficient to the extent of 92% in nitrogen nutrients. However, in the case of phosphates, the availability of domestic raw material is a major hurdle. Roughly 66% of the phosphatic fertilizer requirement is met through domestic production based on either indigenous/imported rock phosphate and imported sulphur or by imports of intermediates, viz., ammonia and phosphatic acid. Since indigenous rock phosphate supplies meet only 5-10% of total requirement they are usually imported. Chambal Fertilizers and Chemicals Limited (CFCL), has the capacity to manufacture 4500 tonnes of urea per day. The company is ideally located close to the fertile belt of Uttar Pradesh, Punjab and Haryana. Though urea is the mainstay of the company, it has diversified into many related/unrelated businesses such as trading in phosphatic fertilizers, manufacturing of potato seeds, software, housing finance and textiles.

Gujarat Narmada Valley Fertilizers Co Ltd.

Urea, ANP and CAN are leading players as the prices boost the profit due to decontrolling by the Government. However the other organic chemicals manufactured are acetic, nitric acids and methanol.

Gujarat State Fertilizers and Chemicals Ltd.

Gujarat State Fertilizer & Chemicals Limited, (GSFC) has varied business interests ranging from fertilizers to industrial chemicals to polymers and LPG bottling. Of the business mix DAP accounts for 52% of the revenue.

DAP was decontrolled in 1992, which resulted into immediate price hike thereby affecting volume off takes. The NPK ratio continues to be skewed in favor of urea however adhoc subsidies provided by government for phosphatic and complex fertilizers has eased the situation for DAP manufacturers. The DAP market is expected to grow at 12% yoy.

Caprolactam, a part of the petrochemical sector is subject to cyclical fluctuations. Initially GSFC reaped rich benefits due to increase in demand from the user industries. However excess capacities both in the domestic and international markets has effected business. Caprolactam is expected to grow at 4% yoy.

Domestic fertilizer industry earnings are regulated, with the Government controlling selling prices and providing subsidies to meet excess costs. In case of urea (the popular fertilizer), Government assures a 12% post tax ROE, based on 90% normative capacity utilisation. This allows companies to earn above normal returns by utilising more than 100% capacity. In case of DAP and other complexes, which were freed from price control in 92, government provides adhoc subsidy. Subsidy is also extended to imported fertilizers. Future prospects of fertilizer companies are therefore closely dependent cumulative Average Growth Rate on Government policies.

Domestic fertilizer consumption is growing at a CAGR of 12% since last 25 years. Usage pattern of N:P:Kis skewed to a ratio of 8.5 : 3.1: 1 as against the ideal 4:2:1, due to the lopsided and halfhearted government policies. The industry continues to face problems such as shortage of natural gas, the key raw materials and delays in receipt of subsidy payments,

Rashtriya Chemicals and Fertilizers Ltd.

Rashtriya Chemicals & Fertilizers Ltd (RCF) is one of the largest fertilizer and chemical company in Asia. The company has its manufacturing facilities at Trombay and Thal respectively. These units produce Nitrogenous, Phosphatic and Potash Fertilizers along with wide range of Industrial Chemicals. The company has good marketing network spread all over the country, which is a prerequisite in a decontrol scenario. The company has earmarked about '25Bn towards expansion and modernization, for the next five years. The company is currently at the threshold of privatization. Of the equity base of '5.10Bn government holding is about 92%. The cabinet committee for disinvestment has decided to offload 51% stake to a strategic partner. The partner would be finalized via a bidding process carried out by a global advisor. RCF has depreciated plants and the cost of manufacturing urea is as low as '4600 per ton while the international prices are about ' 6500/ton. In a decontrolled scenario this would translate into profits of about ' 2000/ton.

Nagarjuna Fertilisers & Chemicals Ltd.

Nagarjuna Fertilizers & Chemicals Limited (NFCL), is the largest urea manufacturer in South parts (an installed capacity of 3000/mt per day). The plant operates at 135-140% capacity but the efficiency is not reflected on to the earnings because of government policies. The company also trades in DAP manufactured by Godavari Fertilizers, Coromandel Fertilizers etc. but in the long run NFCL proposes to be a complete Solution Provider to the agriculture industry. The company believes in enhancing farmer's productivity and to do so it provides a wide range of products from seeds to agro-chems to specially fertilizer and others. Further to widen its reach the company has decided to start a portal dedicated to agriculture.

The company has also ventured into the Oil sector. NFCL is the lead company in the 6 mtpa refinery coming up at Cuddalore in Tamil Nadu. The company decided to enter into this field since growth in fertilizer business has nearly plateaued as it is impossible to increase capacity utilization beyond the current operating levels. The company foresees huge unmet demand for petro products especially in south.

The Department of Fertilizers in consulation with Ministry of Petroleum & Natural Gas has drawn up a concrete action plan for supply of gas to the fertilizer sector till the year 2011-12, considering the fact that the growing production of food grains :

Of the 37 urea units including existing, closed, green-field and brown-field expansion projects, as many as 13 urea units are already based on HBJ pipeline and have adequate supply of gas. For the other 9 gas based units, though the supply of gas is deficient, assurance from Petroleum ministry for the adequate gas supply by 2008-09,

For the three FO/LSHS based plants of National Fertilizers Ltd in North India also the pipeline connectivity and availability of gas is expected by 2008-09.

For the 7 closed units of the public sector Hindustan Fertilizer Ltd and Fertilizer Corporation of India, connectivity and gas supply may be available by 2009-10.

For the unit of Fertilizer and Chemicals Travancore (Cochin) and Madras Fertiliers Ltd (Chennai), gas is likely to be available by 2008-09.

The problem seems to be with the Goa unit of Zuari Industries Ltd, the Tuticorin unit of SPIC and the Mangalore unit of Mangalore Chemicals and Fertilizers Ltd as there is no provision for gas supply. The installed capacity as on 2005 has reached a level 120.6 MMT of Nitrogen and urea capacity is 205 MMT while 56 MTPA phosphorus of India is the third biggest producer of fertilizers while GNFC is the World's largest urea producer. There are 57 large sized fertilisers while 29 are for urea remaining DAP, Phosphorus. There are small and medium size single super phosphate units.

1.1.5 Petro Chemical Industries and Products

Haldia petrochemical complex shall be the second biggest downstream industry in the country. The complex commenced production from April 2001.

Haldia Petrochemicals Ltd. is '5170 crores, with '1979 crores Equity and '3191 crores Debt. Production Started in Aug 2001 by Haldia Petrochemicals Ltd, a resurgence of industrial development is targeted in the region, around a range of downstream plastic processing industries which will be supported by ancillary and service industries.

Plant	*LSTK Contractor*	*Technology /Licensor*	*Capacity*
Naphtha Cracker Unit	Toyo, Japan	ABB Lummus, USA	420 KTPA
Butadiene Extraction Unit	Lurgi, Germany	BASF, Germany	74.5 KTPA
Pyrolysis Gasoline Hydro Unit	Lurgi, Germany	IFP, France	300 KTPA
Benzene Extraction Unit	Lurgi, Germany	Lurgi, Germany	75.5 KTPA
C4 Hydrogenation Unit (EPCM)	IDEA, India	IFP, France	110 KTPA
Polypropylene Plant	Daelim, S. Korea	Montell, Italy	210 KTPA
High Density Polyethylene Plant	Daelim, S. Korea	Mitsui, Japan	200 KTPA
Lin. Low Density Polyethylene Plant	Tecnimont, Italy	Montell, Italy	225 KTPA
Captive Power Plant (BOO)	Larsen & Toubro, India		116 MW
Nitrogen Plant (BOO)	Praxair, India		9,000 NM3

Gandhar Gas Cracker

IPCL's Gandhar Gas Cracker complex at Dahej will have large gas separation facilities. It can crack 3 lakh ton gas to produce 1.6 lakh ton of HDPE and one lakh ton of ethylene oxide/glycol, PVC and is expected to put on full stream by 2000. Thus the investment by IPCL alone accounts for more than 30,000 crores by the end of this century. The production achieved by IPCL alone is amounting to 9.65 lakh ton in 1997-98. The Gas Authority of India Ltd (GAIL) plant at Auriya is to produce ethylene of 9.3 lakh tons by cracking gas.

A recent study reveals that by 2000-2002 India will be surplus in polyolefins and the capacity of ethylene is shown below:

Reliance	1.6 MMTPA
IPCL	1.0 MMTPA
HPCL	0.43 MMTPA
GAIL	0.30 MMTPA
NOCIL	0.3 MMTPA
Others	0.31 MMTPA

In fact most of the South-East Asian countries due to financial crunch desperately sold the polymer stocks resulting in drop in the prices of polymers. Because of Reliance (Hazira) cracker plant of 7.5 lakh ton of ethylene, almost 50% increase in production of polymers and a fall of 30% imports was noticed. Indian polymer scene is better described by the Table 1.1(*f*).

Table 1.1 (*f*): Polymer scene in 1997-98

	Production	*Imports*	*Exports*	*Consumption (Tons)*
LD/LLDPE	3, 31,600	1,20,000	5,000	4,58,600
HDPE	3,63,000	1,35,000	8,700	4,89,300
PP	4,94,600	1,10,000	8,000	5,96,000
PVC	6,81,300	40,000	30,000	6,91,300
PS	1,20,700	20,000	12,000	1,28,700
ABS	27,500	5,000	—	32,500

Polypropylene demand and supply scene is given below:

Year	*1995-96*	*1996-97*	*1997-98*	*1998-99*	*1999-2000 (Thousand tons)*
Demand	320	415	510	620	750
Capacity	115	540	540	890	890
IPCL	115	190	190	190	190
Reliance	-	350	350	700	700

Besides IPCL and Reliance, NOCIL, Haldia, Square-D, and Krishna Chemicals have entered into polymer industry.

The global over-supply position in polymers, the consequent squeeze on margins for producers, the high debt burden of the Company and a difficult liquidity position are real issues, which are being addressed by the management. It is compounded due to economic recession.

High density (HDPE) polyethylene	80,000 tpa
Polyethylene glycol	50,000 tpa
Acrylonitrile	50,000 tpa
Butadiene	20,000 tpa
Styrene	30,000 tpa
Naphtha Cracker capacity	6,00,000 tpa

Additional products like : Vinylchloride/PVC/Propylene glycol will be also available. Next important complex is being installed at Salempur (UP) to produce

Benzene	1,60,000 tpa
0-xylene	53,000 tpa
p-xylene	1,53,000 tpa
Alkyl benzene	17,000 tpa
Cyclohexane	90,000 tpa

Bongaigan Petrochemical Complex is the second complex under public sector industries occupying immediate place after IPCL. Separation of aromatics is the priority.

Xylene	35,000	tpa (0-xylene 10,550 tpa; p-xylene 23,700 tpa)
DMT	45,000	tpa ....
Polyester	30,000	tpa ....

Petrofoils Cooperative Ltd. (IPCL) was setup in 1974 under joint venture. It is to produce polyester yarn at Baroda - 7,000 tpa.

Hindustan Petroleum Corporation Ltd., Vizag (Former Caltex Refinery, after take over by the Govt. has joined the wing of Hindustan Petroleum Corporation Ltd.). The capacity of the refinery is raised to process 7.5 M^2 TA, this has given a fillip to go for petrochemical venture. And the products likely to be obtained are :

Acrylonitrile	6,000 tpa
Butyl rubber	25,000 tpa
Styrene	10,000 tpa
Linear Alkyl Benzene (LAB)	33, 000 tpa
Vinyl acetate	10,000 tpa
Butadiene	18,000 tpa
Ethyl hexanol	30,000 tpa
Maleic acid	3,000 tpa

(Also supplies Naphtha to Fertiliser Plant at Kakinada)

Union Carbide. Union Carbide commissioned 60,000 tonnes naphtha cracker in early 1966, which heralded the birth of petrochemical industry in India. Beyond its activities as a petrochemical producer, it produces batteries, electrodes and agricultural chemicals. Its major products include 20,000 tpa LDPE, 10,890 tpa of benzene and carbon black.

Gujarat State Fertilisers Company. Setup in 1960 as the first joint sector complex in India. Besides, fertiliser plant at Baroda it produces caprolactum, Nylon-6, melamine.

Caprolactum	20,000	tpa
Nylon-6	4,000	tpa
Melamine	4,000	tpa
Polyester Staple fibre	45,000	tpa
MEG	40,000	at Patalganga

LAB	50,000	tpa
Poly olefins plant at Thane produces HDPE	70,000	tpa
Poly propylene (Reinforced)	4,000	tpa

Nagothane complex (Maharashtra Gas Cracking Complex)

LDPE	80,000	tpa
LLDPE and HDPE	1,35,000	tpa
Polypropylene	60,000	tpa
Ethylene glycol	50,000	tpa
Ethylene oxide	5,000	tpa
Acetylene black	1,740	tpa

Hindustan Organic Chemicals Ltd, Rasayani (Maharashtra) incorporated in 1960, comprises over 20 plants to manufacture several chemicals. Its extension plant at Cochin is producing Nonyl-phenol, Dodecyl phenol, Bis-phenol, Polycarbonates, Diphenyl methoane, iso cynates, phenol, acetone, propylene, PTFE.

Formaldehyde	30,000	tpa
H_2SO_4/Oleum	30,000	tpa
BHC	3,000	tpa
Chlorobenzene	4,400	
Nitrobenzene	11,000	
Nitro toluene	1,200	
Nitro chlorobenzene	4,800	
Aniline	6,000	
Nitrophenols	1,350	
Phenol	40,000	
Acetone	24,640	
Propylene	29,900	

IOCL : Bombay incorporated in 1960

Acetic anhydride	600	tpa
Ethyl acetate	1,880	tpa
Ethyl acetoacetate	300	tpa
Butyl acetate	2,000	
Amyl acetate	250	
Propyl acetate	4,000	
Ethyl chloride	1,000	
Benzyl chloride	300	
Plasticisers	10,000	
Polyester fibres	6,100	

ABS Plastics Ltd., 1973, Ahmedabad

Additives for PVC	3,250	tpa
Leather mix items	1,000	tpa
Polyester-polymeric type Plasticisers	3,250	tpa

Adarsh Chemicals and Fertilisers Ltd. (1957) Gujarat

Maleic Anhydride	3,000	tpa
Beirad Chemicals		
ABS Plastics	2,000	tpa

PVC Producers

IPCL (Baroda)	55,000	tpa
NOCIL (Bombay)	21,500	tpa
Chemplast (Mettur)	20,000	tpa
Shriram Chemicals (Kota)	26,400	
Plastic Resin and Chemicals	12,000	tpa
C and P India Madras	16,000	tpa
Calico	6,000	tpa

Southern Hydrocarbons Ltd. 1977, Lakshmanapatti

Aldehydes, Acetic anhydride, H_2 and other petrochemicals Thirumali Chemicals Ltd. 1972, Bombay : (Phthalic Anhydride, 6000 tpa)

Gujarat Petro Synthesis proposes to produce : Polybutene (Raw material will be supplied by IPCL)

Continental Chemicals Syntec Detergents UP and Haryana

FICOM-organics; Dodecyl phenol—2500 tpa.

Coming up petrochemical industrial and the productions are enlisted here.

J.K. Synthetics Ltd. (1993)

Nylon, Polyester	3,000	tpa
DMT	600	tpa
Methylene glycol	1000	tpa

Imperial Chemical Complex

PVC	52,000	tpa
Polyester	13,000	tpa
HDPE	28,000	tpa
LDPE	28,000	tpa
CA Moulding Powder	1,300	tpa
MF Moulding Powder	275	tpa

Indian Organic Chemicals Ltd. (Tamil Nadu) (with the assistance of MRC): PTA; Polyester fibre 30,000 tpa. (MRL expansion to 5.6 M^2 ta has a FCC unit to handle 6 lakh tonne feed stock to produce cycle oils, LPG and petrochemicals include LAB 50,000 tpa and PP 17,000 tpa). Also, it incorporates an Aromatic extraction plant based on Bombay High Naphtha to produce

BTX	21,500 tpa
Butylene	25,000 tpa
i-Butylene	
-OXO alcohols	16,000 tpa

1.1.6.1 Synthetic Detergents

One of the important industries of modern society and industrial development, is the detergent industry. Tendency is to produce cheap and effective cleaning agents. Textile industry is the largest consumer; industries other than household, are leather, paper, food industries. During 2nd Five Year Plan the synthetic detergent industry started in our country. Major producers of synthetic detergents are :

Surfactants demand (units in tonnes per annum)

	Actual			*Projected*	
	2003-04	*2004-05*	*2005-06*	*2006-07*	*2007-08*
LAB					
Capacity	314	434	434	434	431
Production	388	388	465	460	460
Imports	2	11	0	4	5
Exports	54	43	89	65	42
Demand	336	356	376	399	423
EO					
Capacity	107	107	107	107	107
Production	70	78	85	91	96
Imports*	0	0	0	0.4	2.7
Exports*	0	0.2	0	0	0

Hindustan Levers	Bombay, Calcutta, Haldia (LAB)-	40,000 tpa
Tata Oils	Bombay	10,000 tpa
Swastik Oils	Bombay	20,000 tpa
Godrej Soaps		20,000 tpa
Indo Burma Petroleum Co. Ltd., Bombay		
I.P.C.L. Detergent alkylate and about 20 units more		30,000 tpa

Non Soapy-Detergents (NSD) powder market has grown upto the tune of 1.7 million tons and is likely to reach 3 million tons by the end of 2000. Hindustan Levers Ltd. the pioneer in the synthetic detergents industry launched Surf, in 1959, followed by Rin and Wheel powder and Bars. Proctor and Gramble's Ariel and another competitive market item Nirma were introduced in Eighties. The industry by the turn of century will grow up to more than 4500 crore-rupees worth. Major innovations have taken place in detergent manufacturing—like enzyme incorporations to eliminate different stains and even tough dirt and oily residues. Further traditional spray drying technology has also been discarded and concentrate powders of these detergents have been significantly left their presence in the market.

1.1.6.2 Out Look of Olefins

Impact of olefins in the modern world is multifacet. Ethylene and propylene are well known and butene is another important olefin the use of which has not been upto the mark. Plastics and synthetics have carved special place on modern life and the effect is enormous and unparalleled, the day begins and ends with plastics. World wide share of polymer consumption (14 polymers) is shown below and the growth rate of polymers in developed countries is 3%.

N. America	26.6%
Asia Pacific	36.6%
West Europe	23.3%
East Europe	5.3%
Africa/ME	3.6%
S/Central America	4.7%

In 1977 World Ethylene supply was 79 million tons (US ethylene production 29 Million tons). By 2010 ethylene consumption will be 140 million tons demanding a capacity growth of 60 million ton. Asia Pacific and Africa are sighted to be fastest growing regions. Ethylene capacity in India in 1997 was 1.41 million tons which was excess by 4 lakh tons, however by 2005 it will be deficit by 1.2 million ton. Ethylene supply, production, consumption are always complex problems. Economic competitiveness is the main reason, for which mega mergers, improvement in design and effective methods of naphtha cracking are some complements or remedies to the industry. This is a big threat to the SSI and MSI industries, which are the main producers of commercial articles. Large distribution net works, focus on business management are the helpful criteria.

Polystyrene PVC, PET, are some prospering plastics. Even timely pioneering of one product is sufficient to establish the industry as seen in the case of GSFC for caprolactum. Propylene ranks next to ethylene and is being consumed in several ways, propylene consumption never dwindles but may be more expensive because of gasoline manufacturers, if win back from gasoline industry, it gives a considerable relief to petrochemical industry. This may not be true with India as the gasoline industry is irrelevant to propylene production. 46 million tons of propylene was consumed in 1998 through out the world. While US and E Europe are the largest markets, Asia market growth rate is highest. Perhaps manufacturers of propylene chemicals are guilty of building over capacity, as can be seen in 1999 itself about 16 producers added 3.2 million tons and the trend is not checked. Propylene growth rate is expected to be 5.3% while ethylene may be hovering around 4.8%. Because of FCCs in Refinery operations more availability of propylene leading to more investments in industry are justified. Global production of plastics in 1998 was 128 million tons and is expected to touch 160 million ton in 2003. Some times unexpected down turns happen too. In 1977 ASEAN economic crisis lashing out, perhaps India remained unaffected, while PE capacity of 2.2 million tons has been delayed or cancelled. In our country demands of these two polymers are projected to increase in next three years as polyethylene 23%, Polypropylene 34% Butylene production has shot up because of institution of FCCs (presently 9) in refineries.

1.1.6.3 Out look of Fibers and Filaments

During a three year period of 1995-98 Global polyester demands grew by 44%, even then ASEAN crisis has not affected the production of TPA which grew at the rate of 6 to 8%. Polytrimethylene terephthalate a competitor for PET has not been introduced yet. 20 million tons of synthetic fiber produced round the world, against this India's performance is not bad. Synthetic fibers production in last two decades showed a big jump in production. In 1980-81 polyester was only 35 million kgs and nylon was 20 million kgs, by 1997 polyester was 6.85 lakh tons and nylons touched 65,000 tons. Not much change in acrylic filament production was noticed (Max. 45,000 tons). Of late the commencement of Reliance would increase the production of PTA, DMT, Ethylene oxide by 40-50%.

The plastic industry is geared more to take up newer packaging materials which may require 52% of the plastics leaving 15% for consumer goods, perhaps rest shared by Automobile and Electrical industries. PVC is another strong competitor for PP and PE and has many advantages due to selective and versatile applications.

Further the new Millenium will witness a revolution in materials science.

Indian polymer industry was only 2.7 million ton in 1999. Ninth Plan demands of olefins are shown below :

Olefins Demand

	1998	*2001-02*
	('000 tons)	
Ethylene	2002	3419
Propylene	886	1581
Butadiene	140	246
Aromatics	1450	2366

Polypropylene demand in 1995-96 was 320,000 against a capacity of 115,000 tons which was entirely supplied by IPCL; by 2000 the demand will rise up to 750,000 tons while the capacity may be 890,000 tons of which 700,000 tons will be from Reliance itself. Reliance is the largest producer of PE 4,00,000 tpa, the tenth largest in the World, with production capabilities of one million ton polypropylene it is again the biggest in the World and also the highest producer of PTA (350,000) and p-xylene of 1.4 million tons etc. Polymers form 50% of the petrochemical industry while 40% is shared by fibers, the remaining goes for other uses.

Demands of Unsaturates (India)

(in '000 tons)

	1999	*2001-02*
Ethylene	2000	3419
Propylene	888	1582
Butadiene	140	246
Aromatics	1450	2366
World Ethylene	79 mmt (1997)	110 mmt by 2003

USA Alone	30 mmt	World growth rate 4.8%
India Propylene	1.4 mmt 46 mmt (1998)	World Growth rate 23% 55 mmta 2003
World growth rate %	5.3%	
India	0.32 mmt	Growth rate 34%
HDPE (World)	19 mmt	26 mmt (2003)
PET (World growth rate)	15%	India 30%
PP (India)	320,000 (95-96)	750,000(99-00)
CAPACITY	115,000	890,000
IPCL	115,000	190,000
RELIANCE		700,000
PVC	693,000 ('98)	

Synthetic Fibres ('97) (million kgs) (Filament Yarn)

Nylon	63
Polyester	395
(Staple Fibre)	
Polyester	270
Acrylic	55

Thermosettings : (Demand : Tons : 2000)

UF	28,000
PF	89,000
MELAMINE-F	2,700

Engineering Plastics

ABS	38,000
Polyamides	20,000
Polyacetals	4,500
Polycarbonates	15,000

1.2 FEED STOCKS FOR PETROCHEMICALS

Our petrochemical industry started much later than the oil industry; hence there was no coordination between these too. Lately integrated complexes have been installed and are becoming successful. Our country with meager resources in hydrocarbons has to utilize and conserve the hydrocarbons to a great extent. Thus implementing a petrochemical complex project virtually depends upon the feed stock availability and an assured supply for at least 20 years. The feed stocks for an integrated complex will thus be naturally different from an independent installation.

Generally the feed stocks are classified on the basis of the existing from i.e. solid, liquid and gas. These are :

1. Gases : Associated gas (obtained with crude oil), Lean gas (Gas stripped of the condensable; mostly methane), Refinery off gas (gases from crackers, cokers, stabilizers and other units) Natural gas, LPG etc.
2. Light Liquids : Natural gas liquids, Naphthas, wild gasolines, Kerosene, gas oils Reformates, coker fractions, FCC lights etc.
3. Heavy Liquids : Vacuum gas oils, fuel oils, waxy distillates, heavy crudes etc.
4. Solids : waxes, Deasphalter bottoms, bottoms of distillation columns and storage tanks, Residuum.

Table 1.3: Approximate Composition of Different Gases

Component	*Natural gas*						*Process gas*		
	1	2	3	4	5	6	7	8	9
Hydrogen	-	-	-	-	0.2	0.1	6	1	0.5 to 1.0
Methane	80.3	71.4	19.26	78.44	16	1	6	4.1	9.3
Ethane	7.9	5.3	0.43	7.24	17	16.9	6	3.4	4.5
Ethylene	-	-	-	2.6	20.8	-	-	31.3	27
Propane	3	3.8	0.1	4.59	21.4	25.1	17	0.6	0.35
Propylene	-	-	-	-	9	17.4	-	13.1	16.4 + 0.64 + others
Butane (*i* & *n*)	1	1.8	0.05	1.95	20.5	10.2	33	-	0.14
Butenes	-	-	-	-	13.1	8.1	-	7	6.2 + 5.05 dienes
C5+	0.3	0.6	0.02 C6 = 0.01	0.5 C6 = 0.26	-	-	Balance C5+	Saturates 6	4.5
Nitrogen+ Oxygen	3.1	10.9	71.66 + 1.94	0.01	0.01	traces		Balance heavy liquids	C6 + 7.43 BTX 7.9
Carbon dioxide	4.4	6.1	6.14	6.49	0.1	0.2			Feed oil 9.7
H_2S+Mer-captans	1185 ppm	-	He 0.09-0.12	-	-	-	-	-	-

Legend:

1. South Basein gas
2. Cambay
3. Ghotaru (Rajasthan)
4. Bombay High
5. Thermal Cracker (under pressure)
6. Coker
7. Reformer
8. Naphtha
9. Hydrocracker gas

In fact there is no bar in using any hydrocarbon, however the economy, purification, availability of the raw material in abundance and technology for the particular fraction are the only impediments. It is traditional to use some fractions for certain intermediates or products, like naphtha for olefins or for synthesis gas or hydrogen. One can see the production of hydrogen through different processes using different feeds, however the available raw material has to be utilized for pecuniary gains and conservation of the petroleum stocks. When a raw material is used for producing a chemical, the purification of the stock is foremost important and the purification is devised on the state and condition of the feed stock. Each fraction or each destination requiring a special type of purification perhaps in- addition to common purification procedures. The complications of purification and cost of purification usually increase with the molecular weight of the feed stock. Though a maximum purification is anticipated in gaseous feed stocks, the cost may not be that much compared to the purification of heavy stocks, because the gas, either field or process, can be easily purified. Usually the field gases will be rich in saturates though the composition of each is different and the processed gases will be rich in unsaturates. Hence the purification procedure and the degree of purification varies from gas to gas. Table 1.3 shows the composition of different gases.

1.2.1 Purification of Gases

All the gases above listed are used as feed stocks for petrochemicals. However, if facilities for petrochemical industries are not available, then the gases of high heating value are diverted to industrial heating systems or to domestic necessities. All the gases, irrespective of origin, contain some unwanted extraneous constituents named impurities. Most common of the impurities are listed below :

1. Water vapour
2. Mechanical (suspended) impurities
3. Chemical impurities
4. Other than above mentioned impurities

1.2.1.1 Exploration and Development of Gas and Oil Fields

Most of the Country's NG comes from Western Off-Shore area. South Basein gas fields Tapati and onland gas fields in Tripura, Rajasthan, Gujarat are famous. Presently KG Basin has major significance. Next to Bombay High this basin is the leading producer of oil and gas. The ONGC & the OIL have together planned to add 865 million tonnes of oil equivalent in-place hydrocarbon reserves during Ninth Plan. These efforts will be supplemented by addition of some reserves in the exploration blocks awarded to the private parties. An accretion of up to 200 million tonnes oil and oil equivalent of gas (O+OEG) is envisaged in deep water areas where a breakthrough is expected in the Ninth Plan. The estimated recoverable reserves will-be in the range of 245.84 to 342.72 million tonnes of oil equivalent. The cumulative oil and gas production during the Ninth Plan is projected at 335.16 million tonnes of oil equivalent (Oil production is about 180.8 MMT). It may be mentioned that considering the reserve accretion and the production of oil and gas during the Ninth Plan, the Reserve Replacement Ratio (RRR) will be in the range of 0.73 to 1.0. All efforts would be made to bring the RRR above the unity level. Although a New Exploration Licensing Policy (NELP) has already been formulated, certain fiscal and financial incentives would be required under this Policy to attract investment for exploration. Exploratory activity should be given infrastructure status to attract necessary investments from both public and private sector companies. According to The Petroleum Ministry projection oil production during the XIth Plan Period (2007/8-11/12) is about 207.756 MMT, of which ONGC would be responsible for 141.06 MMT OIL will produce 18.99 MMT others would produce the balance. Similarly NG production will be stepped upto 229 BCM 45% higher than the XI th Plan achievement of 159 BCM. Similar is the case with LPG with the current consumption rate of 11 MMTPA by the end of XI th Plan 17 MMTPA will be required to meet the domestic and transport sector needs.

OIL AND GAS PRODUCTION (Ninth Plan)							
Organisations	97-98	98-99	99-00	00-01	01-02	05-06	Total
OIL PRODUCTION (Million Tonnes)							
(*i*) ONGC	27.73	28.09	29.18	29.87	30.02		144,89
(*ii*) OIL	3.10	3.20	3.29	3.38	3.50		16.47
(*iii*) JVC PRODN.	3.59	4.40	4.08	3.93	3.46		19.46
TOTAL	34.42	35.69	38.25	37.18	36.98		180.82

GAS PRODUCTION (MMSCMD)							BCM
(*i*) ONGC	62.75	63.81	65.49	65.29	69.04	51.0	119.04
(*ii*) OIL	5.60	6.33	7.19	7.81	8.28	4.0	12.86
(*iii*) JVC PROD.	6.73	7.26	6.98	6.82	6.79	20.0	12.63
TOTAL	75.08	77.40	79.66	79.92	84.11	75.0	144.53

MMSCMD: Million Metric Standard Cubic Metre per Day @ inclusive of 20 LNG

BCM: Billion Cubic Metre

Refining Capacity Addition During 9th Plan (Million Tonnes)

Refineries	*97-98*	*98-99*	*99-00*	*00-01*	*01-02*	*04-06*
Capacity as on 1. 4.97	61.55	61.55	61.55	61.55	61.55	61.55
1. Gujarat, IOC	—	—	—	3.00	3.00	13.70
2. Panipat, IOC	0.00	6.00	6.00	6.00	6.00	6.0
3. Haldia						6.0
4. Barauni						3.3
5. Guwahati						1.0
6. Mathura						8.0
7. Digboi						0.65
8. Numaligarh, NRL			3.00	3.00	3.00	3.0
9. Visakh, HPCL	—		3.00	3.00	3.00	7.5 (8.4)
10. Mumbai						5.5(7.9)
11. Bina.BPC						9.0(15)
10. BPCL, Mumbai						6.90
11. Bina.BPC	—	—				6.00
12. MRPL Expn			6.00	6.00	6.00	9.50
13. Kochi						7.5
13. Bongaigon						2.35
14. ESSAR, Pvt				9.00	9.00	9.00
15. Reliance				15.00		22.00
16. Paradip						15.00

17. Ennore Petrochemical Complex, CPCL + IOC						15.00
18. Bhatinda IOC						15.00
19. ONGC TAIPAKA						3000 BPSD
20. Bina						3.0
Sub-total: Cap.Addition	0.00	6.00	43-40	46-40		66-69
Total	61.55	67.55	104.95	107.95		>137 MMTPA*
*By the end of 10th Plan						
Eighth Plan To 2005 Petroleum Products-consumption						
Component	97-98	98-99	99-2K	00-01	01-02	04-05
(*i*) Light Distillate	16.87	18.25	19.62	21.03	22.15	36.83
(*ii*) Middle Distillate	50.45	53.43	56.62	60.24	64.38	60.96
(*iii*) Heavy Ends	16.41	16.88	17.43	17.80	18.27	15.51
Total	83.63	88.56	93.67	99.07	104.80	127.00*

*inclusive of boiler fuels

Draft Report of the Working Group on Petroleum and Natural Gas for the Xlth Five Year Plan (2007-08) is to be believed; India will transform into a gas surplus country by the year 2011-12. A supply of 40 MMSCMD of gas from the KG Basin (Reliance Industries Ltd (RIL) and 54 MMSCMD from GSPC's discovery, in the KG Basin will be the first stage. The other scenario takes into account the unlikely possibility of no gas supplies at all from the two KG Basin blocks. Under scenario-I, the domestic gas supply position goes up from 80.54 MMSCMD in 2007-08 to only 108.30 MMSCMD in 2011-12 whereas under the other scenario, the supply touches 202.30 MMSCMD. In this period the imported LNG supply to move up from 30.45 MMSCMD to 83.12 MMSCMD. Thus, the total supply under the two scenarios could be either 285.42 MMSCMD or 191.42 MMSCMD. Also the report has projected total countrywide gas demand at 178.97 MMSCMD in 2007-08 and at 282.55 MMSCMD in 2011-12. So under the scenario where gas is available from the KG basin, the country will have a surplus of 2.87 MMSCMD of gas whereas under the worse case scenario, the deficit will be 91.13 MMSCMD. If the demand projections are right, India may actually end up with a larger surplus than has been projected under one of the scenarios by the Working Group. RIL is planning to produce not 40 MMSCMD, but 80 MMSCMD of gas from its D-6 discovery within the Xlth Five-Year Plan. In fact, the company is informally talking of taking output up to as much as 110 MMSCMD. Similarly, the GSPC's find in the KG Basin is likely to be equally big and gas output could swell well above the 50 MMSCMD. If gas output from the KG Basin goes up, there may, in fact, be no requirement for additional imports of LNG beyond what is brought in by 2007-08 or gas through transnational pipelines.

Total hydrocarbon resources, inclusive of deep waters, are estimated at around 28 billion tonnes oil and oil-equivalent of gas (O+OEG). As on 01.04.2004. initial in-place oil of 7.89 billion tonnes and ultimate reserves of 2.94 billion tonnes have been established. The resources estimated by DGH for its 'internal use', for the country, are 32 billion tonnes (O+OEG). India's sedimentary area (up to economic zone 200 miles) is 3.14 million square kilometers, of which 0.42 million sq.km area is offshore area, spread over 6000 km coastal line and 0.4 million sq.km area is available in the form of continental slope, while the remaining area is land based. India's offshore basins are divided into 8 regions out of total 26 basins, as shown in Fig. 1.1.b. The hydrocarbon potentiality of the entire region is 23 billion tons. The biggest offshore activity started with discovery of Bombay High in 1973. Oil and oil equivalent (OEG) gas available in the country as per survey's of April 2004 is 1650 MMT. According to estimates the reserves of gas are 763 billion cu. Meters while the production has not gone up beyond 90 MMSCD. While the demand goes well over the production schedule.

1.2.1.2 Some Projects Under Implementation

Petronet LNG Limited company was incorporated in April, 1998 and is setting up LNG Terminals at Dahej (5 MMTPA) and at Kochi (2.5 MMTPA), have recently signed the sale purchase agreement for LNG supplies with M/S Rasgas-Mobil of Qatar and work on setting up of the terminals is progressing.

Kandla-Loni LPG Pipeline

GAIL is in the process of laying a 1246 km LPG pipeline network from the Western region (Kandla, Jamnagar in Gujarat) to the Northern Region (Loni in U.P. via Delhi) at an estimated capital cost of '1229.45 crores including a foreign exchange component of '386.73 crores. The line pipe diameter ranges between 8" and 16". LPG will be collected at the import location at Kandla and Jamnagar (refineries of M/s. Reliance and M/s. ESSAR). The initial capacity of the pipeline system would be 1.7 MMTPA. The capacity of the pipeline will be increased to 2.5 MMTPA by augmenting the pumping capacities to match the LPG demand in the Northern Region.

Gas Processing Complex at Gandhar

GAIL is setting up an LPG plant at Gandhar in Gujarat. The plant shall process 5 MMSCMD feed gas will have a capacity of producing 207 TMT of LPG.

1.2.1.3 Water Vapour

This is exceptionally present in all the fractions. Gases from atmospheric column or other refinery units, condensate gases and natural gas do have variable amounts of water vapour. This can be removed by passing through drying agents like silica gel, bauxite, dehydrated salts or even by absorbing with glycols. It is observed that some hydrocarbons have the capacity to form crystalline hydrates, such examples are CH_4 $7H_2O$, C_2H_6 $7H_2O$, etc. These gases exhibit the crystalline forms even above the freezing point of water. These hydrates not only clogg the delivery lines but facilitate other abnormalities like corrosion, malign catalyst, cause hydrolysis in chemical reactions.

In modern literature, hundreds of dehydrating agents are mentioned, of late, the tendency of using polyhydric alcohols[9] or glycols has been on increase.

Liquid absorbents are not as efficient as solid desiccants for absorbing water vapour. Ethylene glycol is a conventional absorbent used to remove the moisture from gasoline as well as gases. However, the amount of glycol required for circulation is very high, i.e. 20-40 times of water vapour removed. Usually 35-40°C drop in dew point is obtained with glycols.

Addition of toluene to glycol helps in reducing the water content to less than 1000 ppm and is used in DRIZO process[10]. Solid desiccants, synthetic gels, calcium chloride can decrease the dew point even by 80°C. Presently, molecular seives of type 4A, 5A have shown a commendable performance. They can lower the dew point even by 100°C. These sieves are vastly used in N.G. and air liquefaction plants.

Obviously drying by means of solid adsorbents is more economical and simple. When bentonite clays are used as adsorbents, regeneration may not be attempted. But, when synthetic adsorbents are used the regeneration is done simply by heating the bed to 200°C to 300°C. The moisture capacity of the clays is usually small-of the order 2.5 to 4%, while that of synthetic alumina gels is 5 to 8%. Similarly, synthetic alumina gels voraciously adsorb moisture up to 22%, when they are fresh, but due to fatigue in long duration of operation the capacity drops to 10%. Each adsorbent has got its own moisture removal capacity and regeneration temperature. The best dehydrating agent, of course, is lithium chloride, but its use is curtailed due to its cost.

1.2.2 Mechanical (Suspended) Impurities

The sources for mechanical impurities are many. Mining operations invariably bring clay, dust etc. Catalytic cracker gases are abound in catalyst dust. Process gases like coker, cracker gases, if not contain any solid particles, likely to bring condensed droplets of tars. Relatively, Atmospheric Distillation Unit (ADU) and stabiliser gases are free from such impurities. Usually all types of mechanical impurities can be removed by washing with water. Suitable solvent is preferred if tar and heavy hydrocarbon oils are present, sometimes, the removal of suspended particles is resorted to by simple settling but it does consume a good deal of time. When a solvent wash is decided to eliminate chemical impurities naturally water washing may be skipped off. The washed solvent is filtered before recycling, this will ensure the continuous removal of suspended impurities too.

1.2.3 Chemical Impurities

Sulphur and its compounds are the foremost of the chemical impurities. Sulphur can exist right in gaseous form to liquid form. The elimination of sulphurous compounds will not only reduce the corrosion problem but indirectly helps quality of fractions too. Such cases are plenty; sulphur free gasoline requires less lead to boost the octane number. Further sulphur defies the activity of catalyst. In some reactions sulphur is very detrimental too.

Sulphur exists in the form of hydrogen sulphide, mercaptans, thiophanes, thiophenes, sulphides, polysulphides etc. These impurities are present in all the gaseous fractions and light petroleum fractions (full range naphtha). Sulphides, polysulphides are present in high boiling stocks. Even natural gas that forms the feed for ammonia has to be desulphurised. This is carried out using activated carbon fortified by zinc oxide and other metallic[11] additives. At a temperature of 400°C, these catalyst preparations are good for removal of H_2S, mercaptans and small percentage of COS.

Cat cracker or coker gases invariably contain carbon dioxide to a small amount, i.e. 1000 ppm. Other impurities like ammonia, oxides of nitrogen mark their presence to an extent of a few ppm.

Many processes are listed for the selective removal of these impurities. The most famous process of all these is ethanolamine[12] (Girbotol). This employs a 20 to 30% aqueous solution of ethanol amine (mono, di, tri). Ethanol amines being basic can absorb all the acidic gases freely as per the reactions cited :

$$(CH_2\,CH_2\,OH)\,NH_2 + H_2S \rightarrow (NH_3\,CH_2\,CH_2\,OH)_2\,S$$
$$\rightarrow (NH_3\,CH_2\,CH_2\,OH)\,HS$$
$$(CH_2\,CH_2\,OH)\,NH_2 + H_2O + CO_2 \rightarrow (NH_3\,CH_2\,CH_2\,OH)_2\,CO_3$$
$$\rightarrow (NH_3\,CH_2\,CH_2\,OH)\,H\,CO_3$$

All the absorption reactions are favoured by low temperatures (25 – 30°C), while temperatures above 100°C favour the desorption reactions.

Other solvents based on amine compositions are diglycol amine (DGA), disopropylamine (DIPA), methyl diethanol amine (MDEA). Some new solvents[13] are credited with taking very heavy loads i.e., about 2 to 3 cubic meters of gas can be absorbed per equal amount of circulation (mole to mole). Further sulphur (mercaptan) removal is upto 55% (C_1-fractions).

The recent innovations suggest methyl diethanol[14] amine (MDEA) is a better solvent when high concentration of CO_2 is present in the feed gas. When concentration of acid gases exceeds 0.45 Mol/Mol amine solution; MEA, DEA, TEA are not suitable. The first plant in the world Dome's North Caroline Amine15 Plant has been converted to MDEA solvent system and is found to operate convincingly at a higher efficiency.

In fact, there are many processes which are currently under operation to scrubb the off gases, a gist of which is presented in Table 1.4. When a high amount of carbondioxide (65%) and fairly a good amount of H_2S are present, the mixture can be handled by Selexol process which uses dimethyl ether of propylene glycol[2].

All the gas treating processes may be classified into three classes depending upon the operational pressure.

(*a*) Low pressure operation (Below 10 bars) Girbotol, Economiser, Flex sorb-PS, Thiolex, Selexol, Sulferex, Sulfolin.

(*b*) Medium pressure (upto 100 bars). MDEA, Adip Benfield, Catacarb, Flexsorb-HP, Giammarco-Vetrocoke, Stretford, Sulfinol.

(c) High pressure (above 100 bars) Rectisol, Selexol, Benfield.

When sulphur removal alone is to be judged, Amine/Claus process is economical for large installations, because of the capital-intensive nature. Such installations are good for recovering more than 25 tpd of sulphur. Locate iron-redox process is suitable for recovering less than 1 tpd of sulphur, however operating costs are much. Sulferox of DOW, employs improved iron chealates, that can work under higher concentrations without much dilution thus decreasing operating and equipment cost. Bender Sweetening is generally useful for NGLs, gasolines, however LPG can also be processed. These plants are considered to be cheap as the mercaptans are converted to disulphides only.

Beavon sulphur recovery (BSR) MDEA process is a purification process for the tail gas from different units. The process works in two steps, initially all sulphur is converted to H_2S by hydrogenation. In second stage H_2 S is absorbed by amine. Merox process is another new one; it can treat gases, LPG, light naphtha and kerosenes. This extracts mercaptans from gases and low volatiles, while heavy stocks respond to oxidising mercaptans to disulphides.

ADIP Process uses Di isopropanol amine or tertiary amine, methyl diethynol amine in 50% aqueous solution and is capable of handling gases of 100 ppm sulphur 10 ppm sulphur in liquids COS is generally oxidised. Similarly Thiolex employing caustic solution extracts Hydrogen sulphide, carbon dioxide and mercaptans. Beavon-others process is peculiar in one respect it alters the composition of Claus tail gas and by first catalytic oxidation of all sulphur compounds over a CuO bed and then hydrogenated to convert the gases into H_2S or NH_3. MDEA extracts sulphur compounds to separate clean gas from acid gas stream and Seleoxl separates clean gas from elemental sulphur[30].

Modop is a process for removal of sulphur from tail gas to meet the pollution regulations. Tail gas is first heated with fresh air and fuel to oxidise all the components of sulphur and then hydrogenated to yield H_2S which is later selectively oxidised to sulphur. Shell Claus off-gas treating (Scot) process also works on the same principle. Selexol is another process for recovering sulphur from H_2S streams, while sulferex is suitable for removing SO_2 from off gases. Stretford process is suitable for operation on H_2S gas streams.

The operation is simple and illustrated in Fig 1.2*a*. Mixture of gases enters at the bottom of absorber, through a gas distributor. The gas goes into the absorber and mixes freely with the down coming liquid, and the purified gas leaves the absorber at the top. The fat liquid is filtered and desorbed in a similar column by steam heating. When DEA solutions are used for gas treating plants filters are required to remove fine particles, which are trapped during contact with liquid. This is to remove foams and inimical deposits. This is however absent with MEA[16] solution. Absorbed CO_2, H_2S are liberated by steam stripping as shown in the figure. The solution is then made to the required concentration and cooled; after which it is fed into the absorber from the top. The absorption, desorption temperatures are kept around 25°C, 110°C respectively. To facilitate easy absorption, the absorber is superimposed with a pressure of 0.5 to 1.0 MPa. It may be thus mentioned the merits of a process is more or less connected with the amount of impurities and nature of these[17]. Beavon-otter process is shown in Fig. 1.2(*b*). The selector X separates sulphur from gas.

Membrane separators[18] along with amine treatment have been tried for efficient and economic operations. Sulphur removal[19] by adsorption or solvent absorption is followed in the same operations where CO_2 or H_2S removal is affected. Molecular sieves are also employed to desulphurise certain stocks, specially gases. In most of the cases, solvents are regenerated by expelling the absorbed gases. Except in certain cases, (the regeneration is done by air-blowing (Merox)) when solutions are employed for removing impurities, the gases are again dried.

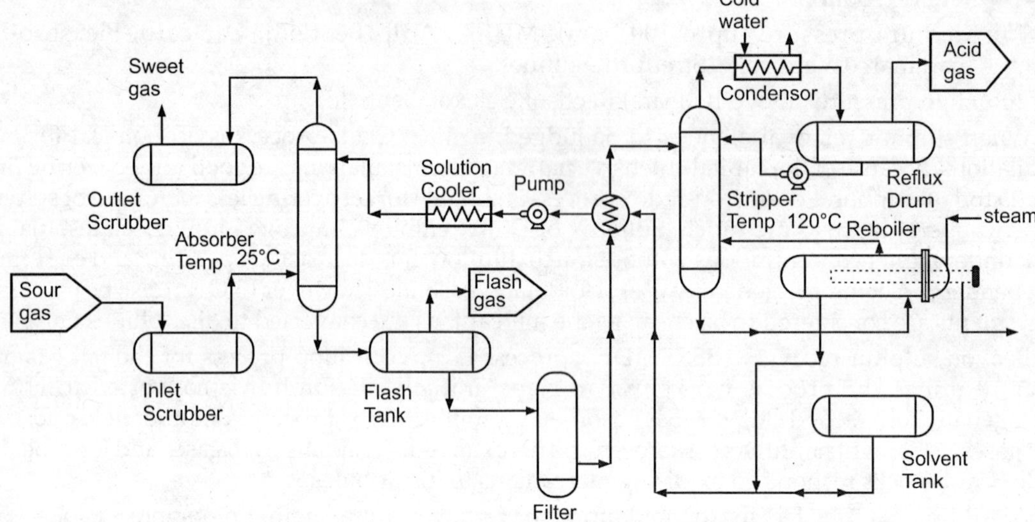

Fig. 1.2 (*a*): Process Flow for Typical Ethanolamine Sweetening Unit.

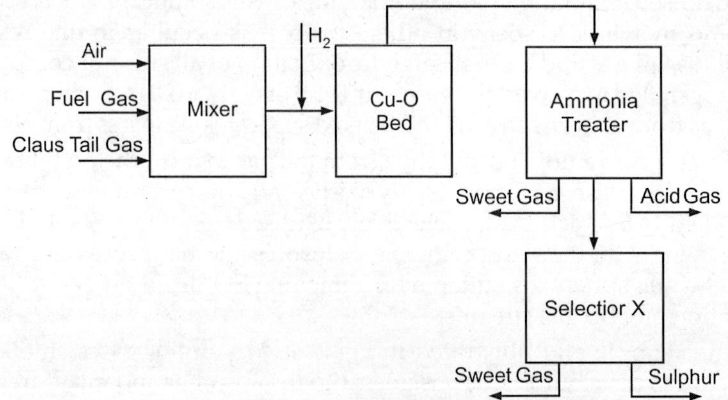

Fig. 1.2 (b): Beavon-otter process.

1.3 SEPARATION OF GASES INTO INDIVIDUAL CONSTITUENTS

Gases ranging from $C_1 - C_4$ are freely available either in cracked, ADU, or stabiliser streams. Further, permanent gases like air, N_2, CO, CO_2, and H_2 are also present in varying extents. The separation of the gases into individual constituents is a costly process, but has to be.

Industrially the separation techniques consist of

1. Absorption-Desorption
2. Compression-Liquefaction
3. Low-temperature fractionation
4. Adsorption
5. Special techniques.

Table 1.4: Survey of Different Processes for Treatment

SI. No.	*Name of the process*	*Chemicals involved*	*Constituents removed*	*Operating conditions*	*Regeneration*
1.	Girbotol	Ethanol amines Mono-di	H_2S, CO_2 from natural, cracked gases	Reaction temp. 25°C 20% concentration	Backward reaction 105°C
2.	Shell solutiser	Tripotassium phosphate	H_2S, from NG	35°C	105-110°C air blowing
3.	Alkazid BASF	Sodium, potassium salts of diethyl/ methyl amino propionic acids/ amino acetic acids	H_2S,CO_2 from NG, liquid hydro carbons	Ordinary conditions	
4.	Benfield Corporation	Hot potassium carbonates containing Benefield reagents	CO_2, H_2S, COS NG, synthesis gas	High pressures, above 3kg/cm^2. Any convenient temperature	
5.	Cat Carb	Potassium salts	CO_2 (NG, H_2, NH_3 or sour gases)	T:100 - 130°C, 20-40 bars	Steam heating
6.	Adip	Dipropanal amine	COS, H_2S, CO_2 mercaptans, acid gases (NG, SG, refinery gas)	Wide flexibility in operation conditions	P:20-25 kg/ cm^2, T:35°C –38°C
7.	Unisol Atlantic	Methanol + KOH	CO_2		
8.	Mercapsol	Cresols + Naphthenic acids	Mercaptans (gases & gasoline)		
9.	Tannin solutizer	Organic salts + Potassium phosphates	H_2S, CO_2 gas		
10.	Vetrocoke	Sodium potassium arsenite sol. ($K_3 As_2 O_3$)	CO_2, NG, RG		
11.	Battersea	Aromatic amines (alkyl salts)	SO_2 (gas)		

12.	Flour Econamine	Aqueous sol. of primary alkyl amine/diglycol-amine ($HO-C_2H_4$ $O-C_2H_4-NH_2$)	H_2S, CO_2 (NG, SG, RG)	Sol. 40-60% P : upto 80 kg/cm^2	
13.	Purisol	N-methyl pyr-riolidone	H_2S/acidic constituent (Gas)	Pressure 80 kg/cm^2 T 30°C	Steam strippings P : 80 kg cm^2 T : 180°C
14.	Rectisol	Methanol + KOH	CO_2, H_2S, NH_3, COS, HCN gum formers (gas)	35 kg/cm^2	
15.	Selexol	Dimethyl ether of polyethylene glycol	SO_2, CO_2 (NG, SG)		
16.	Sulphonil (Shell)	Sulpholane, alkanol amino and water	H_2S, COS, CO_2 Mercaptans (NG, SG, and LNG)	80 kg/cm^2	
17.	Flour solvent	Anhydrous organic compounds propylene carbonate (Diglvcolamine)	O_2 and H_2S from NG, SG	30-40% sol. P : upto 80 kg/cm^2	
18.	SNPA-DEA	Diethanol amine (modified)	$H_2S + CO_2$ (NG)	P : 50-75 kg/cm^2 T : Low (30°C)	Steam stripping at 180°C and 80 kg/cm^2
19.	Takahax	Alkaline sol. (PH 8.5) Sodium 1, 4 naphtoquinone 2 sulfonate as redox catalyst	(NG and SG)		

1.3.1 Absorption-Desorption

The gases are selectively absorbed in suitable solvents and then desorbed. The absorption is usually a physical process, however it may be even chemical absorption. The removal of SO_2, CO_2, NO, etc. is basically by chemical absorption. Hydrocarbon gases are mostly separated by physical absorption; propane, butane are highly soluble in light hydrocarbon oils, permitting the recovery up to 70-80%, with a purity approaching 100%. Some gases like acetylene may be absorbed by solvents like dimethyl formamide. Even aromatics can be separated by using tetra ethylene glycol.

Almost all olefins and diolefins can be separated by selecting a suitable solvent. Ethylene, propylene, butene, butadiene may be separated by a mono ethanol diamine solution of cuprous nitrate/chloride, due to varied solubilities (shown in Table 1.5) under different pressures. The ethanolamine and metal salt mixtures can absorb CO and acetylene in addition to olefins. Absorption is done at low temperature, while desorption is carried out at 50°C. Silver fluoroborate is found to be better than amine, as it is not affected by the presence of acetylene-carbon monoxide mixtures. $Ag BF_4 . 2C_2H_4$ is formed during absorption which easily decomposes.

Table 1.5: Solubilities of Some Unsaturated Hydrocarbons in Monethanolamine-CuCI at 20°C

Hydrocarbon	*Pressure (atm.)*			
	1	5	10	20
Ethylene	8.4	15.8	21.2	24.2
Propylene	1.1	4.3	6.1	—
n-Butene	1.0	—	—	—
Butadiene	10	—	—	—

Thus choosing best operating conditions any single constituent can be separated. Similarly iso-butene from butane mixture may be separated by 65% sulphuric acid at 35°C by absorption, where other olefins do not react with the acid at that condition. Further, where the boiling points of many components are close, fractionation fails hence alternative routes are sought. Sometimes azeotropic or extractive distillation may be useful. Butenes are separated conveniently from n-butane by using extractive distillation; the added component being furfural. By addition of a solvent the relative volatility of one component can be shifted; as an illustration 1-Butene and *n*-butane have relative volatilities 1.046 and 0.871 respectively; in presence of furfural the volatilities shift to 1.718 and 2.02. Paraffins and olefins of C_4 fraction can be separated by simple distillation because of the closeness of boiling points. It is observed relative volatility of saturated hydrocarbons increases more than that of unsaturates in presence of furfural as given below :

	B.P. °C	*Relative Volatility*	*Relative volatility in presence of furfural*
i Butane	– 11.7	1.209	2.6
n-Butane	– 0.6	0.871	2.02
1-Butene	– 6.3	1.05	1.72
iso-Butene	– 6.9	1.07	1.70
Cis-Butene	– 3.5	0.77	1.06
Trans-2-Butene	– 0.9	0.84	1.19
Butadiene	– 4.5	1.00	1.00

Infact the separation of butanes follow a serial order. Firstly, i-butane is removed from mixed C_4 fraction. Later, distillation is recalled to separate t-Butane (BP-11.7°C) from *i*-butene (BP-6.3°C) as overhead product leaving bottom product as a mixture of *n*-Butane (BP-0.6°C) and 2-butene (0.9°C). Both the overhead and bottom products are distilled with third component, furfural, in separate columns. The relative volatility of paraffin increases in presence of furfural, thus aiding the separation. Fig. 1.3 shows such a distillation with furfural as solvent.

Aqueous acetone solutions are also useful in extractive distillation (20% Water + 80% Acetone). Acetone-water mixture is convincingly used to separate butane and butene streams into individual components. Absorption with extractive reflux has been successfully used to split ethane and propane 20 streams. The most useful diene of C_4 fraction is butadiene, and is removed from other olefins by chemical absorption method or using extractive distillation.

Butadiene is separated from C_4 streams by absorbing in a solution of copper ammonium acetate, containing about 17 to 18% of copper. Two or three columns are required for complete absorption of butadiene from butene fractions. These columns operate at less than room temperature (0°C). Iso-butene is easily removed from C_4-fraction by extractive distillation. Butene-butadiene fraction is absorbed in series of absorbers. The solution is heated to remove butadiene vapours, unabsorbed butenes-leave from the absorber top, while regenerated butadiene escapes from the stripper. Perhaps, further purification of butadiene is required. Isoprene is also similarly separated by using acetone solution.

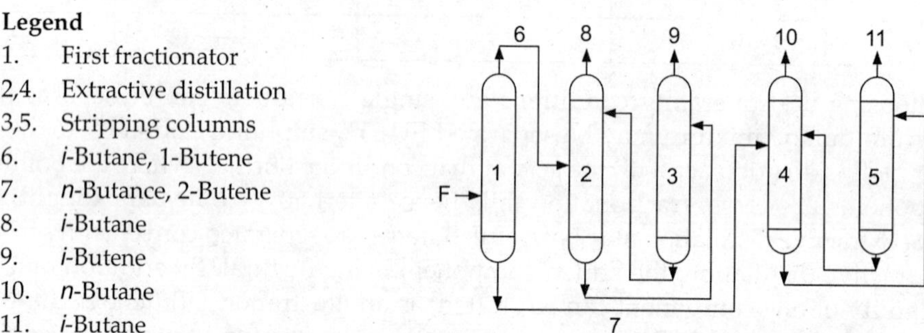

Fig. 1.3: Separation of C_4 Components by Extractive Distillations.

1.3.2 Compression and Liquefaction

This technique is suitable specially if a mixture contains a component, which when compressed and cooled liquefies easily keeping the remaining components unaffected. Gas when compressed at bubble point condenses, of course, if the bubble point is high, say greater than 30°C, it can be easily liquefied by cooling. Depending upon the concentration of the component in the mixture, simple cooling or application of pressure or both have to be resorted to. The dew point of the component is given the maximum credit. This process is mostly used in combination with other operations. Sometimes, casing head gases are compressed and extracted with gas oil fractions to remove the light gasoline fraction, which again reflects the application of the process.

1.3.3 Low Temperature Fractionation

In this operation whole mixture is cooled and liquefied. This liquefied mixture (liquid) is subjected to distillation at low temperature. The production of low temperature can be done by external cooling agencies or by self expansion against resistance or no resistance. The effectiveness of this technique has resulted in cheap production of polymer grade olefins. Different feed stocks ranging from ADU, casing head to cracker gases can be liquefied. Separation of a gas containing C_1 to C_3 components is discussed here.

The dried gas when cooled to 55 to 60°C under a pressure of 35 atmospheres, ethane-ethylene fraction condenses, while methane, nitrogen, hydrogen and carbon-monoxide remain as gases. Hence it is desirable to operate the column more specifically, reflux drum under those conditions. After the non-condensable gases are removed, the liquid fraction containing ethane-ethylene can be distilled separately in a coulmn that contains 60 plates and operates at 20 kg/cm^2 pressure. The reflux is maintained around – 30°C. This way ethylene is separated out as distillate and ethane as residual product. So far, the above mixture has been dealt assuming up to C_2 fraction is available. In case C_3 fraction is also available, this is first liquefied by using a separate column operating at - 10°C under

25 kg/cm^2 pressure. Usually a 35 plate column in desired to separate C_3 fraction from all the more volatile fractions upto C_2.

After collecting the C_3 fraction in liquid state, a separate column of the same order is pressed into service for separation of propylene from propane. From this, it is clear that the increasing pressure naturally elevates the operating temperature or vice versa. Modern method of producing liquefied petroleum gas is by turbo expansion technology. Such technique has been in operation in India at Duliajan for the production of LPG. The plant produces 60,000 tonnes LPG from the natural gas liquefaction. LPG being higher boiling component can be liquefied first and taken out before liquefying methane.

Turbo expansion technology differs from Joule—Thompson expansion in one respect only. In Joule-Thompson expansion, there is no external work done, whereas in Turbo-expansion, the gas is allowed to do some work i.e. driving turbines for compression, while it suffers cooling.

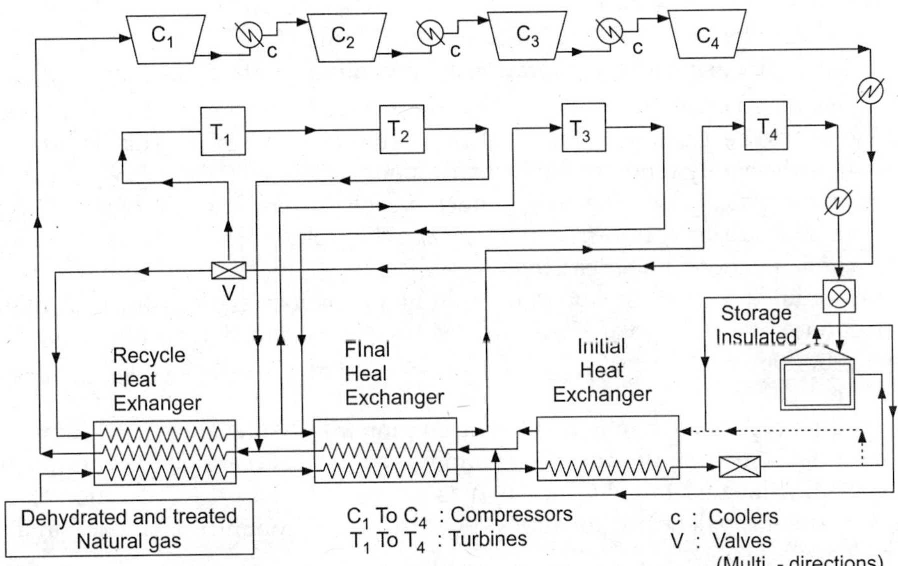

Fig. 1.4: Natural Gas Liquefaction.

The feed gas is first compressed up to 20 MPa in stages and allowed to expand in turbines, as a result, temperature goes as low as – 100°C. Turbo expansion technology or Joule-Thompson expansion effect is invincible where natural refrigerants (may be useful up to –40°C) fail. A more sound method called cascade cooling is economically employed to substantiate the simple principle of natural refrigeration. In this method, the easily condensable component is liquefied first, this acts as a refrigerant for the second liquefaction, the resultant of which is employed for cooling the most vehement one. It may be illustrated further like this, propane is first condensed by using water at not more than 50°C and at a pressure of 1.9 MPa. The boiling point of propane at ordinary pressure is – 43°C. So, this is permitted to act as a refrigerant to ethane fraction, which can be liquefied at – 33°C under a pressure of 2 MPa. Ethylene boils at – 103°C at ordinary pressure, hence this component can safely condense methane, at – 100°C under a pressure of 3 MPa. Thus, methane faction can be liquefied much above its boiling point (–161°C) under pressure. The system is evenly balanced because of the critical temperatures and pressure as shown below:

	B.P. at 760 mm Hg °C	*Critical pressure kg/cm²*	*Critical Temperature °C*
Methane	– 161.5	45.0	– 81.55
Ethane	– 88.6	47.5	+33.3
Ethylene	– 103.8	50.9	9.7
Propane	– 42.0	41.0	+97.8
i-Butane	–11.72	35.0	+135.6
n-Butane	– 0.5	36.4	+153.2
Butadiene	– 4.41	42.0	152.2
i-Pentane	32.3	32.0	177.0
n-Pentane	32.2	33.6	177.4

C_4 and above fractions are usually available in liquid state. Fig. 1.5 illustrates the separation of gaseous constituents by using low temperature condensation and fractionation.

Purified gas is dried to remove moisture and compressed to 35 atmospheres. After compression it is cooled by process heat removers to as low as possible (around 15°C). This mixture is sent into column (1), where the highly condensable gases (Propane and Butane) liquefy, which are drawn out from the bottom of the column. The top product is cooled in a cooler to – 8 to – 10°C again by process coolers and streamed into fractionator (2), where the imposed pressure is sufficient to condense C_3 and C_2 fractions. Overhead fraction is a gas consisting of methane and some liquid and transferred to column (3). From the top of the column (3), methane comes out. The liquid product mainly C_2 fraction joins the stream of ethane-ethylene fraction obtained from column (4). This column (4) takes up the bottoms of column (2) as feed, which contains C_2, C_3 and C_4 fraction. The column (4) fractionates the mixture into C_2 stream as overhead and the bottoms, remaining liquid. Ethane-ethylene mixture is separated in long distillation column (5). Overhead of this column is ethylene, thus ethane forms the bottom product. The bottoms of column (4) are mixed with the bottoms of column (1) and fractionated to get C_3 fraction as top product and C_4–C_5 fraction as bottoms in column (6). Propane-Propylene fractionation is done in another column (7). Ethane-Ethylene splitter is shown in Chapter 3.

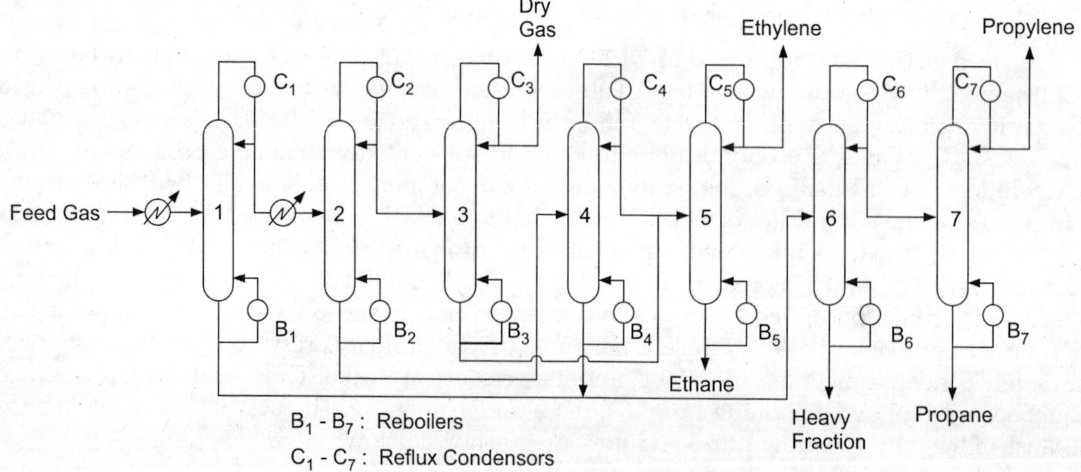

Fig. 1.5: Low Temperature Fractionation

Propane or propylene obtained serves as the initial refrigeration system for incoming feed. Ethane-ethylene fraction cools the top product of column to condense methane. Effective cooling by out going cool gases markedly improves the economy of the process. Propane can be very easily liquiefied at room temperature simply by application of pressure.

1.3.4 Adsorption

In the earlier stages of application, adsorption was considered to be an operation of purification. Various kinds of impurities ranging from colour donating substances to corrosive and toxic chemicals were attended by this process. Use of charcoal in masks at war fields or bone charcoal for bleaching sugar solution were authentic. Tswett's application of selected adsorbents for separation and isolation of pigments and natural colours, has gradually instituted chromatography, a modern tool of chemical analysis. In all these applications natural adsorbents have been gradually replaced by synthetic ones. The use of natural absorbents, like bentonite clay, Fullers earth, bauxite has been received with wild acclaim for decolorisation of lube oils in petroleum industry. Though the hydrodesulphurisation technique has gained momentum to interlace all the petroleum and chemical industries, still orthodox refiners are continuing the old clay treatment process.

Synthetic absorbents like silica gel, molecular sieves, zeolites have been permeated into petroleum industry, the survival of petroleum industry without these has become a question.

Separation of paraffins from aromatics whose boiling range do not permit the conventional techniques, has been successfully done by this technique. Based on the selective or preferential adsorption, Molex process (Molecular sieves) works for removal of normal paraffins from either branched chain paraffins or aromatics. Also detergents manufacturing is based upon the selective removal of paraffins from other constituents (gasolines and kerosenes). The adsorbability is governed by the physical and chemical properties of adsorbates and the nature and structure of adsorbents. As a highlight, ethylene has been successfully separated from cracked gas mixtures, which is described below.

The unit namely, hypersorber is shown in Fig. 1.6 contains 4 chambers. The feed gas enters the section (3) and rises to adsorption chamber, where the heavy unsaturates are adsorbed, the adsorbent, in this system being activated charcoal. Light gases like H_2, CH_4 are not adsorbed and escape as such through top section. The downward travelling catalyst reaches bottom section where heating is done to expel the adsorbed ethylene (265°C). Heating is done by circulating dowthern mixture or high pressure-superheated steam. The outgoing streams usually go through cyclones, which catch any charcoal dust. The desorbed adsorbent once more heated (for complete regeneration) and elevated to the bin above the top of the hypersorber. Proper distribution is done by distributors at the entrance of each section. A highly efficient hypersorber can yield 98% ethylene from a concentration of 6% in inlet feed, at a purity of 98 to 99%.

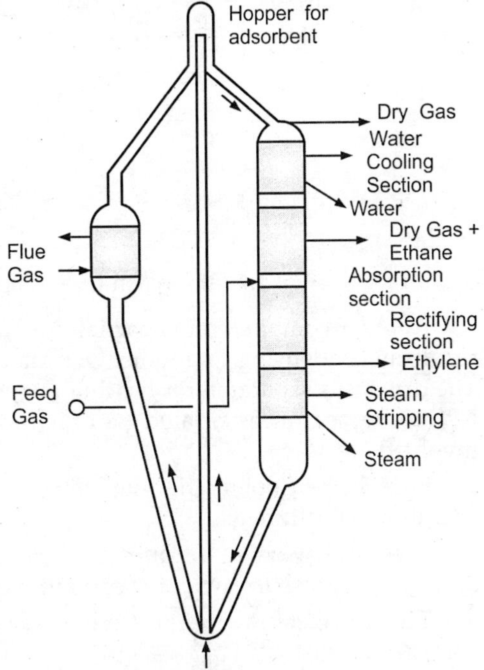

Fig. 1.6: Hypersorber

1.3.5 Special Techniques

Special techniques are invited when the individual techniques fail to prescribe economic operation. For a close economy, two or more methods are congenially joined together. Fig. 1.7 describes a combined absorption and fractionation unit for separation of refinery cracked gases into dry stream and olefins or feed stocks for petrochemical industry.

Feed gas under a convenient pressure (8 to 10 atms) is first sent into depropaniser absorber (1), where a solvent fraction (C_7-C_{10}) trickles down. This solvent rapidly absorbs C_3 and above components. The top gas contains unabsorbed methane, ethane and propane fractions along with other non condensables like air, H_2S, hydrogen, etc. The fat solution goes to the second column called debutaniser, the overhead shall be butane fraction and residue contains high proportions of pentane. Pentane is stripped off in the column (3). The heavier bottoms are recirculated as solvent. The unabsorbed gases are compressed (4) (30-35 atm) cooled and dried and then sent to low temperature fractionation battery. This system has been already described (Fig. 1.5.).

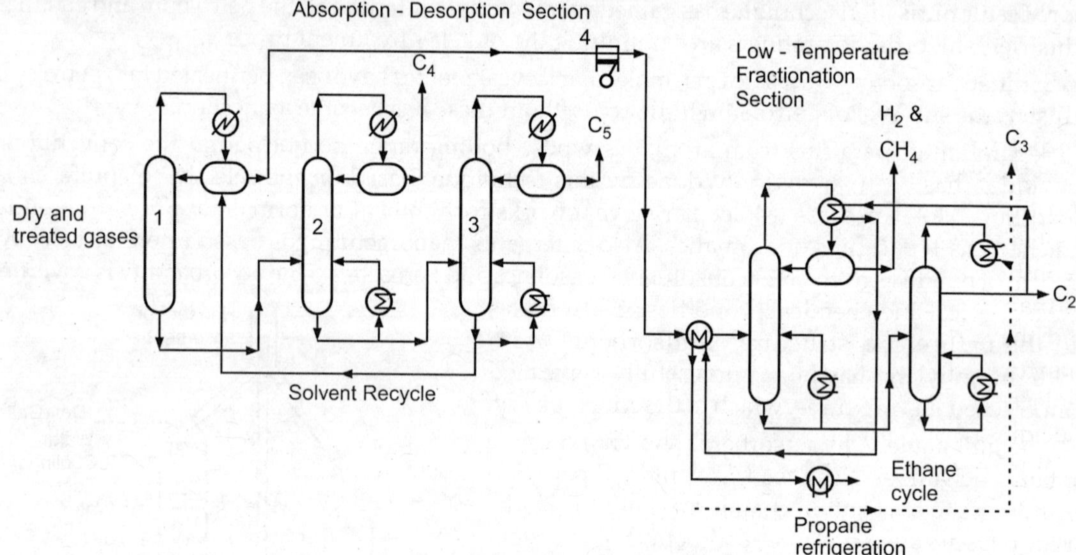

Fig. 1.7: Absorption-Low Temperature Combination for Separation of Gases.

Another combination is partial liquefaction-fractionation. In this process the cooling efficacy results in condensing C_2 and heavier fractions. Thus, hydrogen and methane remain in gas phase. The dry gas is separated for heating purposes and the concentrates are processed in liquid phase. LPG recovery from associated gas or natural gas is a fit example in this direction and the major steps involved are :

First Stage: Cooling the whole mass of gas to 30°C under 35 atmosphere pressure to liquefy C_4 and above constituents.

Second Stage: The uncondensed gas is subjected to severe conditions, i.e. upto +30°C, and 20 atmosphere pressure, whereby propane condenses leaving ethane and ethylene in gaseous form.

Third Stage: Ethane and ethylene are liquefied and fractionated,

Fourth Stage: Dry gas constituting CH_4 and H_2 is separated from ethane/ethylene mixture, and this mixture is used as a refrigerant.

Fifth Stage: The heavy bottoms of first stage unit are processed for C_3 and heavy ends.

1.4 SEPARATION OF AROMATICS

Most useful aromatics are benzene, toluene and xylene. All these being liquids at room temperatures, precise fractionation can separate. However, when they are in fractions of close boiling ranges, paraffins will hurdle the separation. Aromatics from paraffins can be separated by azeotropic distillation or by liquid extraction. So, preferably, aromatics as a group may be separated from these by azeotropic distillation. It is apparent that paraffins are separated by extractive distillation, while aromatics require azeotropic distillation. The principle underlying azeotropic distillation is that added component forms a constant boiling mixture (usually less than the b.p. of individual components) which can be easily condensed and further fractionated. Industrially liquid extraction is used to separate aromatics from cracked feeds. Solvents like liquid sulphur dioxide, dimethyl formamide,(24) dimethyl sulphoxide(25) and ethylene glycols(26) etc. are in much use in industry. All these solvents preferably extract aromatics from non-aromatics.

The most suitable source of light aromatics (BTX) is reformate of naphtha (80-140°C) or cracked naphtha fraction. The description of these methods appear here.

1.4.1 Azeotropic Separation of Toluene

Any thermally cracked or catalytic reformed stocks contain at-least 18% toluene. This toluene fraction in close boiling cut (100-115°C) is first separated from the main stream by distillation. Mostly used azeotropic solvents are methanol and methyl-ethylketone. Methanol has a tendency to form azeotrope with toluene. Azeotropic mixture contains 55% methanol. Fig. 1.8 describes the operation in detail. Methanol and feed mix enter the column (1), where methanol forms a top product with naphthenes and paraffins. This product enters the tower (3). The bottoms of column (1), contain methanol and aromatics as an azeotrope enters the column (2). The top product is mainly methanol and the bottom product is toluene of the purity 96-98%, which is further refined to eliminate unsaturates either by treating with sulphuric acid or by distillation. In the column (3) methanol is replaced by scrubbing with water thus setting free hydrocarbons from the bottom of the column, the liquid wastes is taken into column (4), where fractionation separates methanol as a top product from water. The methanol is sent back to column (1) to begin the operation.

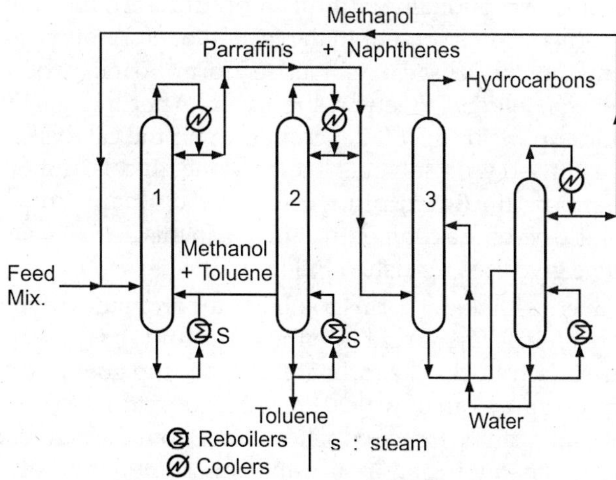

Fig. 1.8: Azeotropic Distillation for Toluene.

1.4.2 Extractive Distillation : Separation of Styrene

Styrene is obtained as byproduct during cracking of naphtha for ethylene and it constitutes about 4 to 6%. The cracked fraction can be used as a raw material without any treatment for recovery of styrene although it is highly advantageous to tailor the fraction to C_8 cut only, to rise the concentration of styrene to greater than 30%, (Boiling range of such stream is 125-148°C). Specially styrene O-xylene separation is a difficult thing because of close boiling ranges of these two. Extractive distillation is successful in separation of these two. Fig. 1.9 explains the operation. Initial separation may be done by extraction or distillation to concentrate C_8 fraction from the rest. Separated C_8 fraction is sent into an extractive distillation unit, where a suitable solvent encounters the feed. The solvent styrene form a high boiling mixture and crude xylenes leave the tower from the top of column (1). The residue obtained from the bottom of this column contains styrene, a high boiling mixture and crude xylenes leave the tower from the top of column (1). The residue obtained from the bottom of this column contains styrene and solvent. This solvent mix is distilled again in a second column (2) to free styrene as a top product. Rich mixture of solvent and heavy ends formed due to polymerisation of styrene go into a separate column (3), where pure solvent is set free from the bottoms of the column. Polymer wastes are discarded. To prevent polymerisation of styrene small amount of inhibitor is added. STEX[27] process uses a special solvent for this operation, though phenol is the best suggested one.

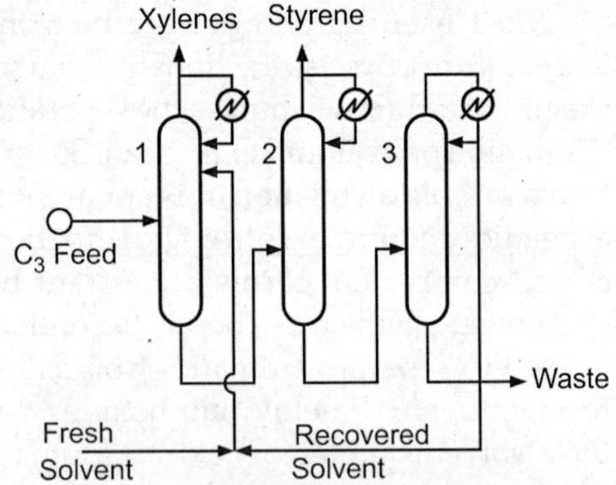

Fig. 1.9: Styrene by Extractive Distillation.

1.4.3 Aromatics separation from Reformates

Full range naphtha when subjected to catalytic reforming paraffins cyclise through dehydrogenation reaction. The cyclic compounds together with isomerised paraffins boost octane number. Hence, this process has become a major refinery operation to produce quality gasoline. The same process has become exceptionally faithful to produce aromatics from paraffin rich stocks. Depending upon the carbon atoms present in the feed, the reformate turns out to contain the same number of carbon atoms, but in cyclised or isomerised form. So it may be suitably chosen $C_6/C_7/C_8$ atoms to produce benzene, toluene, xylene respectively. After the reaction, the reformate may be enriched with aromatics up to 40 to 50% depending upon the catalyst and the process variables. The rich reformate is extracted with solvents like ethylene glycols, dimethyl formamide, dimethyl sulphoxide, sulpholane or liquid sulphur dioxide etc. Snam Progettis[28] new solvent N-formylmorpholine has also become a competitive solvent in this field. A brief description on extraction of aromatics by ethylene glycols is furnished in Fig. 1.10.

Ethylene glycol mixed with water forms a solvent for aromatics. Aromatic rich stock heated to 150°C under a pressure of 8-10 bars is sent into the extractor (1); where the udex solution comes in contact with the aromatic rich stock. The raffinate from the top goes into a water washer (3), where solvent is washed off by a spray can of water. From the top, raffinate is obtained; and by steaming the remaining solvent stains can be removed. The extract from the extractor goes into a fractionator (2), where the top product containing aromatics and small amount of solvent is finally distilled in a column (5) to get aromatics and solvent from the bottom of the column to join the extractor. Raffinate washings are collected and separately fractionated in a column (4) to drive off excess water to maintain the degree of dilution of the solvent, i.e. 80-90%. Solvent recycling is 5 to 6 times to hydrocarbon feed. Ethylene glycol is suitable to extract only light hydrocarbons, while the other solvents such as DMF, DMSO liquid sulphur dioxide are used for improving smoke point of kerosene.

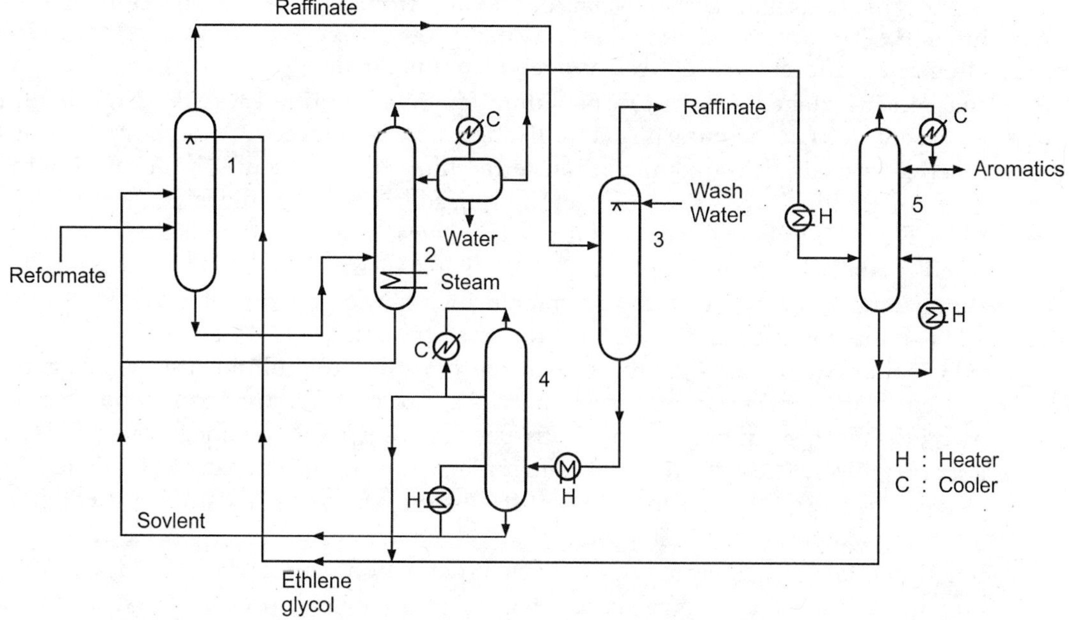

Fig. 1.10: Udex Extraction Process.

1.4.4 Separation of Aromatics by Crystallisation

Azeotropic or extractive distillation is an invincible tool for separation of aromatics from paraffins of the same boiling range. However, these methods fail when separation of C_8 aromatics to individual components is attempted. There are four isomers of C_8 aromatics whose boiling points are in close range as given below :

Component	*Melting point °C*	*Boiling point °C*	*Density gm / cm*
O-xylene	– 25.2	144.4	0.880
p-xylene	+ 13.3	138.4	0.861
m-xylene	– 47.8	139.1	0.864
Ethyl benzene	– 95.0	137.2	0.867
Styrene	– 31.0	145.2	0.902

It may be guessed how difficult it should be to separate these individual components from a mixture by conventional fractionation alone. So a different procedure involving the advantages of difference in melting points and boiling points is taken up. *p* and m-xylenes can be separated by fractional crystallisation, while O-xylene can be separated by distillation from ethyl benzene. As shown in Fig. 1.11, first fractionator (4) separates all the four components to O-xylene as bottom residue and overhead as a mixture of *p* and *m*-xylenes and ethyl benzene. O-xylene is removed as bottom product. The reflux in these columns is maintained at 60 to 70. Usually, such large columns are split into series of 3 columns, to diminish the construction costs, difficulties further smooth maintenance is also possible. Now *p* and *m* xylenes are to be separated in cooler (C_1). *p* and *m*-xylenes are first cooled to – 20°C. This starts crystallising *p*-xylene, while m-xylene remains as liquid. The crystals of p-xylene from the crystalliser (1) are separated in a centrifuge (F_1). The separated crystals still contain some m-xylene carrys p-xylene to 10%, hence an altogether different procedure has been recommended.

P-xylene and m-xylene mixture can be submitted to extractive crystallisation using n-pentane as solvent, so that the eutectic crystallisation may be avoided. Another method is to follow adductive crystallisation using carbon tetra chloride as solvent. The former method was developed by Phillips, while the later method is used in USSR. To improve the purity of p-xylene, the recovered p-xylene is mixed with carbon tetrachloride and cooled to – 78°C in a cooler (C_2) and crystallised (2). A complex crystalline substance melting at – 40°C results, which on distillation yields p-xylene and carbon tetrachloride. The complex from the crystal-User is centrifuged (F_2) and the filtrate is sent into the first crystalliser (1). Carbon disulphide is regenerated in column (6) to join the process. The bottoms from the column are cooled again in a cooler (C_3) to – 70°C and sent into crystalliser (3). The crystals are again separated in a centrifuge (F_3), the mother liquor from the centrifuge contains p-xylene concentrate and is sent into the first crystalliser. The crystals from centrifuge-3 show a solid phase of m-xylene about 85% and this can be purified in a separate unit if desired. Many methods of purification are available as discussed below :

Sulphonation is used to separate m-p-xylenes. When sulphonation is done to m-p-xylenes mixture, only m-xylene is found to be susceptible. The sulphonated product is extracted with water or any solvent and hydrolysed to get back m-xylenes.

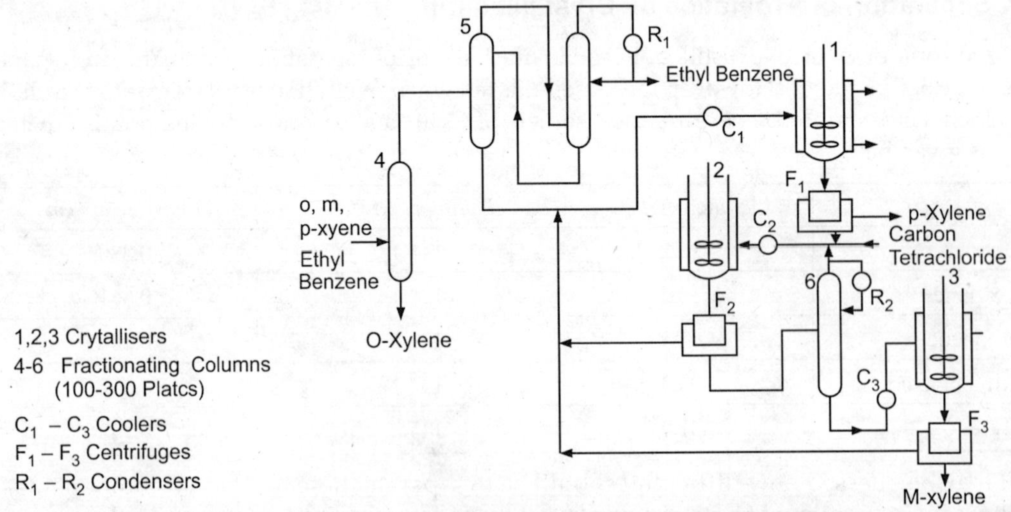

Fig. 1.11: Fractional Crystallisation for Separation of C_8 Isomers.

Another method appears in literature which describes the oxidation of m-xylene to phthalic acid with air at a temperature of 200°C under a pressure of 15 to 30 atm. (Amoco oxidation),

A supplementary to the above methods is understandably isomerisation technique[29]. When ethyl benzene concentration is not favourable for extraction or separation, it may be converted to other xylenes by isomerisation catalysts. This method has been developed in view of economy by ICI[30], using non-noble catalysts at atmospheric pressure and at a temperature of 100 to 500°C maximize p-xylene.

1.4.5 Solid Adsorbents

Specific adsorbents with well defined pore size, like molecular sieves are credited with an excellent performance of separating paraffins from aromatic or a particular isomer from host of isomers of same carbon atom. Parex process has suggested that p-xylene can be effectively separated from other C_8 isomers. Iso Siv[20] process has been in operation for concentrating n-paraffins from kerosene or gasoline fraction. In this process molecular sieve of size 5A is used.

In Parex process adsorption takes place at a temperature of 150-175°C and under a pressure of 10 kg/cm^2, preferably in liquid phase only. By this method purity of p-xylene obtained is found to be 99.5%.

REFERENCES

1. (*a*). Urja Sept 85, 122., (*b*) S.D. Valia and Y.D. Bhatt, Urja, Sept. '85, 137.
   (*c*). N.K. Sharma, Urja, Sept. 85, 121.
2. Sushila Khanna, Economics and Political Weekly, Vol. XIX, No.31, Aug. '84, 1319.
3. Hindu Survey of Indian Industries, 1985, 127, 1983, 121.
4. Biswajit Dhar, Economic and Political Review, Aug. 84,1386 (annual No.).
5. Duleep Singh, Commerce, Dec. 7, 85,1025.
6. HCP, January '78, 141, Feb. '82, 51.
7. CEW, Oct. 85, 39 and CEW, June 1982, 123.
8. B.K. Jain, CEW, Oct. 85, 67.
9. F.C. Reisenfled and A.L. Kolo. Gas purification. GPC, 1974, 490.
10. R.L. Pearse, J.E. Protz and C.W. Lyon. HCP, Dec. 72, 79.
11. J.E. Finneran, L.J.Buividas and N.Walem. HCP, April, 1972, 1227.
12. Joseph W. Holmes, Michael L. Spears and Jerry A. Bullin. C E P May, 1984, 47.
13. K.F. Butwell, O.J. Kubek and P.W. Sigmund. HCP, March, '82,108.
14. C. Blane and J. Elgue. HCP, Aug. '81, 111.
15. G.R. Davies, R. Sandermann and S.T. Donnelly. HCP, May, '84, 79.
16. W.L. Scheirman, HCP. Aug., '73, 95.
17. Volker. H. Mohr and Gerhard Ranke, C E P, May, '84, 27.
18. R.L. Schendel, C E P, May, '84, 39.
19. R.N. Meddex, Chemical reactions as a Means of Separation. Edited by Billy L. Crynes, Marcel Dekker, 1977. N.V. 155.

20. E.J.H. Hoffman. HCP and P R. 1963, Vol. 42, 3, 67.
21. HCP, April, 1982, 164, 106.
22. Luigi Bazzi and Carlo Rescalli. Chemical Engineering World (India) April, 1985, 61.
23. M.F. Symoniak, R.A. Reber and R.M Victory. HCP, May, '73, 1101.
24. B.K.B. Rao and K. Venkateswarlu. Chemical Industry Developments, August, '72, 21.
25. B.K.B. Rao and Devandran, HCP, August, 1976, 127.
26. B.K.B. Rao. HCP, May, 1978, 105.
27. Masakisato HCP, May, 1973,141.
28. Cinelli, E., Noe. S., and Paret, G, HCP, 1972, 141.
29. Nelson, E. Ocker Bloom, HCP, Oct, 1971, 101.
30. HCP, August, '69,109.
31 & 32. Hindu Survey of Indian Industries, 1996 & 1998.
33. HCP, April, 1999, 59.

CHAPTER 2

Chemicals from Methane

2.0 INTRODUCTION

Methane is the simplest of all hydrocarbons and available in nature as a major constituent of natural gas. Besides this, associated gas and gases from crude distillation (atmospheric) and cracking units are also abundant sources of methane. The composition of these gases was shown in Table 1.3.

Methane and ethane are substantially soluble in liquid petroleum and the solubility is around 0.22 cubic metres per standard barrel per kg/cm^2 pressure difference. Thus the reservoir fluids are bound to contain 18 to 25 cubic metres of gas per barrel under the reservoir conditions. Surfacing of these fluids naturally free a huge quantity of gas, in the form of associated gas, at the site of mine itself. Remaining dissolved gas is expelled during distillation of crude.

In our country, no large deposits of gas and liquid hydrocarbons are available. Gas fields are in offshore areas of Maharashtra, Kutch, Godavari and Andamans etc. Land based deposits in Assam, Tripura, West Bengal, Gujarat are explored since the last two or three decades and the concentrated search for hydrocarbons is still on. The activities on ONGC and OIL would likely to result in successful strikes of hydrocarbon deposits in near future to meet major portion of 80 to 90 M^2 TA demands. India is the 3rd largest consumer of natural gas in Asian region though it amounts to less than 10% of the primary energy consumption. Some statistics on production and projected demands are shown in Table 2.1 (a) and (b).

Naturally, the value of any hydrocarbon is the highest when it is converted into petrochemical. The surveys till now show that 50% or more of the gas in our country is marked for fertiliser industry and relatively a small amount is shared by petrochemical industry. However, the Seventh Plan initiated major petrochemical industries, which are likely to be materialised by the end of the Eighth Plan. Considering the fact, that the chemistry of natural gas (Methane specially) has been more recent than the other feed materials, gas based industries are strictly audited.

Originally natural gas was used either in heating installations or in carbon black industry. Presently the uses are many and the chemistry is profusely relevant as depicted in Fig. 2.1. In fact the age old reservation of hydrocarbons of higher molecular weight for petroleum chemicals has been shifting towards methane, specially after the processes like oxidative pyrolysis and electrocracking have come up. The subject large number of commodity chemicals potentially accessible through C1 ¡V C2 chemistry like ethyleneglycols, acetic acid, ethanol and higher alcohols, is drawing constant attention of the manufacturers. A gist of such reactions has been focused by Aquilo et al. and R.K. Gupta et.al. (1) and (2).

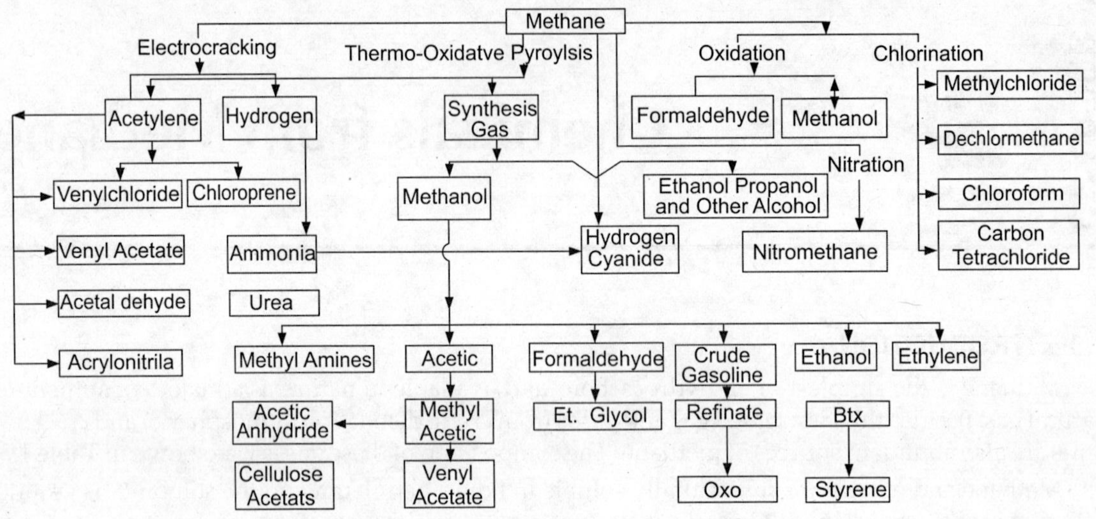

Fig. 2.1: Principal Reactions of Methane.

Some more Methane Reactions

$$\text{Chlorination } CH_4 + Cl_2 \rightarrow CH_3Cl$$
$$\rightarrow CH_2Cl_2$$
$$\rightarrow CHCl_3$$
$$\rightarrow CCl_4$$
$$CH_3Cl + aq.KOH \rightarrow CH_3OH$$
$$2\,CH_3C1 + Ag_2O\,(Dry) \rightarrow CH_3OCH_3$$
$$CH_3Cl + NH_3 \rightarrow CH_3NH_2 + HCl$$
$$CCl_4 + H_2 \rightarrow CHCl_3(+ Cl_2 \rightarrow Cl_4)$$
$$CHCl_3 + O_2 \rightarrow COCl_2\,(Phosgene)$$
$$CHCl_3 + aq\,KOH \rightarrow HCOOK$$
$$2\,CHCl_3 + Ag \rightarrow C_2H_2 + 6\,AgCl$$
$$CCl_4 + H_2O \rightarrow COCl_2$$
$$CH_3CN + H2O \rightarrow COCl_2$$
$$CH_3CN + H_2O \rightarrow CH_3COOH$$
$$CH_2Cl_2 + NaOH \rightarrow CH_2(OH)_2 \rightarrow HCHO$$
$$CH_4 + S(Vapour) \rightarrow CS_2 + H_2$$
$$HCHO + 2H \rightarrow CH_3OH$$
$$HCHO + HCN \rightarrow CH_2(OH)CN$$

Formaldehyde Cyohydrin

$$HCHO + NH_2OH \rightarrow C - C = NOH$$

|
H

Formaldehyde oxime

$$(HCHO)_6 + NH_3 \rightarrow (CH_2)_6 N_4$$

Urotropin Hexamethylene Tetra amine

$$HCN + H_2O \xrightarrow{\text{Acid cat.}} HCOOH + NH_3$$

$$CO_2 + H_2 \rightarrow HCOOH$$

$$CO + NaOH \rightarrow HCOONa$$

210°C/10 atm

$$HCOOH + NH_3 \rightarrow HCONH_2 \text{ (Formamide)}$$

Grignards Reaction

$$CO_2 + CH_3MgI \rightarrow CH_3-\underset{\underset{O}{\|}}{C}-OMgI + H_2O \rightarrow CH_3COOH$$

Acetic Acid

$$CH_3OH + CO \text{ (cobalt acetate)} \rightarrow CH_3COOH$$

$$+ \left(\frac{\text{Rhodium/Iodine}}{\text{65 bars, 250 C}}\right)$$

Table 2.1 (a): Petroleum Industry in India Oil Production

On Shore	85-86 (M^2 T)	Vllth Plan Period (M^2T)	Vlllth Plan
Upper Assam	2.84	17.10	
Cambay	4.10	25.00	65.182
Chachar-Nagaland	0.07	0.89	
OIL			
Upper Assam	3.11	16.49	16.66
Off Shore			
Bombay	20.17	102.31	133.25
Krishna-Godavari	—	1.10	
Other Fields			
Bengal, Kutch etc.	—	3.84	
Kerala, Mahanadi	—	0.30	—
Off Shore			
New Fields	—	8.41	
Total	30.29	175.4 (Actual)	215.092 (Projected)

During Seventh Plan 175.44 million tonnes of oil was produced by ONGC (ONGC's share: 144.49) and Oil India Ltd., falling short by 1.8 million tonnes of the plan target. The major source of these hydrocarbons is through offshore ventures. Even then the terminal year of Eighth Plan may be requiring 75 million tonnes oil, for consumption which may leave a gap of 25 million tonnes to be met by imports. It can be seen from the past performance of OIL industry that about 30% of crude demands must be met from imports only. Although the scene on the gas front is slightly better, the infra structural facilities for handling and consumption of gas have not come up in a proper manner as the discoveries. This will leave flaring of gas as usual i.e. about 30%.

Table 2.1 (b): Prognosticated Reserves-XProduction

Hydro Carbon Deposits		
Geological Reserves		(1-1-'89)
Cambay Basin		932 Million tonnes
Bombay High		3130 Million tonnes
K.G. Basin		98 Million tonnes
Cauvery		52.2 Million tonnes
Pondicherry		24 Million tonnes
Assam - Arunachal		1127 (including Gas)
Andamans	—	465 (M^2T)
Kutch	—	380 Million tonnes
Saurashtra	—	2 10 Million tonnes
Tripura - Mizoram		970 (Gas 13 billion M^3)
Bengal	—	300 M^2 T
Himalayan-Gangetic belt		370 Million tonnes
Rajasthan		340 M^2 T (Gas 2.5 million M^3)
Deep continental shelf		5000 M^2 T.

Gas Production	*Billion cu. metres 85-86*	*VIII Plan period*	*VIII Plan targets*
O.N.G.C. (on shore)			
Krishna-Godavari	0.01	0.093	
Upper Assam	0.42	2.56	34.102
Cambay	1.06	5.49	
Chachar-Nagaland	0.033	0.32	
OIL			
Upper Assam	1.66	7.50	11.967
New Fields			
ONGC	—	0.31	
OIL	—	0.02	
ONGC OffShore			
Bombay High	4.20	27.75 (Free Gas)	92.545
		21.68	
Krishna-Godavari	—		
OIL	7.50		
Kaveri basin (New Field)	1.66	1.10	
Total	9.28	71.60	138.614

Table 2.1 (c): Refinery Capacity

Public Sector	*Capacity (M^2 T) 1989*	*LPG availability 84-85*
IOC, Gauhati	0.85	7
Barauni	3.30	27
Baroda	9.50	—
Haldia	3.0	31
Mathura	9.0	153
Digboi	0.50	
BPCL, Bombay	6.00	160
HPCL, Bombay	3.50	71
HPCL, Vizag	7.0	85
CRL, Cochin	7.5	91
MRL, Madras	6.5	181
BRPL, Assam	1.55	
O.N.G.C.		288
OIL India Ltd.		47
Swing Capacity HPCL, Bombay	2.99 81.85	1237 (thousand tonnes)
Assam 4th Refinery	3.0	47
Mangalore	9.0	8th Plan Additions
Karnal	6.0	

(More information is given in Appendix).

2.1 OXIDATION OF METHANE

Considerable difficulty is experienced in oxidation of low molecular weight hydrocarbons, and is found to be the maximum in case of methane.

The following are the oxidation reactions that proceed via free radical mechanism.

$$\begin{aligned} CH_4 + 1/2\,O_2 &\rightarrow CH_3OH \\ CH_4 + O_2 &\rightarrow CH_2O + H_2O \\ CH_3\,OH + 1/2\,O_2 &\rightarrow CH_2O + H_2O \end{aligned} \quad \text{...(A)}$$

$$\begin{aligned} CH_2O &\rightarrow CO + H_2 \\ CH_2O + 1/2\,O_2 &\rightarrow CO + H_2O \\ CH_2\,O + O_2 &\rightarrow CO_2 + H_2O \end{aligned} \quad \text{...(B)}$$

All the reactions in group A are valued, while in group B are waste reactions. Thus the main product of reaction may be treated as methanol, though formaldehyde forms a major side product. In all oxidation reactions of paraffins or olefins, oxygen-hydrocarbon mixture must be kept outside the combustion limits.

2.1.1 Production of Methanol by Oxidation

Oxidation of methane is carried out by oxygen. Methane to oxygen ratio is maintained at 9 :1 and under a pressure of 100 atmospheres the oxidation of methane takes place at a temperature of 360°C. This gives a product stream containing 17% methanol, 0.8% formaldehyde and the rest is mixture of oxides of carbon and steam. It is found that the pressure and ratio of methane to oxygen favour the formation of methanol over formaldehyde. With addition of ethane to methane stream the situation has been changed in favour of methanol resulting upto 30%, this helped in a considerable reduction in temperature of reaction. Catalysts like nitrogen-oxides, hydrogen chloride, metal phosphates are in use for this reaction. This is said to be an obsolete route and the new route namely through oxidative pyrolysis has become attractive.

2.1.2 Methanol from Synthesis Gas

Carbon monoxide and hydrogen obtained from steam reforming operation are relatively free from detrimental components, if carbon monoxide should be obtained from metallurgical industries, it has to be purified vigorously. The reaction between carbon monoxide and hydrogen is simple catalytic and resembles closely ammonia synthesis

$$CO + 2\,H_2 \longleftrightarrow CH_3OH + 125.5\ kJ/mol$$

Equilibrium yields are shown in Table 2.2 for different temperatures and pressures.

Table 2.2: Equilibrium Yield of Methanol

Temp. °C	*Ke*	*Mole fraction of Methanol at*		
		1 bar	100 bar	300 bar
200	0.34	12×10^{-2}	0.95	0.98
300	0.0011	5×10^{-4}	0.66	0.85
400	negligible	—	0.07	0.33

Because of less conversion, many times recirculation of the products is unavoidable. Basically there are three technologies well explored for methanol synthesis. These being high pressure, low pressure synthesis and liquid phase technology. High pressure synthesis offered by ICI, Lurgi, TOPSOE etc. operates in the pressure range of 30 to 35 MPa and temperature between 360-400°C. Catalyst used being zinc oxide-chromium oxide mixture.

Low pressure synthesis is also offered by Topsoe, ICI, Lurgi and others, and has been readily accepted by the industry. The conditions are milder, i.e. pressure in the range of 5-10 MPa and temperature in the range of 220-350¢XC. Zinc-chromium catalyst has been replaced by copper based catalyst.

Liquid phase synthesis developed by Chem Systems, uses fluidised bed reactor. The catalysts like the oxides of zinc, chromium resist high temperature but poisoned easily. Some times, oxides of alkali metals are doped on these oxides to promote higher alcohols. Oxides of nickel and iron are found to be active chain generators. Though vanadium - zinc - chromium catalysts are selective to reaction, not susceptible to poisons and high temperature. The synthesis takes place over a bed of heterogeneous catalyst in adiabatic beds or in tubes packed with catalyst. As the oxidation generates a lot of heat, heat removal assists the forward reaction, hence cooling of the reactants at intermediate stages is required.

Fig. 2.2 gives the outlines of the process. The purified gas mixture, carbon monoxide and hydrogen in the required mole ratio is compressed to 200 bars and mixed with hot recycle gas and fed into the converter from the top. Multiple bed converter easily facilitates the heat disposal. Initial heating of the reactor through external agency will sustain the reaction. The product gases are heat exchanged with the feed to be cooled and these are separated in a high pressure separator.

The liquid products from the bottom of the separator are taken to purification unit, where a battery of fractionating columns separates formaldehyde and methanol. The purge gas from separator is recycled. A high space velocity or contact time as small as 10 seconds is preferable. Industrially a rate of 600 cubic metres per hour per kg. catalyst is maintained. About 2500 cubic metres of gas mixture is required per ton of methanol. Present day capacity of the plants is raised to 2,500 TPD per reactor for economy.

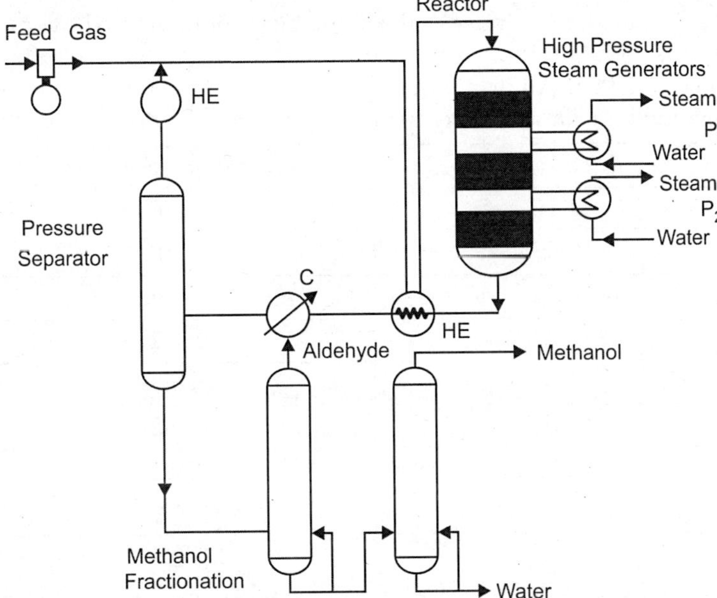

Fig. 2.2: Methanol Production.

Convincingly, fluidised bed reactors, (The fluidisation is done by inert hydrocarbon like methane or ethane) developed by Chem System, give a high conversion of 15 to 20% against the conventional fixed beds of 5 to 6%. Presently, the tendency of using both carbon monoxide and carbon dioxide is gaining significance after the significant disclosure by CPI-Vulcan process. Mitsubishi Chemicals employ Zinc-Boron-Copper catalyst that functions under a pressure of 50 to 200 bars and a temperature of 300¢XC.

2.1.2.1 Uses of Methanol

Methanol is mostly used as a chemical intermediate and 50% of it is usually converted into formaldehyde. Dimethyl terephthalate, methylamines, methacrylites, methyl halides are some of the important derivatives of methanol. Presently, methanol has gained importance as a component to gasoline-alcohol mixtures for petrol engines, due to the fact of rising prices of oil and impending pollution laws. Chang and Silvestri, Dejajaifve et. al (4) & (5) have used ZSM series of catalyst of small pore to study the conversion of methanol to hydrocarbons. Later on the well known process for methanol to hydrocarbons and gasoline known as Mobil process has come into operation.

2.1.3 Mobil Process

Methanol can be converted to light olefins over an acid catalyst, which are then oligomerised or cyclised to aromatics over zeolite catalysts. This process can limit the chain growth to a maximum C_{10}, thus fitting in the range of gasolines. Dehydration of methanol is successfully carried over- A_12O_3 to yield olefins. About 75% of methanol is converted to C_{5+} gasoline. Fischer Tropsch's process has also given its share to convert methanol to hydrocarbons of high octane value, through the following reaction :

$$n\,CH_3OH \rightarrow (CH_2)n + nH_2O$$

Conventionally, the source for such reaction is a mixture of carbon monoxide and hydrogen rather than methanol. According to Inui & Takequani, methanol can also be converted to olefins in a flow reactor over zeolite catalysts at atmospheric pressure in presence of nitrogen. With a high space velocity of 1000 to 4000 hr¡V1 at a temperature range of 300 to 400¢XC a dilute mixture of 10 to 20% mole per cent methanol and rest being nitrogen gives the following spectrum of products.

	Per 100 kg of methanol
CH_4	1.176
C_2H_4	6.168
C_3H_6	13.396
C_3H_8	0.452
$1 + i - C_4H_6$	6.153
$2 - C_4H_8$	6.152
C_4H_{10}	0.444
C_5	5.779
H_2O	56.223

Methanol seems to yield 30-60% ethylene, while naphtha can yield not more than 30% ethylene, a definite advantage to be utilised in future. Further silicate catalyst developed by Union Carbide with Mn-Fe deposits(7) produce at moderate conditions, 30-35% olefins of $C_2 - C_4$ range and 30% aromatics. Medium pore zeolites of Mobil also contribute aromatics because of the acidity of zeolites.

2.1.4 Iso-butylene from Methanol[(8)]

A stream containing 8 to 15% by weight of i-butylene and 15-50% i-butane can be produced from methanol by using silicate catalysts impregnated with oxides of thorium zirconium, titanium etc. Iso-butylene is an important raw material in the manufacture of methyl-t-butylether, butyl rubber, polybutenes etc. MTBE is used as an additive to gasoline.

Methanol conversion is pronounced at 380 °C and HSV of 10 – 11.

2.1.5 Formaldehyde

Formaldehyde is obtained mainly as a constituent in the oxidation products of paraffins, specially methane. However its separation in pure form becomes uneconomical. It can be cheaply produced either by dehydrogenation or oxidation of methanol. Oxidation is carried over a silver gauge or finely made copper gauge. Methanol to air ratio is kept in 0.3 to 0.5. The reaction is carried in vapour phase at 450-600¢XC under atmospheric pressure. Lumuus and Montecatni developed iron-molybdenum combinations for smooth oxidation in fixed bed reactors. The catalysts can withstand poisons and can give higher yields.

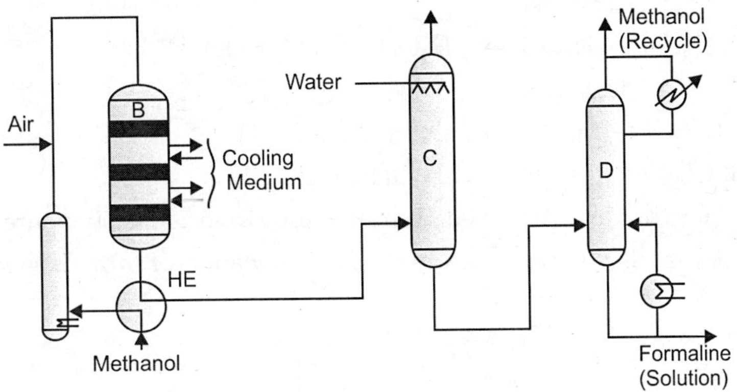

Fig. 2.3 Formaldehyde from Methanol.

Fig. 2.3 shows the flowsheet on oxidation of methanol. Methanol is first heat exchanged and sent into a vaporiser (A). The vapours are mixed with required amount of air and sent into reactor (B) where the silver gauge is spread in three or four levels which can help in cooling the gases. The outgoing gases exchange heat and goes into an absorber (C), where a stream of water absorbs formaldehyde and unreacted methanol. The dilute mixture is distilled to separate formaldehyde in the form of solution; methanol in the form of vapour in a fractionator (D). Methanol vapours are cycled back to the process. The exhaust gas contains carbondioxide and air.

2.1.6 Acetic Acid

In the Monsanto process, methanol is made to react with carbon-monoxide under mild conditions of temperature and pressure to yield acetic acid. This reaction is carried over a catalyst, rhodium coated with iodine. A high yield over 90% is expected.

Rh-I_2 catalyst. Temperature 200°C : Pressure 6.5 MPa

$$CH_3OH + CO \rightarrow CH_3COOH$$

$$(CH_3ONa)$$

$$CH_3OH + CO \rightarrow HCOOCH_3$$

The reaction in liquid phase with basic catalyst CH3ONa gives methyl formate along with acetic acid. Cobaltous iodide, developed by BASF is very useful in production of acid at high pressure (6.5 mPa). Acetic anhydride can also be produced by carboxylation of methylacetate on rhodium catalyst.

$$CH_3COOCH_3 + CO \rightarrow \begin{matrix} CH_3CO \\ | \\ CH_3CO \end{matrix} \!\!>\! O$$

2.1.7. Ethylene Glycol

Ethylene glycol is the newest attempt from methanol, formaldehyde; it was developed by Celanese and Redox Technologies. In the process, radical initiator, ditertiary butyl peroxide is used as the catalyst. The reaction sequence follows:

Initiation

$$(CH_3)_3CO.CO\,(CH_3)_3 \rightarrow 2\,(CH_3)_3\,CO$$

Transfer

$$(CH_3)_3CO + CH_3OH \rightarrow CH_2OH + (CH_3)_3\,COH$$

Condensation

$$CH_2OH + 2\,CHO \rightarrow HOCH_2CH_2O + CO$$

$$HOCH_2\,CH_2O + CH_3OH \rightarrow (CH_2OH)_2 + CH_2OH$$

Reaction conditions are reported to be : Temperature 125 to 200°C ; Pressure 5 to 20 kg/cm^2.

Ethylene glycol can also be produced by hydroformylation of formaldehyde in presence of cobalt/rhodium catalysts

$$HCHO + CO + 2H_2 \rightarrow \begin{matrix} CH_2OH \\ | \\ CH_2OH \end{matrix}$$

2.1.8 Hexamethylene Tetramine[10]

Formaldehyde reacts with ammonia to form hexamethylene tetramine, which is used in plastic industry and in pharmaceutics too.

$$6CH_2O + 4\,NH_3 \rightarrow$$

CH$_2$ — N
N – CH$_2$ CH$_2$
N
CH$_2$
CH$_2$ CH$_2$
N

$$+ 6H_2O$$

Formaldehyde solution reacts with anhydrous ammonia in a cooled reactor. The product is cooled purified by evaporation and centrifuging. After centrifuging the crystals are dried and the mother liquor is recycled back to the reactor.

2.1.9 Ethanol

The synthesis of ethanol from methanol provides an example of reductive carboxylation chemistry.

$$CH_3OH + CO + 2H_2 \rightarrow CH_3CH_2OH + H_2O$$

The technical limitation of methanol homologation to ethanol is found in catalyst activity and selectivity. Cobalt carbonyls mixed with ruthenium have been found to be useful in enhancing ethanol production rather than simple cobalt carbonyls.

2.2 HALIDES OF METHANE

All the halides of methane can be indirectly produced from methanol. The direct halogenation of methane, leads to the higher halogenated products than the simple methyl halide. Methane is the most difficult of all hydrocarbon for any reaction, however Chlorination and bromination are conducted at elevated temperatures or photo-chemically. A fixed ratio between methane and chlorine can give rise to preferred chloromethane at higher yields, while other chloromethanes also accompany. As an example methyl chloride can be produced by choosing very great ratios of methane to chlorine and passing over a catalyst like cupric chloride deposited on pumice at 450°C.

$$CH_4 + Cl_2 \rightarrow CH_3Cl + HCl$$

As the Chlorination increases, higher chlorides appear due to overriding resonance effect over polar effect of chloride.

The substitution reactions go through free radicals. Free radicals being active, can bring the chain reaction mostly in an uncontrollable way. The initial stage of formation of free radicals may be supposed to be the only difficult step in operation as shown below:

$$Br_2 \rightarrow Br + Br$$

$$CH_4 + Br \rightarrow CH_3 + HBr$$

$$CH_3 + Br_2 \rightarrow CH_3\,Br + Br$$

$$CH_3\,Br + Br \rightarrow CH_2Br + HBr$$

$$CH_2\,Br + Br_2 \rightarrow CH_2Br_2 + Br$$

and so on, finally termination of free radical takes place

$$CH_3 + CH_3 \rightarrow C_2H_6$$

$$Br + Br \rightarrow Br_2$$

$$CH_2Br + Br \rightarrow CH_2Br_2$$

Presence of a catalyst is not necessary, as the chlorination reaction is light sensitive one; further the fixed ratio of reactants does yield the specific chlorinated product. Thus, it is easy to produce carbon tetra chloride by recycling the intermediates, rather than to confine on any single chlorinated intermediate. A product distribution of Monochloromethane 58.5%, Dichloromethane 2.3%, chloroform 29.5%, carbontetrachloride 9.7% is more common in such reaction.

A process for manufacturing chloromethanes from methanol and chlorine has been designed by Valcun Materials & Co. Methanol and hydrogen chloride are reacted in presence of a catalyst like zinc chloride at a temperature of 350°C to yield mainly chloride and small amount of other chloromethanes.

Methylchloride is primarily consumed in silicons production and tetramethyl lead. Other uses are in manufacturing rubbers, herbicides, amines and it can be used as solvent too. Methyl chloride is used in the production of dimethyl sulphoxide as shown by the reaction.

$$CH_3\,Cl + Na_2S \longrightarrow (CH_3)_2S \longrightarrow (CH_3)_2\,SO$$

Methylene dichloride is a good paint removal solvent; also it is a good propellant for aerosols because of its non-flammable nature. The other products of chloro-fluorination of methane are gaining importance as refrigerants and aerosol propellants.

2.2.1 PTFE

The monomer of teflon, tetra flouro ethylene is an odourless, colourless and non toxic gas (B.P. — 76.3°C). The heat of formation is — 152 k. cal. per mole, while heat of polymerisation is characteristically higher than other monomers i.e. — 41.12 k.cal.

Teflon has become a very useful engineering plastic because of its high thermal stability, insolubility in solvents, outstanding electrical insulating and dielectric properties and low frictional coefficient (a property suitable for coatings with waxes). It is very tough and not easily attacked by chemicals below 300°C. Further it is resistant to heat; this quality has permitted its use in a temperature range of 300 to minus 200¢XC; infact it is as much resistant as platinum12. Flouro-chloro carbons can be decomposed over platinum at 100¢XC as shown by the following reaction :

$$2\,CHClF_2 \rightarrow 2\,CF_2 + 2HCl$$

Chloro-fluorination of methane can take place in a single stage according to the following equations :

$$CH_4 + 4Cl_2 + 2HF \rightarrow CCl_2F_2 + 6HCl$$

$$CH_4 + 4Cl_2 + HF \rightarrow CCl_3F + 5HCl$$

or

$$CCl_4 + 2HF \rightarrow CCl_2F_2 + 2HCl$$

Montecatini Edison process uses chlorine and hydrogen fluoride mixture to react with methane in presence of a catalyst (chlorides of metals) in a fluidised bed reactor. The mixture is recycled to allow a close control over reaction temperature which is in the range of 370 to 470°C at a pressure of 4 to 6 bars. Outlines of this method is shown in Fig. 2.4.

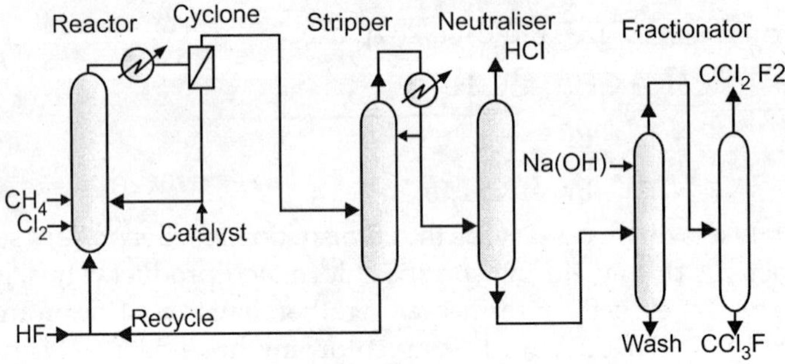

Fig. 2.4 Flouro-Chloro Methanes.

Fluorocarbons are purified after absorbing the acid gas by water wash and later sodium hydroxide wash. The yield of carbons on the basis of methane is well above 99%.

2.3 METHYL AMINES

Like chloromethanes, ammonia can give rise to three amines namely primary, secondary and tertiary. A continuous process for the manufacture of mono, di and tri-methylamines consists of the reaction between ammonia and methanol in a fixed bed reactor containing dehydration catalyst. Catalysts like silica, alumina gels function at a pressure of 6-12 atmospheres and a temperature of 380-450°C. The ratio between methanol and ammonia has a great influence on the product pattern. A ratio of 2 : 1 (methanol to ammonia) gives a product containing mono, di and tri-methylamines in a ratio of 43 : 24 : 33 respectively. Yield seems to be higher than 98%.

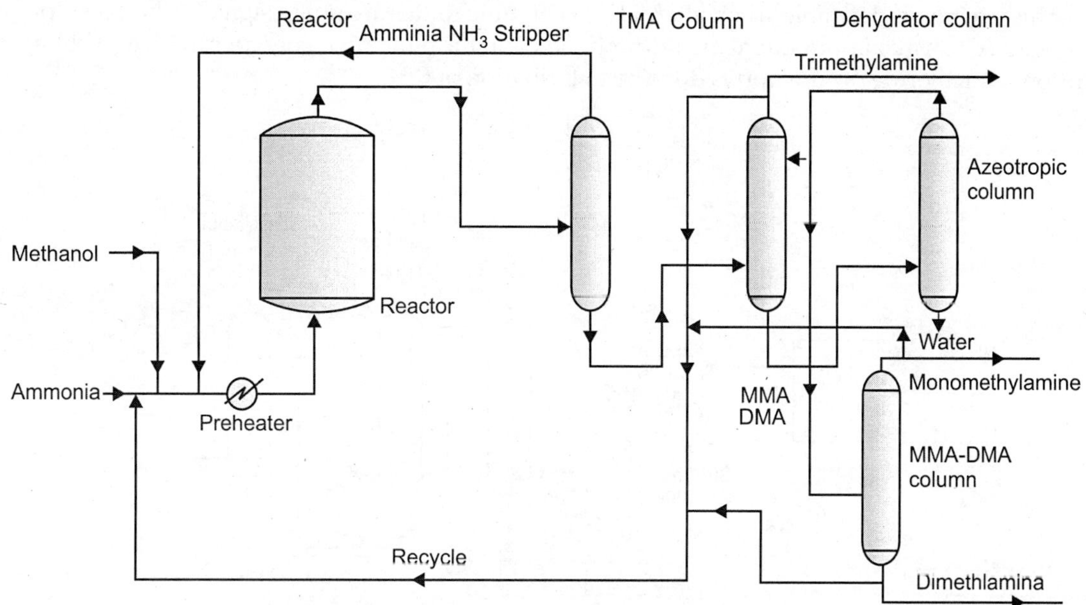

Fig. 2.5 Methylamines Production.

The production sequence is shown in Fig. 2.5. The product of all these three amines is separated in a battery of distillation columns. Trimethylamine-ammonia form an azeotrope; hence go back as recycle mixture to the first column itself. It can also be separated by extractive distillation with water). The reaction water may be separated immediately or can be distilled after tri-methylamines distillation. From the next column excess of TMA comes out as a top product while the bottoms of the column contain DMA and MMA. The mixture is fractionated in two distillation columns, while the liberated vapours are cycled into the process.

Amines are used as intermediates. Dimethyl amine is used for producing dimethyl formamide, a solvent and a substance used in plastics according to the following reaction :

$$(CH_3)_2\,NH + CO \rightarrow (CH_3)_2\,N-\underset{\underset{\displaystyle O}{\|}}{C}-H$$

2.4 CARBON DISULPHIDE

Carbon disulphide is produced by reacting sulphur in vapour form with methane over a catalyst like alumina gel or synthetic clay. Drastic heating is foremost important. Lower conversion rates demand recirculation of exhausts. The reaction occurs in different ways.

$$CH_4 + 2\,S_2 \rightarrow CS_2 + H_2S$$

$$CH_4 + S_2 \rightarrow CS_2 + 2H_2$$

$$CH_4 + 2H_2S \rightarrow CS_2 + 4H_2$$

Another route is by oxidation.

$$CH_2 + 2S + O_2 \rightarrow CS_2 + 2H_2O$$

This reaction is desirable when alternative economic routes are not in sight. Carbon disulphide is consumed to a maximum extent in rayon industry and the remaining goes for carbon tetrachloride. Carbon tetrachloride production is done by chlorination of CS_2.

$$CS_2 + 3Cl_2 \rightarrow CCl_4 + S_2Cl_2$$
$$CS_2 + 2S_2Cl_2 \rightarrow CCl_4 + 6S$$

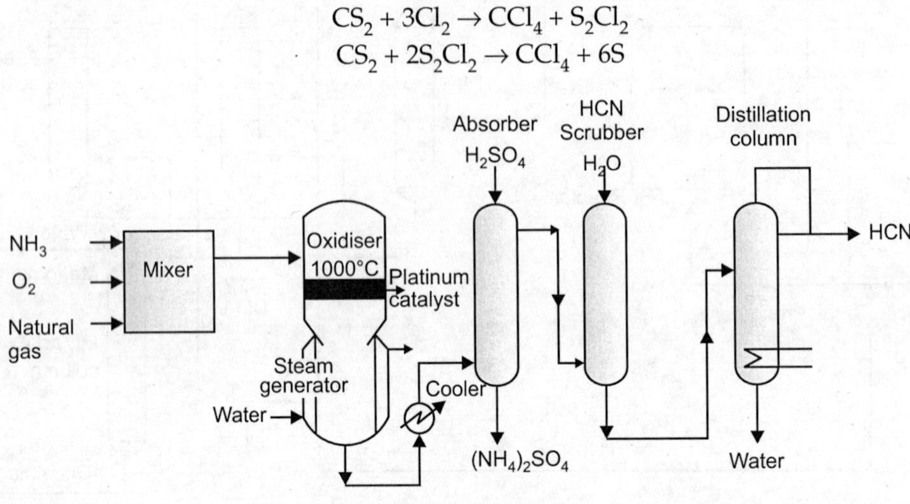

Fig. 2.6: HCN Synthesis.

2.5 HYDROGEN CYANIDE

Hydrogen cyanide seems to be an unconventional product of methane. It is produced by oxidation of natural gas or methane in presence of ammonium over a platinum gauge, which acts as catalyst. The oxidation raises the temperature to 1000°C. The effluent gases cooled in heat exchangers and then scrubbed with dilute sulphuric acid to remove ammonia. The remaining gas is cooled, compressed and absorbed in water. Dilute gas solution is heated to expel the gas in pure form. Ethylene cynohydrin can be obtained by reacting hydrogen cyanide with ethylene oxide. Fig. 2.6 shows the outlines of hydrogen cyanide production.

2.6 LIQUID FUELS FROM METHANE

A novel method of conversion of methane to liquid fuels, through free radical oligomerisation over ZSM-5 catalyst has been presented by Fox14 et al. Methane is selectively oxidised in a circulating fluid bed reactor, with reducible metal oxide catalyst to produce free radicals, which can later combine in a reactor over ZSM-5 catalyst to yield various olefins and oligomer products. With oxygen, at elevated pressure 4:1 methanol and formaldehyde can be produced, which can also be oligomerised. Third method consists of methane oxychlorination under pressure over $CaCl_2$ – KCl catalysts to yield mono and dichloro methanes. These are oligomerised to liquid fuels mainly gasolines. Thus oxidation, partial oxidation, oxychlorination products can be compelled to give rise to gasoline type fuel.

REFERENCES

1. A. Aquilo, J.S. Alder, D.N. Freeman & R.J. Voorthoeve. HCP, March, 83, 58.
2. Tuhin K. Roy & S.K. Gupta, Chemical Age of India, Vol. 35, No. 3, 1984, 144.
3. R.A. D alla Betta & K.K. Ushiba. HCP, Nov. 1980, 157.
4. Chang, C.D. & Silestri A.J.J. Catalysis 47, 1977, 49.
5. Dejaifve P., Vedrine J.C., Boils V. & Derouane E.G.J. Catalysis, 63 (2), 1980, 337.
6. T. Inui & Y. Takegami, HCP, Nov. 82, 117.
7. V.U.S. Rao & R.J. Gormley. HCP, Nov. 1980,141.
8. R.G. Anthony, & P.E. Thomas, HCP, Nov. '84, 95.
9. Harold A. Wittcoff, & B.G. Reuben, Industrial Organic Chemicals in Perspective Part-I; John Wiley & Sons, N.Y. 1973, 62.
10. Albert V. Hahn. The Petro Chemical Industry Market and Economics, McGraw-Hill, 1970, 80.
11. D.L. King, K.K. Ushiba & T.E. Whyte, HCP, Nov. 82, 135. .
12. F.T. Spencer, HCP, July, 1966, 91.
13. J. Ramioulle & Andre David, HCP, July, '81, 117.
14. J.M. Fox, T.P. Chen & B.D. Degan; CEP, April 1990, 42.

❒

CHAPTER 3

Chemicals from Ethane-Ethylene-Acetylene

3.1 ETHANE : OCCURRENCE

Ethane is associated with methane as a second major component (upto 10%) in natural gas. In casing head gas and associated gas its presence is high upto 20%. Even in major secondary processes like thermal cracking or reforming and catalytic reforming, it is associated with other gases to a nominal extent of 5-15%. Practically faced with separation difficulty from other components, the possible utility for a maximum spectrum of derivatives from ethane is cut into smaller size. The chemicals obtained from ethane are very few and really much interest is not shown towards exploitation of ethane. However, its value when converted into an olefin (ethylene) is remarkably high. Ethane is indeed the best preferred feedstock for ethylene. In fact, what methane to reforming is, ethane to ethylene is. Some countries like USA, recover ethane from natural gas to a substantial extent (even 25%). Though chloro and nitro derivatives are cited as direct derivatives of ethane, the spectrum can be increased if an attempt to utilise ethane from natural gases is systematically made.

3.1.1 Halides of Ethane

The most important halide of ethane is ethyl chloride. This is obtained in a similar way as methyl chloride, though ethane chlorination is easier than methane. Conspicuously, the chlorination of ethane results only in monochloroethane; as ethyl chloride resists further replacement of hydrogen atoms. The costly side product of direct chlorination of alkanes is hydrogen chloride, which has no sufficient marketability, hence manufacturers proceed through ethylene and hydrogen chloride reaction. Ethyl chloride, commonly, is produced by reacting ethane and chlorine over a catalyst, usually a metal chloride at 300 to 500°C.

Most of the ethyl chloride (90%) is consumed in the following way for the production of tetraethyl-lead:

$$4\,NaPb + 4\,C_2H_5Cl \rightarrow Pb(C_2H_5)_4 + 4\,NaCl$$

Certain aldehydes and ketones catalyse the above reaction. The remaining chloride is used as refrigerant or as anaesthetic. It has been found that fluoro carbons developed by using a mixture of methane and ethane are less toxic and more stable than methane fluorocarbons alone.

Yet, another chlorine derivative of a great significance is through a recent process TRANSACT(1); that involves the reaction between chlorine and ethane in the presence of oxygen to give rise to vinyl chloride. The streams of these gases are sent into molten salt bath, where a high temperature and less residence time favours the synthesis of vinyl chloride. This method has earned considerable interest as it dispels the use of purified ethylene stream.

3.1.2 Nitro-Ethane

Nitration of ethane has no significance when compared with nitration of aromatics. Majority of the nitro compounds, find their way into explosives : Some are offered as a source of amines. Some applications such as lacquers, solvents are common.

The nitration reaction is very smoothly conducted by spraying nitric acid into a hot stream of gas which is kept under pressure (10-12 bars) and around 400°C. Direct nitration with a nitrogen peroxide and oxygen is also cited in literature. Recycling is common in such reactions.

3.1.3 Oxidation of Ethane

While methane oxidation primarily yields methanol and small percentage of formaldehyde, ethane oxidation gives a complex mixture of products. Though the composition of products vary depending upon the conditions of oxidation, still the components are same. When ethane is oxidised at a pressure of 100 bars the following products are expected:

Product	*Temperature 262°C Time 40 Min.*	*278°C 5 Min.*
Ethyl alcohol	24.3%	28.2%
Methyl alcohol	15.5%	43.6%
Acetaldehyde	6.3%	20%
Acetic acid	52.3%	5.5%
Formaldehyde	1.6%	2.7%

It is easy to oxidise higher paraffins, though getting a single product is remote. Often a mixture is obtained. In all these oxidation reactions, the gaseous mixture is supposed to be made up of a ratio outside the combustion limits to avoid explosions. Shown below, are the combustion limits for different gases at normal conditions :

Hydrocarbon	*Lower combustion limit*	*Higher combustion limit*
Methane	5.35	14.85
Ethane	3.12	14.45
Propane	2.37	9.05
Butane	1.60	8.50
Pentane	1.42	8.00
Ethylene	3	17

Ethane, propane and higher alkanes can be oxidised by air instead of oxygen. It is preferred to introduce steam alongwith air when oxidation of higher paraffins is undertaken to moderate the reaction.

Propane oxidation products at different pressures[(2)] show:
(propane to Air ratio 1 : 3.6)

Product	*Pressure (Bar).*	*1*	20	60
	Temp. °C	373	281	252
Aldehydes		20.5	21.8	13.5
n-alcohols		19.7	21.0	17.5
Iso-propyl alcohol		1.3	2.8	6.2
Acetone		0.5	4.3	12.5
Acids		4.3	17.0	19.0
Carbon-dioxide & Carbon-monoxide		28.6	33.1	31.3
Propylene		25.1	0.0	0.0

With increase of pressure, lower oxidation temperature is sufficient. Formidable problem of separation of reactor effluent into pure components at economical rates forms a major hurdle liberal utilisation of this technology. Like propane, butane yields on oxidation acetaldehyde, formalin and acetic acid in a weight ratio of 3.2 : 3.3 : 2.0 respectively.

3.2 ETHYLENE PRODUCTION

3.2.1 Cracking (Dehydrogenation of Ethane)

The gigantic and invincible growth of petrochemical industry owes principally to one single intermediate ethylene. In the last three decades the production of ethylene round the world increased several folds and continues to do so. During 1950 the production of ethylene was 0.67 million tons only, while it touched 40 million tons by 1980. However the world capacity, in 1990 against a demand of 48.4 MMT, the installed capacity was 51.5 MMT. By 1996 this figure changed to 79.3 MMT and production of polymers like polyethylene (LLDPE 33.9, HDPE 28.6, LDPE 38.2, VCM 25 MMT) was witnessed.

The scene is quite different in India. During 1990-91 the production ethylene was 2.4 lakh tons while the demand was 7 lakh tons: the demand for ethylene, propylene, butadiene, during 2001-2002 would be 3.119,1.561,0.24 MMT against an installed capacity of 1.517,0.722,0.153 respectively. By 2005 India is expected to have ethylene capacity of 4 MMT as shown below :

Existing Naphtha /gas Cracker		
State/Name of Unit	*Feedstock*	*Capacity (Ethylene)*
	GUJARAT	
IPCL, Vadodara	Naphtha Cracker	130,000 tpa
RIL, Hazira	Naphtha/NGL Dual Feed	750,000 tpa
IPCL, Gandhar	Gas	300,000 tpa
	MAHARASHTRA	
OSWAL Agro, Mumbai	Naphtha	23,000 tpa
NOCIL, Mumbai	Naphtha	63,000 tpa
IPCL, Nagothane	Gas	400,000 tpa

Existing Naphtha /gas Cracker		
State/Name of Unit	*Feedstock*	*Capacity (Ethylene)*
	UTTAR PRADESH	
GAIL, Auriya	Gas	300,000 tpa
	WEST BENGAL	
Haldia Petrochemicals Ltd., Haldia	Naphtha	466,000 tpa
Panipat Refinery	Punjab	300,000 tpa

Proposed/planned Olefinic Cracker for which, industrial approvals have been approved.		
Project	*Feed Stock*	*Ethylene Capacity ('000 tpa)*
Reliance Assam Petrochemicals Ltd., Brindavan, Shri Ram Mill Compound Ganpathrao Kadam Marg, Mumbai.	Gas	300 200
Karnataka State Industrial and Investment Development Corp. Ltd.	Naphtha	300
Kerala State Industrial Development Corp. Ltd.	Naphtha	500
National Organic Chemicals Industries Ltd., (NOCIL), Mumbai	Naphtha	700
Punjab State Industrial Development Corporation Ltd., (PSIDC)	Naphtha	300
Tamilnadu Industrial development Corporation Ltd., (TIDCO), Chennai	Naphtha	500

3.2.1.1 Thermal Cracking Reactions and Theory

Heat sensitive, high molecular weight paraffins fragment when the temperature exceeds 400°C. A molecule of C_nH_{2n+2}, where 'n' ranges from and above 25 easily splits into two, almost at the middle, resulting in one saturated molecule and other unsaturated molecule.

$$C_nH_{2n+2} \rightarrow C_{n/2}H_{n+2} + C_{n/2}H_n$$

The above reaction is foremost and conclusive at a moderate temperature around 400°C, but fragmentation continues with increasing temperature, giving rise to often simple products and occasionally complex molecules. Appearance of different intrinsic reactions is common depending upon the molecular size. For a paraffin of less molecular order, the following reactions illustrate the mode of cracking.

$$C_{12} \rightarrow C_6 + C_6$$

(One is always unsaturated)

$$C_6 \rightarrow C_4 + C_2 \text{ or } C_3 + C_3$$

Unsaturates (olefins) obtained in the process crack again

$$C_4H_8 \rightarrow C_2H_6 + H_2 \text{ or } C_2H_4 + CH_4 + C$$
$$C_4H_8 \rightarrow CH_4 + C_3H_4 \text{ (Diolefin/Alkyne)}$$
$$C_2H_4/CH_4 \rightarrow C + H_2$$

Olefin cracks or dehydrogenates to diolefin or an alkyne. Further severity in conditions results in production of methane, hydrogen and carbon. These are regarded as stable, end products of cracking operations. Unsaturates being active, under congenial thermal cracking conditions form dimers, trimers, etc. *i.e.* condensation to bigger molecules.

$$2C_2H_4 \rightarrow C_4H_8$$
$$C_4H_8 + C_2H_4 \rightarrow C_6H_{12}$$
$$2C_3H_4 \rightarrow C_6H_6 \text{ (ring)} + H_2$$
$$3C_2H_2 \rightarrow C_6H_6$$

Hydrogenation also occurs but to a less extent

$$C_3H_4 + H_2 \rightarrow C_3H_6$$

Aromatics, saturated ring structures follow a different pattern of cracking. Chain detachment is frequent, followed by dehydrogenation as shown below.

$$C_6H_5{-}CH_2.CH_2.R \longrightarrow C_6H_5{-}CH_3 + RCH_2 \text{ (Olefin)}$$

$$C_6H_5{-}CH_2.CH_3 \longrightarrow C_6H_5{-}CH{=}CH_2 + H_2$$

Ring saturates are converted to unsaturates

$$C_6H_{11}{-}CH_2.CH_3 \longrightarrow C_6H_6 + C_2H_5 + 2H_2$$

Ring opens in the extreme conditions of cracking

$$C_6H_{12} \longrightarrow CH_2{:}CH_2 + CH_2 = CH - CH = CH_2 + H_2$$

$$C_6H_6 \longrightarrow 3C_2H_2 \text{ or } CH_2 = CH\,CH = CH_2 + 2C$$

$$C_6H_6 \longrightarrow C_2H_4 + 4C + H_2$$

Above reactions illustrate the carbon and hydrogen formation. Light paraffins when subjected to cracking dehydrogenation reaction shall be imminent and it dominates over other cracking reactions :

$$C_4H_{10} \rightarrow C_4H_8 + H_2$$
$$\rightarrow C_3H_6 + CH_4$$
$$\rightarrow C_4H_6 + 2H_2$$
$$\rightarrow C_2H_6 + C_2H_4$$

The tendency of dehydrogenation becomes less, when the chosen molecule is above C_4. Dehydrogenated molecule normally exhibits isomeric forms, such as

$$C_4H_{10} \rightarrow CH_3CH_2CH = CH_2 + H_2 - 131.5 \text{ kJ/mole}$$
1-butene
$$\rightarrow CH_3CH = CH.CH_3 + H_2 - 116.8 \text{ kJ/mole}$$
Cis-2-butene
$$\rightarrow CH_3CH = CH - CH_3 + H_2 - 115.8 \text{ kJ/mole}$$
Trans 2-butene

A plausible explanation based on all these reactions concedes that the product of cracking may have higher molecular weight than the feed itself, obviously due to polymerisation or condensation.

Industrial thermal cracking operations are classified based upon temperature parameter as shown in Table 3.1.

Table 3.1: Thermal Cracking Operations

Cracking Temperature °C	*Nature of Operation*	*Products*
425-460	Visbreaking	Fuel Oil
460-520	Thermal Cracking	Gas, liquids
520-600	Low temp., coking	Gas, liquids, coke.
600-800	Gas Production	Gas and unsaturates.
800-1000	High temp, coking	Gas, heavy aromatics, coke
above 1000	Decomposition	H_2, gas, carbon-black.

Theory of Thermal Cracking. Cracking proceeds *via* free radical mechanism(3). Free radicals are atoms or group of atoms with bare unpaired electrons, thus differing from other types of charged particles known as ions. The following reactions explain the difference.

$$NH_4Cl \rightarrow NH_4^+ + Cl^- \quad \text{ions}$$
$$H_2 \rightarrow H^+ + H^- \quad \text{charged particles}$$
$$CH_4 \rightarrow CH_3^+ + H^-$$
$$H_2 \rightarrow \dot{H} + \dot{H} \quad \text{radicals}$$

In a molecule of hydrogen there are two electrons sharing the spin in opposite directions. For hydrogen paired electrons are arranged in S-shell like | H | ↓ | H | ↑ |. When atomic hydrogen is formed without charges unpaired electrons result in S-shell; $H_2 \rightarrow$ | H | ↓ | + | H | ↑ |

Similarly,

$$R.\dot{C}H_2CH_3 \rightarrow R\dot{C}H_2 + \dot{C}H_3$$

These unpaired electrons are very active, and try to form stable compounds by acquiring unpaired electrons from sources available. Motive power of free radicals to combine with other radicals or metallic bodies is immense; generous disociation into small radicals is also high.

Hydrocarbon (charged) ions are usually referred as carbonium ions or carbanium ions depending upon the charge they carry.

$$CH_3.CH = CH_2 + H^+ \rightarrow CH_3 - CH^+.CH_3 \text{ carbonium ion}$$
$$CH_3CHO + OH^- \rightarrow CH_3CHOOH^- \text{ (carbanion ion)}$$

The reactions with free radicals are fast; indeed only one radical is sufficient to bring about a series of reactions as shown below :

or

$$
\left.\begin{array}{c} H_2 \rightarrow H + H \\ O_2 \rightarrow O + O \\ C_5H_{12} \rightarrow C_5H_{11} + H \end{array}\right] \text{Radical production}
$$

$$
\left.\begin{array}{c} C_5H_{12} + H \rightarrow C_5H_{11} + H \\ C_5H_{12} + C_3H_5 \rightarrow C_7H_{16} + H \\ C_5H_{12} + C_2H_5 \rightarrow C_5H_{11} + C_2H_6 \\ C_3H_7 + C_5H_{12} \rightarrow C_4H_{10} + C_4H_9 \end{array}\right] \text{Propagation}
$$

$$
\left.\begin{array}{c} C_3H_7 + H \rightarrow C_3H_8 \\ C_5H_{11} + C_3H_7 \rightarrow C_8H_{18} \\ H + H \rightarrow H_2 \end{array}\right] \text{Termination}
$$

Different free radicals have different lives. Hydrogen free radicals have the highest stability and so a higher life. The stability or life of a radical decreases with the increasing size of molecule. Minimum bond energies invoke the scission at a convenient position during the reactions. The bond energy of C-C is 262.7 kJ/mole while C-H bond requires 358 kJ/mole to break open.

Due to the high stability of tertiary carbonium ions, the asymmetric carbon bonds are easily ruptured than the symmetrical ones. This is amply sufficient to explain why isomerisation does not result in thermal cracking operations. The free radical mechanism offered by Rice, though it cannot correctly predict the composition in kind and quantity of the products, still, is the best tool in view of the complexity of reactions.

Excerpts of S. Barendregt(4) et al's work on cracking oils for olefins indicate a spyro kinetic scheme which involves 2,000 intervening leading to C 25 cracking. 500 reactions, based on radicals were proposed for steam cracking.

3.2.1.2 Discussion on Important Parameters

Earlier methods provisionally used Soaker line system where 1-2 sec. time is required for completion of reaction in presence of steam. In such systems ethylene yield was around 19-22% from light distillates which was considered to be artifice. In all respects ethane cracking is similar to naphtha cracking; requiring a range of temperature 750 to 930°C at 4-6 bars pressure.

Presently tubular reactors of 5 to 14 cm dia. and 120 metre long are in service. The residence time of the reactant is very important which is customarily measured in the radiant section of the furnace (above 750°C) and it is of the order of 0.8 to 1.6 seconds. Outlet gases must be cooled to 300°C almost instantaneously lest the free radical reaction should go to foster thermal polymers. Obviously for any feedstock residence time, temperature, hydrocarbon partial pressure, the quench time are regarded as important parameters.

3.2.1.3 Feedstock-Ethylene Yield

Chemical nature of feedstock dictates very much the yield of ethylene even at the fixed parameters like temperature, pressure and severity etc. For maximum production of ethylene, recycling of products to extinction is often opted. By recycling, ethane can be completely cracked to produce ethylene; which is higher than 35% ethylene obtained from naphtha. Tables 3.2 (*a*), (*b*), (*c*) show the effect of recycling for various stocks. When lower molecular weight feedstocks are cracked more hydrogen is produced. 10 to 16% hydrogen is available if the feedstock is shifted from heavy residue to naphtha. Even for the same stock (naphtha) the partial pressure changing from 1.2 to 0.3 bars raises ethylene formation from 9 to 15%. This signifies the importance of partial pressure. Further aromatic rich stocks are more refractory, hence efficiency of cracker decreases. Ethylene yield is connected to the hydrogen content of the feedstock. Thus ethane with 20.2% of hydrogen content would yield 80.5% ethylene while naphtha of 16.2% hydrogen content can give 37.8% ethylene.[5] The growing availability of natural gas liquids would thus enable ethylene production to shift from heavy stocks to light stocks. One more advantage to be gained by using these light stocks is that unnecessary flaring of product gases till the ethylene specifications are met, can be avoided.

Table 3.2. (*a*): Typical Products obtained through Thermal Cracking of Various Feedstocks

	A Feedstock				*B*		*C*
Composition of product	*Ethane*	*Propane*	*Butane*	*Naphtha*	*Ethane (at 60% conversion)*	*Propane (at 90% conversion)*	*Residuum (Iranian Crude) K = 11.6*
Hydrogen	3.3	1.2	0.7	1.0	3.7	1.19	1.0
Methane	5.1	25.3	23.3	15.0	4.1	24.00	$C_1 + C_2 + G_3 = 16$
Acetylene	0.2	0.3	0.5	0.6	0.3	0.41	—
Ethylene	47.7	36.6	31.2	31.3	48.91	34.47	– 15
Ethane	37.7	6.5	7.3	3.4	40.0	5.82	
Propene	2.1	14.1	17.8	13.1	1.12	14.6	5
Propene	0.4	8.1	0.9	0.6	0.08	9.28	
(3 - acetylenes)	—	—	—	—	0.09	0.53	3
Butadiene							
Butenes & Butanes	1.7	2.9	8.2	7.0	0.86	1.57	—
C_5^+	—	—	—	6.0	0.78	6.16	
Gas oil/Fuel oil	1.8	5.0	10.0	Balance	—	0	Balance

A : Peter Wiseman : An Introduction in Industrial Organic Chemistry (Applied Science), 1979, p. 67.
B : R.L. Espino & M.B. Sherwin : HCP, Nov. 72, p. 100.
C : J. Cusrack & R.H. Mash Ford : HCP, Jan. 74, p. 127.

Table 3.2. (*b*): Condensed Information on Cracking of Different Feedstocks (with ethane recycle)

Product wt %	*Feedstocks*					
	Ethane	*Propane*	*n Butane*	*FRN*	*LGO*	*HGO*
Ethylene	80.0	42.0	42.8	31.2	28.7	25.3
Propylene	1.8	16.2	19.3	16.1	12.8	12.8
Butadene	1.9	3.2	3.1	4.5	3.8	3.6
Aromatics	0.9	3.4	2.4	12.1	11.6	13.6

Data : HCP. Nov. 1973, 127

Table 3.2. (*c*): Different Feedstocks - Cracker Yields[6] (Recycling)

Feedstock	*Ethylene*	*Propylene*	*Butadiene*	*BTX*	*Others*
Ethane	80	1.4	1.4	0.4	17.0
Propane	44	15.6	3.4	2.8	34.2
n-butane	44.4	17.3	4.0	3.4	30.9
Light naphtha	40.3	15.8	4.9	4.8	34.2
Full range naphtha	31.7	13.0	4.7	13.7	36.9
Reformer reffinate	32.9	15.5	5.3	11.0	35.03
Light gas oil	28.3	13.5	4.8	10.9	42.5
Heavy gas oil	25.0	12.4	4.8	11.2	46.6
Waxy distillates	28.3	16.3	6.4	4.5	44.5
Crude residue	21	7	2	11	59
Crude oil	32.8	4.4	3.0	14.4	45.4

Table 3.2. (*d*): Ethane and Propane Thermal Cracking Yield % (Single pass)[6]

Conversion %	C_2 60	C_3 90.75	C_2 70	C_3 93.5
H_2	3.7	1.19	4.34	1.20
CH_4	4.10	24.00	5.95	25.55
C_2H_2	0.3	0.41	0.4	0.43
C_2H_4	48.91	34.47	53.73	36.07
C_2H_6	40.00	5.82	30.00	5.89
C_3 insaturates	1.21	15.13	2.10	14.71
C_3H_8	0.18	9.25	0.29	6.50
C_{4+}	1.6	9.73	3.09	10.74

Table 3.2. (*e*): Varying Yields on Naphtha for 453,000 tpa Ethylene (Chemical Industry Developments)*

Process	*Low severity no. recycle of C_2H_6*	*Moderate severity no recycle of C_2H_6*	*High severity no recycle of C_2H_6*	*High severity C_2 and C_3 recycle*	*High severity C_2, C_3, C_4 recycle*
Once through yield Wt %	26	28	30	30	30
Naphtha feed rate					
Million tonnes	1.76	1.63	1.53	1.08	0.96
Hydrogen rich gas 1000 tonnes	17	17	17	9	0.4
CH_4 rich gas 1000 tonnes	238	252	259	248	250
Ethane, 1000 tonnes	88	81	73	-	-
Propylene, 1000 tonnes	309	276	235	-	-
Butadiene 1000 tonnes	82	75	68	56	-
Butene/Butanes	130	95	63	48	-
BTX(1000 tonnes)	168	173	180	134	124
Fuel oil (1000 tonnes)	52	61	72	52	48

* (CP&E) India, March '74, 30-31

Table 3.2 (*e*) shows the amounts of gases and others obtained from a naphtha cracker that is able to produce 453,000 tpa of ethylene. Table 3.2 (*a*) shows the products of cracking for different feedstocks. Table 3.2 (*b*) shows the same, however the produced ethane is recycled to extinction. Table 3.2 (*c*) shows the effect of recycling. Table 3.2 (*a*) shows the yield of unsaturates at different conversions for ethane and propane fraction for once through.

3.2.1.4 Effect of Residence Time, Severity and Selectivity Factors

A high residence time and high partial pressure affect adversely the yield of ethylene. Ethylene yield with residence time and kinetic severity function (KSF) is shown in the Table 3.3(*a*). The concentration of other components is also affected by KSF; and is shown in Table 3.3(*b*). Partial pressure of feedstock has its positive influence on ethylene yields as shown in Table 3.3(*c*).

Table 3.3 (*a*): Residence time(8) Vs. Ethylene yield at Constant Kinetic* Severity Function (KSF)

Sec.	KSF	2.5	3.0	3.5
0.2		27.5	29.1	30.8
0.4		24.5	27.0	28.2
0.6		23.8	25.9	27.2
0.8		23.0	24.5	26.5
1.0		22.2	24.2	26.2

* Kinetic Severity Function can be determined analytically by finding the time required for disap-pearance of n-pentane over a coil length keeping temperature and hydrocarbon partial pressure constant.

$$\ln \frac{C_1}{C_2} = \int_{\theta_2}^{\theta_1} Ae^{-E/RT} \cdot d\theta$$

where C_1 & C_2 are pentane concentrations at θ_1 and θ_2 time

E - Activation energy A - Constant in cracking.

Next to severity, selectivity[(9)], a complicated term is often mentioned in the literature. In gas cracking it may be termed as the amount of ethylene obtained per unit weight of feed. In liquid products, where enormous by-products appear, the ratio of methane or propylene to ethylene is a more meaningful indicator of cracking selectivity; lower methane to ethylene ratio for given ethylene yield would give higher yields of valuables. Usually selectivity is dependent upon the boiling point of feedstock. Pyrolysis coil development has followed the usual trend of increasing ultimate yields of ethylene by designing more selective coils. A millisecond furnace[(12)], as discussed by John Dluzniewski and Raymond Orris, has become an attractive alternative to get olefins from liquid feedstocks with much less consumption of these, for the same yield of ethylene. The reasons are high temperature operation and a very short residence time (0.01 to 0.15 second only.)

Table 3.3 (*b*): Component Yield versus KSF[(4)]

Component	*KSF = 1*	2	3	4
Ethylene	12.5	23.8	26.5	28.6
Gasoline	32	23.7	21.0	22.2
Propylene	11.9	17.1	15.2	11.9
Methane	6.8	5.2	7.3	16.5
Benzene	4.0	5.2	7.3	9.8
Butadiene	2.8	4.2	4.0	3.8

Table 3.3 (*c*): Ethylene yield versus Partial Pressure(5) at KSF 3.0

HC Partial Pressure	*Ethylene yield %*
100 KPa	30.4
128	28.8
156	27.8
184	25.6
212	25.0

3.2.1.5 Coil Geometry-Heat Flux

Obviously the coil length to diameter and heat flux also play a determining role hence always are dependent as shown below :

Temp. °C	*Coil length Metres*	*L/D*	*Pressure kg/cm²*	*Avg. Heat flux - 1-2K.cals. hr.*	*Residence time sec.*
950	58	700	1.2	43,490	0.46
1000	43	520	0.98	64,010	0.35
1050	35	310	0.8	74,900	0.27

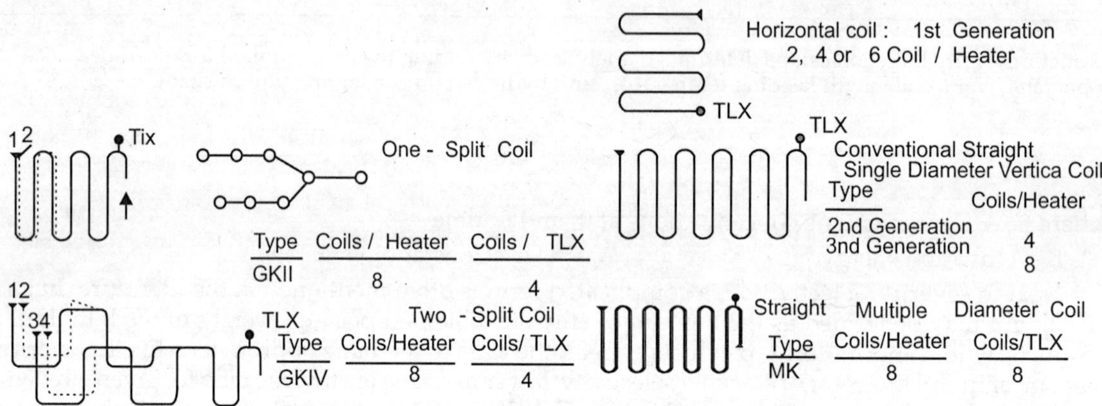

Fig. 3.1: Cracking Coil Designs.

Coil Designs. Coils of different designs are available for cracking operations :

(*a*) Single diameter coil consists of uniform diameter throughout the coil. The coil placement in the furnace may be vertical or horizontal. Most of the modern designs are inclined in favour of vertical assembly only.

(*b*) One split coil consists of two-4 pass tubes combined with two-2 pass tubes and further appended with two-1 pass tubes (shown in Fig. 3.1).

Table 3.4: Coil Diameter versus Ethylene yield (%)

Coil dia (Metres × 10^{-3})	*Temperature °C*		
	927	982	1032
50.8	28.8	30.1	30.5
101.6	26.9	28.0	28.7
152.4	25.7	26.7	27.4

Table 3.5 (*a*): Cracking Coil Performance*

Coil type	*Straight single dia*	*Straight tapered*	*One split*	*Two split*
Relative on stream time	1	1	1	1
Coil temperature °C	870	866	866	865
Propylene to ethylene ratio	0.42	0.42	0.42	0.42
Relative radiant duty	1.02	1.01	1.00	1.00
Relative coil weight	1.00	1.00	1.00	0.93
Residence time second	0.62	0.54	0.46	0.34
Relative HC throughput	1.02	1.02	1.00	0.99
Ultimate ethylene (Wt%)	33.24	33.40	33.95	34.19
Total C_3 (Wt%)	13.83	13.85	14.14	14.36
Benzene (Wt%)	9.22	9.14	8.84	8.66
Toluene (Wt%)	4.64	4.60	4.51	4.46
Xylene (Wt%)	1.07	1.07	1.05	1.04
Butadiene (Wt%)	4.22	4.29	4.48	4.58
Total C_4 (Wt%)	7.04	7.12	7.40	7.58
Total C_5 (Wt%)	2.51	2.58	2.70	2.81
H_2 + Tail gas (Wt%)	18.21	18.13	17.74	17.48

*For identical lengths and high severity
Feed-Naphtha (sp. gr. 0.707 ; Outlet P : 0.6 kg/cm^2)
Steam dilution = 6:10
Steam to HC

(*c*) Tapered coil has its own role to play (Tables 3.4 *a* and *b* & Tables 3.5 *a* & *b*). Compared with equal furnace lengths the same severity, an increased selectivity towards light olefins is indicated in a two split coil. Hydrocarbon partial pressure profile of the split coil is lower than the straight and tapered coils. Residence time is also found to decrease in split coils.

Evidently, size, shape, length to diameter ratio, material of construction and not to reckon the operating conditions, all adequately influence the cracking. Performances of split coils in place of a single dia coils is convincingly commendable, and such novelty and innovation are expected to elate profits.

Table 3.5 (*b*): Comparison of Pyrolysis Heater Coils[9]

Coil	*Naphtha MT/HR*	*Temp. C*		*Ethylene production T/Year per coil*
		Wall	*Outlet*	
10 cm	1.1	990	872	2,290
SRT II HS	2.27	990	840	4,700
SRT I	2.75	945	835	5,700
SRT II HC	6.0	980	830	12,500

SRT : Short Residence Time; H.S. : High Severity; HC : High Capacity.

3.2.1.6 Effect of Partial Pressure (Steam)

Steam helps in cracking operations mainly as a diluent, *i.e.* to reduce the partial pressure of hydrocarbon. Reduction in pressure increases the cracking consequently more light ends are formed. Further steam can oxidise the carbon deposited in tubes. Table 3.3 (*c*) shows the effect of steam on cracking.

Tables 3.2 (*a*) to (*e*) show the composition of outlet gas from a cracker for different feedstocks. Inquisitive look into these tables reveals that the nearest saturated compound would give highest amount of olefin of that carbon number. Ethylene produced from ethane at 60% conversion can be as much high as 50% per pass[11], and Complete extinction by recycling would result in 80% ethylene. Earlier designs used stainless (310) steel or inconel alloys, but now HK 40 (25 Cr : 20 Ni) alloys[13a] are safely prescribed upto 980°C. The wall thickness[13b] plays a vital role, as carbide formation increases, chromium will be withdrawn from the matrix of Nickel-chromium-iron and encourage failures. By sacrificing tube life and resorting to high severity and high temperatures[13c] with a variety of feedstocks; high yield of ethylene can be obtained: Above this temperature Cr, Ni modified with 2% silicon alloys are found to be the best. Other modifiers, like tungsten, Columbian are also used under the trade names G4868, G 4857 etc.

When ethane and propane are chosen as feedstocks the following points may be taken into consideration.

Methane. It enjoys a free ride in the reactor, hence increases the load on reactor and purification system.

Propane and Butane. Both produce olefins of the same carbon atom in the beginning, then diolefin production increases with severity. Hence hydrogenator for converting diolefins to olefins is required. Diolefins in production of mixed polymers and alkylates do not pose any problem.

Carbondioxide. In all cracked stocks its presence upto 20 ppm is inevitable. It is better if it is eliminated in the beginning itself.

Hydrogen sulphide. Present in all feedstocks though the concentration may vary. Sulphur compounds are generally good in cracking operations for they are good inhibitors of coke. Of course, contamination, corrosion should be taken care of.

Oxygen. Dangerous component as explosions are common.

Nitrogen and inerts. No problems.

It may be apprehended even the material of construction can increase the conversion upto 10%. The yields of ethylene, benzene at different severities and contact times are predicted and tested too as shown in Fig. 3.2, usually the ethylene yield shoots up from 25 to 40% as the feedstock varies from gas oils to light naphtha. Also propylene to ethylene ratio increases as feedstock molecular weight increases. High severity favours yield of ethylene. Gas oils from atmospheric columns and vacuum columns when cracked (low and high severity) gives the following pattern[12].

	Low severity		*High severity*	
	AGO	*VGO*	*AGO*	*VGO*
Methane	8	6	3.7	9.4
Ethylene	19.5	19.4	26	23
Ethane	3.3	2.8	3.0	3.0
Propylene	14.0	13.9	9.0	13.7
Butadiene	4.5	5.0	4.2	4.9
BTX	10.7	18.9	12.6	16.9
C_5^+	10	-	8.0	-
Fuel oil & others	23.6	26.4	21.5	23.8

3.2.2 Ethylene Furnaces and Naphtha Cracking

The cracking furnace consists of a separate convection section for preheating and a radiant section exclusively for cracking. Burners are placed along the side walls or at the bottom of the furnace. The temperature on the outside walls of tubes will be around 950°C. Specially for high cracking operations (more ethylene) may be a temperature of 1000°C is required. Pressure in the tubes may touch 5 bars. In such circumstances, the coke formation is increased.

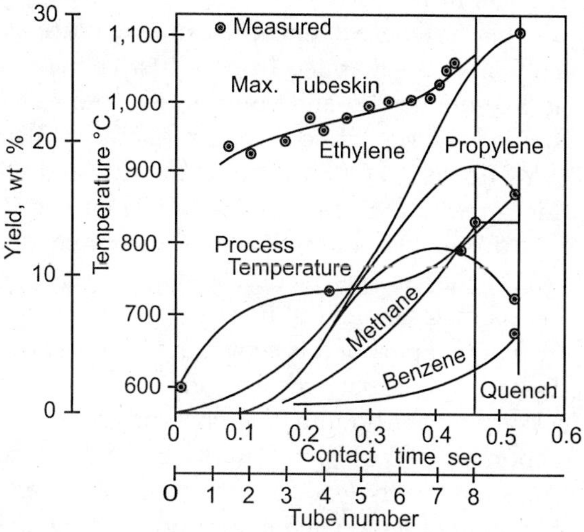

Fig. 3.2: Prediction vs. Measurements

3.2.2.1 Naphtha Cracking

The schematic Fig. 3.3 shows typically the production of ethylene and co-product by the pyrolysis of naphtha. Feedstock is preheated through various process heat exchangers and then passed through the convection section of the furnace whereby the temperature goes upto 300 to 400°C. Here the feedstock is mixed with steam (700-800°C) and routed into the radiant section of the tabular furnace. The coil arrangement, the material of the construction, and dimensions are discussed already. The type of furnace is the propriety of the processor, for example : Lummus employs short residence time (SRT) furnace containing entirely separate pyrolysis radiant section; Steam dilution of feedstock to an extent of 0.6 kg/kg HC is done prior to entering the cracking zone. The temperature of the furnace of these types is usually higher than 900°C and the effluents of furnace record a temperature of 800-850°C, at just emerging point. The gases are fed through high pressure steam (10 to 14 MPa) generators and transfer line heat exchangers (not shown in figure). The products are later fractionated to separate (heavy) fuel oil from volatiles. In other designs direct quenching by means of available process-oils is successfully substituted. Heat saving and extraction of solubles compact plant size are some of the advantages in oil quenching.

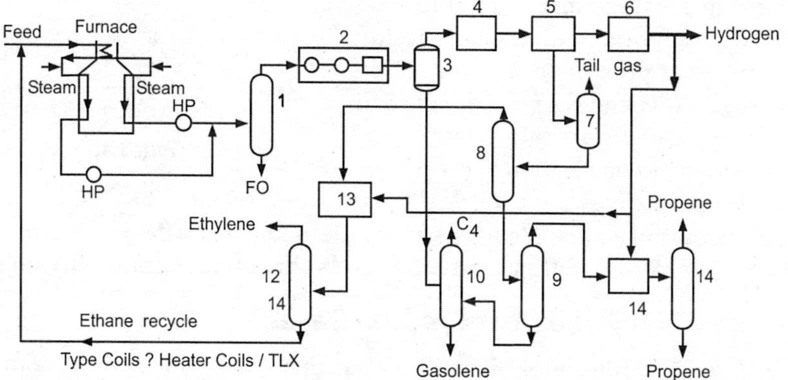

1. Product Fractionator, 2. Coolers and Compressors Section, 3. Liquids-Separator, 4 Acid Gas Removal Unit, 5. Demethaniser Feed Preparation Unit, 6. Hydrogen Purification, 7. Demethaniser, 8. De Ethaniser, 9. Depropaniser, 10. Debutaniser, 11. C_3-Splitter, 12. C_2-Splitter, 13. Hydrogenators for Acetylene, 14. Propadiene

Fig. 3.3: Ethylene Production (Based on Kellogg Technology).

Naphtha vapour and steam in required ratio enter the radiant section of the furnace. The hot products from the furnace are passed through a High Pressure (HP) generator and sent into a fractionator where quenching can take place along with fractionation to a certain degree, *i.e.* only heavy oils are separated from lights. The volatiles issued from the primary fractionator (Refer to Fig. 3.3) are cooled and compressed to 5 to 6 MPa. Part of heavy hydrocarbons (above C_3) are liquefied and separated through separator (3) and goes into the debutaniser (10) and depropaniser (9) fractionators. Often vapours are infested with acidic constituents like CO, CO_2, H_2S, SO_2, etc. which are eliminated in conventional manner. After the treatment and drying the main constituents in the product vapour will be, methane, hydrogen, ethylene, ethane and slight amount of acetylene (1%). All the gaseous fractions are liquefied, except hydrogen which is separated in the gaseous form. This is being carried out in a unit (5) which supplies feed to demethaniser. Hydrogen gas is later purified and part of it is sent to hydrogenation units (13), to convert acetylene and propadiene to ethylene and propane respectively. Tail gas from demethaniser is abundant in methane, and used in fuel system. The bottom fractions from demethaniser are fractionated in column (8) that separates C_2 fractions from the heavier products. This C_2 fraction after mild hydrogenation is fed into C_2 splitter (12), whereby ethylene and ethane are obtained. Ethane is recycled back for cracking. Demethaniser bottoms are fractionated in depropaniser (9) to separate C_3 fractions from C_1 and heavy. Finally C_3 splitting takes place in a column (11). Debutaniser takes feed from two sources; heavy fractions separated from the coolers, compressors units in preliminary liquefaction of vapours; and the other source being the bottoms of depropaniser, C_4 fraction is obtained as top product, while gasoline fraction is obtained from the bottom.

3.2.2.2 C_2 Splitter

Ethane/ethylene separation is achieved mainly by low temperature fractionation. The splitter is usually operated under high pressure or low pressure.

Fig. 3.4 shows the outlines (12) of splitter. Feed gas enters the splitter at + 2.8°C under a pressure of 2 MPa. Usually a small pressure drop of 0.8 to 0.9 kg/cm² is encountered in a plate column containing 125 to 135 trays. The top fraction is condensed in a reflux condenser by refrigerant propylene, and the reflux enters the column at – 30°C under a pressure of 18 to 19 bars. In the stripping section steam is employed and cooling is provided by evaporation of propylene fraction. Bottom fraction leaves the tower at 20 bars pressure and –20°C. A low pressure splitter operates under a pressure of 8 to 9 kg/cm², where by the temperature during operation goes to as low as – 40°C to – 60°C. A very high reflux is essential to obtain polymer grade ethylene.

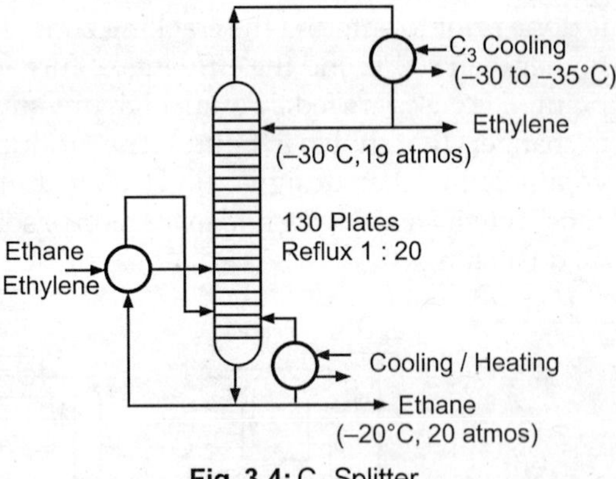

Fig. 3.4: C_2 Splitter.

3.2.2.3 Recovery of Ethylene from Cracked Gases

Irrespective of feedstock, the pyrolysis gases contain many compounds uneconomical for recovery, hence all recovery plants are more or less similar and extract the major components.

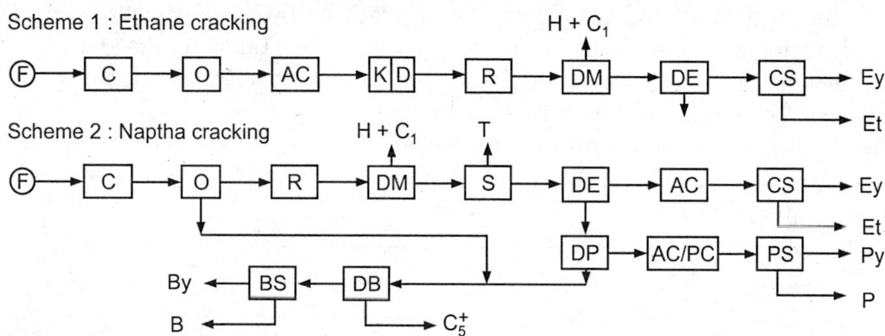

F : Pyrolysis Gas, R : Refrigeration, PS : C_3 Splitter, C : Compressors, DM : Demethaniser, DB : Debutaniser, O : Coolers, DE : DE Ethaniser, BS : C_4 Splitter, AC : Acetylene Converters, CS : C_2 Splitter, BY : Butadiene + Butane, K : Caustic Wash, DP : Depropaniser, B : Butane, D : Dryers, AC/PC : Methyl Acetylene Propadiene Converters, C_5^+ : Gasolines, S : Tail Gas Separator, T : Tail Gas

Fig. 3.5: Separation Schemes for Olefin Streams.

Hydrogen, methane, acetylene and dienes are common diluents; hence the active acetylene/dienes are hydrogenated at a convenient point and extracted alongwith olefins. When ethane is cracked, the composition of outlet gas is simple; hence small amounts of acetylene formed is hydrogenated to ethylene. During propane cracking, acetylene hydrogenators are kept in the same position as in the previous case, C_3 splitter separates propylene from the remaining heavy components. Analogously naphtha cracker requires a debutaniser to recover butadiene, butane fraction and pyrolysis gasoline. In this case, the entire mixture is never subjected to hydrogenation as butadiene is converted to olefin, hence hydrogenation is carried out in up stream of C_2 and C_3 splitter. A schematic diagram showing the recovery of ethylene and olefins is presented in Fig. 3.5 for crackers using ethane and naphtha.

In ethylene recovery the condensation and vaporisation technique is aptly used (Chapter 2), thus high pressure-low temperature or triple cascade system becomes a major tool in separation of different volatiles.

As mentioned early, a small amount of acetylene is present in cracked products, which has to be eliminated. Acetylene is converted to ethylene by mild hydrogenation at 350°C using palladium or cobalt, nickel, or molybdenum catalyst at 2 MPa pressure.

3.2.3 Other Developments

A late comer in olefin production technology is hydropyrolysis[(15)], developed in the laboratory of Pierrefitte-Auby/CECA. This is basically a hydrocracking process without any catalyst and differs in operational technique too. A high temperature and shorter processing time are helpful in getting greater yields of ethylene, while feedstsocks of medium and heavy oils are safely cracked. Due to the high temperature of 800-900°C at short reaction time, less than 0.1 sec., the chain reaction terminates at unsaturated level. Though greater yields of ethylene (over 40%) result, it is at the expense of by-products, which are available in conventional cracking operations. Cracking severity has now been increased in most of the modern plants so that naphtha can yield 34.5% ethylene. Ethylene yield of 25% is now commonly referred to as low severity cracking operation. Phillips high severity cracking process is credited with yielding 42.7% ethylene, while Stone & Webster Process employing an integrated design of pyrolysis furnace furnishes propylene to ethylene ratio in the range of 1:1 to 0.3:1, yielding over 30% ethylene. As the severity increases propylene yield falls while butadiene yield remains unaffected (around 4%).

Another interesting route is partial combustion cracking (PCC), a direct cracking process, in which heat of cracking is supplemented effectively by burning a portion of feedstock. This process seems to be quite appropriate for direct conversion of crudes to ethylene in one step as illustrated by Richard-O.Kirk[16]. The results of which are shown here :

Pressure in the system 35 to 448 K Pa
Temp. of combustion 1600-2000°C
Steam temperature 700-1000°C

Products	
Methane	14%
Acetylene	2.3%
Ethylene	25.2%
Ethane	1.9%
Propylene	8.3%
C_4	6.6%
C_5	3.0%
C_6 - heavy	35.9%
Balance	$CO + CO_2 + H_2$

Ethylene yield from following gist[17], shows the advantage of this process over regular tubular furnace cracking for different feeds.

	PCC	*Tubular*
Naphtha EP 175°C	35%	28%
Gas oil 175-345°C	32%	23%
30 API crude	27%	.
Residual oil	25%	-

Triolefin process is another advancement in the production of ethylene. It was commercialised by Phillips Petroleum. It is highly attractive because propylene is disproportionated to ethylene and butene. This enhances the yield of ethylene to as high as 42.7%. The process flow sheet is presented in Fig. 3.6. Salient features of the process include products of high purity ethylene, butylene that can

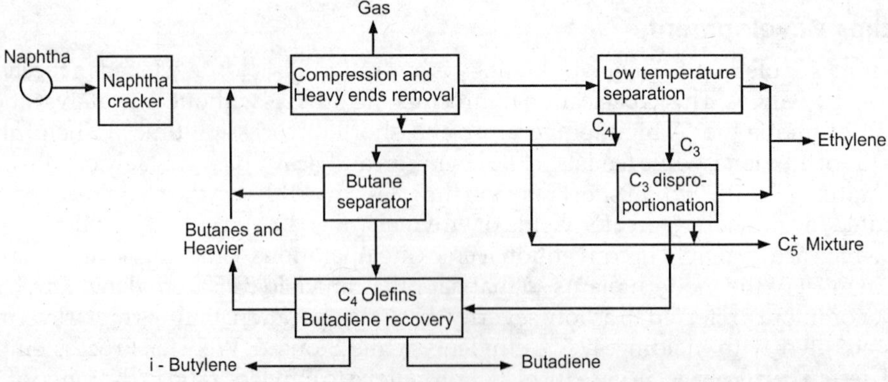

Fig. 3.6: Ethylene Triolefin Process.

directly go for polymer and alkylation units respectively, by relatively inexpensive fractionation rather than by difficult ethane-ethylene, butane-butylene separations. Propylene conversion per pass is approximately 40%, while unconverted propylene can be recycled back, there is no bar in choosing a mixed fraction of propane-propene. The pressure in the reactors can be varied in the range of 1 to 30 bars at WHSV in the range of 10 to 100.

Advanced cracking reactor (20, 19*a*, *b*) developed by Union Carbide, Kureha Chemical Industry Co. and Chiyoda to overcome the deficiencies of conventional ethylene technology. This innovation increases C_2 yield to 40% while reducing the cost of production by 20 to 30%. The product mix can be varied from 2 to 6 in ethylene propylene ratio. The process can work on number of feedstocks including untreated sour crudes. Fig. 3.7 shows the outlines of the process. The reactor is the most critical element of the process; designed on extensive theoretical work to insure the fluid dynamics of penetration, atomisation and vaporisation maximise the operability without sacrificing the severity or selectivity. The cracking of oil takes place at a burner temperature of 2000°C, developed by burning fuel (natural gas) with oxygen and steam at 800°C. Quenching of the products takes place at the bottom of the burner and heat recovery is carried out effectively at the bottom end of the burner. The quench oil products are separated from the quench oil and lighter fractions are fractionated to get light oils and gases. The gases are then treated to remove acid constituents and subjected to usual gas separation technique.

This process represents the first significant advance in olefins technology, and offers a major petrochemical alternative route.

Another innovation uses fluidised bed cracking of crude oil, using coke particles as the catalyst in the operation. The residence time lasts for about 0.2 to 0.3 sec. A ratio of ethylene, propylene, C_4 unsaturates 1 : 0.4 : 0.3 respectively in the products is available.

To supplement the stupendous demands of olefins specially ethylene and propylene, direct cracking of crude oil by Pebble heaters has also come into[21, 22] practice. Homogeneous catalytic cracking of gases is likely to enhance the yield of ethylene and propylene by 5%. The main obstacle in dehydrogenation catalysts in homogeneous phase is the loss of catalyst activity. According to Espino & Sherwin[6] usual problems are successfully overcome and the catalyst can be recovered from the outlet gases, and the same reactor system employed in steam cracking can be instituted.

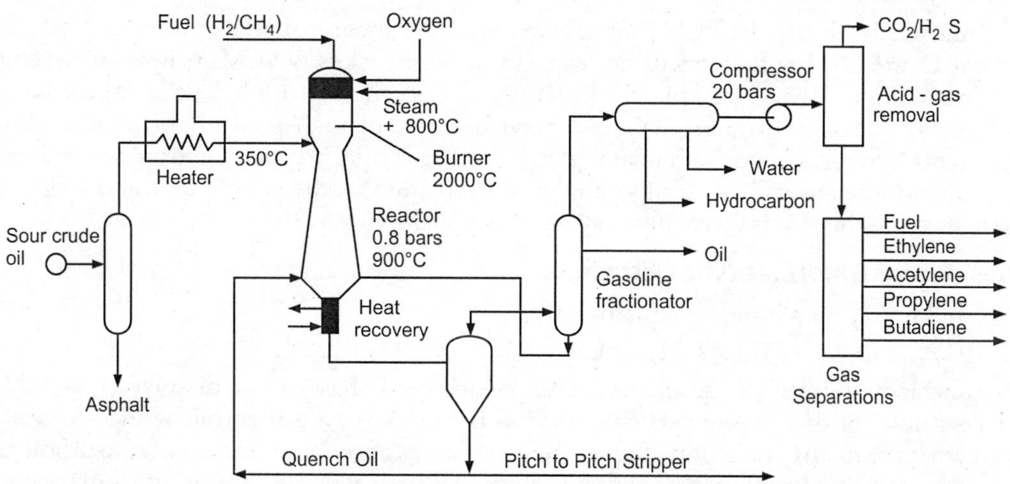

Fig. 3.7: The Advanced Cracking Reactor (ACR) for producing Olefins from Crude Oil.

3.3 ETHYLENE DERIVATIVES

Ethylene is a single olefin that created epoch history in the chemical industry and continues to do so. Switch over of acetylene chemistry to ethylene has further given an incredible scope for its demands. Fig. 3.8 shows the important derivatives of ethylene. A detailed description of production of major industrial products is presented in Fig. 3.8.

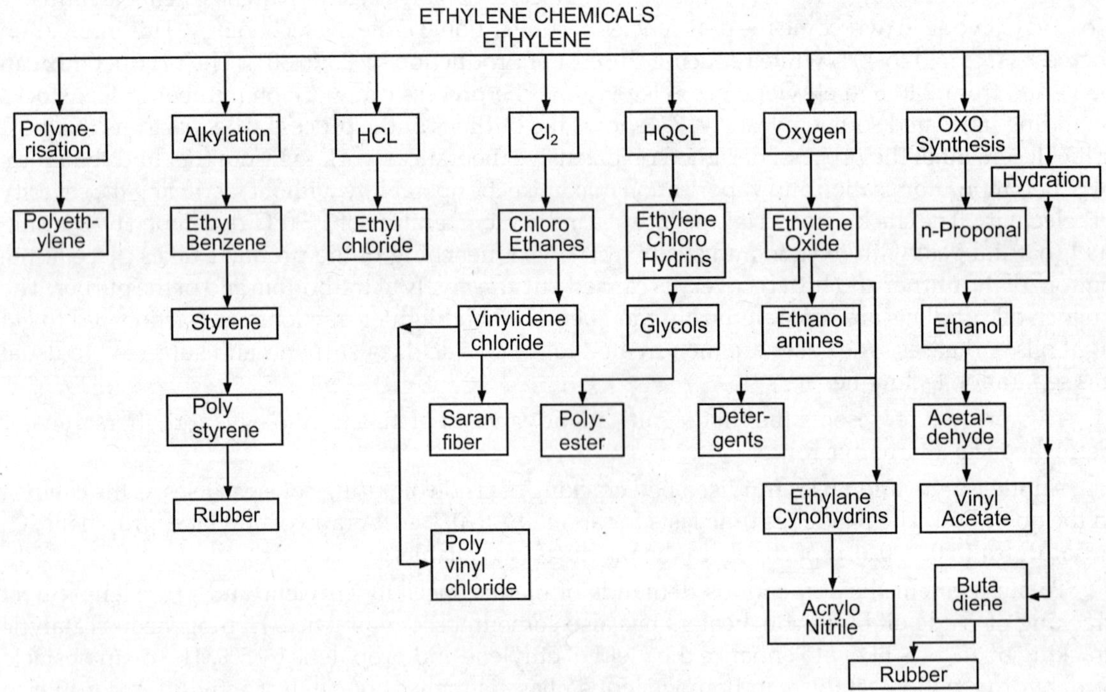

Fig. 3.8: Ethylene Chemicals.

3.3.1 Vinyl Chloride Monomer (VCM)

Commercially three major VCM processes are available, these are based on acetylene, ethylene or mixed gas. Over 90% of world's production capacity of 20 billion kgs. of VCM on the balanced ethylene in based route. Since 1965, several plants based upon oxychlorination of ethylene have been born, and have become a real cause to route out the old acetylene-based plants. About 85% of modern plants are based on this technology and the remaining on mixed-base chlorination. There are two major steps involved in production of VCM. Firstly, ethylene is converted to ethylene chloride and in the second phase diethylene chloride is thermally cracked to give VCM. (B.P. -13°C)

3.3.1.1 Direct Chlorination of Ethylene

The reaction is simple and straight forward

$$C_2H_4 + Cl_2 \rightarrow C_2H_4Cl_2$$

In stoichiometric ratio, the reactants are reacted in liquid phase, over a dissolved metal chloride ($FeCl_3/CuCl_2/C_2H_2Br_2$)[(23)] catalyst. The reaction is found to be exothermic, and heat has to be withdrawn for smooth conduction of reaction *i.e.* without appreciable amount of side products. For this purpose EDC itself can be employed as a solvent. Oxygen also aids the selectivity of the catalyst, as this reaction is drawn to completion by free radical mechanism.

The reaction is conducted below 50°C and at this temperature the homolysis of chlorine may not be occuring; however, it seems to be aided by reactor walls or catalysts like $CaCl_2$ and some organic oxides.

The mechanism goes like this:

$$Cl_2 \rightarrow 2\dot{Cl}$$
$$CH_2 \,.\, CH_2 + \dot{Cl} \rightarrow CH_2 \,.\, CH_2\, \dot{Cl}$$
$$CH_2 \,.\, CH_2\, \dot{Cl} + Cl_2 \rightarrow CH_2Cl \,.\, CH_2\, Cl + \dot{Cl}$$
$$\dot{Cl} + \dot{Cl} \rightarrow Cl_2$$
$$CH_2CH_2\dot{Cl} + \dot{Cl} \rightarrow CH_2Cl \,.\, CH_2Cl$$

In liquid phase, most obviously

$$CH_2 : CH_2 + Cl_2 \rightarrow \underset{\underset{Cl^+}{|}}{CH_2}CH_2 + Cl \rightarrow \underset{\underset{Cl}{|}}{CH_2}\;\; \underset{\underset{Cl}{|}}{CH_2}$$

The above reaction goes *via* carbonium ion mechanism. In vapour phase, the same catalysts are used and found to exhibit identical activity and selectivity too. The unwanted side products are common as listed below :

$$CH_2\,Cl - CH_2\,Cl + 2Cl_2 \rightarrow CCl_2 : CHCl + 3\,HCl$$
$$CH_2\,Cl - CH_2\,Cl + 3Cl_2 \rightarrow CCl_2 : CCl_2 + 4\,HCl$$
$$CH_2Cl - CH_2Cl + 5Cl_2 \rightarrow 2\,CCl_4 + 4\,HCl$$
$$CCl_4 + C_2H_4 \rightarrow 3\,C + 4\,HCl$$
$$CH_2Cl\,CH_2Cl + 7\,Cl_2 \rightarrow 2C_2Cl_6 + 4\,HCl$$
$$2\,CCl_4 \rightarrow CCl_2 : CCl_2 + 2\,Cl_2$$
$$CH_2 : CHCl + HCl \rightarrow CH_3CHCl_2.$$

Unless, the liberated hydrochloric acid vapour is successfully consumed in industry, its frugality is in question. Hence, alternative route for oxychlorination of ethylene or direct addition to acetylene is essential. As seen in Staffer Technology[(24)] an additional reactor in final stages is instituted to minimize ethylene losses in vent (less than 20 ppm) and complete utilisation of ethylene is in sight. Other products like dichloroethylene and ethane, trichloroethane, vinylidine choloride; tetra, penta, hexa chloroethanes accompany. Tetra and hexa chloroethanes are subjected to cracking. Remaining products are commonly directed as solvents. This chlorination step being active reaches over 95% conversion in single stage itself.

In recent methods[(25)], the liquid phase is kept around the boiling point of ethylene dichloride (83.7°C), which when boils consume the heat liberated. Thus EDC solvent is useful in the chlorination step. It is seen that direct chlorination process yields relatively pure EDC[(26)], may be slightly contaminated with ferric chloride, which can be easily removed.

3.3.1.2 Oxy Chlorination of Ethylene

Fig. 3.9 shows the outlines of combined processes of oxychlorination and direct chlorination of ethylene. In the reactor (2) ethylene and chlorine react where the temperature is kept below 40°C by circulating cold water; the pressure in the system is maintained around 2 to 3 atmospheres. The off gases from the reactor are cooled to 5°C, whereby the uncondensed gases can be let off and the liquid product can be fractionated (3). The bottom product EDC, is cracked in a furnace at 500°C.

The product is quenched (4) and cooled; HCl gas is separated and sent into oxy chlorination unit. The product is now distilled (5) and further purified. The heavy ends from column (5) are sent back either as quencher liquid or to crude EDC unit.

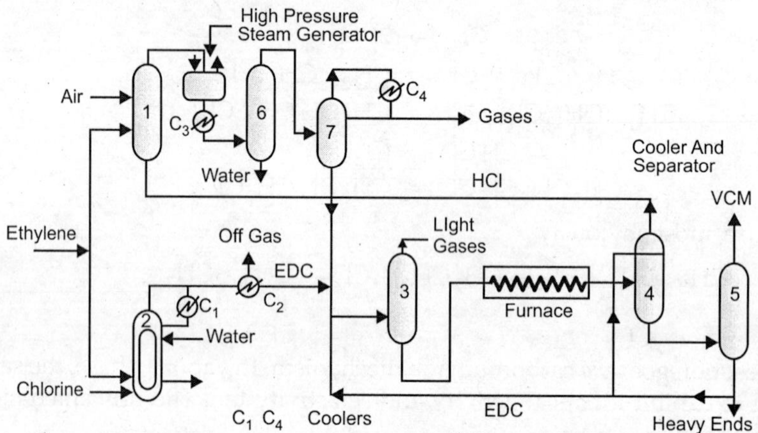

Fig. 3.9: Vinyl Chloride by Oxychlorination/Balanced Ethylene.

HCl from this plant enters the oxy chlorinator reactor (1) where the exothermic heat is removed by outside steam generation equipment. The gases are then cooled (6) and water separated out. The product is distilled (7) to separate unconverted gases and crude EDC.

Commendability of this technology owes to Deacon's principle of generating chlorine in-situ, in presence of copper chloride catalyst, (Mercury chloride is also used in some cases). The reaction can be conducted in liquid phase or in vapour phase, also in fixed beds or in fluidised beds. The heat of reaction is very high and should be efficiently disposed. Fluidised[(18–b)] beds are quite worthy in that respect. Most of the processes operate in vapour phase, except Kellog process that operates at 20 atm pressure and 180°C. In this process aqueous solution of cuprous chloride acts as catalyst. In all these cases the reactions are same and follow the pattern as described below.

$$2HCl \xrightarrow[\text{at } 450°C]{CuCl_2/KCl/O_2} H_2O + Cl_2$$

$$C_2H_4 + 2HCl \xrightarrow[Cu\,Cl_2/KCl]{O_2} CH_2Cl\,CH_2Cl + H_2O$$

The reaction proceeds *via* intermediate cuprous chloride formation

$$C_2H_4 + 2\,CuCl_2 \rightarrow CH_2Cl\,.\,CH_2\,Cl + 2\,CuCl$$

$$2\,CuCl + O \rightarrow CuO + CuCl_2$$

$$CuO + 2HCl \rightarrow CuCl_2 + H_2O.$$

Increase in temperature in all the above cases leads to complicated series of reactions. Fluidised beds give an yield of 94-97% based on ethylene, while fixed bed reactors operate with excess of ethylene to give an yield of 95%.

Other reactions anticipated are :

$$CH_2Cl\,.\,CH_2Cl + C1_2 \rightarrow \underset{\text{Trichloroethane}}{CH_2Cl\,.\,CHCl_2}$$

$$CH_2\,Cl\,.\,CH_2Cl + Cl_2 \rightarrow \underset{\text{Tetrachloroethane}}{CH_2ClCCl_3}$$

$$CH_2ClCCl_3 \rightarrow CHC_1:C\,Cl_2$$

Trichloroethylene.

In fixed bed reactors, the finely deposited copper chloride with some alkali metal chlorides on inert material acts as catalyst. The catalyst is packed in tubes, heat removal is affected through shell side heat transfer salts or steam generation system. The system works at a temperature below 350°C and a pressure of 15 atm. Instead of oxygen, air is also used, this permits lowering of operation temperature, but it has an inherent detrimental problem of recycling with excess of nitrogen contaminant. Hence a device of wedding oxychlorination and direct chlorination will benefit the manufacturers. Acetylene production by modern methods includes a substantial amount of ethylene; such kind of mixed gas stream is hailed by industry. Mixed gas stream processes based on balanced oxychlorination system and balanced ethylene/acetylene system are achieving industrial importance. Different schemes for VCM production are shown in Fig. 3.10.

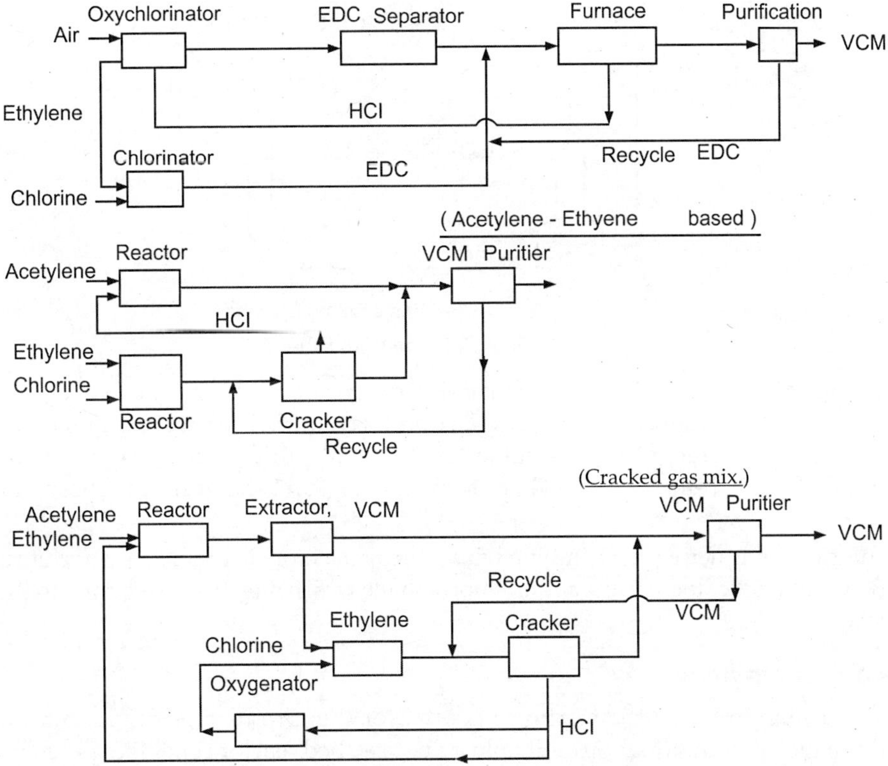

Fig. 3.10: Schemes of Balanced VCM Plants (Ethylene based).

Next step involves in separation of moisture and purifying EDC. This purified EDC is thermally cracked to yield vinyl monomer. The cracking takes place over a catalyst, kaolin/pumic or activated carbon and the temperature is kept around 500°C and the pressure at 25 atm. Conversion during cracking is only 50% at a residence time of 10 seconds. Hydrochloric acid vapour liberated during cracking is separated from VCM and sent to oxychlorination system or acetylene based reactor.

The crude VCM free of HCl is distilled in a column, and VCM is obtained as volatile product. The bottoms contain mainly uncracked EDC, which is recycled to the purifying section.

The by-products of main interest are C_1, C_2 chlorinated organic compounds that are used as solvents. These are obtained as light ends from EDC column. Heavy ends comprise other highly chlorinated products of ethane and butanes.

Yet another process, though not commercialised, is known by the name TRANSCAT process(27), a single step process based on ethane. All the above reactions are allowed to intricately take place in molten salt (copper chloride) catalyst at 650-700°C. Reaction and cracking conditions of the system are expedient to reach the goal.

$$C_2H_6 \xrightarrow{600-700°C} C_2H_4 + H_2$$
$$2\,HCl + O \rightarrow Cl_2 + H_2O$$
$$Cl_2 + CH_2{:}CH_2 \rightarrow CH_2Cl.CH_2Cl$$
$$CH_2Cl.CH_2Cl \rightarrow CH_2{:}CHCl + HCl$$
$$4\,HCl + 2\,CuOCl + KCl \rightarrow CuCl_2 + KCl + CuCl + 2H_2O + 3/2\,Cl_2$$

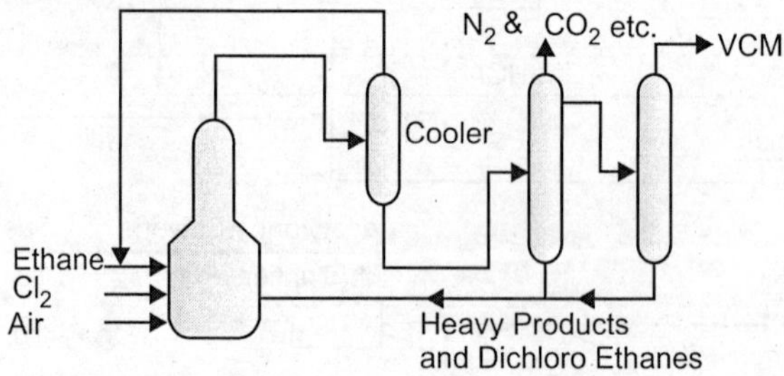

Fig. 3.11 Vinyl Chloride from Ethane.

The reactants comprise of ethane, oxygen and hydrogen chloride and the catalyst being molten salts of copper and potassium chlorides. Often, cuprous chloride is converted into cupric chloride by blowing air. The product stream is subjected to usual purification. The ethyl chloride and unconverted ethylene and ethane are sent back to reactor. Fig. 3.11 shows the outlines of TRANSCAT process.

Vinyl chloride being active can combine with hydrogen chloride in presence of chloride catalyst even at 25-30°C giving dichloroethane and trichloroethane which may be mixed up with the incoming feed to the reactor cycle.

3.3.2 Perchloroethylene

Ethylene dichloride forms the starting material for this product. A single step operation of dehydrochlorination and chlorination of EDC in a fluidised bed reactor concludes perchloroethylene. The evolved gas (HCl) is separated and utilised for generation of chlorine. The product is neutralized and distilled to get pure perchloroethylene. Instead of using EDC, chlorinated organics of C_2, which are considered to be merely waste, can be routed into this system for producing per or trichloroethylenes. Adjustment of reaction conditions can direct the preferential yield of either tri or per chloroethylene. The reactions are shown here :

$$2\,C_2H_2Cl_2 + 5Cl_2 \xrightarrow{\text{Chlorination}} C_2H_2Cl_4 + C_2HCl_5 + 5\,HCl$$

$$C_2H_2Cl_4 + C_2HCl_5 \xrightarrow{\text{Cracking}} C_2HCl_3 + C_2Cl_4 + 2\,HCl$$

$$2HCl + O_2 \xrightarrow{\text{Chlorine regeneration}} H_2O + Cl_2$$

$$C_2H_4Cl_2 + O_2 \xrightarrow{\text{Single stage oxychlorination}} C_2H_3Cl + C_2Cl_4 + H_2O$$

Production of trichloroethylene is traditionally done by chlorination of acetylene in tetrachloroethane solvent at 80°C using chloride of a metal as a catalyst. Chlorinated product being tetrachloroethane which forms the intermediate and on cracking yields trichloroethylene.

Thermal hydrochlorination is the present practice. This can effectively regenerate chlorine. A high temperature above 250°C is required for hydrochlorination. Barium chloride on charcoal acts as the catalyst. From trichloroethylene, perchloroethylene can be produced by chlorination and dehydrochlorination. Chlorination is done over as metal chloride catalyst at 800°C in pentachloro ethane solvent.

$$C_2H_2 + Cl_2 \rightarrow C_2H_2Cl_4 \rightarrow C_2HCl_3 + HCl$$

$$C_2HCl_3 + Cl_2 \rightarrow \begin{matrix} CCl_3 \\ | \\ CCl_3 \end{matrix} \rightarrow \begin{matrix} CCl_2 \\ | \\ CCl_2 \end{matrix} + HCl$$

Production of perchloroethylene is done by either pyrolysis of CCU or by propane chlorination. Decomposition of CCU at 800-900°C gives rise to perchloro ethylene. Usually ethylene dichloride, propane chlorination products are first reacted with chlorine and then cracked. The recovered HCl gas is converted into chlorine. Perchloroethylene and carbon tetra chloride are separated by distillation from the heavy products. Production of perchloroethylene can be done by using EDC and chlorine as reactants. The reaction is conducted in a fluidised bed reactor. The products from the reactor are separated from catalyst dust and quenched, later neutralized. Any carry over of catalyst is separated from the slurry. The crude perchloroethylene is then distilled to separate the light ends. The light ends, trichloroethylene and carbon tetrachloride are cooled to condense. The condensate is recycled. Perchloroethylene is purified after neutralisation and mixed with inhibitor before storing. Economy depends on HCl utilisation. HCl gas obtained from quench tower, is compressed and sent to oxidation unit for getting chlorine.

Perchloroethylene is mainly used as an intermediate in the production of a range of fluorocarbons. This is mostly used as a dry cleaning solvent. It is considered to be toxic, hence alternative solvents are preferred.

3.3.3 Production of Vinylacetate (VAM) Monomer[(28)] (B.P. 73°C Density 0.932g/cc)

The worldwide production of vinylacetate is more than 1.3 million tonnes; shows its importance (next to vinyl chloride). The Indian capacity for this monomer is about 30000 tonnes per year. The antique process was based on the catalytic reaction of acetylene and acetic acid in liquid phase. Owing to the cost of production and handling difficulties of acetylene, this process has been abandoned to give way to ethylene based plants. Later on modified liquid phase process, using acetaldehyde and acetic anhydride had come into operation since early 70s. The chemistry involving the acetylene based process is shown below. Zinc acetate deposited on charcoal acts as catalyst and the reaction proceeds at 170°C-200°C.

$$CH_3COOH + C_2H_2 \rightarrow \underset{\text{Vinyl acetate}}{C_2H_3COOCH_3}$$

Acetal dehyde and acetic anhydride react to give ethylidine acetate CH_3CO

$$2CH_3\,CHO + \begin{matrix} CH_3CO \\ CH_3CO \end{matrix}\!\!>O \xrightarrow{O_2} CH_3COOH + CH_3\,.\,CH(OOC.CH_3)_2$$

$$CH_3\,.\,CH(OOC\,.\,CH_3)_2 \rightarrow CH_3COOH + C_2H_3\,.\,COOCH_3$$

Ethylidine acetate is decomposed into acetic acid and vinyl acetate. ICI developed a process based on the reaction of ethylene and oxygen catalytically in liquid phase. In this process, recycled acetic acid and a mixture of ethylene and oxygen are fed into a reactor, where an aqueous solution of Copper chloride and acetate with palladium chloride acts as catalyst[29, 30]. Palladium ion concentration is kept in the range of 30 to 50 mg per litre of solution. Complex reactions are expected here. Palladium chloride is supposed to oxidise the reactants, during which, it is reduced to metallic state; but the copper chloride present in the solution incessantly oxidises palladium to palladium chloride, as depicted in the following way:

$$2\,C_2H_4 + CH_3COOH + O_2 + PdCl_2 \rightarrow C_2H_3OOCCH_3 + CH_3CHO + 2HCl + Pd$$

$$4\,C_2H_4 + 3\,O_2 \rightarrow 2\,C_2H_3OOCCH_3 + 2H_2O$$

$$CH_3CHO + O \rightarrow CH_3COOH$$

$$Pd + CuCl_2 \rightarrow PdCl_2 + CuCl$$

$$CuCl + HCl + O_2 \rightarrow CuCl_2 + H_2O$$

Obviously the role of Palladium catalyst may be ascribed to the formation of acetaldehyde in the initial stage. An important feature of the process[18–c] is that mole ratio of acetaldehyde to vinylacetate can be varied from 0.3 to 2.5.

$$C_2H_4 + PdCl_2 + 2\,HCl + H_2O \rightarrow CH_3CHO + Pd + 4\,HCl$$

Acetaldehyde being active usually ends up as n-butanol *via* aldol condensation. Acetal-dehyde can be easily oxidised by catalysts such as acetates of Mn/Co/K/Na/Cu.

The whole process is accomplished in a single reactor (usually a fractionator-tray type will suffice) at a temperature of 110 to 150°C and 30 to 50 bars pressure. The process may be described with the aid of Fig. 3.12. Ethylene, oxygen, recycled acetic acid enter the column (1) to contact the catalyst solution in a number of stages. Unreacted ethylene, oxygen along with reaction products aldehydes, acetic acid, acetate and water are the main effluents. Effluent is thoroughly cooled in a cooler (C_1), where the liquid products condense. In the following stage, the condensates are separated (S_1) and routed to purification section which takes place in number of stages. Rectifier (2) separates acetic acid from the rest of the volatiles. From the volatiles

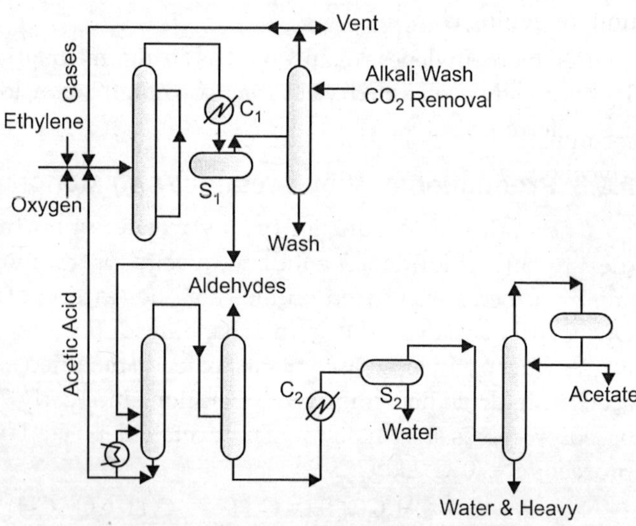

Fig. 3.12: Vinyl Acetate Manufacture.

aldehyde is fractionated as over head product in column (3), the aldehyde depending upon need be oxidised to acid in a separate unit. The bottoms from the column, are cooled (C_2), whereby water separates out in separator (S_2) and catalyst solution is recycled back. It is necessary to up-grade VAM by fractionation in column (4), where vinyl acetate monomer appears as overhead product.

Exhaled gases from separator (S_1) contains unreacted stream of ethylene, oxygen alongwith oxides of carbon. Gas stream is purified by removing carbondioxide by any suitable absorbent and the enriched gas is fit for recycling.

The chemistry involving Palladium chloride and copper chloride impeccably tried for producing acetaldehyde alone, and now it seems to outstrip the old process involving the oxidation of ethanol.

3.3.4 Ethylene Oxide

Ethylene oxide has become a significant chemical after the direct oxidation of ethylene has been commercialised since 1950. The production of ethylene oxide round the world in 1983 is around 8.091 $\times 10^6$ tonnes of which about 47% is produced in USA alone. Once, ethylene oxide was produced only through chlorohydrin process. First ethylene is chlorinated in presence of water to yield ethylene chlorohydrin. This is neutralised with calcium hydroxide to give ethylene oxide and calcium chloride. Except for the losses, the yield expected is high.

$$CH_2:CH_2 + H_2O + Cl_2 \rightarrow CH_2OH\,.\,CH_2Cl + HCl$$

$$2CH_2OH\,.\,CH_2Cl + Ca\,(OH)_2 \rightarrow 2\,\underbrace{CH_2:CH_2}_{O} + CaCl_2 + 2H_2O$$

Modern method of ethylene oxide production is a revolutionary step in the direction of enhancing its uses as an important intermediate. The process involves a single step, direct oxidation of ethylene over silver catalyst. Ethylene being explosive must be oxidised outside the limits of explosion. Simple representation of the reaction is given below :

$$\begin{matrix} CH_2 \\ \| \\ CH_2 \end{matrix} + \tfrac{1}{2}\,O_2 \quad O \quad \begin{matrix} CH_2 \\ | \\ CH_2 \end{matrix}\!\!>\!O + 133.9\ \text{kJ/mole}$$

Exothermic heat has to be removed to control the reaction. The conversion per pass is small, below 10% hence many times the effluents are recycled. Most of the modern industries are based on this technology and old ones also gradually switching over to this process. It is common to bud ethylene glycol plants with these plants.

The process, as shown in Fig. 3.13 starts with a mixture of ethylene (polymer grade or at least 95% pure) and dry oxygen. Mixture is passed over silver catalyst in multi tubular reactor. The pressure and temperature are 1 to 3 MPa and 200 to 300°C respectively. Being a free radical reaction, the oxygen is adsorbed silver surface and this adsorbed oxygen (O_2) is supposed to initiate the reaction as shown below :

$$3/2\ O_2\,(\text{ads}) + C_2H_4 \rightarrow \begin{matrix} CH_2 \\ | \\ CH_2 \end{matrix}\!\!>\!O + O_2 + 106.3,\ \text{kJ at } 250°C$$

Vigorous oxidation reaction leads to formation of carbondioxide.

$$C_2H_2 + O_2 \rightarrow CO_2 + H_2O + 323\ \text{kJ at } 250°C$$

Oxidation of ethylene to CO_2 and H_2O also takes place if excess oxygen is made available, to avoid this, the concentration of oxygen in the reaction mixture should be kept below 9%.

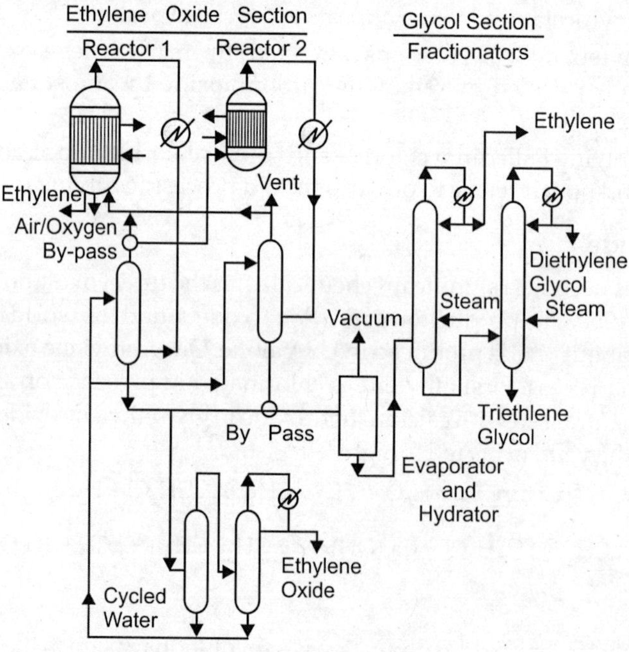

Fig. 3.13: Ethylene Oxide and Glycols.

As a rule hydrocarbons form an explosive mixture with air or oxygen and the explosive limits of these mixtures vary with temperature too.

°C	*Methane*		*Ethylene*		*Acetylene*	
	L	*U*	*L*	*U*	*L*	*U*
100	5.95	13.3	3.2	14.1	2.68	65
200	5.5	14.6	2.95	14.9	2.39	81
300	5.1	15.5	2.75	17.9	2.19	-

L stands for lower combustion limit. U stands for higher combustion limit.

It may be expounded further that ethylene air mixture explodes with the concentration of ethylene increasing from 1.7 to 78 vol % as the temperature decreases from 540 to 460°C. The presence of oxygen too increases the limits of combustion for ethylene. The increase is marked from L : 3 to U : 22 for air to L : 4.1 to U : 62 for oxygen. So in the said reaction temperature range may be chosen so that 30% ethylene concentration is allowed.

While Halcon SD process operates with a mixture of 15% ethylene and 7% oxygen O_2, remaining being nitrogen. Shell & ICI processes operate at 30% ethylene and 8% 62 concentration[32]. The catalyst is usually 8% silver aided by other ions like cesium, potassium on alumina beads. Insert gas or nitrogen is employed for diluting oxygen stream, this is mainly done by adjusting recycled gases, where nitrogen concentration is very high. Air based plants use nitrogen as diluent. Oxygen based plants can use methane ethylene mixture to increase the selectivity. A residence time of 1 sec is prelude to maintain a high space velocity 3000 to 15,000 hr^{-1}.

Reactor is multitubular type in which catalyst is packed[33] in tubes, shell side heat removal is done by circulating heat transfer salts or by generating steam. Sometimes two reactor system is also used; in first reactor the conversion is maintained low to get high selectivity; selectivity* also is found to increase as the temperature decreases. Side reactions lead to carbondioxide and water, while traces of acetaldehyde and formaldehyde are common. It is significant to note that ethylene carbonates may be obtained as intermediates for production of ethylene glycol.

As shown in Fig. 3.13, the products after exchanging heat are absorbed in water. Ethylene oxide goes into water, other gases recycled or partly rejected out. When two system is employed, the waste gases are sent into the second reactor, till all ethylene is oxidised. Usually, after absorption from the second unit, the gases are exhausted to atmosphere. The aqueous solution from scrubbers is fed to stripping section of a fractionator, where ethylene oxide, is obtained as overhead product (ethylene oxide B.P. 10°C). It is further purified to remove the last traces of carbondioxide. Ethylene oxide is a highly wanted chemical in detergent industry. It serves *as* the basic chemical for glycols, amines and serves in adhesives too.

3.3.5 Ethylene glycols

Ethylene glycols (mono, di, tri, tetra) are manufactured by the hydrolysis of ethylene oxide at 70°C under atmospheric pressure or it can be done at elevated temperature and elevated pressure. Depending upon the amount of water and conditions of reaction all the polyglycols appear.

$$\begin{matrix} CH_2 \\ | \\ CH_2 \end{matrix} \!\!>\! O + H_2O \rightarrow \begin{matrix} CH_2OH \\ | \\ CH_2OH \end{matrix} \quad \text{Ethylene glycol}$$

$$2\begin{matrix} CH_2 \\ | \\ CH_2 \end{matrix} \!\!>\! O + H_2O \rightarrow \begin{matrix} CH_2OH \\ | \\ CH_2 \\ | \\ CH_2 \\ | \\ CH_2OH \end{matrix} \quad \text{(the two inner } CH_2 \text{ groups bridged by O) Diethylene glycol}$$

Teijin[49] process for the production of ethylene glycol is centered around Oxy-Chlorination of ethylene in presence of thallium salts in aqueous chloride ions.

IPCL, NOCIL, and India Glycols (Okhla), Reliance (Patalgange) are some of the producers of Glycols.

After hydration reactions the aqueous solution is evaporated and then dehydrated under vacuum. Yield of ethylene glycol is around 85%.

3.3.6 Dithioglycols

Ethylene oxide when allowed to react with hydrogen sulphide produces thioglycols.

---

* Selectivity is defined as the amount of Et. O. produced per unit of ethylene. Usually it ranges 70-75% for oxygen and 68% for air based plants. The recent reports show even 84% selectivity for oxygen based plants.

$$\begin{array}{c} CH_2 \\ | \\ CH_2 \end{array}\!\!\!>O + H_2S \rightarrow \begin{array}{c} CH_2OH \\ | \\ CH_2SH \end{array}$$

$$\begin{array}{c} CH_2 \\ | \\ CH_2 \end{array}\!\!\!>O + H_2S \rightarrow \begin{array}{c} CH_2CH_2OH \\ | \\ S \\ / \\ CH_2CH_2OH \end{array}$$

These glycol are mostly used as solvents in printing industry.

3.3.7 Ethanol Amines

Ethylene oxide reacts with ammonia to give a mixture of mono, di and triethanolamines.

$$\begin{array}{c} CH_2 \\ | \\ CH_2 \end{array}\!\!\!>O + NH_3 \rightarrow \begin{array}{c} CH_2NH_2 \\ | \\ CH_2OH \end{array}$$

$$2\begin{array}{c} CH_2 \\ | \\ CH_2 \end{array}\!\!\!>O + NH_3 \rightarrow (CH_2CH_2OH)_2NH$$

$$3\begin{array}{c} CH_2 \\ | \\ CH_2 \end{array}\!\!\!>O + NH_3 \rightarrow (CH_2CH_2OH)_3N$$

Above reactions resemble hydration reactions. A wide range of temperature, pressure, a varying molal ratio of ethylene oxide and ammonia; give rise to these series of products. As an example, 40°C and 1.5 atmospheres pressure at 10 : 1 ratio of ethylene oxide and ammonia give rise to 70% monoethanol amine, followed by diethanol amine, a little triethanolamine. Decreasing the ratio of ethylene oxide, ammonia and severe operating conditions result in more triethanolamine. All these amines are useful in absorbing H_2S and CO_2. Besides this, useful applications are noted in detergent and solvent industry.

3.3.8 Acrylonitrile

If the cost of production is not accounted then ethylene oxide can produce acrylonitrile as per the below cited reactions. Even with a high velocity of 2000hr^{-1} a trifle conversion 6% per pass results. A small percentage of ether and aldehydes also appears, but the maximum yield is towards absolute alcohol.

$$\underset{\diagdown O \diagup}{H_2C \quad CH_2} + HCN \rightarrow HOCH_2CH_2CN$$

Cyanohydrin

$$OHCH_2CH_2CN \rightarrow CH_2CHCN + H_2O$$

Acrylonitrile

3.3.9 Ethylene Diamine

Ethylene dichloride is used as the starting material for producing diamine

$$\begin{array}{l} CH_2Cl \\ | \\ CH_2Cl \end{array} + 2NH_3 \rightarrow \begin{array}{l} CH_2NH_2 \\ | \\ CH_2NH_2 \end{array} + 2HCl$$

By changing the ratio of ethylene dichloride, tri and tetra ethylene amines are formed.

$$3 \begin{array}{l} CH_2Cl \\ | \\ CH_2Cl \end{array} + 3NH_3 \rightarrow \begin{array}{l} (CH_2)\,NH_2 \\ | \\ (C_2\,H_3)_3\,(NH_2)_3 \end{array} + 6HCl$$

$$4 \begin{array}{l} CH_2Cl \\ | \\ CH_2Cl \end{array} + 4NH_3 \rightarrow \begin{array}{l} (CH_2)\,(NH_2) \\ | \\ (C_2H_3)\,(NH_2)_4 \\ | \\ (CH_2)\,NH_2 \end{array} + 8HCl$$

Ethylene diamine is used for the production of ethylene diamine tetra acetic acid (EDTA), and many emulsifying agents etc.

3.3.10 Ethanol

Ethanol is mainly produced by fermentation of molasses in our country. For academic reasons it may be mentioned that ethanol can be produced by hydration of ethylene, as it is done in some countries. The basic reaction is hydration of ethylene in presence of a catalyst. Hydration can be done in a liquid phase or in vapour phase.

3.3.10.1 Liquid phase hydration of ethylene

A stream of ethylene is sent into concentrated sulphuric acid, which exothermically absorbs ethylene. Next stage consists of hydrolysing this mixture for the production of ethanol and acid separates out below :

$$C_2H_4 + H_2SO_4 \rightarrow C_2H_5HSO_4 + 142\ kJ/mole$$

$$C_2H_5HSO_4 + H_2SO_4 \rightarrow (C_2H_5)_2SO_4 + 93\ kJ/mole$$

$$C_2H_5HSO_4 + H_2O \rightarrow C_2H_5OH + H_2SO_4$$

The sulphuric acid acts as a catalyst lending carbonium ion. The reaction is carried out at room temperature, and pressure increases the solubility of ethylene in acid. Too much concentration of ethylene may give rise to ether.

$$2C_2H_5HSO_4 + H_2O \longrightarrow \begin{array}{l} C_2H_5 \diagdown \\ \quad\quad\quad O \\ C_2H_5 \diagup \end{array} + 2H_2SO_4$$

Absorption reaction is carried out in any packed or plate column. Effluent from the column is sent to hydrolysis unit. Hydrolysis yields ethanol, ether and heavy polymerised products of ethylene. It may be augmented that sulphuric acid being good polymerizing catalyst, oligomers may be expected. In fact the chain length does not exceed more than four monomers. It is also necessary to protect other olefins entering into the system.

A 95% acid can take up 1.2 moles of ethylene per mole of acid; time required for this operation is about 4 hours. Industrially, 1 to 1.5 hrs is given for absorption in 98% acid; the partial pressure of ethylene also has got a favourable effect, hence the pressure is kept upto 20 bars.

After absorption, the second phase reaction, *i.e.* hydrolysis is called for. Hydrolysis requires a higher temperature of 90 to 100°C and a longer period of time if the dilution is high. The hydrolysed mixture is first neutralised and distilled in a column, the top product is ether-alcohol a 60% -water mix. For concentration of alcohol, two or three columns are used. The spent acid from the column is concentrated and recycled. Usually 95% alcohol results in this process. Anhydrous alcohol is obtained by azeotropic distillation with benzene.

3.3.10.2 Direct Hydration of Ethylene

Inherent difficulties posed by handling acid and huge expenditure involved in purification of ethanol had made the above process obscure. New ventures are based on direct hydration of ethylene. The best catalyst for direct hydration as cited in literature is phosphoric acid on diatomaceous earth; or on silica-alumina base. Like methanol synthesis, pressure has favourable effect on equilibrium conversion, while temperature has an adverse effect, as shown below:

Equilibrium Yield of Alcohol

Ratio (Mole) H_2O/C_2H_4	*Yield %*	
	T°C250 Pr: 50atm	*T°C 150 80 atm*
0.2	4.6	6.5
0.6	10.6	15.4
0.9	11.2	20.5
1	14.1	20.5

Reactants ratio, pressure and temperature of the reaction play a vital role in determining the yield at equilibrium. Low conversions usually less than 4%, necessitates recycling of reactants. Further, the acid concentration on the catalyst surface should be kept above 85%, otherwise inadequate conversion results. In practice a contact time of 18 to 20 sec., arid water to ethylene mole ratio of 0.6 to 0.7 with a high space velocity of 2000 hr^{-1} are used to give a trifle conversion 6% per pass. A small percentage of ether and aldehydes also appear, but the maximum yield is absolute alcohol.

Fig. 3.14 shows the outlines of the process. Ethylene and steam are mixed and heat exchanged with the out going reactants, if required heated separately to raise to the reaction temperature. The mixture enters the fixed and reactor packed with catalyst. After the completion of the reaction, the gases are allowed to enter a separator where the unreacted gases are separated. These gases are sent into a scrubber where the alcoholic vapours are absorbed in a water spray. These washings along with

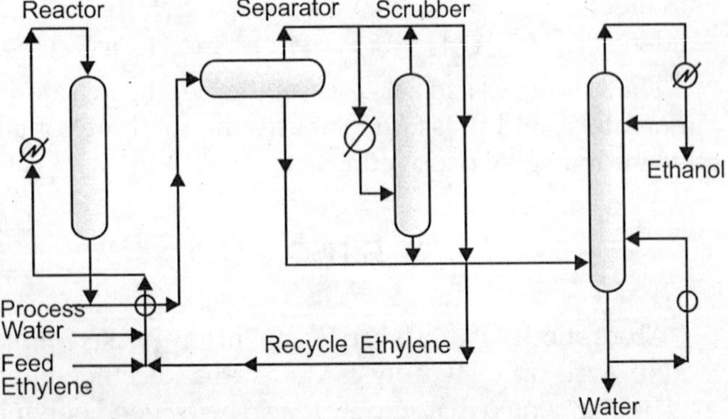

Fig. 3.14 Ethanol Production from Ethylene.

liquids recovered from separator are mixed and distilled in a fractionator for getting alcohol. Unconverted ethylene is recycled.

3.3.11 Acetaldehyde

It is mobile, colourless, inflammable liquid with pungent[35], choking odour, boils at 20.8°C. It is now-a-days produced from ethanol by oxidation or dehydrogenation. Catalysts like oxides of Mn, Ni, V and Fe have been found to oxidise ethanol vapours. Presently silver catalysts 9 : 1 copper-silver alloys or copper granules coated with silver are used in industry. Exothermic heat of 40.55 kcal has to be removed for progressing the reaction with silver catalysts. Often ethanol-water mixture is oxidised to regulate the rate of reaction, thus sacrificing high yields per pass. 40-50% conversion is common. With copper catalysts the temperature of the reaction reaches up to 350°C, while with silver it can go up to 500°C.

Dehydrogenation of alcohol over copper catalyst also results in aldehydes

$$CH_3CH_2OH \rightarrow CH_3CHO + H_2 - 17.28 \text{ kcal}$$

Dehydrogenation commences at 225°C and becomes rapid at 250°C. The conversion is usually low (20-25%).

Acetylene hydration also gives aldehyde, the reaction is catalysed by $HgSO_4$ in presence of H_2SO_4

$$CH:CH + H_2O \rightarrow CH_3CHO + 33 \text{ kcal.}$$

Chisso Engg. Co. developed a reaction between acetylene and steam in presence of 20-25% H_2SO_4 containing ferric and mercuric ions. Gas phase addition of steam takes place over phosphates of calcium, cadmium and zinc. It is suitable for dilute acetylene streams. By product acetone is also obtained.

$$2\,C_2H_2 + 3\,H_2O \rightarrow CH_3COCH_3 + CO_2 + 2\,H_2$$

3.4 ACETYLENE PRODUCTION

Ethyne or acetylene, as is known, is the simplest compound of triple bond. This was first discovered by Admond Davy in 1836, and it was produced from K_2C_2. Berthelot in 1860, named it as acetylene and pronounced its usefulness as an illuminant, which became the incentive for commercial exploitation. Ever since its discovery many laboratory methods were developed for its production; to name a few : fusing potassium ethylene disulphonate with potash; dehydration of vinyl chloride with alcoholic potash etc. Industrial production was relatively unknown until T.L. Wilson (1882) accidentally obtained calcium carbide lumps from the electric arc furnace, which was constructed and operated for the production of calcium from raw materials lime and coke. This was the real birth of acetylene industry through inorganic raw materials. Calcium carbide played an invincible role till 1960, during which time the importance of acetylene as a chemical feedstock reached pinnacles and a slow decline started since mid sixties. Today it is unable to withstand the competition from other olefins, since they are produced with much ease and almost at half the cost. Further, the chemicals obtained from acetylene are economically produced from the olefins.

Today it is industrially produced by the following processes:

1. Calcium carbide
2. Electrolytic decomposition of methane
3. Wulff's Regenerative Cracking
4. Oxidative pyrolysis of hydrocarbons.

3.4.1 Calcium Carbide Route

The basic factors in the manufacture of calcium carbide have not undergone any marked changes since inception. The reaction is simple, but progressive at very high temperatures only, *i.e.* above 1600°C and close to 2000°C, calcium carbide is formed by the reaction, may be through an intermediate step of calcium.

$$CaO + 3C \rightarrow CaC_2 + CO\text{-}464.5\ kJ$$

The reaction is concluded at subatmospheric pressures. The impurities in lime and coke naturally mark their presence in finished calcium carbide to be followed only in acetylene when generated:

$$CaC_2 + H_2O \rightarrow CaO + C_2H_2$$

$$Mg_3N_2 + 3H_2O \rightarrow 3MgO + 2NH_3$$

$$CaS + H_2O \rightarrow H_2S + CaO$$

$$2SiC_2 + 2H_2O \rightarrow SiO_2 + 2C_2H_2$$

$$Ca_3P_2 + H_2O \rightarrow PH_3 + Ca(OH)_2$$

The energy consumption in large modern furnaces is 3-3.5 kWh per kg carbide, *i.e.* (2600-3000 Kcal/kg). About 2.5 to 3 times carbide is required for a unit of acetylene produced. Hence the power consumed on the basis of acetylene is about 10 to 11 kWh plus massive labour charges which deterred the growth of acetylene chemicals. Many of the plants throughout the world are either closed or exchanged for the new technology. In our country, however the picture is different; about 48,000 to 60,000 tonnes of acetylene is primarily produced through carbide route only, due to the non-availability of the petroleum stocks in those days, as well petrochemical industries based on acetylene were meagre. Presently IPCL has ventured to separate acetylene from cracker gases. Roughly the production is around 20,000 tonnes yearly.

3.4.1.1 Thermodynamics of Acetylene

The remaining processes are based on thermal decomposition of hydrocarbon feedstocks, mostly methane, ethane or light fractions. The huge energy requirement for such thermal pyrolysis may be guessed from the following reactions :

$$\underset{\text{(graphite)}}{2C} + H_2 \rightarrow C_2H_2\text{–}226kJ \qquad ...(1)$$

$$2CH_4 \rightarrow C_2H_2 + 3H_2\text{–}380\ kJ \qquad ...(2)$$

The free energy change for the reactions 1 and 2 respectively is

$$\Delta Z° = 223.86 - 0.0533\ T\ kJ/mol$$

$$\Delta Z° = 404.86 - 0.2717\ T\ kJ/mol$$

Evidently thermodynamic potential decreases with increase of temperature. Being an endothermic compound, acetylene is relatively more stable at higher temperatures (> 1200°C) than paraffins *i.e.* even methane; and another threshold stability is below 200°C. Energy requirement and equilibrium yield of acetylene from elements are given below.

Temp °K	*Δ H kJ/mol*	*Equilibrium yield*
298	226.8	2×10^{-35}
1000	222.2	0.0006
1200	222.0	0.0018
1300	220.0	0.002
1500	218.0	0.013
2000	216.0	0.4
2500	214.0	3.2

Large amount of heat is required to produce acetylene. Heat of formation and free energy formation for different hydrocarbons are shown in Fig. 3.15 (*a*) and (*b*). It may be observed that heat requirement decreases with increase of molecular weight of feed. Equilibrium considerations show that at high temperatures, unavoidable side products appear. Bearing all these in mind, the tough job of production of acetylene has to be undertaken by quickly rising the temperature of the reactants with the minimum amount of energy and cooling down the products rapidly to freeze acetylene. Decrease in partial pressure by inerts or by steam or by vacuum is essential, comparing all these, steam seems to favour reaction thermodynamically.

$$C_2H_2 + 2H_2O \rightarrow 2CO + 2H_2$$

The reaction involves many intermediate steps as shown below :

$$CH_4 \rightarrow CH_2 + H_2 \text{ (380 kJ activation energy)}$$

$$2CH_2 \rightarrow C_2H_2 + II_2$$

From C_2H_6, it follows $\quad C_2H_6 \rightarrow C_2H_4 + H_2$

$$C_2H_4 \rightarrow C_2H_2 + H_2 \text{ (360 kJ activation energy at 1000°C)}$$

The maximum concentration of acetylene is found at higher temperatures only. In case of ethane the activation energy is sufficient to delink C-C bonds but not C-H bonds, so the initial radicals are likely to be CH_3. There is butene formation too which easily dehydrogenates to acetylene.

$$C_2H_4 + C_2H_4 \rightarrow C_4H_6 + 2\dot{C} + 2\dot{H}$$

$$2\,C_2H_6 \rightarrow C_2H_2 + C_2H_4 + 3H_2$$

When acetylene decomposes huge energy is released, so the reaction becomes self sustaining leading to explosion. In methane-thermal decomposition, the equilibrium yields are shown below. (See Fig. 3.15 (*c*) and (*d*)).

Temp. °K	*Eq. yield at 0. 1 bar*	*Acetylene in mixture 1.0 bar*
500	0.00	0.00
1000	4.83	1.54
1200	35.00	11.7
1400	85.00	46.7
1600	97.8	84.0
1800	99.0	96.0
2000	99.9	99.0
2200	100	99.7

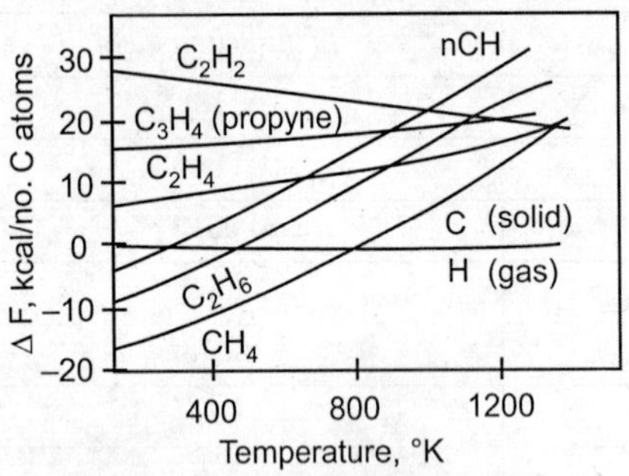

Fig. 3.15 (*a*): Free Energies of several Hydrocarbons.

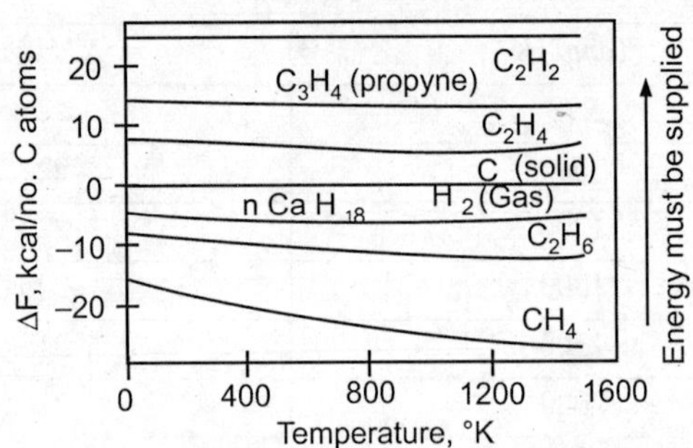

Fig. 3.15 (*b*): Heat of formation of several Hydrocarbons.

In all the processes to be described, the production of acetylene is achieved strictly without attaining thermodynamic equilibrium. Hence a short time is allowed for pyrolysis and immediate quenching of the reactants to freeze acetylene formed. The maximum quenching temperature is around 400°C.

3.4.2 Electric Cracking of Methane[36]

The conversion of methane by means of electric arc discharges into acetylene was known even in 19th century. The principle was later transformed into industrial scale. The electrical energy supplied is meant for two reasons, one for raising the reactants to decomposition temperature, second for supplementing heat of decomposition. Heating of the feedstock saves part of energy, however, the energy consumption in electric arc furnaces is around 6.45 to 7 kWh per kg acetylene, this energy does not include losses at electrodes, the energy for separation and purification of acetylene etc.; all energies if accounted touch around 12 to 13 kWh/kg acetylene. As the molecular weight of the feedstock increases, energy for decomposition decreases, energy requirement for propane decomposition is only 5 to 6 kWh while for methane it touches 7 kWh.

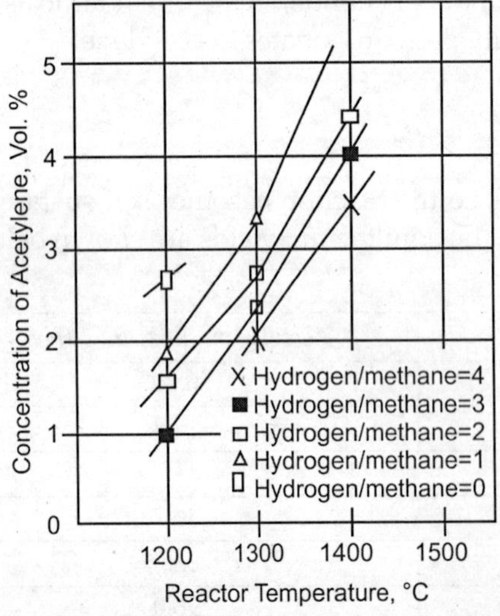

Fig. 3.15 (*c*): Optimum conc. in the cracked gas Vs. Temperature

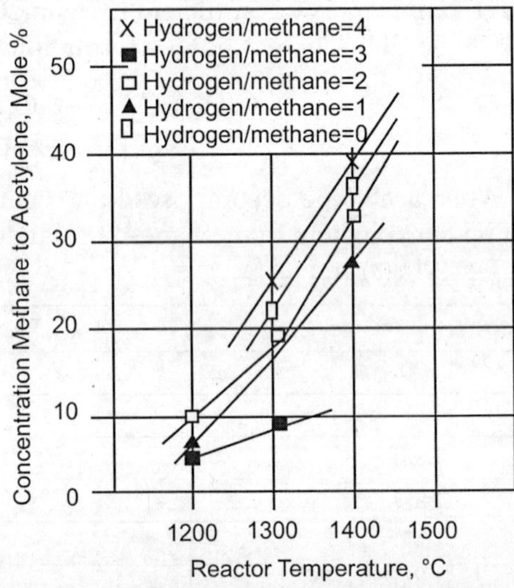

Fig. 3.15 (*d*): Optimum Acetylene yield Vs. Reactor Temperature

3.4.2.1 Kinetics of Methane Cracking

Electric arc process exhibits first order reaction kinetics. Other hydrocarbons do follow the same pattern. The following data shows the energy consumption for different hydrocarbons and yield of acetylene[37]; as the molecular weight of feed increases the energy consumption decreases.

	Methane	*Ethane*	*Propane*	*Propylene*	*Ethylene*
% Acetylene	20	24	26	30	30
Reaction heat kWh/cu.m of C_2H_2	4.7	3.8	3.7	2.2	2.7

There are two well known arc processes, Hulls and Du Pont. Hulls furnace uses high D.C. voltage 8000 V at 1000 amps. Du Pont furnace works on D.C. 3500 V at 3100 amperes. Hulls arc process consists of the furnace that was developed for nitric oxide production from air. The rapid quenching of effluents by liquid hydrocarbons is an advantageous thing, as ethylene produced due to cracking of the hydrocarbons forms a by-product. About 0.5 kg ethylene, 0.3 kg carbon black and 0.25 kg of hydrogen accompany each kg of acetylene. The requirements of methane is 2.5 to 2.8 kg per kg acetylene; the power requirement is of the order of 9 to 9.2 kWh/kg acetylene.

3.4.2.2 Du Pont Process[37, b]

The arc furnace is shown in Fig. 3.16. An arc is struck between a Cathode of soderberg type 1.25 cm dia. The anode is water cooled concentric copper tube of 9 cm.i.d. Anode is surrounded by a magnetic coil that spreads the arc. Hydrocarbon gas enters the arc zone and passes through the anode. The gases coming out of the anode are given a thorough wash with water. This quenches the gases to about 300°C. Recent innovations suggest that plasma jets are likely to replace the present arc furnaces. Argon plasma at 12000°C gives a yield of 90-99% acetylene resulting in no further recycling of hydrocarbons.

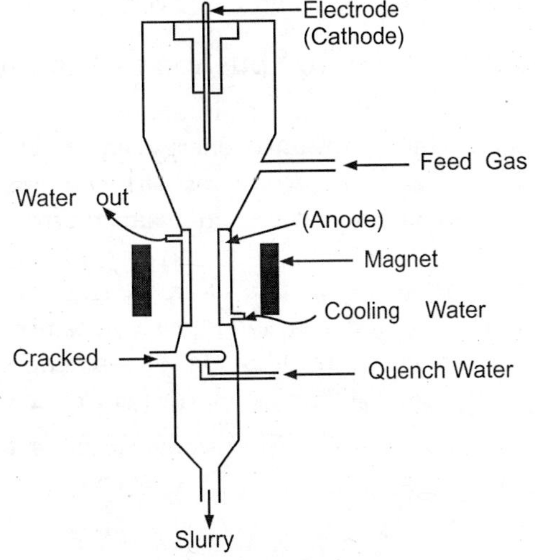

Fig. 3.16. Du Pont Arc Furnace.

3.4.3 Wulff's Process[38, 39]

It is essentially based upon regenerative principle. The thermal decomposition is accomplished in a pair of regenerative heat exchangers, situated on either side of the combustion chamber. The reactor furnace is constructed of heat resisting, nonspalling fire bricks workable in oxidising and reducing atmosphere; perfectly aligned intrinsic channel labyrinth. Air and fuel are preheated in the left regenerator and allowed to complete combustion in the central chamber. Combustion products exit through right regenerator exchanging the heat to chequer brick work. This heat is effectively availed by the pyrolysis gas, that enters through right regenerator and escapes through common duct to quench section. After pyrolysis cycle is over, the fresh air and fuel enter through this regenerator and consume the deposited carbon and tarry material. Thus alternatively heating and pyrolysis take place in each of these regenerators, with central combustion chamber. Pyrolysis is envisaged at a temperature of 1300°C and above, for methane; while for propane it is around 1200°C; products leave at 800°C and regenerators cool down to 400°C. Two-makes constitute one cycle, lasts about a minute, cracking itself is of the order of 0.04 to 0.1 sec. A subatmospheric pressure (0.5 bar) is maintained during pyrolysis to enhance the yield of acetylene. These reactors are usually operated to produce adjustable ethylene-acetylene ratios. In case of methane the ratio is 2 to 5 :1 while for propane it is 3 to 5:1 -1:3 to 5.

Wulff's process has some advantages like no pure oxygen[40] requirement; products of combustion and cracking do not mix. Further no soot or carbon formation is observed. This does not mean that it is free of disadvantages, main one being operation under subatmospheric condition.

The outlet composition of the gases is given below for methane and propane feedstocks :

	CH_4	C_2H_2	C_2H_4	CO_2	CO	H_2
Methane	28.3	7.1	0.9	1.3	7.8	49
Propane	15	10	3.8	1.6	6.7	56

Rest being nitrogen and oxygen.

It can be judged from this that propane should be a better feedstock than methane. Steam is also introduced along with feedstock in a ratio of 7 to 1. The residence time lasts less than 0.04 sec. and the pressure is kept below atmospheric usually 0.5 to 0.6 atm. It is likely to be cheaper if naphtha feedstock is used and may even become the cheapest of all the processes.

3.4.4 Partial Combustion or Thermo Oxidative Pyrolysis

Presently, acetylene is manufactured by several companies by partial oxidation route. Fig. 3.15 (*c*) shows the[41] amount of energy required for formation of methane (exothermic compound); it requires highest energy for conversion to acetylene. Numerous undesirable side reactions appear at such high temperatures, hence a well balanced fortuitous combination of chemical kinetics and burner design may foredain the yield of actylene. Partial oxidation burners for the production of acetylene starting from methane or natural gas operate on the principle utilisation of combustion heat, of a sizable portion of feedstock and to raise the temperature of the remaining portion to formation temperature of acetylene[42]. The following set of reactions explain the yield of acetylene:

$$CH_4 + O_2 \rightarrow CO + H_2O + H_2$$

$$CO + H_2O \rightarrow CO_2 + H_2 \text{ (Shift)}$$

$$CH_4 \rightarrow C_2H_2 + H_2$$

$$C_nH_m + H_2O \rightarrow CO + H_2 \text{ (Reforming)}$$

$$2\,C_2H_2 \rightarrow C_4H_2 + H_2$$

Though reforming and shift reactions are doubtful, diacetylene results as the main product due to acetylene dimerisation.

Acetylene is also likely to be oxidised by water vapour.

$$C_2H_2 + 2H_2O \rightarrow 2CO + 3\,H_2$$

It is observed that by preheating feed and increasing mole ratio of oxygen to methane raise the reaction temperature as shown in Fig. 3.15 (*b*) and (*c*).

Preheating the feed to 900°C and a mole ratio of 0.6 (O_2 to CH_4) give a temperature of 1700°C, where the maximum conversion of CH_4 takes place, *i.e.* nearly 37%.

Increase of oxygen to methane ratio and preheating the feed can definitely elevate the available reaction temperature but no useful purpose is served, because of the reverse and side reactions. 95% of the unburned methane can be converted to acetylene when the temperature crosses 1900° K.

The effect of dilution of the feedstock, methane with hydrogen was investigated by Taiseki Kunngi[43] *et.al.* during thermal decomposition. Figs. 3.15 (*c*) and (*d*) show the conversion of methane to acetylene and concentration of acetylene in product gases. The maximum yield is at the ratio of 2 and 1200°C, while for 1300°C–1400°C, it is about 3. It may be seen that dilution has a prominent effect in prolonging the contact time and the yield of acetylene during pyrolysis of methane.

Recent developments in acetylene[44] production include hydrogen plasma jets, where by 10% acetylene is achieved at 10 kWh/kg of acetylene power consumption.

Partial combustion burners are credited with furnishing extremely rapid high temperature heating as required by acetylene.

The total hydrocarbon and oxidant feeds are routed to reaction zone, where the combustion takes place. Amount of oxidant is usually much less than the stoichiometric ratio for complete combustion. The combustion heat is sufficient to crack the unburnt portion of hydrocarbon. Fig. 3.17 shows schematic cross-section of a burner developed by BASF. Preheated oxygen and natural gas enter separately but in the mixing throat complete mixing takes place after which it is passed through a plurality of holes in a burner block (2). Combustion occurs in the space marked (3). The burner block prevents flash back and acts as a distributor of the gas for proper flame. The flame initiation begins in the mixing zone (3). The high velocity gases (30 m/sec) create a turbulent combustion flame. The decomposition of methane progresses from the flame initiation zone to the fully developed combustion zone. The total burner gas mixer is immediately quenched just beneath the reaction zone (4) with water. The product gases are separated from tarry slurry and purified. Usually 6.5% concentration of acetylene is reported for a weight ratio of oxygen to methane 1.4 to 1 when no preheating was done. However, by preheating the gas mixture the concentration of acetylene can be raised to 9 - 9.3% by volume. The other components to be followed are hydrogen 42.5%, carbon monoxide 37.3%, carbondioxide 3.8% and methane 5.5%. SBA-Kellogg process employs two types of burners. Type I burner is metallic and useful for methane-oxygen system. The preheating of the gases to 650°C by external agencies achieves an acetylene yield of 8.1%. The gases are thoroughly mixed before entering the combustion chamber. This is aided by multitubular mixing sections. The injection of oxygen transversely into methane to successive locations along the mixing throat aids in increasing yield of acetylene. The residence time in the reaction zone, and reaction volume is adjusted by manual control. The reaction chamber which is a pair of alloy steel tubes sliding one over the other without any gap for escape of gases, while the top tube is fixed to the burner. The lower one is adjusted by raising or lowering, this fixes the residence time, water quenching takes place at two places, just inside the reaction in chamber and outside the reaction chamber. Time of reaction is of the order of 2 to 3 milli seconds. Quenching brings down the reaction gases (at 400°C to 500°C). The outlet gas composition shows acetylene 8.1%, hydrogen 57.1%, carbon monoxide 28.9%, methane about 5%, other hydrocarbons and nitrogen balance the rest.

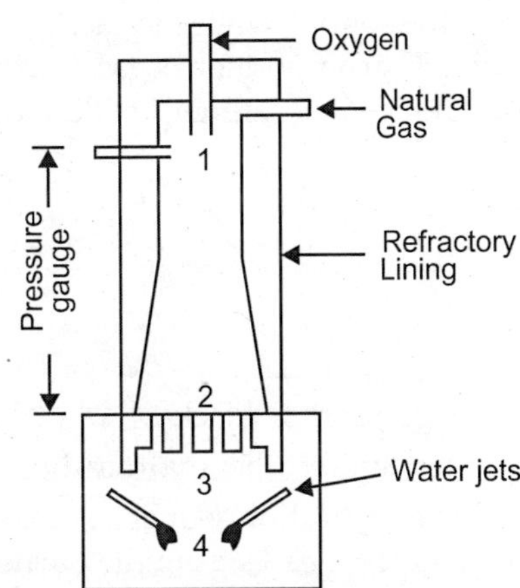

Fig. 3.17: Partial Combustion Burner for Acetylene production.

In type II burner, preheated oxygen and fuel gas are introduced at a velocity of 100-200 m/sec into combustion chamber[45] consisting of a water cooled refractory ring (carborundum). Fine holes are drilled through this block, these are separate for oxygen, fuel and steam. Through different channels these enter the burner holes. Steam at 600°C is admitted for better mixing and it eliminates carbon deposition and reduces partial pressure. Preheated naphtha, propane are injected just below the burner holes into combustion chamber as shown in Fig. 3.18. The thermal decomposition of these stocks takes place due to high temperature of flame. The cracked gases are immediately quenched with huge jets of water. The water slurry carrying carbon particles leaves from the bottom, while the gases leave through a side outlet. The composition of gas from naphtha feedstock after removal of carbondioxide shows:

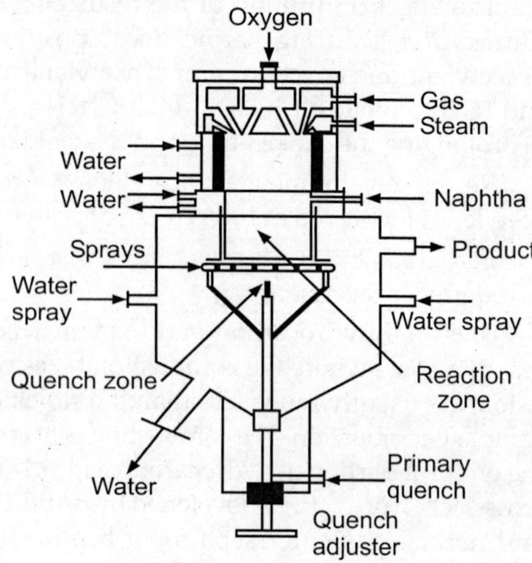

Fig. 3.18: SBA Burner.

C_2H_2	9.51%
C_2H_4	16.34%
H_2	38.8%
CO	15.64%
CH_4	13.12%
Carbon black	2.56%

Remaining other hydrocarbons.

The separation sequence of acetylene produced from hydrocarbons is shown in Fig. 3.19. The prequenched gases from the burner are again given a water wash in column (1). Any carbon, tarry material is carried through washings to a common tank where water is separated for recirculation. The gases are then dried and compressed (not shown in figure). Acetylene is separated from the mixture by selective solvents; the best being dimethyl formamide followed by N-methyl pyrrolidine[42].

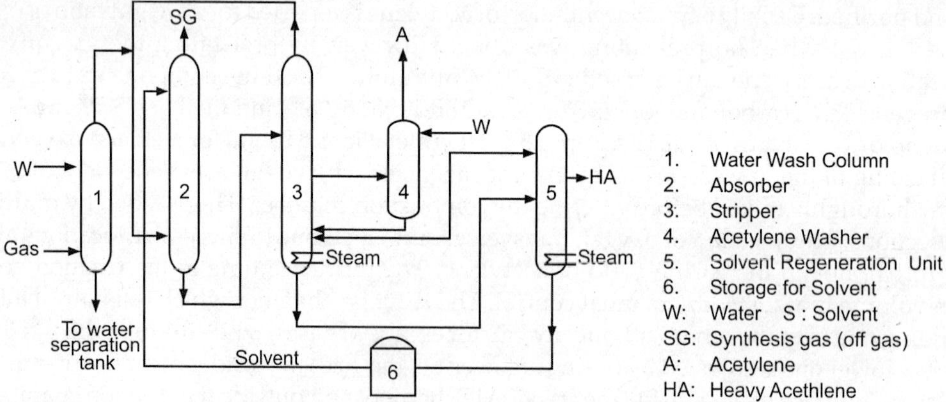

Fig. 3.19: Acetylene separation by Selective Absorption.

Mixture of compressed gas at about 2 bars, is sent into absorber (2) where counter current absorption takes place. The off gas from the top of the column is usually synthesis gas. The fat solution is stripped in column (3) for liberating acetylene from solvent. The off gas from the top of the column is recycled back into the absorber. Acetylene is again washed with water to remove solvent and the washings sent into stripper (3). From the top of the column acetylene is taken off. From the bottom of the stripper washings are taken into solvent regenerator (5) where solvent is collected from the bottom of the column and sent into storage tank (6). Heavy acetylenes are obtained from the middle of the column. The top products from the column are sent into column (3) for stripping.

3.5 CHEMICALS FROM ACETYLENE

The most important products of acetylene are acrylic acid and acrylonitrile. Most of the compounds obtained from acetylene during the first half of 20th century are now economically obtained from other sources. However, the chemistry of Reppe's synthesis still continues, one such synthesis is acrylic acid.

3.5.1 Acrylic Acid

There are many routes for the production of acrylic acid, one route considered to be important, is based upon acetylene and synthesis gas as shown below.

$$CH = CH + H_2 + CO \rightarrow CH_2CHCOOH + 220kJ$$

Under pressures acetylene is reacted with a mixture of carbon monoxide and water over a catalyst, nickel (nickel bromide). The reaction is carried out at 225°C and 100 bars pressure in liquid phase, solvent used being tetra hydrofurane. Alcohol, instead of water, if substituted in the above reaction corresponding acrylic ester results. The reaction product contains 18% acrylic acid and is distilled to separate tetra hydrofurane and acrylic acid.

3.5.1.1 Direct Additions

Acetylene being very active can be directly added to many reactants, one such reaction leads to acrylonitrile. Presently these methods have become obliterated or nearing obsolescence, as ethylene, propylene chemistry offers economic routes.

3.5.1.2 Acrylonitrile

Direct addition of HCN at acetylene results in acrylonitrile.

$$CH = CH + HCN \rightarrow CH_2CHCN$$

Reaction goes in presence of a catalyst ammonium and cuprous chlorides in hydrochloric acid at 70-90°C. Addition of cynide radical increases the rate of reaction. Acetylene diluted with nitrogen upto 50% is mixed with HCN in a ratio of 6:1 and injected into catalyst solution. All HCN is consumed and the reaction products are scrubbed with water to obtain a 2% solution of acrylonitrile. The solution is concentrated to get acrylonitrile.

3.5.1.3 Vinyl Chloride

Vinyl chloride is formed by addition of one mole of HCl to acetylene; addition of two moles of acid leads to the production of unsymmetrical dihalide. The reaction is carried out in the vapour phase using mercuric chloride impregnated on charcoal. The reaction temperature range is 120 to 200°C.

3.5.1.4 Vinyl Acetate

This is also an additive product.

$$CH{:}CH + CH_3COOH \rightarrow CH_3COOCH{:}CH_2 + 122kJ$$

In liquid phase, acetylene is reacted with glacial acetic acid in presence of mercury sulphate at 60°C. Alternatively in vapour phase, it is reacted with acetic acid at 210°C over a catalyst zinc or cadmimum acetate deposited on charcoal. The exit gases are cooled to 0°C, where by acetylene is separated, to be recycled. The liquid is fractionated to separate acetate from acid. Mercuric salts did not yield high amount of acetate, but the side product ethylidine diacetate was an important product. Mercuric othophosphate was found to be a superior catalyst in this reaction. With mercuric salt of acetyl sulphuric acid, operating below 40°C, the conversion to acetate was found to be very high, increase of temperature resulted in more of the side product.

Wacker-Hoechst produced vinylacetate[46] by reacting acetic acid vapour with excess of acetylene over zinc acetate on carbon base catalyst at a temperature of 170-222°C. The conversion was found to be around 70%. (See Fig. 3.20).

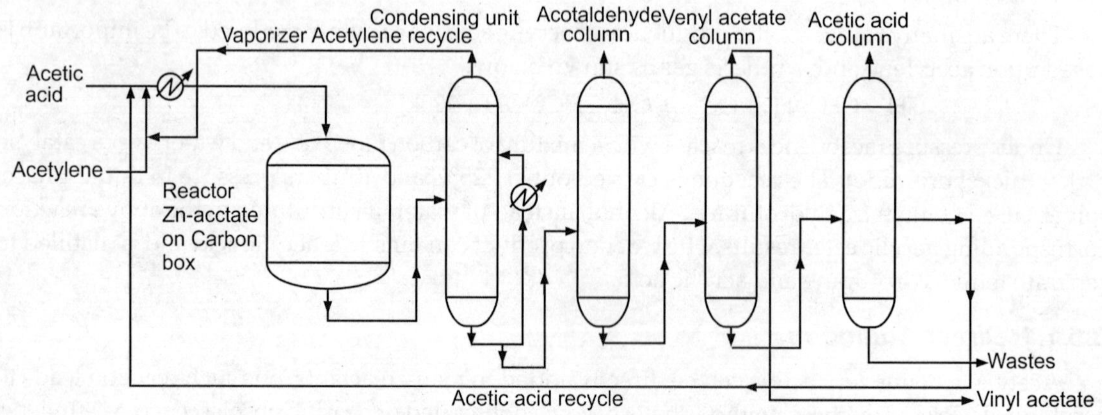

Fig. 3.20: Vinyl Acetate (Wacker process).

3.5.1.5 Acetone

The commercial production of acetone form acetylene is based on the reaction between steam and acetylene over zinc-iron oxide catalyst at 460°C.

$$2\,CH:CH + 3H_2O \rightarrow (CH_3)2CO + CO_2 + 2H_2$$

Acetaldehyde and acetic acid are found to be the intermediate compounds in the production of acetone.

$$CH{:}CH + H_2O \rightarrow CH_3CHO$$
$$CH_3CHO + H_2O \rightarrow CH_3COOH + H_2$$
$$2\,CH_3COOH \rightarrow (CH_3)_2CO + CO_2 + H_2O$$

3.5.1.6 sAcetaldehyde

Acetylene conversion into acetaldehyde can also be considered as an addition reaction, *i.e.* hydration. Hydration is carried out in dilute acetic acid[43a, b] consisting of 0.1 to 2% H_2SO_4; 0.05-4% $HgSO_4$ 25% acetic acid and remaining water. Sometimes $Fe_2(SO_4)_3FeSO_4$ and mercuric oxide in H_2SO_4 are also used.

Optimum conditions of hydration are found to be 70-80°C and 11 bars. Steam and acetylene are admitted into the rubber lined reactor containing the catalyst solution. The conversion is found to be 50 to 55%. The out going gases when scrubbed with water gives 7% acetaldehyde solution. The remaining gas is cycled back. The catalyst solution has to be degassed and treated with 30% nitric acid to convert Fe^{++} to Fe^{+++}; the catalyst solution is then regenerated and recycled. Acetaldehyde solution is distilled to get pure aldehyde. Side products like crotonaldehyde and acetone are common. Fig. 3.21 shows the outlines of the process.

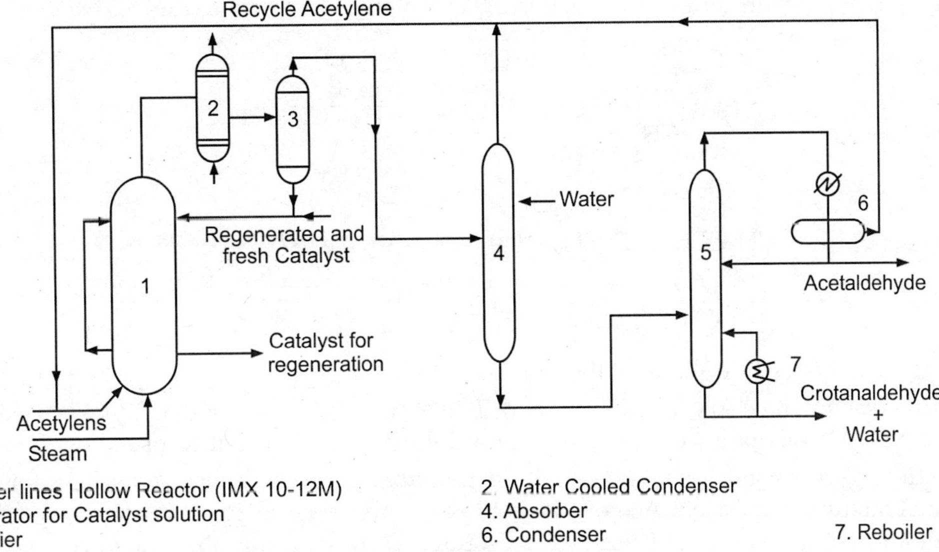

Fig. 3.21: Acetaldehyde from Acetylene.

Vapour phase hydration though possible in practice, has not been instituted. A great variety of catalysts have been used, including copper-nickel, copper-palladium, molybdates, vandates and chromates. A temperature of 300 to 400°C is needed in such reactions.

3.5.1.7 *Trichloroethylene/Perchloroethylene*

These are used as degreasing solvents and in production of fluorocarbons.

Traditionally trichloroethylene is produced from acetylene and still persists. In presence of $FeCl_3$ at 80°C in tetrachloroethane solution, acetylene reacts with chlorine to form tetra-chloroethane

$$CH \vdots CH + 2Cl_2 \rightarrow CHCl_2CHCl_2$$

Tetrachloroethane is reacted with lime to remove HCl from a molecule, which results in trichloroethylene.

$$2\,CHCl_2CHCl_2 + Ca(OH)_2 \rightarrow 2\,CHCl{:}CCl_2 + 2H_2O + CaCl_2$$

Thermal dehydrochlorination is the present practice, which can effectively save costly chlorine, at high temperature of 250°C.

$$CHCl{:}CCl_2 + Cl_2 \rightarrow CHCl_2\,.\,CCl_3$$

Dehydrochlorination may be done by lime or by thermal decomposition, which results in perchloroethylene as shown below:

$$\begin{array}{c} CHCl_2 \\ | \\ CCl_3 \end{array} \rightarrow \begin{array}{c} CCl_2 \\ | \\ CCl_2 \end{array} + HCl$$

3.5.2 Isoprene

Many processes are available for the production of isoprene. One such commercial process is based upon the reaction between acetone and acetylene which yields methyl butynol which on hydrogenation gives butenol, which on further dehydration yields isoprene. No doubt it is a tortuous route compared to direct dehydrogenation of isopentane. The reactions are :

$$C_2H_2 + \begin{matrix} CH_3 \diagdown \\ \quad CO \\ CH_3 \diagup \end{matrix} \rightarrow CH_3\overset{\overset{\displaystyle O}{||}}{C}CH_2CH : CH_2 \quad ...(1)$$

$$CH_3\overset{\overset{\displaystyle O}{||}}{C}CH_2CH : CH_2 + H_2 \rightarrow CH_3CH(OH)CH_2\,CH : CH_2 \quad ...(2)$$

$$CH_3CH(OH)CH_2CH{:}CH_2 \rightarrow H_2O + CH_2 : \overset{\overset{\displaystyle CH_3}{|}}{C}CH : CH_2 \quad ...(3)$$

Fig. 3.22 shows the outlines of the process[18.e].

First reaction takes place at 20 bars and 10 to 40°C in liquid ammonia : KOH is used as catalyst (Reactor 1). Always excess of acetylene is desirable to avoid by product formation.

After the reaction is over solvent and excess acetylene are flashed in a separator (2). All the gaseous mixture is recycled, Acetone is distilled (3) off from the liquid products, which are again distilled to separate butynol from heavy products formed during the reactions, which are to be rejected (4). It is seen butynol forms an azeotrope with water, the separation of which is not important.

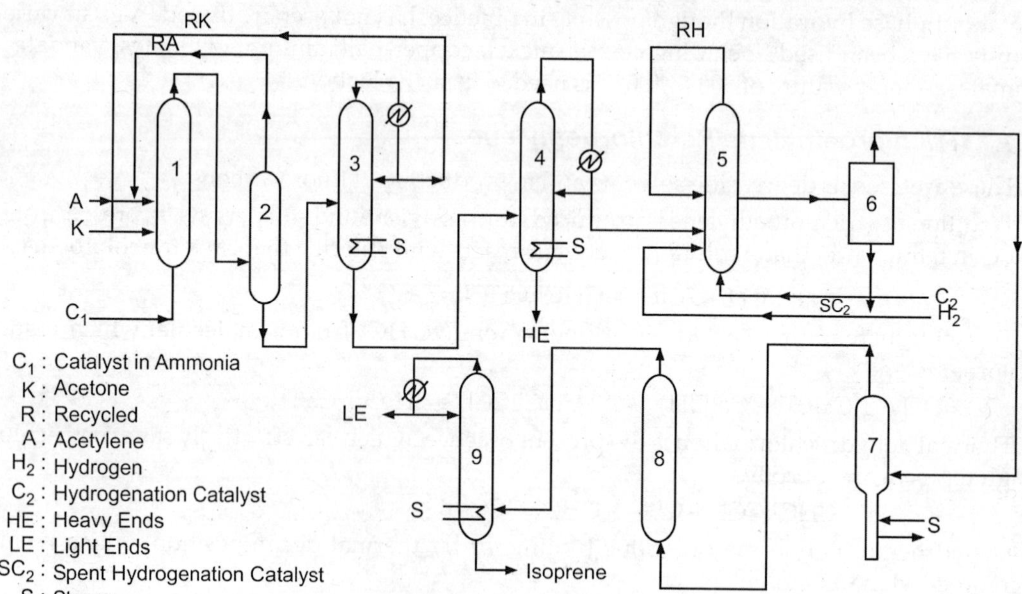

Fig. 3.22: Isoprene Production

As a whole, azeotropic mixture of water and butenol (MBI) is subjected to hydrogenation in a catalytic reaction (5). The catalyst is of significance, as it should not give rise to the formation of 2-methyl butenol-2. The third (3) reaction is easily carried out by alumina at 250-300°C under normal pressure. The hydrogenerated product is separated from the catalyst in a catalyst recovery unit (6).

After catalyst separation MBE-water azeotrope, goes into a preheater or evaporator (7). Preheated vapour is dehydrated in a catalytic reactor (8). The dehydration reaction is carried out by alumina at 250-300°C at normal pressure. The product is distilled (9) to separate isoprene and heavy ends. Polymer grade isoprene results in this process, hence no further purification is attempted.

3.5.3 Vinyl Ethers

Butlerov and Reppe studied in detail the addition of alcohols to acetylene in presence of potassium hydroxide. The product, vinyl ether is extensively active, because of double bonds. Most of the ethers are polymerised or used in polymers for making plastics and synthetic rubber etc.

$$CH{:}CH + ROH \rightarrow CH_2{:}CH\,.\,OR$$

Vinyl ethers are produced by action of acetylene on alcohols, in presence of alkali catalysts. The temperature of the reaction lies in the range of 150–170°C, while the required pressure increases with decrease of carbon atom in the alcohol; to elaborate methyl alcohol requires 20 to 25 bars, while butyl alcohol needs only 5 bars. The reaction is carried out in packed columns, where 1-2% potassium hydroxide in alcohol circulates. Vinylation reactions can be carried out in vapour phase also using sodium acetate catalyst[47]. Yield of ether is expected to be 50-70% on acetylene basis. Vinyl ethers when hydrolised in presence of acids produce aldehydes:

$$CH_3O\,.\,CH{:}CH_2 + H_2O \rightarrow CH_3CHO + CH_3OH$$

An interesting reaction seems to be the addition of formaldehyde to acetylene to produce propargl alcohol

$$CH{:}CH + HCHO \rightarrow CH{:}C\text{–}CH_2OH + 158\ kJ/mole$$

By mild hydrogenation of propargl alcohol allyl alcohol is obtained, severe hydrogenation results in n-propyl alcohol

$$CH{:}C\text{–}CH_2OH + H_2 \rightarrow CH_2{:}CH\text{–}CH_2OH$$

$$CH{:}C\text{–}CH_2OH + 2H_2 \rightarrow CH_3CH_2CH_2OH$$

3.5.4 Acetonitrile

Acetylene and ammonia when passed over zinc sulphate deposited on inert material at 400°C, under moderate pressure, acetonitrile results. Equimolal mixture gives a conversion of 90%. The reaction is shown below :

$$CH{:}CH + NH_3 \rightarrow CH_3CN + H_2$$

Acetonitrile has to be separated from hydrogen after liquifaction of the reactor products. This being toxic is consumed as an insecticide. It is also used as a solvent for butane-butadiene mixtures.

3.5.5 Dimerisation of Acetylene

Acetylene with a triple bond can easily be dimerised or trimerised. Most important compounds, monovinyl and divinyl acetylenes are obtained by this way

$$2\,C_2H_2 \rightarrow CH_2 : CH.C \vdots CH$$

$$CH_2{:}CH\,.\,C{:}CH + C_2H_2 \rightarrow CH_2 : CH.CH{:}CH.C \vdots CH$$

By passing acetylene through acid catalyst solutions at 60-70°C under moderate pressure monovinyl acetylene results. Usually ammonium copper chloride solution in hydrochloric acid is used as catalyst. Divinyl acetylene also follows to some extent. The reaction is highly exothermic (about 130 kJ per mole of acetylene reacted) necessitating effective heat removal. The reaction products are brought to – 30°C to liquefy the products, unconsumed acetylene is cycled back. To reduce the highly polymerised products acetylene is diluted with nitrogen. Monovinyl acetylene is routed to the production of chloroprene, a very valuable monomer in rubber synthesis. Monovinyl acetylene is reacted with hydrogen chloride over copper chloride catalyst at 40-60°C. The reaction is fast.

$$CH_2{:}CH.C \equiv CH + HCl \rightarrow CH_2{:}CH.CCl{:}CH_2$$

Being active acetylene polymerises in presence of finely polished copper at elevated temperatures, thus copper contact is forbidden for the sake of avoiding these crystalline polymers of $(CH)_n$ nature. Acetylene is further found to possess the capacity to cyclopolymerisation, *i.e.* yielding cyclised structures. The principal chemical produced by cyclopolymerisation is benzene.

3.5.5.1 Acetylene Cyclisation

Acetylene when passed through activated charcoal heated to 600-650°C, cyclises giving a large number of polymers. Benzene is the main product. Another interesting compound is cyclic octatetrene which is accomplished by Reppe reaction.

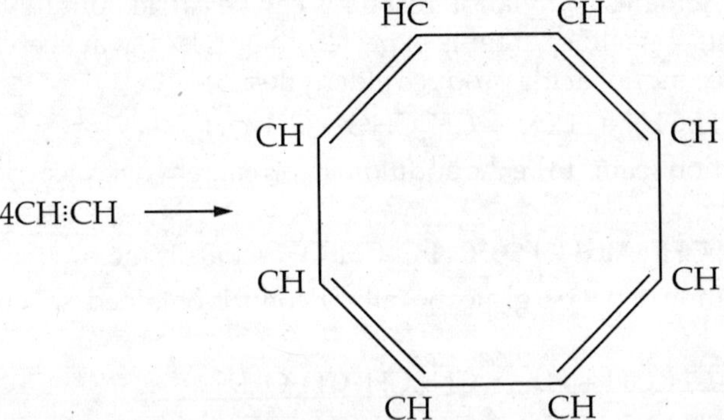

This reaction is carried out in presence of nickel cynide suspension in tetrahydrofurane at 60-70°C at 20 bars. This compound can be hydrogenated to cyclo octane. Cyclo octatetrene can be easily oxidised to phthalic acid.

3.5.5.2 Polyacetylene

When Hideki Shirakawa, a graduate student at the Tokyo Institute of Technology, attempted to prepare polyacetylene, he obtained not the expected dark powder, but a silvery film which coated his glassware. The unexpected product was found to be the result of an error in which the student used a thousand times more catalyst than required. The metallic appearance of the product caused chemists to see if it had metallic properties. In its original form it wasn't a very good conductor, but in 1976 Alan Heeger and Alan MacDiarmid found that when polyacetylene films were treated with iodine (doped), the material became gold coloured and was an excellent electrical conductor-an organic metal. Doped polyacetylene also changes colour with temperature (thermochromism) — red at low temperature, blue at higher temperatures.

REFERENCES

1. A.L. Waddams, Chemicals From Petroleum, GPC.
2. P. Belov, Fundamentals of Petroleum Chemical Technology, Mir Publications, Moscow, 1970,195.

3. B.K.B. Rao, Modern Petroleum Refining, Processes, Oxford-IBM Calcutta, 1984, 285.
4. S. Barendregt, M. Dente, E. Ranzi & Duim, Oil & Gas J. April 6, 1981, 90.
5. Reno. S. Lack & Robert, O. Skamser, HCP, Nov. 72, 99.
6. R.L. Espino & M.B. Sherwin, HCP, Nov. 72, 99.
7. L.N. Ng, C.N. Eng. & R.S. Zack, HCP, Dec. 83, 90.
8. L.E. Chambers & W.S. Potter, HCP, Jan. 1974, 121.
9. R.K. Dorn & M.J. Haddock, HCP, Nov. 1972, 80.
10. S.B. Zdmoik & E.J. Green, Oil & Gas J. Feb. 19 & Oct. 16, 1967, 118.
11. Chemical Engineering News, Dec. 8, 1975, 29.
12. John Dlugniewski, R. Orris & J. Wallace, Oil & Gas J. April, 6, 1981,121.
13. (*a*) CM. Schillmoller, Ulrich. W. Vander Bruck, HCP, Dec. 84, 59. (*b*) R.H. Krikke, J. Moving & K. Smit, HCP, Aug. 76,161. (*c*) W.G. Ashbaugh, HCP, Aug. 76,162.
14. H.Z. Kister & R.W. Townsend, HCP, Jan. 84,108.
15. C. Barre, E. Chanvekilian & R. Dumon, HCP, Nov. 76,176.
16. Richard O. Kirk, CEP, Feb. 83, 78.
17. Mitwing, Chem. Tech., Jan. 80, 21 & Chem. Engg. Sept. 4,1972, 50.
18. (*a*), (*b*), (*c*), (*d*), (*e*). HCP, Nov. 1971,140, 221, 219,118,170.
19. (*a*) Y.C.HU; HCP, Nov. 1982, 109. (*b*) R.L. Baldwin & G.R. Kamm, CEP, Jan. 83, 68.
20. T. Ishikawa & R.G. Keister, HCP, Dec. 78, 112.
21. S.A. Miller, Ethylene and its Industrial Derivatives, Earnest Benn. Ltd., London, 1969, 73.,
22. R.A. Duckworth, C & PE, Feb. 68, 68.
23. Indian Chemical Engineer, Oct. - Dec. 1977, Nov. (1-4).
24. R.W. Mcpherson, C.M. Starks, G.J. Frayer, HCP, March 79,85.
25. C.M. Schillmoller, HCP, March 79, 90.
26. C.M. Dick, HCP, March 79, 93.
27. David P, Keane, Robert B. Stobaugh, Phillip L. Townsend, HCP, Feb. 73, 101.
28. R.B. Stobaugh, W.C. Alien & R.H. Sternberg, HCP, May 1972,153.
29. Ludwig Kniel, Olaf winter, Karf stork; Ethylene, Marcel Dekker.
30. HCP, April 1967,146.
31. Krekeler, H. Kroneing W. Seventh WPC, Vol. 5, London (1967) 41.
32. Brian J. Qzero, Joseph V. Procelli, HCP, Mar. 84, 50.
33. L.F. Hatch, Samir Matar, From Hydrocarbons to Petrochemicals, GPC, 1981 (Houston) 95.
34. Industrial Chemist, 1963, 39, 65.
35. Richard Devon, Moutel L. Schwatz, Chem. Engg., Sept. 4,1972, 50.
36. S.A. Miller, C & PE, March 68, 79.
37. Walter E. Lobo, CEP, Vol. 57. No. 11,1961, 37.
38. (*a*), (*b*) S.A. Miller, Acetylene, Vol. 1,1965, Earnest Bunn Ltd., London, 381, 397.
39. CPE, 1966, 47 (2), 71.
40. CPE, 1968, 49 (3), 101.

41. Kenneth Kobe, Advances in Petroleum Chemistry, Inter Science, Vol. Ill, 399.
42. W.B. Howard, B.H. Wood & E.J. Kaltenbacher, CEP, Vol. 57, No. 11,1961, 51.
43. P.J. Leroux & P.M. Mathieu, CEP, Vol. 57, No. 11,1961, 54.
44. Taiseki Kunugi, Takaaki Tanura, Teisuka Natio, CEP, Nov. 1961, Vol. 51, 43.
45. L.S. Opolak, 8th WPC (Moscow) Vol. 4, 367, Applied Science, London, 1971.
46. WPC 3rd, Section IV.
47. S.A. Miller, Acetylene—Its Properties and Uses, Vol. 2, Ernest Benn Ltd, London, 1965,135, 206.
48. O & G.J. May 1996, 65-66.
49. Technology in Indian Ethylene Oxide and Glycols. Ministry of Science and Technology, GOI. New Delhi.

CHAPTER 4

Chemicals from C_3, C_4 & Higher Carbon Atoms

Paraffins of C_3 and higher carbon atoms have no significance either as finished products or as chemical intermediates. Most usually paraffins are either cracked or reformed. The value of these compounds is best judged by the yield of olefins or diolefins they produce during dehydrogenation. Next to ethylene, propylene takes the leading in chemical industry, specially in the production of polymers, copolymers, alkylates etc. Though propylene should yield variety of chemicals because of its complexity of molecule, still its demand is surprisingly much less than ethylene; obviously the complexity of molecule may result in wide spread by-products, which may not be recoverable at commercial rates or of no direct significance. Most of the propylene obtained is a by-product of refinery operations or as a sister of ethylene it is born in steam cracking of naphtha.

4.1 PRODUCT FROM PROPANE

Propane being gas, is available from flash columns of crude or atmospheric distillation unit. It forms the major constituent of LPG. Hence LPG, though manufactured, may be termed as the major source of propane. In intensive hydrocracking operations 10-15% propane is likely to be formed. Propane like other paraffins is subjected to oxidation, that results in aldehydes, alcohols and acids. The trend is similar to ethane oxidation, but the complexity increases with increase in carbon atoms and difficulty is experienced in isolation of oxidative products. Oxidation products of propane were discussed in the previous chapter. Other products like nitro propanes and chloropropanes follow the reaction pattern of ethane derivatives.

4.2 Dehydrogenation of Propane and Higher Paraffins

Olefins and diolefins are generally produced by dehydrogenation of paraffins usually greater than C_4. Propane does yield propylene only; however C_6 and above give rise to aromatics on dehydrocyclisation; which form starting materials for hexa and higher carbon compounds.

In dehydrogenation reactions, the structural carbon atom is not changed[(1)] on the other hand, thermal cracking produces methane onwards to bigger molecules even bigger than the feedstock itself, due to thermal polymerisation. Dehydrogenation is basically a catalytic reaction. Alumina, alumina-silicates, oxides of chromium and heavy metals, some times promoted by potassium hydroxide, magnesium and zinc oxides; are used as catalysts. The reaction is extremely fast, specially with high molecular compounds; low pressure and higher temperatures (500-600°C) are desirable. Dilution with steam or inert gas may not affect the dehydrogenation reactions. For ethylene and propylene, dehydrogenation of corresponding paraffins is not attempted as pyrolysis of all petroleum fractions yield these. Infact separation of individual components is very much practised in the steam cracking of naphtha.

Dehydrogenation scheme is not much different from cracking reactions as shown below :

$$C_3H_8 \rightarrow C_3H_6 + H_2 - 108\text{ kJ}$$

$$C_4H_{10} \rightarrow C_4H_8 + H_2 - 120\text{ kJ (average)}$$

$$C_4H_8 \rightarrow C_4H_6 + H_2 - 132\text{ kJ}$$

Assertively the reaction takes place in a fixed bed reactor though fluidised beds are also assigned. The product contains olefin, paraffin, and hydrogen, as seen in the case of ethane and propane. But with C_4 and higher paraffins, inclination for going to diolefins is great. Infact butadiene, isoprene are produced by dehydrogenation of butane and i-pentane respectively, which show the significance of dehydrogenation reactions for higher paraffins. In this chapter dehydrogenation of butane is described (see Fig. 4.15).

4.3 CHEMICALS FROM PROPYLENE

Propylene is known as a sister of ethylene and if properly planned it can replace ethylene in chemical industry. Though cheaper than ethylene it has to establish a firm seat in polymer industry. The main source of propylene is thermal cracked gases in which it is largely associated with ethylene. Higher boiling stocks usually produce more propylene than ethylene. Significant change in cracking pattern has resulted even 50-55% propylene from heavier feedstocks. No special separation techniques are required as it is preceding ethylene recovery. An extra C_3 splitter is contrived to separate polymer grade propylene from a mixture of propylene and propane. In most of the countries the demand of propylene is half of ethylene, and a day may come it may supersede ethylene, because of its ease of production and storage. Lighter feedstocks for ethylene production are likely to be less available in future hence the contribution from heavier stocks will naturally weigh the propylene industry; *i.e.* to say any increase in ethylene demand divides the responsibility in equal shares. Important industrial chemicals obtained from propylene are presented in Fig. 4.1.

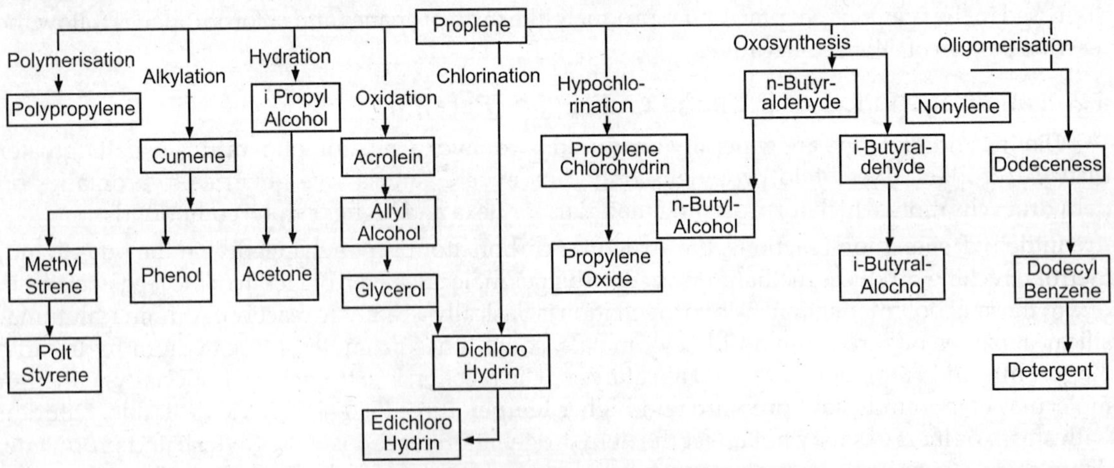

Fig. 4.1: Chemicals from Propylene.

4.3.1 Isopropyl Alcohol

The process is analogous to ethylene hydration. The hydration can be carried out in two ways, either by sulphuric acid (intermediate compound formation) or by direct hydration in presence of a catalyst.

4.3.1.1 Sulphuric Acid Hydration[2]

Propylene hydration is relatively easier than ethylene and the rate of polymerisation is also high. These two factors lead to the formation of other side products, hence contamination increases. The reaction proceeds *via* carbonium ion formation as shown below :

$$CH_3.CH:CH_2 + H_2SO_4 \rightarrow CH_3CH.^+CH_3 + HSO_4^-$$

Again addition takes place

$$CH_3CH.^+CH_3 + HSO_4 \rightarrow \underset{\displaystyle HSO_4}{CH_3.\overset{}{CH}.CH_3}$$

Hydrolysis separates the acid.

$$\underset{\displaystyle HSO_4}{CH_3.CH.CH_3} + H_2O \rightarrow \underset{\displaystyle OH}{CH_3.CH.CH_3} + H_2SO_4$$

Propylene may be absorbed in highly concentrated acid (92%) at ordinary pressure and temperature (20°C) or at slightly higher temperature under an elevated pressure of 25 atmospheres for 80% acid. The feed may be either pure propylene or a mixture of propene and propane containing 50 to 80% olefin. When mixtures are used the pressure is raised to give a partial pressure of propylene 10-15 bars. This ensures the acid saturation of 1.5 to 2.0 moles of gas per mole acid. Though the time of absorption is kept around 4 to 5 hours for all concentrations of acid, it is desirable to use less concentrated acid, for the sake of prevention of side polymers, further dilute acid assists hydrolysis. Hydration by sulphuric acid can be instituted with pure propene or a mixture of propene (20-50%) - propane; often polymers result as side products. By using some solvents which are unaffected by acid, the polymer formation can be greatly reduced. Propene absorption in acid is conducted in plate absorbers fitted with cooling coils on each plate. Acid resisting material like rubber lined-m.s. or karbate is used. Usually 15-20 stages are necessary for absorption.

In the second stage, the reacted mass is steamed keeping the temperature at 70–80°C and acid strength not more than 35%; which is achieved by diluting the mixture with water. The time of hydrolysis depends upon the strength of the acid. Higher concentration of acid leads to less time; however in propanol production the hydrolysis is very easily conducted. Fig. 4.2 shows the brief outlines of the propene hydration. Usually, ether is produced in the reaction which is adjuncted to the reactor as reflux; this decreases the production of ether. Spent acid is regenerated for use. This mixture is steam stripped, the overhead product contains steam and isopropanol vapours to be condensed. The condensate is given a wash with dilute alkali to remove an acid stains and further distilled to get pure alcohol. Dilute acid withdrawn from the stripper is concentrated and recycled. Isopropanol forms an azeotrope with water (12.3%) hence pure alcohol can be obtained by azeotropic distillation only.

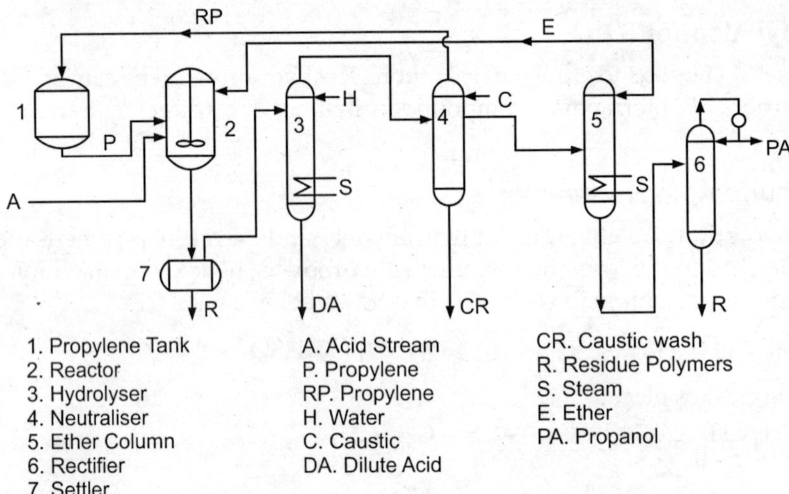

Fig. 4.2: Isopropanol Production (Sulphuric acid process).

4.3.1.2 Direct Hydration

Direct hydration of propylene employs a catalyst phosphoric acid (Veba process) on inert support. The reaction between steam and olefin occurs over the catalyst at a temperature of 180°C. A pressure of 2,000 KPa is necessary for a conversion of 10%. Conversion proceeds at much slower rate than acid hydration. Direct hydration using cation catalyst results in higher yields with other advantages.

Flow sheet in Fig. 4.3 (*a*) depicts the operation as developed by Deutsche Texaco[3a] which uses cation exchange resin. In this process liquid propylene from tank (1) goes under pressure and mixes with preheated water (2). This evaporates propylene and maintains the ratio of propylene and steam in a super critical state. This mixture trickles down the fixed bed reactor (3). The catalyst is usually ion exchange resin, kept at a temperature of 130-160°C and a pressure of 80-100 bars. The effluents are sent into a H.P. separator (4). Heat is removed by cold water. Then the effluent goes into a LP separator (5), where unconverted gas is collected for compression and recycling. The alcoholic solution is neutralised with alkali (6) and then fed into the distillation battery (7-8-9-10). The first column (7) in the battery separates ether, formed in the process. The second column gives the overhead product isopropanol and excess water is taken out as bottom product. Water is ion exchanged to remove all the salt and sent into process. Dehydration of azeotropic mixture is done in columns (9 and 10) to get pure propanol. Fig. 4.3 (*b*) shows equilibrium conversion to i-propanol at various pressures and temperatures over the ion exchange catalyst.

In Tokuyama process[3b], liquid phase hydration takes place over a zirconium-tungsten combination in water at a temperature of 240-280°C and 150-200 bars pressure. This catalyst has very high selectivity, converting as high as 60% propylene per pass. The liquid mass is passed into a flasher where propylene is flashed off. The catalyst solution along with i-propanol mixture is separated in an azeotropic column. Top product primarily contains alcohol-water azeotropic mixture, while the bottom product is mainly catalyst solution for recycling.

Tungsten oxide promoted with zinc oxide on silica supports gives 50% conversion at 230-240°C under a pressure of 200 bars. With cation exchange resin catalyst, as shown in Texaco Trickle bed process[4a] the conversion of 75% is optimum even under mild conditions of operation (temperature 130-160°C and pressure 60-100 bars).

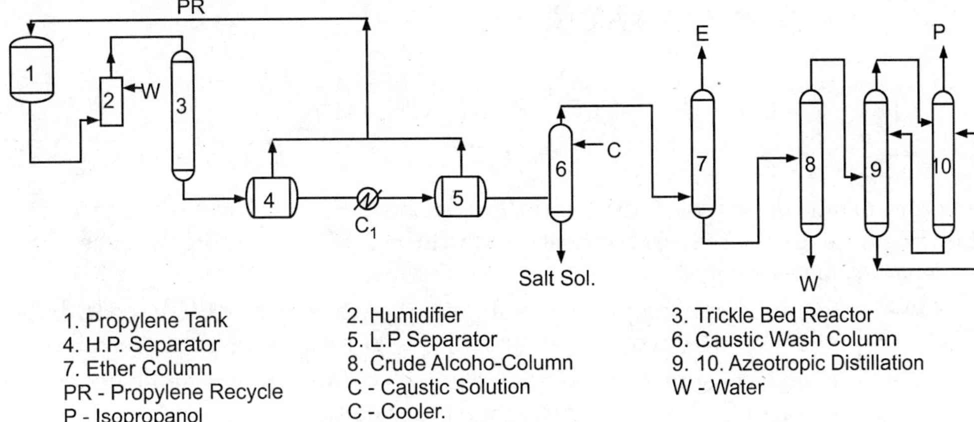

Fig. 4.3 (*a*): Isopropanol Production by Direct Hydration.

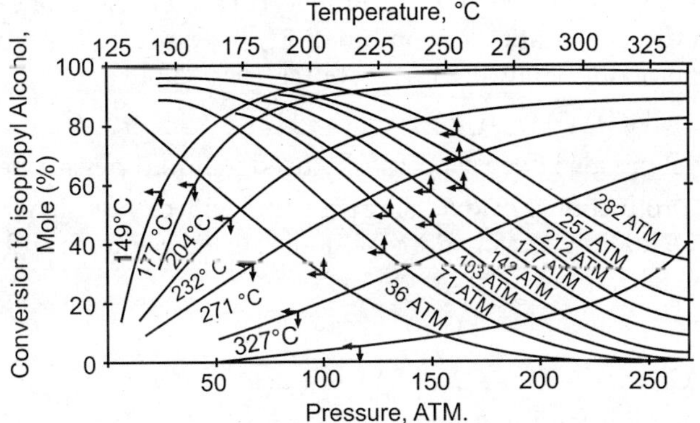

Fig. 4.3 (*b*): Shows Equilibrium Conversion to Isopropanol at various Pressures and Temperatures.

4.3.1.3 Deutsche Texaco AG (DTA) process[5]

This trickle bed process, employs ion exchange resins based on styrene cross linked with varying amounts of divinyl benzene as shown below :

–CH–CH_2–CH–CH_2–CH–CH_2–CH–CH_2.

O O O

SO_3H SO_3H SO_3H

From the literature it is known that low temperature and high pressures favour the reaction. The reaction goes by carbonium mechanism as shown below :

$$H_3C-\underset{\displaystyle H}{\underset{|}{C}}=CH_2+H^+\rightarrow H_3C\,{}^{+}\underset{\displaystyle H}{\underset{|}{C}}-CH_3$$

$$H_3C-\underset{\displaystyle H}{\underset{|}{C^+}}-CH_3+HO^-\rightarrow H_3C\,.\,CH\,.\,OH\,.\,CH_3+H^+$$

The optimum ratio of propylene to water is found to be 12.5-15 : 1. There is a chance of polymer formation if the water is less. The reaction heat is found to be 65 kJ per mole of acid, hence the catalyst bed has a separate quenching system.

The catalyst containing sulphonic groups are found to be active for acid catalysed hydration. The activity is also proportional to H^+ ion concentration; further these catalysts resist vehemently hydrolysis.

Isopropanol is mainly consumed in the production of acetates, amines, xanthates etc. It is also used as a solvent in cosmetic industry, hydraulic and brake fluids.

4.3.2 Acetone

Over 50% of isopropanol is utilised in the production of acetone. Acetone can be produced either by dehydrogenation or direct oxidation or combination of these two. Dehydrogenation is achieved over copper, silver or zinc oxide catalyst. The reaction is carried out at 500°C and 3 bars.

$$CH_3.CHOH.CH_3 \rightarrow (CH_3)_2\,CO+H_2$$

Raney nickel catalyst has also been successfully used in liquid phase operation at 150°C.

Another method of production is by oxidising propanol with oxygen, in which hydrogen peroxide forms as a co-product. Oxidation is easily done at moderate temperature and pressures without any catalyst.

$$CH_3\,.\,CH_2\,.\,CH_2OH+O_2 \rightarrow H_2O_2+(CH_3)_2\,.\,CO$$

$$2CH_3CH_2CH_2\,OH+O_2 \rightarrow 2(CH_3)_2CO+2H_2O$$

Air oxidation reaction is a combination of oxidation and dehydration. The catalyst is copper or silver and the oxidation is carried out at 500-600°C under 5 bars pressure. The drawback of this process is no byproduct; hydrogen peroxide. In dehydrogenation reaction the product is a mixture of acetone and alcohol, while in oxidation reactions the above mixture is obtained in a dilute form with water.

As a co-product, it accompanies several interesting reactions as shown below :

(*a*) Propanol and acrolein react to produce acetone and allyl alcohol.

$$CH_3CH_2CH_2OH+CH_2:CH.\overset{\displaystyle O}{\overset{||}{C}}.H$$

$$\rightarrow\; CH_2:CHCH_2OH+CH_3.\overset{\displaystyle O}{\overset{||}{C}}.CH_3$$

(*b*) Oxidation of propane/butane

$$CH_3CH_2CH_3\longrightarrow CH_3.\overset{\displaystyle O}{\overset{||}{C}}.CH_3+H_2O$$

$$CH_3:CH_2\,CH_2\,CH_3\xrightarrow{O_2}CH_3.\overset{\displaystyle O}{\overset{||}{C}}.CH_3+CO_2+H_2O$$

During oxidation of propane; maximum concentration of acetone produced is 12.5% at 60 bars and 252°C when propane to air ratio is 1:3.6; while 7% acetone can be produced by oxidising butane at 400°C under a pressure of 20 bars. Acetone is now produced directly from propylene. The reaction conditions are very much similar to the production of acetal-dehyde by catalytic oxidation of ethylene using Wacker catalyst[(4b)] system ($PdCl_2$) $CuCl_2$.

4.3.3 Propylene Oxide

Propylene oxide is extensively used in production of propylene glycol which is consumed in polyester resins production. Other applications are insignificant. This has generated new interest in expansion of propylene oxide. Unlike ethylene oxide which can be produced by direct oxidation, the only route for propylene oxide so far used is through intermediate compound chlorohydrin. The reactions involved in chlorohydrin production are given below :

$$CH_3 . CH = CH_2 + Cl_2 + H_2O \rightarrow CH_3 . \underset{\displaystyle CH_2Cl}{\underset{|}{CH}} . OH + HCl$$

$$CH_2Cl - \overset{\displaystyle CH_3}{\overset{|}{CH(OH)}} + Ca(OH)_2 \rightarrow 2H_2C - \overset{\displaystyle CH_3}{\overset{|}{CH}} + CaCl_2 + H_2O \quad (\text{epoxide ring: } C{-}O{-}CH)$$

4.3.3.1 Chlorohydrin Route

Fig. 4.4 furnishes the outlines of the process.

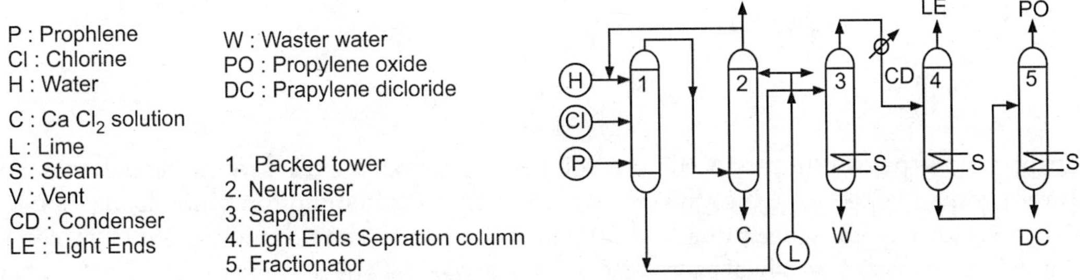

Fig. 4.4: Production of Propylene Oxide *via* Chlorohydrin Route.

The reactions are simple, chlorine and water are introduced into a tower (1) usually packed with inert material where chlorine is converted into hypochlorous acid. A stream of propylene is passed into this tower, and the temperature is closely guarded not to exceed 50°C, by admitting cold water. Propylene should not be allowed to come in contact with chlorine, an equilibrium mixture of hypochlorous acid must be always present. Temperature and concentration of propylene, *i.e.* twice chlorine are important parameters for preventing the side products like propylene dichloride and other chlorinated products. Chlorohydrin obtained must be in very dilute form (6%) to prevent such side products. In fact, propylene dichloride when collected in large amounts may be cracked to perchloroethylene and carbontetrachloride. Usually 10% dichloride is available for unit weight of chlorohydrin.

The liquid stream containing chlorohydrin is drawn from the bottom of the tower and neutralised (3), with lime. The unreacted propylene from the top of the tower after scrubbing with alkali is recycled, occasional purging is also done. The saponification (3) is conducted at slightly moderate temperature (80°C) with excess of lime after which the mixture is steam stripped to remove propylene oxide into the overhead condenser, where it is condensed. The condensate is usually mixture of 26% oxide and the rest water, with a small amount of propylene dichloride and other organics. Purification is done by further distilling under vacuum (5) after separation of light ends. Propylene oxide yield, based on chlorohydrin is 95%, while on entering propylene it is 90%. The difficulty with the process is in getting chlorine and disposing wastes. Incessant pursuits resulted in oxirane process in 1969 (a joint venture of ARCO and Halcon International).

4.3.3.2 Direct Oxidation Process[6]

By oxidising propylene with oxygen, only acrolein results

$$CH_3.CH{:}CH_2 + O_2 \rightarrow CH_2{:}CH\,CHO + H_2O$$

Even organic peroxygen carriers were found to be ineffective, because of the variety of side products not economically recoverable, *e.g.* formaldehyde, acetaldehyde, acetone etc. Normally peroxygen carrier is formed by liquid phase air oxidation of the feed material which is converted directly to the epoxidation product. Oxirane process operates with isobutane or ethylbenzene as the feed for peroxidation. The reaction sequence is given below :

$$4\,(CH_3)_3\,CH + 3O_2 \rightarrow \underset{\text{Tertiarybutyl hydroperoxide}}{2\,(CH_3)3\,C\text{–}O\text{–}OH} + \underset{\text{Tertiarybutyl alcohol}}{2(CH_3)_3\,COH}$$

$$(CH_3)_3\,C-O-OH + CH_3-CH=CH_2 \rightarrow CH_2.\overset{CH_3}{\overset{|}{CH}} + (CH_3)_3\,COH$$

$$\underset{\text{Propylene Oxide}}{\backslash\;\;/\atop O}$$

In this process peroxide is produced from isobutane by air oxidation in a reactor, and the conditions are carefully maintained to reduce the side product alcohol, although formation is inevitable and some times shows its presence more than 50% in the product. The product is sent to epoxidation reactor where a constant supply of propylene is made available. The reaction products are sent into a stripper, where unreacted gases are vented off. The bottom product containing alcohol, oxides are split in a separate column to collect propylene oxide at the top of the column. Alcohol and propylene oxide are purified in a number of subsequent steps. Corresponding olefin, i-butylene, is also accomplished as a side product, which is separated along with alcohol. Since t-butanol is a solid (MP 25.8°C) it is advisable to convert it to i-butene which may be economically consumed in alkylation industry.

Another process[7] duly commercialised yields methyl methacrylate as a side product. In this isobutyraldehyde is employed as an oxygen carrier. The aldehyde is converted to per-iso-acid, which on reacting with propylene yields, propylene oxide and methacrylic acid. The reactions are shown below :

$$\underset{\text{1-butyraldehyde}}{CH_3-\overset{\overset{\displaystyle CH_3}{|}}{C}H_1-CHO} + O_2 \rightarrow \underset{\text{per (iso) butyric acid}}{CH_3.\overset{\overset{\displaystyle CH_3}{|}}{C}H.COOH}$$

$$CH_3\,\underset{\underset{\displaystyle CH_3}{|}}{CH.COOOH} + CH_3.CH = CH_2 \rightarrow \underset{\text{O}}{CH_2.CH.CH_3} + \underset{\text{Methacrylic acid}}{CH_3.\underset{\underset{\displaystyle CH_3}{|}}{CH}.COOH}$$

Methacrylic acid is condensed with methanol to produce methyl methacrylate. Process on the similar lines using peracitic acid as oxygen carrier has been developed by Daicel (Japan).

In case of ethyl benzene oxidation ethyl-benzene hydroperoxide is formed at 90°C and 16-60 bars, in presence of molybdenum naphthenate catalyst. The resulting phenylmethylcarbonil is dehydrated to give styrene, which is a valuable side product. It appears from all these reactions, the basic economy rests completely on the saleable by-product.

4.3.3.3 Per Acetic Acid Process[8]

This process owes its birth extensively to Daicel Japan, that led to rapid commercialisation of the process, from an initial installed capacity of 12000 tonnes/year since 1969. This has been received with acclaims because of high yield and selectivity. The process has been shown in Fig. 4.5. The first step of the process is the production of peracetic acid. Peracetic acid is produced by reacting acetaldehyde (flash or recycled), ethylacetate, a metal ion catalyst and air in a specially designed reactor. Oxidation takes place at 30-40°C under *a* pressure of 30-40 bars. A stabiliser is also added to increase the life of peracetic acid. The peracetic acid is distilled off under vacuum. Acetaldehyde recovered is fedback after purification 30% solution of peracetic acid in ethyl acetate solution is later used in the reaction. The reactions are given below:

(on metal catalyst)

$$CH_3CHO + O \rightarrow CH_3COOH$$
$$\rightarrow CH_3OH + CO_2$$
$$CH_3\,COOH + O \rightarrow CH_3\,COOOH$$

All the information is not shown in the Fig. 4.5.

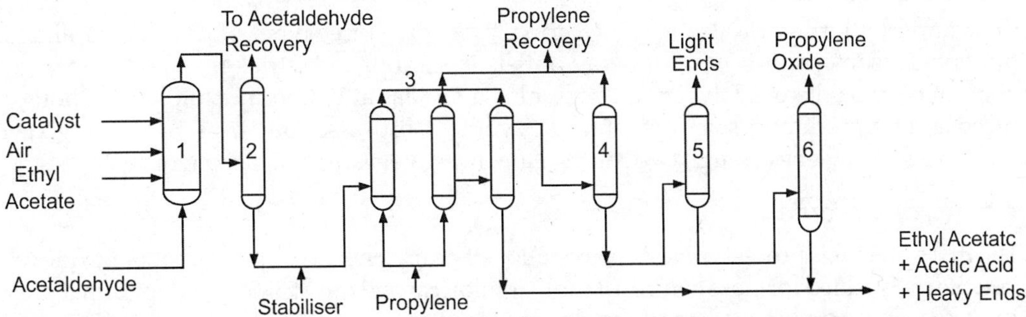

Fig. 4.5: Production of Propylene Oxide *via* Direct Oxidation Route (Daicel process).

A mixture of acetaldehyde, ethyl acetate and a metal ion catalyst are continuously fed into a specially designed reactor (1). Compressed air is continuously supplied to the reactor. The reactor is kept in the temperature range of 30-50°C under a pressure of 20-40 bars for facilitating the smooth oxidation. The reaction products are distilled to separate acetaldehyde in a fractionator (2) for recycling to reactor (1). The peracetic acid in the form of a solution with ethyl acetate is first dosed with a stabiliser to increase the life of peracetic acid and then transferred into the reactor.

Peracetic acid enters the first reactor of three reactor series (3), where it encounters a stream of propylene as shown in Fig. 4.5. The unreacted propylene is collected from the top of the reactors, while the reaction mixer travels from first to third reactor. The overhead product from the last reactor is sent into a battery of distillation columns (4, 5 & 6). In the first column still unreacted propylene goes out to join the recycling propylene. The light ends consisting of aldehydes are taken out and the remaining products are distilled in the last column (6) where propylene oxide is collected as overhead product. The bottoms contain ethyl acetate and acetic acid. These are separately processed for recovery of ethyl acetate and acetic acid as a by-product. Acetaldehyde recovered from the distillation column (2) is purified and recycled. From the last reactor, the bottom product, which is rich in acetate and acetic acid is processed separately.

4.3.4 Propylene glycol

About 30% of propylene oxide goes for production of propylene glycol. Like ethylene glycol, it is also a hydration product of propylene oxide, usually, epichlorohydrin obtained in the process of propylene oxide is converted to glycol by hydrolysing with steam in a stirred reactor at 150°C under pressure 10-15 bars

$$\underset{\diagdown O \diagup}{CH_2 - CH}\, CH_3 + H_2O \rightarrow \begin{array}{c} CH_3 \\ | \\ CHOH \\ | \\ CH_2OH \end{array}$$

Dilute glycol solution is evaporated to required concentration. Propylene glycol, in many cases can form a substitute for ethylene glycol, because of the similar properties of the two compounds.

Many types of propylene glycols are manufactured and most of them are used in polyester resins, plasticisers and detergent industry. Di and polypropylene glycols are used for making rigid polyurethane foams.

Most important trihydric alcohol is glycerine. It can give three types of derivatives, all of them exhibit, optical isomerism. It has got many outlets in various industries like food, paper, textile, pharmaceuticals, paints etc. Glycerine commands a special attention in explosives. Though it is produced as a by-product in soap industry, the synthetic processes outstrips. Synthetic glycerol is produced from propylene through two routes, namely acrolein and epichlorohydrin.

4.3.4.1 Acrolein route

The first step in this process consists of oxidation of propylene on supported copper oxide catalyst with air. Steam is added for close control of the oxidation carried out at 350°C under normal pressure.

$CH_2 : CH\,CH_3 + O_3 \rightarrow CH_2 = CH\,CHO + H_2O + 324\ kJ$

Acrolein is reduced to allyl alcohol by treating with *excess* of isopropyl alcohol in presence of catalyst like magnesium *oxide* or zinc *oxide.* The temperature is kept around 400°C.

$CH_2{:}CH.\,CHO + (CH_3)_2\,CHOH \rightarrow CH_2{:}CH.\,CH_2OH + (CH_3)_2\,CO$

Allyl alcohol is treated with hydrogen peroxide to give glycerine.

$$CH_2{:}CH\,CH_2\,OH + H_2O_2 \rightarrow \begin{array}{c} CH_2OH \\ | \\ CHOH \\ | \\ CH_2OH \end{array}$$

Tungustic acid acts as catalyst in this reaction carried out in liquid phase at 60-70°C. Alternatively, acrolein is reacted with hydrogen peroxide to get glyceraldehyde. Hydrogenation of aldehyde in presence of a catalyst yields glycerine :

$CH_2{:}CH\,CHO + H_2O_2 \rightarrow CH_2OH - CHOH - CHO$

$CH_2OH - CHOH - CHO + H_2 \rightarrow CH_2OH - CHOH - CH_2OH$

Another process once in use was isomerisation of propylene *oxide* to allyl alcohol.

$$\underset{\diagdown\ O\ \diagup}{CH_3\,.\,CH\,.\,CH_2} \rightarrow CH_2{:}CH\,CH_2\,OH$$

The reaction is brought out in vapour phase in presence of lithium phosphate catalyst at a temperature of 250°C.

4.3.4.2 Allyl chloride route

Olefins under specific conditions can give rise to substitution products with chlorine. Dry propylene at 500°C reacts with chlorine to give allyl chloride. The pressure in the reaction is kept around 2.5 bars.

$$CH_2{:}CH\,.\,CH_3 + Cl_2 \rightarrow CH_2{:}CH.\,CH_2Cl + HCl + 110\,kJ$$

The reaction products are separated in a fractionating column. By-products like dichloropropanes, dichloropropylenes are also formed. To increase the yield of allyl chloride 5 : 1 ratio of propylene to chlorine is used.

Hoechst's process employs oxychlorination method. Propylene and a mixture of oxygen and hydrogen chloride are reacted at 240°C over tellurium bearing catalyst.

Allyl chloride is made to react with hypochlorous acid to dichlorohydrin at 280°C.

$$CH_2{:}CH\,.\,CH_2Cl + HOCl \rightarrow CH_2OH.CH.Cl.\,CH_2Cl$$

Another way of getting allyl alcohol is by hydrolysis of allyl chloride with 5% sodium hydroxide solution.

or $CH_2{:}CH.CH_2\,Cl + NaOH \rightarrow CH_2{:}CH.\,CHOH + NaCl$

This allyl alcohol is hypochlorinated to chlorodihydrin.

$CH_2{:}\,CH.\,CHOH + H_2O + Cl \rightarrow CH_2OH.\,CH\,OH.\,CH_2Cl$

Chlorodihydrin is hydrolysed with sodium hydroxide to give glycerine

$CH_2.\,OH.\,CH\,OH.\,CH_2\,Cl + NaOH \rightarrow CH_2OH.\,CHOH.\,CH_2\,OH + NaCl$

Allyl chloride is treated with preformed hypochlorite solution to give dichlorohydrin which by hydrolysis with lime yields epichlorohydrin.

$2CH_2OH.CHCl.CH_2Cl + Ca(OH)_2 \rightarrow 2\,CH_2\,OH.CHOH.CH_2Cl + CaCl_2 + H_2$

$CH_2OH.CHOH.\,CH_2OH$

Epichlorohydrin is again hydrolysed with 5% sodium hydroxide solution to give glycerine.

Production of glycerine through different routes is shown in Fig. 4.6. Acrolein and allylchloride processes use propylene as feedstock, while peracetic acid process developed by Daicel[9], may use propylene oxide directly instead of propylene. However, the advantage of using this process is: peracetic acid produced is useful for propylene oxide production too. High yields of glycerine through epoxidation of allyl alcohol with peracetic acid is the main feature, further the absence of chlorine in the reaction is taken as a great advantage. The chemistry involved in the process is given below:

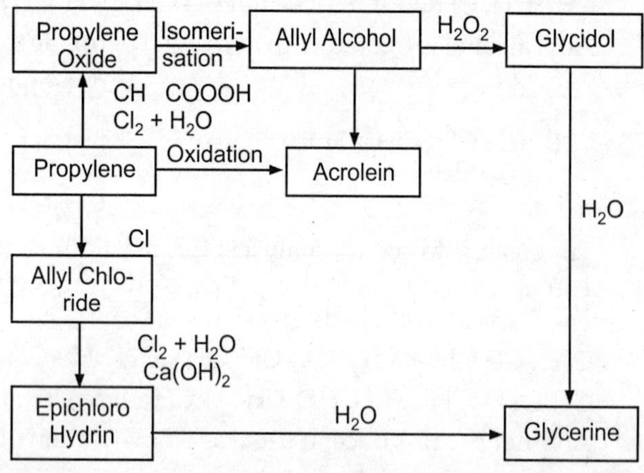

Fig. 4.6: Different Routes for Glycerine from Propylene.

$$CH_2:CH-CH_2OH + CH_3COOH \rightarrow \underset{\diagdown O \diagup}{CH_2.CH}.CH_2OH + CH_2COOH$$

$$\underset{\diagdown O \diagup}{CH_2 . CH}\,CH_2\,OH + H_2O \rightarrow \begin{array}{c} CH_2OH \\ | \\ CHOH \\ | \\ CH_2OH \end{array}$$

Allyl alcohol is produced by the isomerisation of propylene oxide, with a catalyst like lithium phosphate. Peracetic acid is industrially obtained from a lower aliphatic carboxylates, such as ethylacetate or methylacetate. Peracetic acid and allyl alcohol react in equimolar ratio at 40 to 50°C. Addition of water is advantageous as the formation of hydrogen peroxide is slowed down. In the FMC process[10a], propylene oxide is also produced using propylene and oxygen carrier. Fig. 4.7 shows the features of FMC process. The basic steps include:

1. Propylene oxide is converted to allyl alcohol over a catalyst in an isomerisation reactor (1).
2. Acetaldehyde and air react to produce peracetic acid (2).
3. Peracetic acid and allyl alcohol react in-reactor-cum-distillation column (3). The column operates at low pressure of 0.2 to 0.3 bar and is maintained at 40-70°C. The reaction takes place in a solvent usually a ketone. In Daicel process, ethyl acetate-water-peracetic acid solution is fed into the column. This seems to be the conspicuous difference between the two processes, other than starting materials.
4. Glycidol then mixed with water and hydrolysed in an autoclave (4).
5. Hydrolysis product is fractionated in two or three distillation columns (5) to separate acetic acid and water.

In this process allyl alcohol is produced by isomerisation of propylene oxide according to the technique developed by Rhone-Progil. Propylene oxide is bubbled into the reactor-liquid. A high boiling solvent is used as the liquid phase reaction medium as the reaction takes place at 280°C. Lithium phosphate suspended in the solvent acts as catalyst. The reaction products are vaporised and separated (not shown in Fig. 4.7).

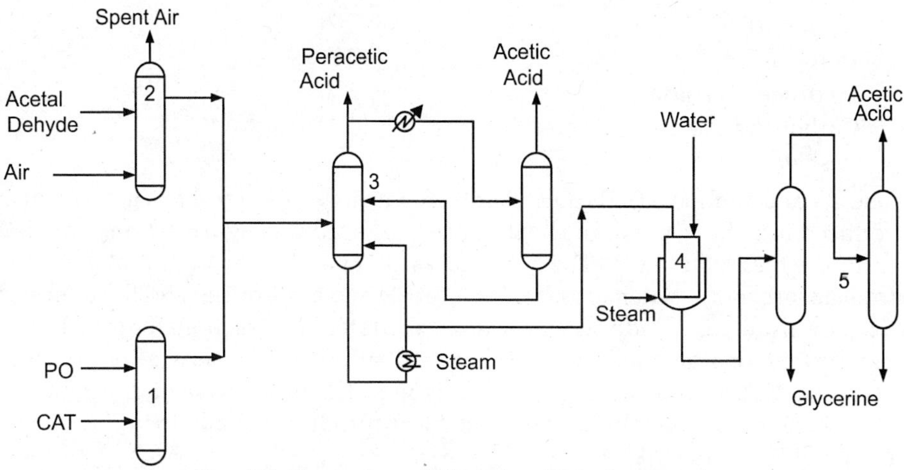

Fig. 4.7: Production of Glycerine (Daicel process).

4.3.5 Acrylonitrile

Throughout the world, acrylonitrile is produced by propylene ammoxidation. It is well known that oxidation of propylene results in acrolein, further oxidation of acrolein results in acrylic acid. Acrylonitrile process is closely related to direct oxidation but in the presence of ammonia acrolein is converted to acrylonitrile, however, the reactions are concluded in a single step and in a single reactor. The reaction between propylene, ammonia and air is shown below:

$$CH_3 . CH : CH_2 + NH_3 + 1.5\,O_2 \rightarrow CH_2 : CHCN + 3H_2O$$

Side reactions[11]

$$CH_2 : CH . CH_3 + NH_3 + 2.5\,O_2 \rightarrow CH_3CN + CO_2 + 3H_2O$$

$$CH_2 : CH . CH_3 + O_2 \rightarrow CH_2 : CHCHO + H_2O + 330\text{ kJ}$$

$$CH_2 : CH . CH_3 + NH_3 + 2\,O_2 \rightarrow HCN + CH_4 + CO_2 + 2H_2O$$

$$CH_2 : CH. CH_3 + 4.5\,O_2 \rightarrow CO_2 + H_2O$$

$$CH_2{:}CHCHO + O_2 \rightarrow CH_2 : CH\,COOH + 250\text{ kJ}$$

Ammoxidation reaction has called for exciting attention, for there was no previous set theory to show the bonding between C and N atoms. Before 1960, the only way of production of acrylonitrile was addition of HCN to acetylene. Sohio process developed bismuth phosphomolybdate catalyst to progress the ammoxidation reaction, later catalysts were based on uranium, which were less efficient.

In the catalyst containing bismuth and molybdenum, the atomic ratio of bismuth to molybdenum is not to exceed one; as the increase would result in complete oxidation to carbon dioxide.

4.3.5.1 Production of Acrylonitrile

Propylene, air and ammonia are reacted catalytically at a temperature of 425-500°C under a pressure ranging from 1 to 3 bars. Fig. 4.8 presents the process developed by Sohio, the widely adopted one, throughout the industrial world. The products of oxidation at optimum conditions are:

Acrylonitrile	71 %
Acetonitrile	2 %
Acrolein	0.7%
Hydrogen cyanide	2 %
Carbon dioxide	4.8 %
Rest water	

Propylene, air and ammonia (99% purity) are sent into a fluidised bed reactor (1). Catalyst in fine form is introduced into the reactor. The heat that is produced during oxidation is carried away by steam generators in the reactor itself. The outlet gases after passing through a catalyst dust collector (2) goes into an absorber (3). Water absorbs acrylonitrile and acetonitrile and the unabsorbed gases leave at the top of the tower. Crude acrylonitrile is separated in a fractionator (4) while the bottom product from the fractionator is sent into another distillation unit for separating crude acetonitrile produced up to 1 to 2% acrylonitrile. Acrylonitrile is again distilled and further purified in columns 5 and 6. HCN obtained from the column (5) is a salable by product (usually 15 to 20% of product). The yield of acrylonitrile *is* about 80%.

Acrylonitrile is a colourless liquid boiling at 78.5°C. It dissolves in organic solvents and moderately in water, it forms explosive mixtures with air. It is a valuable monomer in use for nitro fibers and nitrile rubber. Other markets for acrylonitrile are adiponitrile and acrylamides.

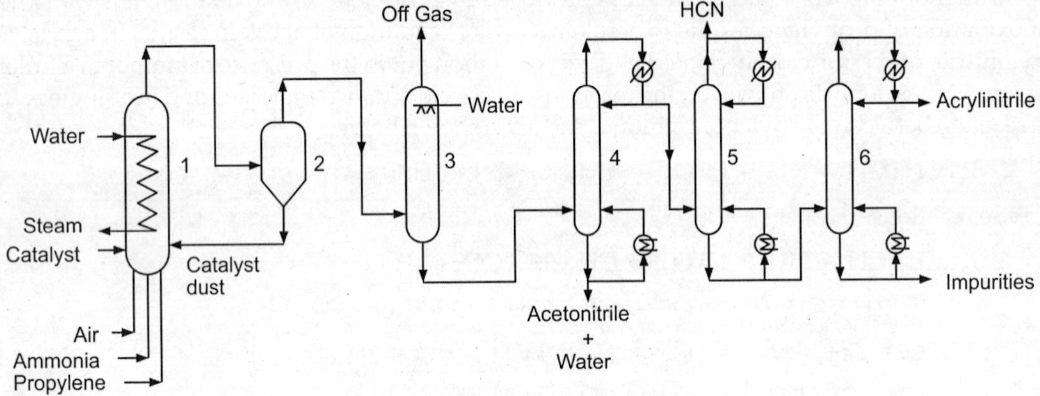

Fig. 4.8: Production of Acrylonitrile.

Monetedison process uses a fluidised bed reactor having the following conditions[12]:

Temperature	420-460°C
Pressure	2 bars
Feed composition	
Propylene	6.7 to 8%
Propane	0.2 to 0.6%
Ammonia	7.5 to 9
Air	Balance
Contact time	2 to 4 sec.

Catalyst: Oxygenated compounds of tellurium, cerium and molybdenum supported on silica 40-100 micron size.

It differs from Sohio process in having one scrubbing unit. The reactor products are first absorbed with lean ammonia sulphate-sulphuric acid solution to recover unconverted ammonia after which the operation is same.

4.3.6 Acrolein

Yet another oxidation route of propylene yields acrolein, rather than propylene oxide, significantly different from oxidation product of ethylene. There are two processes, Shell and Sohio for the production.

In shell process the reaction is accomplished by supported copper oxide catalyst and at temperature of 350°C, and ordinary pressure. Acrolein when further oxidised gives acrylic acid. In Sohio process, propylene, air and steam are introduced into a 2-reactor fixed bed multitubular system. Despite the operation is carried out at temperature of 300-400°C, huge amount of heat has to be taken out because of high exothermic nature of the reaction. Usually heat of reaction is removed by circulating a coolant in shell side of the reactors, and heat is consumed in steam raising operations.

By products like acetaldehyde and acetic acid invariably accompany the main products.

4.3.6.1 Production of Acrolein

Both Shell and Sohio process differ fundamentally in choosing the source of oxygen. In Sohio process the catalyst is bismuth oxide on molybdena and air is employed as the source of oxygen. Shell process uses direct oxygen and the reaction is accomplished over copper oxide catalyst. The reaction conditions of both the processes are given below[13]:

Sohio process	Shell process
Temperature 300-360°C	350°C
Pressure: 1-2 kg/cm	2 kg/cm^2
Catalyst: Bi-Mo system	CuO supported
Conversion : Excellent	20%
Source of oxygen : Air	Oxygen

During acrolein production, acrylic acid and acetaldehyde appear as by products. Sohio process for the production of acrolein is outlined in Fig. 4.9.

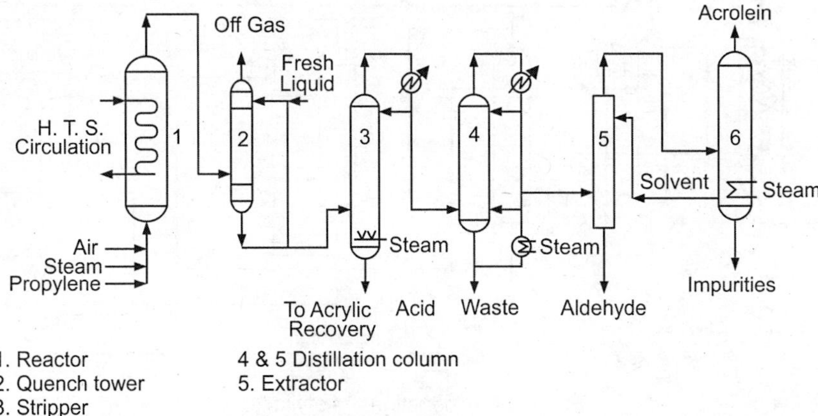

Fig. 4.9: Production of Acrolein.

The reactor is equipped with all facilities to remove large amount of heat by directly extracting with cold water or by heat transfer salt. The products are quenched (2) to remove the off gases from reaction mass; acrylic acid and acetaldehyde along with water and acrolein are drawn off from the bottom of quench tower for separation. In strippers (3 & 4) acrylic acid is obtained as the bottom product, while the top product is distilled to separate aldehydes from water. Acrolein is extracted (5) from acetaldehyde by suitable solvent and regenerated (6) separately. The off gases from the quencher are incinerated.

4.3.7 Acrylic Acid and Esters

Oxidising propylene to acrylic acid and esterifying it, has alluring effect on propylene raw material to discard the existing Reppe's synthesis. Acrylic acid is mainly consumed in methyl methacrylate production. This can be very easily produced in a similar form through ammoxidation reaction, excluding ammonia from the feed mixer; catalyst and other conditions remain the same. Actually this is a two step process, in the first step acrolein is produced, in the second step acrylic acid is produced.

4.3.7.1 Reactions

The reactor may employ catalyst in fixed or fluidised condition. Usual catalysts are metal oxides like cobalt, bismuth, antimony. Fig. 4.10 shows the outlines of the process. Oxidation being fast, it has to be controlled by admission of steam. The reaction mixture contains propylene, air and steam in the ratio of 1 : 5 : 4 respectively, and the reaction proceeds at 300-370°C in two reactors (only one reactor is shown in figure) arranged serially. The temperature in the first reactor shall be higher than the second reactor. Otherwise the reaction can be conducted in a single reactor at 260-360°C. The effluents are showered with water in an absorber (2) to dissolve the acrylic acid and the spent gases are let out. Acrylic acid in aqueous phase is extracted by means of a solvent (3). Light ends are let off or recycled. The acrylic acid is recovered and purified in a fractionator (4). Acrylic acid is made to react with alcohols like methanol or ethanol in a reactor (5) for corresponding esters. The reactor effluents mainly ester and alcohol are separately processed in distillation column (6). Recovered alcohol is reused.

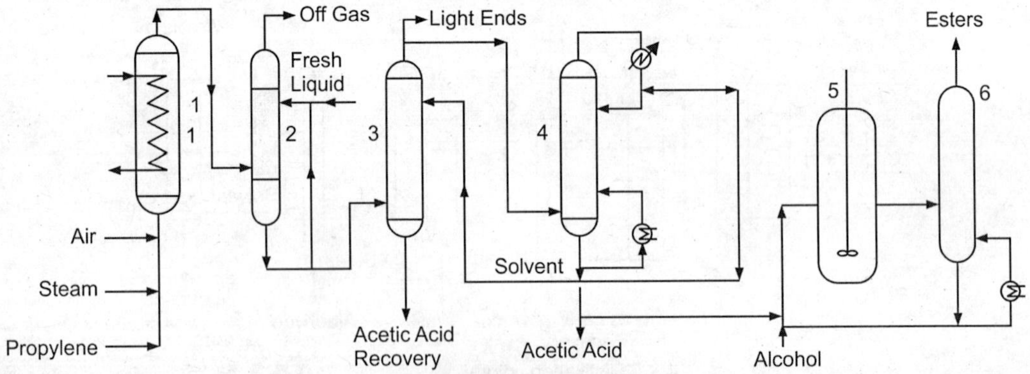

Fig. 4.10: Production of Acrylic Acid.

Table 4.1: Reaction Conditions of Two Step Process[10b]

	First	*Second*
Propylene concentration % (volume)	56%	—
Steam concentration % volume	20-50	—
Space velocity hr^{-1}	1300-1600	1800-3000
Temperature°C	300-400	200-300
Pressure (Bars)	1	1

Oxidation catalysts are synthesised either to suit a single step process or two step process.

4.3.7.2 One step catalyst

It is a multicomponent system containing molybdenum as the main component with other polyvalent metal oxides. Tellurium oxide acts as a promoter in the complex. Though the activity of catalyst is high, its reactivity and life are not satisfactory. The system suffered because propylene conversion increases along with carbondioxide formation; on the contrary conversion of propylene decreases, acrolein to acid is also suppressed.

4.3.7.3 Two step catalyst

The catalyst system is similar to previous one. The reaction conditions are shown in Table 4.1. In the first reactor steam dilution helps in conversion of propylene to acrolein upto 95% with high selectivity. Similarly the second reactor completely converts acrolein to acid without side products. No recycling of propylene or acrolein is attempted.

4.3.7.4 Methyl Acrylate

IPCL produces about 5000 tpa of methyl acrylate utilising acrylonitrile available from the plant. It is colourless liquid with a flash point of – 28°C boiling at 100.6°C. Its sp. gravity is 0.949 and forms explosive mixture with air. Acrylonitrile is first reacted with sulphuric acid of 84% strength as shown below:

$$CH_2.CH = CN + H_2O + H_2SO_4 \rightarrow CH_2 : CHCONH_2\ H_2SO_4 + \Delta H$$

The temperature is closely guarded during the reaction (below 90°C). After the formation of amide, the solution is allowed to maintain 85°C for few hours. After maturing, the solution is reacted with methanol. Though methyl methacrylate is the main product, side products like methyl propionate, acrylic acid ethers are also formed.

$$CH_2 = CHCONH_2\ H_2SO_4 + CH_3OH \rightarrow = CH_2CHCOOCH_3 + NH_4HSO_4$$

The product is separated by distillation.

4.3.8 Disproportionation of propylene (Meta Thesis)

By disproportionating propylene over catalysts like Lewis acids, salts of tungsten, molybdenum or rhenium, a mole of ethylene and butene result

$$2\,CH_2 = CH\text{-}CH_3 \rightarrow \underset{\displaystyle CH_2}{\overset{|}{CH_2}} + CH_3.CH_2.CH = CH_2$$

or

$$CH_3-CH : CH.CH_3$$

or 2-butene

4.3.9 Oligomers and polymers of propylene

Propylene can easily form self additives. However tri, tetra polymers (oligomers) are industrially important, as they can form the basic chains in detergents. Oligomerisation is conducted by sulphuric or hydrofluoric or phosphoric acid.

Phosphoric acid deposited on inert material acts as a good catalyst. Oligomerisation takes place around 200-240°C under a pressure of 20 bars. Usually trimers and tetramers result under those conditions.

The Dimersol[(14)] G. process dimerises C_3 fraction, by using nickel co-ordination complex with alkyl aluminium, as catalyst.

4.3.9.1 Propylene tetramers

Propylene polymerisation does not take place significantly with sulphuric acid, but with phosphoric acid the polymerisation is fast. The catalyst is moulded into pellets by mixing orthophosphoric acid and diatomaceous earth. The pellets are roasted at high temperature. Orthosphoric acid is catalytically active Fig. 4.11 shows the propylene tetramers production from propylene or mixed stocks. Purified propane-propylene mixture is heat exchanged with the products and then sent into the reactor from the top. The reactor is packed with catalyst. Because of exothermic heat, the reactor temperature rises to around 230°C. The reaction products are heat ex-changed, cooled and sent into depropaniser, part of the gas may be recycled as it contains 4-5% of propylene. The heavier fractions are fractionated in a number of columns, to separate light polymers, tetramers and higher ones. In the reactor, the pressure is kept around 60-70 bars. The light polymers may be fed back into the reactor. Tetramers and pentamers boil in the range of 175-250°C. Propane should not be less than 50% in the feed mixture as the heat liberated during reaction leads to wayward situation in the reactor.

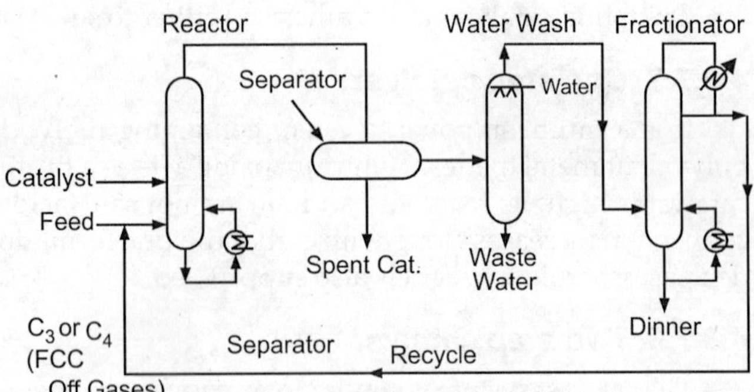

Fig. 4.11: Dimer sol-G Process.

Dimersol-G. Process is connected with dimerisation of propylene, butylene; while Dimersol-X deals with copolymers of propylene and butylene. Dimersol-E produces gasoline out of refinery of gases. The flow sheet for all these processes is similar to Fig. 4.11.

4.3.10 Polymers and copolymers of propylene

Propylene is as much interesting and important chemical as ethylene for the production of polymers. Nearly 20-30% propylene goes for propylene polymers. The remaining is shared by the products like propylene oxide, acrylonitrile, alkylated and oxo products. Polymerisation of propylene has been dealt in detail in chapter 6. Some of the propylene copolymers are discussed here.

Propylene can be copolymerised either with ethylene or butylene. Propylene and ethylene are being economic monomers can be successfully converted into a product of acceptable quality, to form a group of rubbers, EPR or EPDM. These rubbers have a lower density than most of the industrial rubbers. Elastomers of this type do not possess double bonds necessary for cross linking. Ethylene propylene terpolymer (EPT) or ethylene propylene diene monomer rubber (EPDM) can be cross linked using sulphur. These elastomers have good abrasion, oxidation and heat resistance characteristics.

The development of such polymers stems from the polymer techniques of employing family of zieglers catalysts used in polymerisation of ethylene or propylene. Conditions of polymerisation are quoted as moderate temperature about 60°C and pressure 20-25 bars (DSM-Stamicarbon process). Alternatively a low temperature of –20°C and pressure (2 to 3 bars) are used, when vanadium-aluminium complexes are opted as catalysts. Solution polymerisation is quite imperative. Depending upon the properties of elastomers the proportions of ethylene and propylene are varied as 5 to 10 of ethylene, rest propylene. This gives rise to copolymer of good plastic properties having heat, frost resistance and higher melting point (137°C) than polyethylene; films are also made out of such polymers. The saturated nature of these polymers offer high resistance to aging. Ethylene-butylene copolymers are noted for very high resistance to cracking and possess good elongation properties. The difficulty comes across in processing for subsequent uses which has been reduced by using three monomers in the ratio of ethylene 55%, propylene 40% and the third component 5% in polymerisation. Since the molecule is long chain of saturation, Sulphur vulcanization is impossible, however peroxide or any other treatment would substantiate the olefin bond. The kind of treatment is said to be an introduction of diolefin into the molecule, as is the case with butyl rubbers. Among the dienes in commercial use, are dicyclopentadiene, haxadiene. This does not mean that ethylene-propylene terpolymers, (EPDM), can complete with the elastomers, as the basic disadvantages like slow curing rates, poor adhesion and compatibility still prevail. If these problems are overcome economically, they may cope up with the other elastomers. Production of EPDM rubbers is dealt under polymerisation chapter.

In solution polymerisation ethylene and propylene monomers are absorbed in a solvent, to this stream, third monomer diolefin (3 to 8%) is added and sent into a polymer reactor. The catalyst solution (some times two different catalysts are used) is introduced into the reactor. The reaction proceeds under mild conditions. The products are flashed, whereby the monomers are separated and routed back. The polymer is washed with water or alcohol by decomposing the catalyst. The polymer is separated from the solvent. The solvent after purification is sent to the absorber.

4.3.11 Phenol

Phenol is a well known chemical, abundantly available in mixed form in coal tars. Phenol is a starting material for production of caprolactum and adipic acid. Phenol-formaldehyde thermosetting resins, oldest in the chemical history, are manufactured on a large scale. Further it is used in dyes, drugs and germicidal preparations. It is produced from petroleum stocks to a large extent through two routes namely cumene and chlorobenzene. Both the processes use benzene as raw material. Description of phenol production through cumene is furnished below.

4.3.11.1 Production of phenol through cumene

The cumene process is regarded as the industrially significant. In the first stage benzene is alkylated with propylene, either in liquid phase or in vapour phase; the resulted propylbenzene is oxidised in the second stage to yield phenol and acetone as shown by the reactions.

$$C_6H_6 + C_3H_6 \rightarrow C_6H_5C_3H_7 + 101 \text{ kJ/mole} \qquad ...(1)$$

$$C_6H_5C_3H_7 + O_2 \rightarrow C_6H_5OH + CH_3.CO.CH_3 \qquad ...(2)$$

Reaction (1) is alkylation which is carried out in presence of a catalyst. The catalyst may be sulphuric, phosphoric acid on diatomaceous earth, HF, $AlCl_3$, BF_3, aluminosilicates, and many chlorides of metals. Obviously depending upon the catalyst, the process becomes homogeneous or heterogeneous. Preferably in industry benzene alkylation with propylene is conducted over Friedel-Crafts catalysts.

The reaction conditions vary with the catalysts as shown below :

Catalyst	$AlCl_3$	H_3PO_4	H_2SO_4 (90%)	*Alumina[15] Silicates*
Benzene to propylene (mole) ratio	2.5	4	4	3
Pressure, bars	1-5	15-25	5-10	20
Temperature°C	70	250	30	350
Propylene conversion % per pass	complete	complete	complete	50-70%
Polyalkyl benzenes %	3 to 6	3	2	3
Propylene concentration % in feed	80	40	80	60

In all alkylation reactions, the end utility of the product is to be specified; for example 1-propyl benzene produced with phosphoric acid catalyst is not suitable for production of phenol, as it oxidises poorly for the fact it contains some unsaturated compound. Production of isopropyl benzene is shown in Fig. 4.12. It may not be essential to use pure propylene fraction, and it can be a mixed stream of propane-propylene but purified from other con-stituents like O_2, CO_2, H_2S and water. Benzene and olefin stream (3:1 ratio) are mixed before entering the heat exchanger (1) Preheating (2) if required is done, otherwise the feed goes into the reactor (3) where the catalyst ($AlCl_3$) is packed in tubes. The temperature is closely guarded by circulating oil. The hot oil from outlet gives away the heat to feed mixture in the exchanger (2). The reaction products at 250°C exchange heat to the preheater and find the way to the fractionation units 4, 5 and 6. The pressure in the fractionator is usually brought down to 10 bars whereby propane fraction is easily recovered. The remaining liquid product is stripped off in column (5) where benzenes escape as over head product and recycles. The remaining alkylate is distilled in column (6) to set free alkyl benzene as top product and polyalkyl benzenes as bottom product. It is seen that alkylation being fast not only the desirable alkyl benzene is formed, but polyalkyl benzenes are also formed; however, the dealkylation of polyalkyl benzenes is equally fast, hence at equilibrium more alkyl benzene shall be available. Sometimes these polyalkyl benzenes (specially di-isopropyl benzenes) are mixed with the feed to shift the equilibrium in favour of isopropyl benzene.

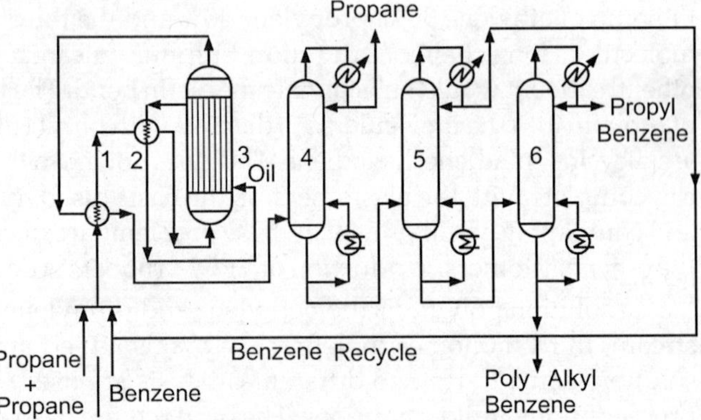

Fig. 4.12: Alkylation of Benzene with Propylene.

Sulphuric acid alkylation is also preferred, when liquid phase alkylation is sought. The acid concentration in such cases is kept around 80-90% and the reaction temperature is lower than 40°C. The second stage in the production of phenol consists of oxidation of cumene.

4.3.11.2 Oxidation of Cumene of Phenol

Cumene is oxidised with air under controlled conditions. This is almost non-catalytic process. The reaction consists of two steps : firstly cumene hydroperoxide is formed, next it is decomposed to phenol and acetone as shown below:

$$C_6H_5{-}CH(CH_3)_2 + O_2 \longrightarrow C_6H_5{-}C(CH_3)_2{-}O{-}O{-}H + 122\ \text{kJ}$$

$$C_6H_5{-}C(CH_3)_2OOH \longrightarrow C_6H_5{-}OH + \begin{matrix} CH_3 \\ CH_3 \end{matrix}\!\!>\!CO + 255\ \text{kJ}$$

Oxidation can be carried out in liquid phase quite conveniently. The pressure and temperature in the oxidiser are kept within 3-5 bars and 110-120°C respectively. The reactor is usually a plate column, where cooling coils are kept on plate for assisting heat removal, Fig. 4.13 shows the process outlines. Isopropyl benzene, after getting heated in a heater (2) up to 110°C is sent into the plate column (1) from the top, where it encounters a rising gust of air.

The overhead vapours are condensed (3) and uncondensable exhaust gases are stripped off (4). The liquids are cycled back with fresh cumene.

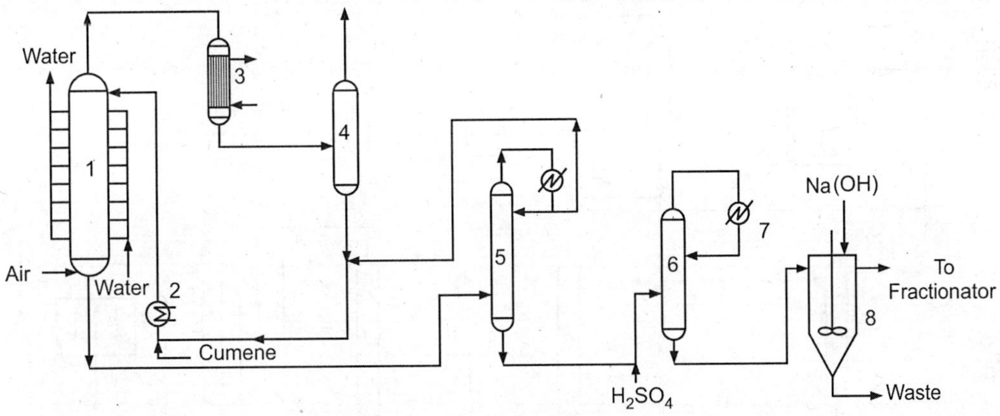

Fig. 4.13: Phenol Through Cumene.

The bottom product from the reactor column which usually contains 20-25% of hydroperoxide. It is concentrated to 80% by distillation under vacuum (less than 30 mmHg) in a column (5). Overhead product from this column (5) contains cumene for recycling. The bottom product is decomposed with an acid catalyst (H_2SO_4) in a plate column (6). Very large amount of heat is liberated during this cleavage reaction, hence only dilute acid (5-10%) is used. The heat removal is carried out by circulating the reaction mass through an outside heat exchanger (7). The decomposed product is neutralised with sodium hydroxide (8) and the organic mass is sent into multicolumn fractionator to separate acetone and phenol.

4.3.12 Bisphenol A

Bisphenol-A (2, 2 bis 4-hydroxyphenol) is mainly used for the production of epoxy resins followed by polycarbonates. Other insignificant uses are resins and chemicals.

It is produced by condensation of a molecule of acetone with two moles of phenol (Fig. 4.14).

$$2\,\underset{\text{(O)}}{\overset{OH}{|}} + CH_3.\overset{\overset{O}{\|}}{C}.CH_3 \rightarrow OH - \text{(O)} - \underset{CH_3}{\overset{CH_3}{\underset{|}{\overset{|}{C}}}} - \text{(O)}\ OH + H_2O$$

The reaction takes place in presence of an acid catalyst like hydrochloric or sulphuric acid. Phenol and acetone are saturated with acid stream and are sent into a reactor (1). The reactions products are distilled off (2) to remove HCl and water vapours. Bottom product contains bisphenol and phenol. The mixture is separately distilled in another column (5). Crude bisphenol from the distillation column is passed into isomer still (6) where the impurities having lower boiling point than bisphenol are distilled off. The heavier product is again distilled (7) to separate bisphenol vapours from the tarry residues. Bisphenol is further purified by dissolving in a solvent and crystallization (8, 9). The extract containing impurities comprising mainly ortho, para isomers goes into the solvent recovery unit. All the fractions from 5, 6, 4 units are recycled to the reactor.

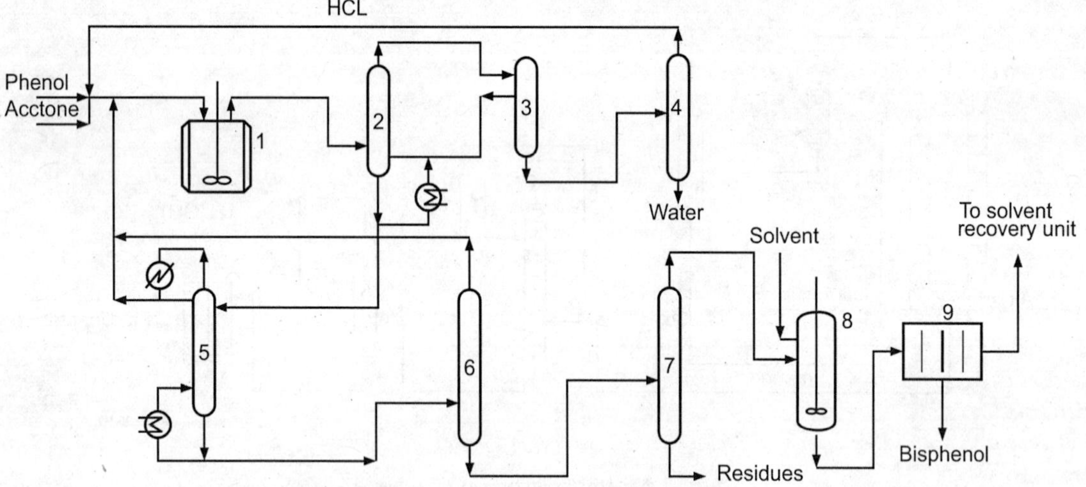

Fig. 4.14. Production of Bisphenol-A.

4.4 DEHYDROGENATION OF BUTANES

Though dehydrogenation of ethane and propane for production of olefins has not been pronounced in industry, higher paraffins do, example being butane, pentane etc. Like cracking, it proceeds through homolytic radicals; often it is conducted over catalysts to minimise the unnecessary light gases, which form in large quantities during thermal cracking or dehydrogenation. Usually a single stage dehydrogenation yields olefins and two stage dehydrogenation leads to dienes. Depending upon the catalyst; pressure, temperature, space velocity are judiciously choosen.

Table 4.2 Dehydrogenation Catalysts of Butane/Butene

Catalyst Types	*Al_2O_3 90% Cr_2O_3 8% K_2O*	*Cr-K Salts on support*	*Fe_2O_3 18% K_2O 4% MgO 72% CuO 4.6%*	*Bi + MO Oxides on silica gel Bi:MO = 0.8*	*Alkali earth bromide and a transition metal on Al_2O_3*
Feed	Butane	Butane	Butane (i & n)	n-Butane	n-Butane
Product	Butene	Butenes C_3/C_2	Butadiene Butenes/ C_3	n-Butene	n-Butene
Reaction Temperature°C	575-600	580	580-650	520-550	520-550
Reactor type	Vertical Tubular	Fluidised Bed	Vertical adiabatic	adiabatic	adiabatic
Conversion	20%	33%	45-50%		80%
Space velocity	200 m^3/m^3 h	120 m^3/m^3 h	150 m^3/m^3 h	200 m^3/m^3 h	430 m^3/m^3 h

4.4.1 Catalysts

During dehydrogenation of a n-butane, five major products appear.

$$CH_3.CH_2.CH_2.CH_3 \rightarrow CH_2:CHCH_2CH_3 \quad \text{n-butane}$$
$$CH_3CH:CHCH_3 \quad \text{cis-2-butene}$$
$$CH_3CH:CHCH_3 \quad \text{trans-2 butene}$$
$$CH_3CH:CH_2 + CH_4$$
$$CH_3CH_3 + CH_2:CH_2$$

Many dehydrogenation catalysts are developed[17] (Table 4.2); these are usually based on metal oxides; $Cr_2O_3 - Al_2O_3$ (Houdry type), $Ca_3Ni(PO_4)_6Cr_2O_3$ on supports (Dow type) and many mixed oxides like Fe_2O_3 (18.4%) K_2O (4.6 %) MgO (72.4%) CuO (4.6%) etc. Catalysts (Houdry type) are usually promoted with oxides of potassium, magnesium, zinc etc.

4.4.2 Dehydrogenation for butenes

Dehydrogenation of butanes for production of butenes is carried out in single stage either in fixed beds or fluidised beds as shown in Fig. 4.15. Because of the isomers, the separation of individual butenes becomes tedious rather than impossible; various separation schemes are shown in Fig. 4.16. When cracker gases are handled which contain some butadiene, it has to be selectively removed either by extraction with a suitable solvent or by mild hydrogenation to convert butadiene to butane. The separation of C_4 streams is shown in chapter 1. Specifically, when butane is dehydrogenated; resulting butenes and butanes have to be separated economically either by distillation or by combination methods i & n butenes. Butenes can be separated by using molecular seives, a technique developed by Union Carbide; but this fails if n & i-butanes are also there in the feed stream. Isobutylenes are produced by dehydrogenating isobutanes in similar fashion.

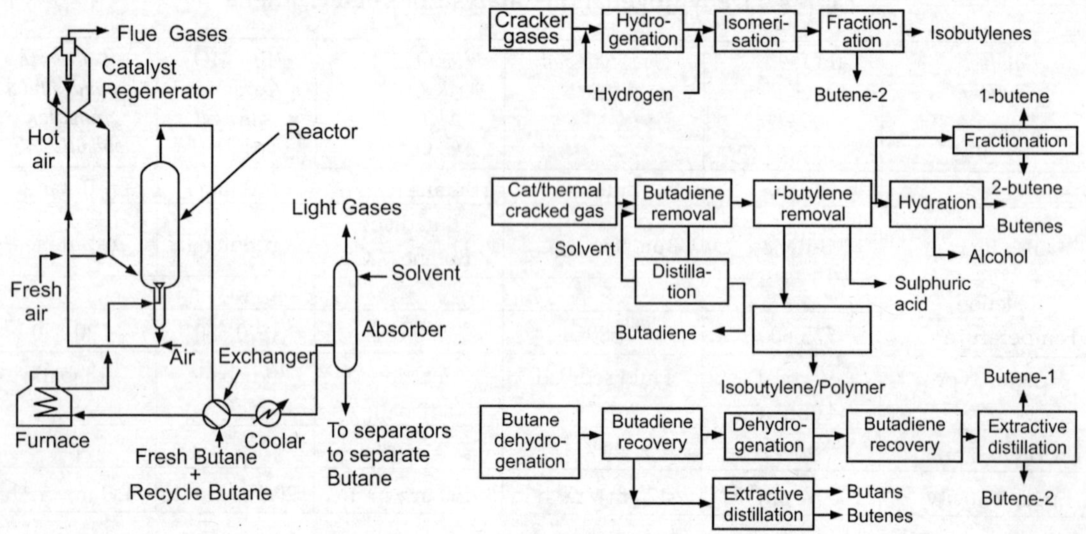

Fig. 4.15 Fluidised Bed Reactor for Butane Dehydrogenation.

Fig. 4.16 Schemes for Separation of C_4-Streams.

4.4.2.1 n-Butylene products : Iso and n-butanol

The hydration of butylenes to secondary butyl alcohol bears a resemblance to production of isopropyl alcohol by sulphuric acid process. The feed must be free from i-butene and butadiene. Saturates do not affect the reaction. Sulphuric acid of 75-85% concentrations at 20-35°C would bring the hydration. All the n-butenes give secondary butyl alcohol only. Direct hydration of butenes has not been successful. Isobutylenes through sulphation and hydrolyses yield tertiary butyl alcohol in this reaction. In this reaction much dilute acid and lower temperature are used.

Butene-2 is converted into secondary butyl alcohol: this when hydrogenated yields MEK. Conversion takes place in presence of copper catalyst at higher temperature. This process is going to be adopted by Cetex Petrochemicals Ltd. to produce 4000 T of MEK near MRL.

These alcohols have limited use as solvents and chemical intermediates.

An alternative process for the production of n-butanol is through oxo route, where propylene is converted to butyraldehyde, which on hydrogenation gives n-butanol.

4.4.2.2 Oxo Route (n-butanol)

Propylene hydration is accounted for maximum, (about 80%) of the oxo products. Hydroformylation habitually uses a catalyst tetrahydrocarbonyl of cobalt $HCO(CO)_4$. This type of catalyst is later modified with alkyl phosphines[(18)] such as $HC(CO)_3PR_3$; (PR_3 tribuity phosphines). Shell uses these modified oxo catalyst and product with such catalyst shows 90% straight chain compounds. Cobalt has been replaced with rhodium or iridium which is found to be more selective and reactive.

A comparison between the cobalt and rhodium catalyst for hydroformylation of propylene is given below :

	Cobalt	*Rhodium*[(19)]
Temperature°C	145-180	60-100
Pressure (bars)	100-45	6-25
LHSV	0.5-1.0	0.1-2.5
n/isodistribution in product	3-4:1	8-16:1
Product	aldehyde/alcohol	aldehyde
Yield	67%	70%

Rhodium carbonyls are found to be 10^2 to 10^4 times more active than cobalt catalysts, hence the reaction conditions are very mild and the product pattern can be easily regulated.

Fig. 4.17 shows the process based on BASF technology[(20)]. The reaction proceeds according to the equation.

$$C_3H_6 + 3CO + 2H_2O \rightarrow C_4H_9OH + 2C0_2 + 215 \text{ kJ}$$

The catalyst is iron pentacarbonyl and butyl pyroledine. Water and n-butanol act as solvent to the catalyst system. Propylene and carbon monoxide along with the catalyst are fed into the reactor and the reaction concludes very fast producing by products like hydrogen, propane and carbon dioxide. After purifying the out going gases, propylene and carbon monoxide are recycled. Catalyst is separated along with butanol, excess butanol collects as a separate layer, which is taken for distillation.

In another version where cobalt catalyst[(21a)] is employed (BASF), crude oxo product containing a mixture of more butyraldehyde and less alcohol results. Butyraldehyde is separated and hydrogenated. Butyraldehyde can be cracked at 300°C to reverse the reaction.

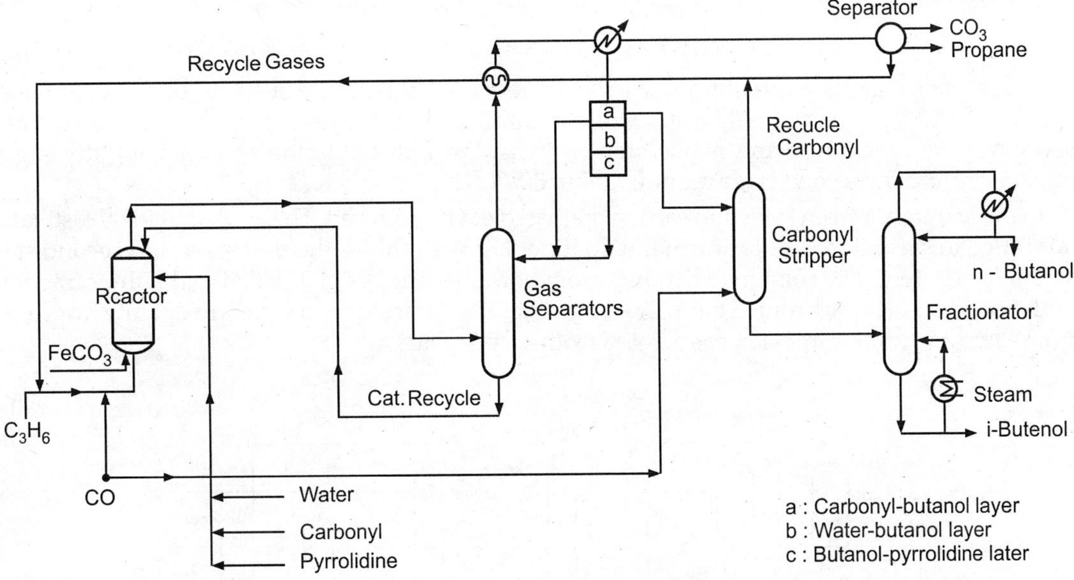

Fig. 4.17: Butanol Production.

4.4.2.3 Oxidation of butane

Products of butane oxidation differ little in composition from those propane oxidatives. Normally butane is oxidised in presence of steam and oxygen from air. A ratio of butane, air and steam of 1: 5 : 30 is maintained at a temperature of 400°C under a pressure of 2-30 bars. The products contain acetaldehyde 18-20%, acetone 7%, methanol 18-19%, propanol and butanol about 1%, carboxylic acids 11-12% and aldehydes about 15%, followed by carbon dioxide and carbon monoxide. Generally higher pressure produces high amount of alcohols and less amount of aldehydes and acids.

4.4.2.4 Acetic acid

Catalytic liquid phase oxidation of n-butane results in acetic acid. The reaction conditions are

Temperature	:	150-250°C
Pressure	:	50 bars
Yield	:	75-80%
Catalyst	:	Cobalt acetate

Other secondary products include formic acid, methyl ethyl ketone and ethyl acetate. Cleanse process *is* based on Cobalt Catalyst, which gives more acetic over formic acid.

4.4.2.5 Methacrylic acid[22]

Isobutylenes can be oxidised in two stages as shown below, to produce methacrylic acid.

$$CH_2 : \underset{}{\overset{CH_3}{\overset{|}{C}}} - CH_3 + O_2 \rightarrow CH_2 : C\,(CH_3) - \overset{O}{\overset{\|}{C}}H + H_2O$$

$$2\,CH_2 : C\,(CH_3) - C = OH + O_2 \rightarrow 2\,CH_2 : \overset{CH_3}{\overset{|}{C}} - \overset{O}{\overset{\|}{C}} - OH$$

The reaction conditions are supposed to be moderate, a temperature of 400°C, under a pressure of 1-2 bars would yield about 80% acid. Molybdenum oxide with several oxides of heavy metals pronounce the reaction. Being a two step process, the first oxidation requires slightly higher temperature and the second stage requires about 300°C.

In the first step, as seen from the reaction methacrolein is produced. This is separated by distillation and routed to second reactor for further oxidation. Fig. 4.18 shows the first stage and second stage reactors separately. The combined product from both the reactors is distilled and lighter fractions, butylene and acrolein; are routed back. The acrylic acid is extracted with water to separate from acetic acid. Water, acetic mixture is later separated from acrylic acid.

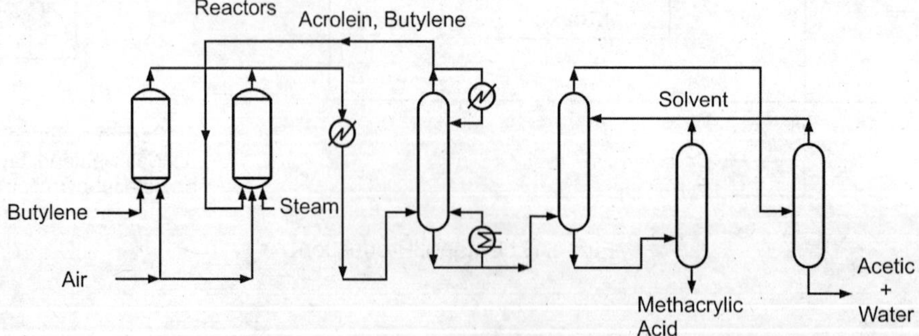

Fig. 4.18: Methacrylic Acid Production.

Methacrylic acid is also produced from acetone-cynohydrin process. A large amount of ammonium sulphate is obtained as a waste product.

$$(CH_3)_2CO + HCN \rightarrow (CH_3)_2COHCN$$

$$(CH_3)_2COHCN + H_2SO_4 + H_2O \rightarrow CH_2 : C(CH_3) . COOH . NH_2 . H_2SO_4$$

$$(CH_3)_2CO . HCN + CH_3OH + H_2SO_4 \rightarrow CH_2 : C(CH_3) . COOCH_3 . NH_3 . (H_2SO_4)$$

The above reactions show the production sequence.

4.4.3.1 Methacrylonitrile

Ammoxidation of butylene is similar to propylene ammoxidation and the product lethacrylonitrile is consumed in the industry as acrylonitrile. Another method to produce lethacrylonitrile is to react ammonia with isobutyl alcohol at a temperature of 55°C over dehydration catalysts.

$$(CH_3)_2 CH_2CH.OH + NH_3 \rightarrow CH_2 = C(CH_3) - CN + H_2O$$

Ethylene oxide when reacted with acetonitrile in presence of alumina at 600°C yields nitrile.

$$CH_3 CN + CH_2 - CH_2 (O) \rightarrow CH_2 = C(CH_3) - CN + H_2O$$

4.4.4 Methyl-tert-butyl ether (MTBE)

MTBE is the fastest growing (growth rate more than 20%) chemical in the world. Because of the pollution stipulation through out the world (though not much in Third world countries) lead replacement has become unavoidable. MTBE is the best substitute for lead in gasoline. Started with a world wide production of 3.5 MM tonnes in 1986 it crossed over 10 MM tonnes 1989-90 till now there is no plant in India, but expected soon. Major amount of i-butylene utilised for the production of MTBE. This is produced by the addition reaction between methanol and i-butylene.

$$CH_3 . C(CH_3) = CH_2 + CH_3 OH \longrightarrow CH_3 - C(CH_3)_2 - O - CH_3$$

The reactions conditions are very mild, but require an acid catalyst; sulphonated polystyrene resin is found to be very active. The reaction between alcohol and butylene can occur in liquid phase also. Usually it is conducted in vapour phase. The reaction products are distilled to separate butylene and unreacted methanol, if any.

4.4.4.1 Production of MTBE

i-Butene and methanol are reacted over an ion exchange catalyst at mild operating conditions. The feed is mainly refinery butene, hence the economy of the process depends upon the availability of butene. Usually not more than 8% butane will be available from the crackers. The produced ether should be cheaper as it is blended with gasoline which itself is cheap. The better solution will be to isomerise all butenes resulted from dehydrogenation of butane and utilise them.

Fig. 4.19 shows the utilisation of isomerisation during synthesis of MTBE. Mixed butenes along with methanol enter into the fixed bed reactor (1) where almost 90% of i-butene is consumed. The products are cooled to 20°C (c) whereby unreacted methanol and MTBE are condensed. Butenes are separated first from the reactor-mix by distillation in a column (2). Methanol and MTBE mixture is obtained from the bottom of the column is again distilled in a separate column to collect MTBE. The gas phase containing methanol vapours is washed with water (3) and dried (4). This gas is now heated in a heater to 300°C (5) and admitted into isomerisation unit, (6) where 2 and n - butenes are converted into i-butene. This stream after getting cooled (c) to required temperature mixes with incoming streams and finally goes into the reactor (1). This way all the butenes available during dehydrogenation of butane can be availed. This ether is mostly used as an octane booster.

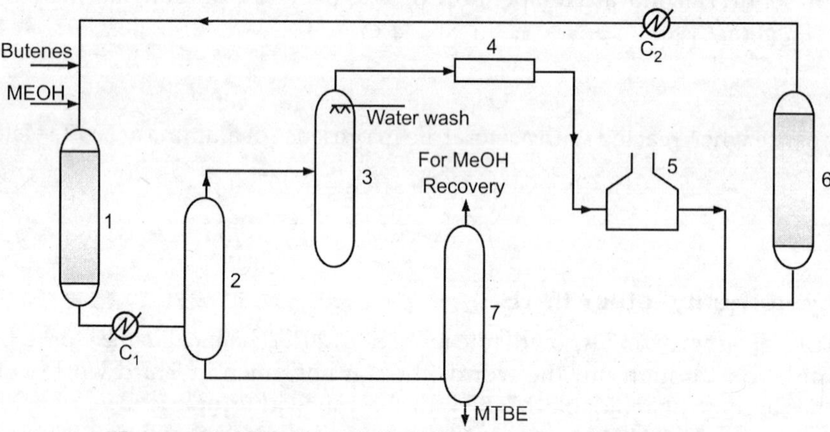

1 Reactor, C1, C2, Cooler, 2. De butanization, 3. De methanolisation, 4. Dryer, 5. Heater, 6. Isomerisation unit, 7. Distilation unit.

Fig. 4.19: MTBE production.

4.4.5 Maleic Anhydride

Maleic anhydride, is produced industrially by the catalytic vapour phase oxidation of C_4 unsaturated hydrocarbons. Catalysts like, Molybdenum, vanadium pentaoxide activated with phosphorous oxide on silica gel are used. Oxidation is carried out either in the fixed beds or in the fluidised beds. Maleic anhydride is also manufactured by the vapour phase oxidation of benzene at 400-500°C over vanadium pentaoxide catalyst. Other method is based on oxidation of cyclopentene over vanadium - molybdenum oxide catalyst at 450-500°C.

Recent attention is however towards butanes as feedstock.

The suggested catalyst for this oxidation is vanadium pentoxide activated with phosphorous oxide. Addition of oxides of chromium, iron and zinc results in higher yields of maleic anhydride.

Butane oxidation is carried out at atmospheric pressure over iron, $V_2O_5 - P_2O_5$ on silica catalyst at 500°C.

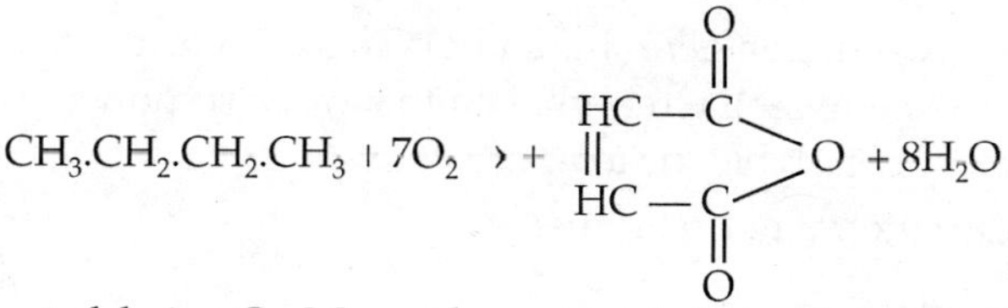

$$CH_3.CH_2.CH_2.CH_3 + 7O_2 \rightarrow \begin{array}{l} HC-C(=O) \\ \| \qquad\qquad \rangle O \\ HC-C(=O) \end{array} + 8H_2O$$

Other catalyst suggested, being Co-Mo oxides supported on silica. n-Butene, oxidation is carried over Mo-V-P oxides[23] on silica at 450°C under 2 to 3 bars pressure. The development of process is credited with Bayer.

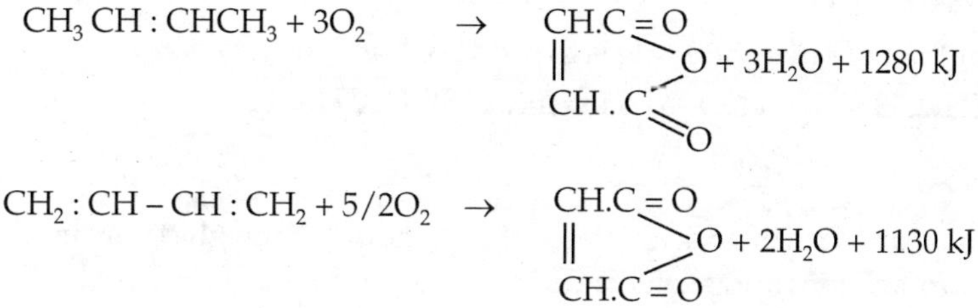

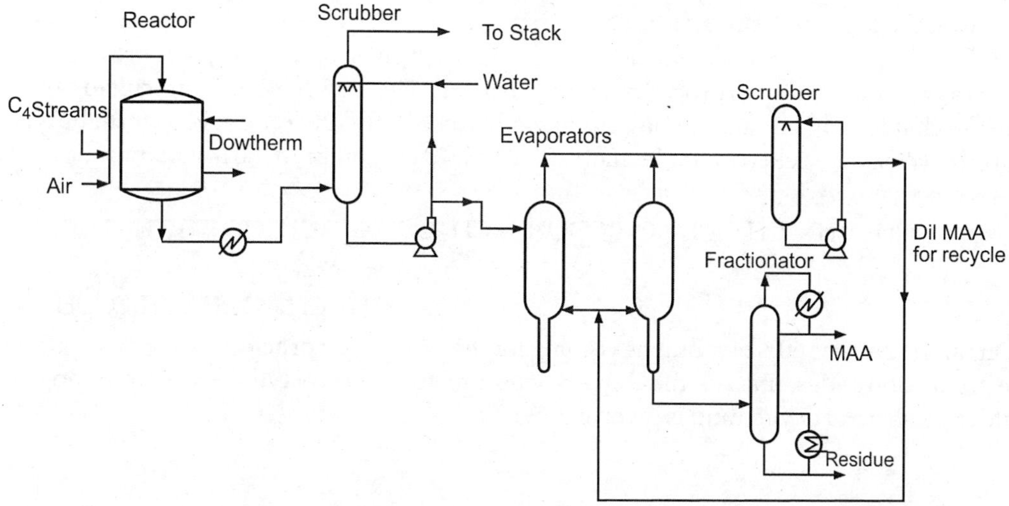

$$CH_3\,CH:CHCH_3 + 3O_2 \rightarrow \begin{array}{l} CH.C=O \\ \| \qquad\quad \rangle O \\ CH.C=O \end{array} + 3H_2O + 1280\text{ kJ}$$

$$CH_2:CH-CH:CH_2 + 5/2O_2 \rightarrow \begin{array}{l} CH.C=O \\ \| \qquad\quad \rangle O \\ CH.C=O \end{array} + 2H_2O + 1130\text{ kJ}$$

Fig. 4.20: Maleic Anhydride from C_4 Unsaturates.

Fig. 4.20 shows the technology based on Bayer's Process. Bayer's process employs a mixture of butane and butylene, and catalyst vanadium pentoxide in pellets. These pellets are packed into a large number of tubes in a reactor. The reaction heat is dissipated by circulating dowtherm to keep the reaction temperature at 300°C. The hot gases pass through heat exchangers and finally through coolers. The cooled gases are scrubbed with dilute solution of maleic anhydride in water. The concentration of the solution in the scrubber reaches to 30%, which is then further concentrated in falling film evaporators. The concentration of the solution in the scrubber reaches to 30%, which is then further concentrated in falling film evaporators. The vapours of anhydride are cooled and further fractionated for purification. Though the yields are low (30%) compared to benzene oxidation, still the raw material availability at cheap rates encourages the process. Maleic anhydride is used as a modifier in plastics, as it can copolymerise with other monomers. As an insecticide and as a stimulator for plants growth, it serves in agriculture.

Maleic anhydride hydrogenation gives butendiol. This can react with terephthalic acid to give butylene terephthalate. Like ethylene terephthalate this can also polymerise to produce a sort of polyester in the name of PBT or polybutene terephthalate.

With ethanol it produces tetra hydro furane.

4.4.6 Butylene Oxide-Glycol

Butylene oxide, though an oxidation product of butylene, cannot be produced by direct oxidation, hence chlorohydrin route is followed. The reaction conditions and sequence of reactions follow as in the case of propylene oxide.

Most important use of butylene oxide is to produce butylene glycol.

$$\underset{\substack{\backslash\,/ \\ O}}{CH_3\,CH_2.CH.CH_2} + H_2O \rightarrow CH_3.\,CH_2.\,CHOH.\,CH_2OH$$

The hydrolysis is carried over an acid catalyst. Butylene oxide and glycol are used for the production of plasticisers, pharmaceutics and surfactants.

4.4.7 Acetic Acid from Butane

Studies on direct oxidation of butene of acetic acid were given up, as many side products resulted, of which a great number led to products tending to form resins thus making it difficult to isolate acid. The olefin character of butane can be eliminated by reacting with acetic acid to form butylacetate which on oxycracking yields two moles more than the acid consumed in the reaction, as shown below: (Bayer's Process)[(21b)]

$$CH_3.\,CH_2-CH:CH_2 + CH_3COOH \rightarrow CH_3.COO.CH\,(CH_3).\,CH_2\,.\,CH_3$$

$$\text{(on oxycracking)}$$

$$CH_3COOH + CH_3COOH + CH_3COOH$$

On carbon balance 60% of n-butene is going for the production of acid. The major steps involved in the production as described by the above reaction are formation of butyl acetate, cracking of butyl acetate in presence of oxygen and is given below.

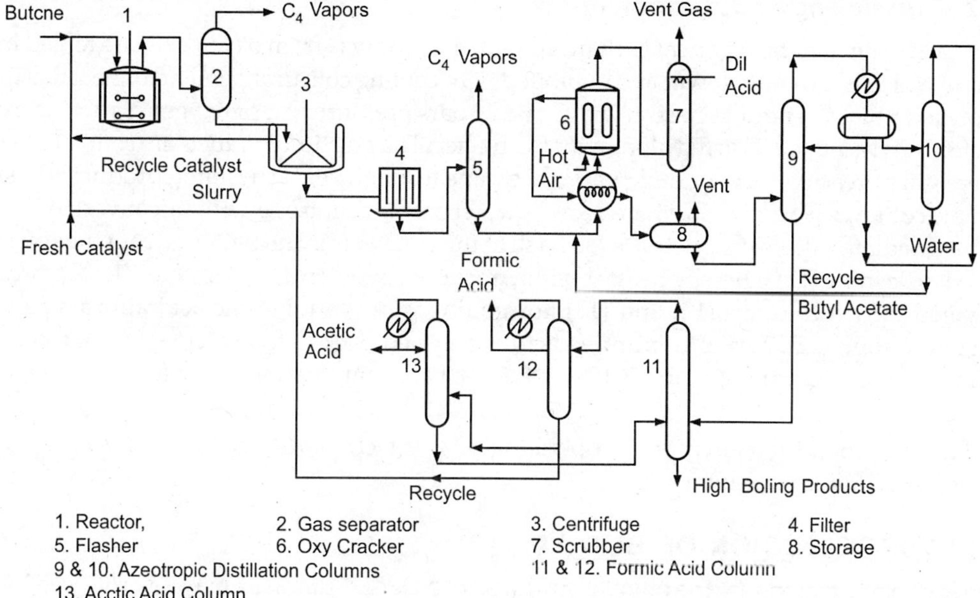

Fig. 4.21: Production of Acetic Acid (Bayer's Process).

4.4.7.1 Production of butyl acetate[23] (Refer. Fig. 4.21)

Acid and butylene are reacted with excess of acid (1.5 mole per/mole olefin) in presence of acidic ion exchange catalyst in liquid phase at 110-125°C under a pressure of 15-25 kg/cm^2. All butylene react to give normal and secondary products. Stirred tank reactor (1) is used for this purpose. Catalyst concentration is kept at about 10%. The catalyst is separated from the reaction mixture by a centrifuge (3) and recycled to the reactor. Clear solution from the centrifuge is filtered (4) and flashed (5) to remove n-butenes; acid and ester from the bottom product.

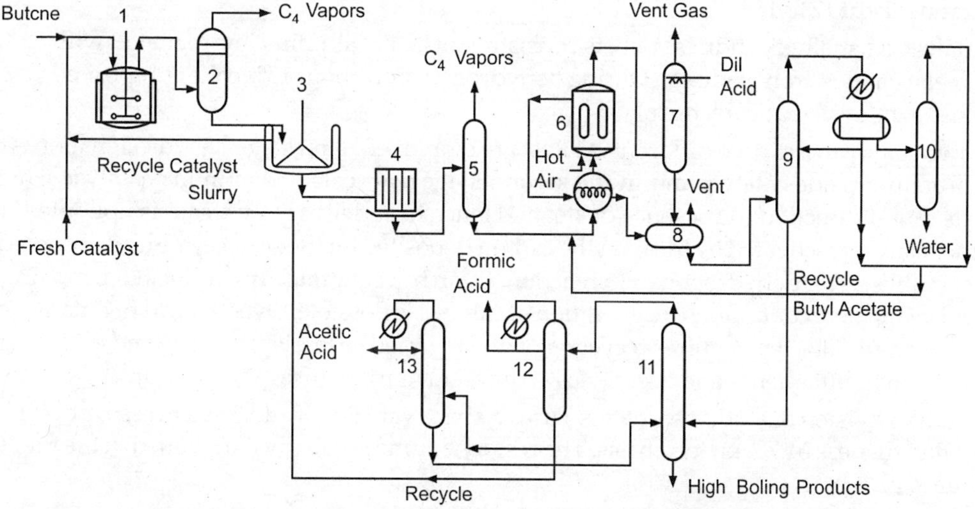

Fig. 4.22 (*a*): Production of Butadiene from Butane.

4.4.7.2 *Oxicracking of butylacetate*

Acid and ester after having heat exchanged with the products from oxycracker are forced into the reactor (6) which is an empty tower and equipped with cooling coil arrangement for heat dissipation. Reaction takes place in liquid phase at 200°C and 60 bars pressure. Air or oxygen is used as oxidant. Unreacted butyl acetate is also fed alongwith fresh ester. The reaction products are sent into a storage (8) where still unreacted gases or other vapours escape through vent. The liquid product is flashed in azeotropic columns (9) and (10) where recycle water forms azeotrope with unconverted ester and other intermediates, these products are fed back to the cracker (6). The bottoms from the column are subjected to flashing (11) whereby heavy boiling fractions are obtained as residue. The top product is fractionated in two columns (12) and (13), for getting pure acid. Formic acid appears as the top product in column (12). Part of unpurified acid is recycled back to the reactor (1). All the vapours including formic acid vapours (about 10% is formed) are sent into a furnace for steam generation (incineration).

In fact, it is doubtful if any plant is operating on this technology, as it is always cheaply made available by oxidation of alcohol.

4.5 DEHYDROGENATION OF BUTANE

Butene and butadiene are the obvious products of butane. Butadiene may be produced by direct dehydrogenation in one or two stages, intermediate chemical being butene. Fig. 4.22 (*a*) shows the dehydrogenation flow sheet of butane. Feedstock is heated to 530-600°C and sent into 3 or 4 reactors in parallel. The catalyst is loaded in tubes, necessary heat may be augmented by external heating of the reactors or some times heat carrier like steam is used. The catalyst in such cases must be steam resisting. The products are quenched, compressed and sent into absorber. The absorber uses a solvent that can be specific with diene only, hence olefin and paraffin are unabsorbed and directed to mingle the feedstock. The extracted butadiene is stripped off from the solvent and mixed with a stabilizer before storing.

Butadiene concentration goes as high as 43.5% during dehydrogenation at 700°C and 10 kPa or less pressure. Partial pressure reduction is aided by sending steam into the reactor (steam to butane motal ratio is kept 12 to 1).

Catalyst is usually cylindrical pellets consisting of active alumina impregnated with chromium oxide, metal halides, iron oxide etc. During dehydrogenation about 15% coke is deposited on catalyst which has to be systematically burnt off.

In two stage process, two different catalyst are employed. First stage dehydrogenation is carried over chromium oxide catalyst and in the second stage over calcined alumina catalyst. Chromium catalysts cannot function in presence of steam. Hence heat delivery by steam is not opted. In such cases the catalyst is activated by burning the carbon deposited on the surface; the heat absorbed by the catalyst is utilised for dehydrogenation reaction. It can be economically conducted in moving bed or fluidised bed, where continuous regeneration of catalyst is possible. Overall yield of butadiene on the basis of butane is 50%, this requires recycling of unconverted gases. Figs. 4.22 (*b*) and 4.22 (*c*) show the yield of butanes, butadienes and acetylenes at different temperatures and pressure.

Houdry Division (Catadiene process) uses a space velocity of 1 to 3 at a pressure of 0.1 bar and yield of butadiene is 64.7% on feed basis. The residence time is usually maintained in the range of 5 to 15 minutes.

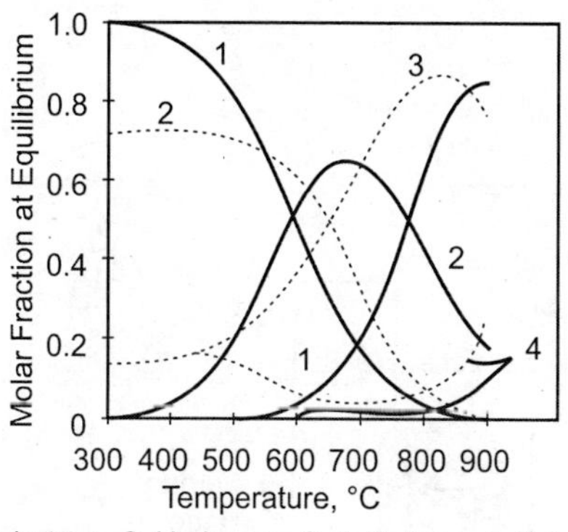

1, n-butene, 2. Hydrogen, 3. 1. Butene, 4. Butadiene
solid lines - 1 atm, Broken lines-1 atm, 0.1 atm

Fig. 4.22 (b): Equilibrium compositions for Butane dehydrogenation reactions as a function of temperature and pressure.

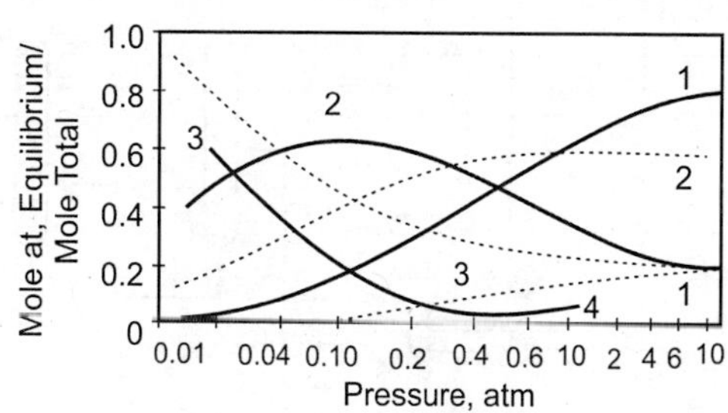

Fig. 4.22 (c): Equilibrium compositions for butane dehydrogenation reactions as a function of temperature and pressure.

In Nippon Zeon Process n-butene[(16)] butane mixed stream is processed for butadiene. Initially recovery of butadiene from C_4 fraction of a naphtha cracker leaves n-butene and butene as main components. These off gases form the feedstock in this process. n-Butene in the feed is oxydehydrogenated to produce butadiene in a multi-tube reactor, packed with catalyst. The conversion of n-butene reaches above 80% and the butane stream is relatively inactive. Air oxidation in presence of steam is common, part of the off gas from the solvent extraction unit is recycled with fresh feed. The high boilers, aldehydes and acids are eliminated from the reactor effluent. Fig. 4.23 present the salient features of the process. A metal oxide catalyst brings about the conversion of butene to butadiene; the catalyst in pellet form is packed into the tubes, the reaction heat is used to raise steam. The reactor product is cooled, quenched and scrubbed with water to remove water soluble impurities. Aldehyde absorber removes aldehydes as a solution in water. The aldehydes are stripped in a stripper and vented.

Excess ethylene[28] can be selectively disproportionated over titanium catalysts to C_3 and C_4 olefins (Oleflex process). i-Butane obtained by this method is a good source for MTBE.

4.5.1 Butadiene Recovery

The unabsorbed gases from the aldehyde absorber are compressed and passed into a C_4 absorber. Unabsorbed gases leaving from the C_4 absorber enter a solvent removal column, where a small amount of solvent present in the off gases is removed. A part of the off gas is recycled to the reactor as a diluent to the feed mixture and remaining left into the atmosphere. A small amount of oxygen, carbondioxide absorbed in the solvent with crude butadiene are stripped in a stabiliser. The crude butadiene is stripped in a C_4– stripper. Butadiene from the stripper is again absorbed in a solvent and stripped to get purified butadiene.

For high conversion (85%) of butene to butadiene, O_2 to n-butane ratio is kept 0.9 to 1.00 mole/mole, that yields about 75% butadiene. The concentration of butane in feed mixture is kept around 8 to 10 mole per cent and the contact time lasts between 1.8 to 2.0 seconds.

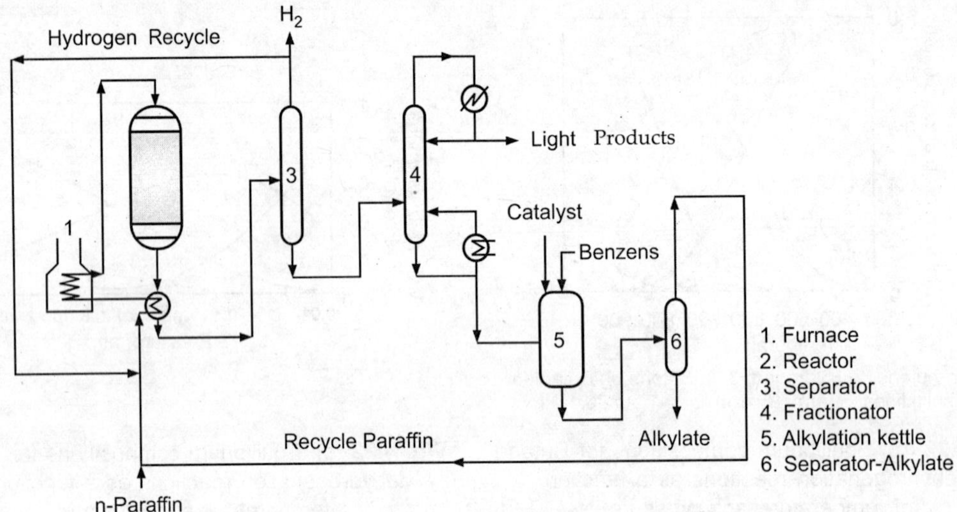

Fig. 4.23: Dehydrogenation Coupled with Alkylation Process.

The UOP route[24] to the production of linear alkyl benzenes consists of dehydrogenation of n-paraffins and alkylation with benzene. Paccol[25] process is highly gifted with such selective dehydrogenation reactions of n-paraffins of C_{10}-C_{15} range for linear olefins. Paccol dehydrogenation combined with UOP alkylation unit is shown in Fig. 4.23. In this process a linear paraffins is first mixed with recycle paraffin and the mixture is again mixed with some portion of hydrogen, generated in the process. The mixed gas is then heated to the reaction temperature and charged into the reactor. The reactor is fixed bed type, where low pressure and high space velocity favours dehydrogenation reaction. Most of the reactant is converted into olefin. From the products, first hydrogen gas is separated. Liquid products are fractionated to separate the light gases and liquids, perhaps formed during cracking, Paraffins are usually obtained from alkylation unit, where olefins and paraffins are fed into the reactor. This simplifies the production pattern as olefins are consumed and paraffins are unaltered and recycled back. Paccol dehydrogenation is conducted with a catalyst, DeH-5 or DeH-7. Alkylation of benzene with olefins is carried with HP catalyst. Unreacted n-paraffins are separated from alkylate in a fractionator. The n-paraffins are routed through an alumina treater to remove any fluoride contamination before dehydrogenation is done.

4.5.1.1 Butadiene Products

Butadiene was supposed to be one of the important starting materials for synthetic rubbers. However, it has not kept its high sounding applications as an intermediate. Future for this chemical seems to be alluring.

Hexamethylene diamine (HMDA) was produced from butadiene in earlier days as shown by the following reactions :

$$2\,CH_2:CH.CH.=CH_2+Cl_2\rightarrow CH_2\,Cl.\,CH:CH.\,CH_2Cl+CH_2Cl.\,CHCl.\,CH:CH_2 \quad ...(a)$$

$$CH_2Cl.\,CH:CH.\,CH_2Cl+2Na\,CN\rightarrow NC\text{–}CH_2.\,CH:CH.CH_2CN+2\,NaCl \quad ...(b)$$

$$NH.\,CH_2\,CH:CH.\,CH_2\,CN+H_2\rightarrow CN\,(CH_2)_4\,CN \quad ...(c)$$

$$CN\,(CH_2)_4\,CN+H_2\rightarrow NH_2\,(CH_2)_6\,NH_2 \quad ...(d)$$

Butadiene vapour phase chlorination results in dichlorobutenes, the reaction is allowed to proceed at 150°C with dry chlorine. As shown in reaction 1,4 and 1, 2 chloroderivatives are formed 1, 2 dichlorobutylene is isolated and recycled back to intercede its further formation. With sodium cyanide in presence of $CuCl_2$, 1, 4 dichloro- 2-butylene yields 1, 4 dinitrile. Dinitrile is hydrogenated over palladium catalyst at 300°C at moderate pressure to adiponitrile.

Adiponitrile can be produced by adding HCN to butadiene in two stages.

Adiponitrile is further hydrogenated to hexamethylene diamine in presence of cobalt catalyst at 200°C and 35-40 bars pressure as shown by reaction (*d*).

4.5.2 Adipic Acid

Another important ingredient for Nylon 6, 6, adipic acid is also produced from butadiene as shown by the carbonylation reaction.

$$CH_2:CH.CH:CH_2 + 2CO_2 + 2H_2O \rightarrow HO-\overset{\overset{\displaystyle O}{\|}}{C}-CH_2-CH_2-CH_2-CH_2-\overset{\overset{\displaystyle O}{\|}}{C}-OH$$

Reaction conditions are

Temperature 220°C; Pressure 80-90 bars; Catalyst RhC12; Yield 45%.

4.5.3 Sulpholane

Sulpholane is one among the recent and important derivatives of butadiene. Butadiene reacts with sulphur dioxide in liquid phase at a temperature of 90°C under a moderate pressure of 8 to 15 kg/ cm^2. The reaction is carried out in a series of stirred tank reactors. The reactions are shown below.

$$CH_2:CH.CH:CH_2 + SO_2 \rightarrow \begin{array}{ccc} CH & = & CH \\ | & & | \\ CH_2 & & CH_2 \\ & \diagdown SO_2 \diagup & \end{array} + H_2 \rightarrow \begin{array}{ccc} CH_2 & — & CH_2 \\ | & & | \\ CH_2 & & CH_2 \\ & \diagdown SO_2 \diagup & \end{array}$$

A mixture of butenes and butadiene can be used as the starting material, as sulphur dioxide exclusively reacts with diene. It may also help in easing the separation problem of butadiene-butene mixture.

4.6 CHLOROPRENE (2-CHLORO-1, 3 BUTADIENE)

Chloroprene production involves two steps. In the first step butadiene is chlorinated to produce 3, 4 dichloro-1 butene ($CH_2:CH\ CHCl.CH_2Cl$). This dichloro butene is dehydrochlorinated to yield chloroprene.

Chlorination is carried out in vapour phase at 300-375°C at normal pressure. A mixture 3,4 dichloro and 1, 4 dichloro butylenes results.

$$CH_2:CH–CH:CH_2 + Cl_2 \rightarrow CH_2.CH.CHCl.CH_2Cl$$
$$\text{or} \quad CH_2Cl.CH = CH.CH_2Cl$$

1, 4 dichlorobutene is isomerised over cuprous chloride catalyst to convert into 3, 4 dichloro butene. Later the mixture is dehydrochlorinated to the product chloroprene. Dehydrochlorination is a sort of partial neutralisation with lime.

$$2CH_2:CH.\ CHCL\ CH_2Cl + Na_2O \rightarrow CH_2:CHCl.\ CH:CH_2 + 2NaCl + H_2O$$

Isoprene and chloroprene are used extensively in rubber making.

4.7 DERIVATIVES FROM HYDROCARBONS HIGHER THAN BUTANE

Paraffins of C_5 and higher homologues do not have much ready uses in the chemical industry. All these compounds exhibit profound isomerism; further cyclisation starts from C_5 onwards. Apparently, effective separation of such fractions into individual constituents, though not impossible, is practicable with remote economy. Except for limited applications neither the paraffins nor the olefins of higher order are directly consumable in the industry. The main service of these fractions are fuel ingredients. Only those fractions or components that directly do not go into fuel fractions, or when these are produced in excess of requirement, are alternatively chosen for petrochemicals; for example, naphtha supplies aromatics, which are useful feedstocks for petrochemicals, hence processed for aromatics. Remaining naphtha is not suitable for fuel blend, hence employed either for production of olefins or for synthesis gas. In fact any paraffin can be cracked for olefin. The heavy fractions are economically utilised by partial oxidation to synthesis gas rather than cracking.

4.7.1 Separation of Paraffins

Paraffins are usually separated by extractive distillation, while n-paraffins from i-paraffins are separated by molecular seives[26].

Molecular seives are alkali metal aluminium silicates (or synthetic zeolites of finite pore size and are used for selective separation. Adsorption under pressure drives the n-paraffins into intra crystalline cavities of these silicates, while other components are unaffected. By decreasing the pressure or by purging with a suitable solvent, the trapped n-paraffins can be extracted. n-Paraffins of $C_{10} - C_{14}$ range can be isolated from kerosene fractions, which when de-hydrogenated yield olefins suitable for detergent making.

Similarly oxidation, chlorination, nitration can be done, but to a limited extent only, as there is no demand.

4.7.1.1 Wax Products

Wax *is* extensively produced in petroleum processing. Except for a few applications like candle making, coatings etc. hitherto not much petrochemical activity is cited. However, wax can be oxidised to high molecular fatty acids and fatty alcohols. Most of these fatty acids are consumed in soap industry.

Higher fatty primary alcohols are obtained by hydrogenating fats, secondary alcohols are produced by wax oxidation. Oxidation is carried over catalysts like boric acid, permanganate in liquid phase at 130 to 200°C. The mechanism is supposed to be free radical, where oxygen radicals attach to linear paraffin forming a hydroperoxide radical. This attachment conveniently takes place on tertiary carbon atom as shown below :

$$R\ldots\ldots CH_2CH_3 \rightarrow R\ldots\ldots \underset{\displaystyle OOH}{\underset{|}{CH}}\ldots\ldots CH_3$$

Hydroperoxide decomposes quickly into alcohol and a ketone

$$R\ldots\ldots CH_2\,\underset{\displaystyle OOH}{\underset{|}{CH}}\ldots CH_3 \rightarrow R.\,CH_2\,CH_2\,OH\,.\,CH_3 + \begin{matrix} CH_3 \diagdown \\ \quad CO \\ CH_3 \diagup \end{matrix}$$

The ketone is further oxidised to ketohydroperoxide which decomposes into acid and aldehyde

$$\underset{\substack{\|\quad|\\ O\quad OH}}{R.R.C.CH.CH_3} \xrightarrow{O} RCH_2 - C\begin{matrix} \nearrow O \\ \searrow OH \end{matrix} + RCH_2 - C\begin{matrix} \nearrow\!\!\!\nearrow O \\ \searrow OH \end{matrix}$$

As a result of these complicated reactions, the product becomes a complex mixture of oxygen derivatives; aggravating separation problems. Fatty acids can be economically produced by oxidising wax and the products are saponified with alkali. Products not saponified are stripped off. The unoxidised wax is separated from soap solution by settling; which is recycled. The soap solution is treated with sodium sulphate and sulphuric acid at 100°C under vigorous stirring. Sodium sulphate and acid rnixer is distilled sodium sulphate in water comes at the bottom of the column, while fatty acids appear as top products.

4.7.1.2 Wax Cracking

Wax can be easily cracked for the production of olefins. The resultant olefins are straight chain and known as α olefins, *i.e.* the double bond preferably is situated at the end of the chain. The products are rarely separated. The whole mixture is usually sent to alkylation units or oxo units. They are useful as detergent feedstocks.

4.8 ISOPRENE

Isoprene can be produced in a number of ways; one method is extending dehydrogenation of isopentane. The i-pentane dehydrogenation follows exactly, the same way and uses the same catalyst. The reaction product is separated and purified.

Isopentane on dehydrogenation gives three isomers as shown below :

$$1.\ CH_2 : CH - \overset{\overset{CH_3}{|}}{C}H - CH_3 \quad ; \quad 2.\ CH_2 = \overset{\overset{CH_3}{|}}{C} - CH_2 - CH_3 \quad ; \quad 3.\ CH_3 - \overset{\overset{CH_3}{|}}{C} = CH.CH_3$$

All these methyl butenes yield isoprene on further dehydrogenation.

In the first stage dehydrogenation at 525-550°C, about 50% of the i-pentane is de-hydrogenated. In the second stage almost at the same temperatures gives about 70% of isoprene based i-pentenes.

4.8.1 Production by Shell Process

Isoprene in shell process is obtained by dehydrogenating t-amylenes (methyl butenes). τ-Amylenes are usually gathered from naphtha cracking C_5 cut is extracted with aqueous sulphuric acid to remove the unsaturates. The extract fraction is then neutralised with alkali and alkali separates out. The olefin is stripped and sent to dehydrogenators. Fig. 4.24 shows the process[(21c)].

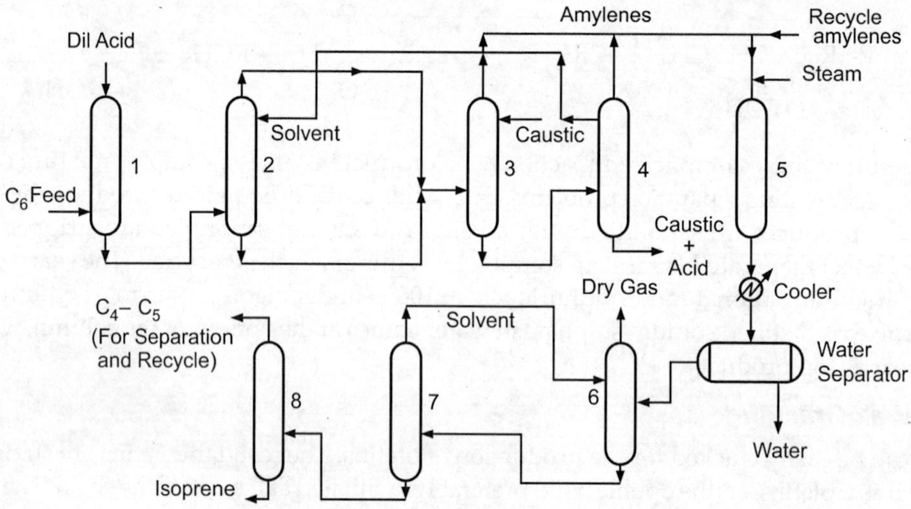

Fig. 4.24 Production of Isoprene.

C_5 feed is contacted in an extractor (1) with dilute sulphuic acid. The lean extract may be treated with another solvent to transfer the hydrocarbons (2). The solvent rich in olefin fraction is then washed with dilute caustic (3) to remove acid stains. In this process the olefin is set free (4) from the solvent stream. t-Amylenes obtained are sent into a dehydrogenator (5), where recycle C_5 stream and steam also join. The product is purified by extractive distillation using acetonitrile as the solvent. The solvent comes into contact with water free reactor product in an extractor (6). The dry gas leaves the extractor from the top. The rich solvent stream is separated by distillation (7 & 8). The overhead product is $C_4 - C_5$ fractions, which can go to butane dehydrogenation unit. The crude isoprene obtained from the distillation unit (8) is again purified with acetonitrile solvent for polymer grade product. Preliminary extraction of reactor products with lean hydrocarbon oil followed by acetonitrile is more convincing, because of the less quantity of acetonitrile as shown in shell process.

4.9 DINITRILES[27]

Acrylonitriles and isobutylenes react to produce C_7– nitriles on further reactions with acrylonitrile, C_{10} nitriles result. These dinitriles can be easily hydrogenated to diamines, a raw material for nylon production.

REFERENCES

1. Richard O. Kirk, CEP, Feb. 83, 78.
2. A.L. Waddams; Chemicals from Petroleum; 4 Ed. GPC; 155.
3. 3 (a) - (b) H.C.P, Nov. 1973; 141, 143.
4. 4 (a) - (b) H.C.P, Nov. 1979; 99, 122.
5. W. Neier, J. Woellner. HCP. Nov. 72, 113.
6. RB. Stobaugh, Vincent A. Calarco, RA Morris & L.W. Stroud; HCP, Jan. 1973, 99.
7. H.A. Wittcoff & Brayan G. Reuben; Industrial Organic Chemicals in Perspective, John Wiley, Part-I, 77.
8. Kazuo Yamagishi Osanu Kageyama, Hiroshi Haruki & Yashiaki Numa; HCP, Nov. 76; 103.
9. Kazuo Yamagishi & Osanu Kageyama, HCP, Nov. 76; 139.

10. (a) - (b) HCP, Nov. 71; 161, 117.
11. T. Ohara, M. Hirani & N. Shirizu, HCP, Nov. 72; 85.
12. Giorgio Caporali; HCP, Nov. 72; 144.
13. Lewis, F. Hatch & Samar Matar, From Hydrocarbons to Petrochemicals; CPC; 1982; (Houston Texas) ; 116.
14. Y. Chauvin; J. Gollilard & J. Leonard; HCP, May 82, 111.
15. P. Belov, Fundamentals of Petroleum Chemicals, Mir Publishers, Moscow; 125.
16. Akira Yoshioka, Hiroshi Hokari, Ryoji Sato, Haruhisa Yamamoto & Kinichi O-kumura; HCP, Nov. 84, 97.
17. Theodore Dumas & Walter Bulani: Oxidation of Petrochemicals; Chemistry and Technology; Applied Science, Essex. 1974; pp. 113,128-56.
18. Kurt Christian Traencker & Rita Payer, HCP, June 1975; 85.
19. Fowler, R & Heng Connor & K.A. Baehl, HCR, Sept. 1976, 249.
20. HCP, Nov. 71; 138.
21. (a) - (b) - (c) HCP, Nov. 73; 107, 93, 143.
22. Oda, Y.I. Gotoh, K. Uchida, T. Morimoto, J. Endoh & T. Ueno, HCP Oc. 1975; 115.
23. HCP April 1967; 143.
24. M.J. Gaffuso; B.V. Vora & T. Imai; Advances in Catalysis Science & Technology. Edtd T.S.R. Prasada Rao; Willey Eastern, N. Delhi, 1985.
25. B.V. Vora; P.R. Pujad & J.B. Speinner; HCP, Nov. 84; 86.
26. H C P, Sept. 1972, 214.
27. C.A. Drake, R.W. Campbell, H.W. Will, J.W. Vauderveen & S.J. Marwill, CEP, Dec. 82, 83.
28. P.R. Piyado & B.V. Vora HCP, March, 1990; 69.

CHAPTER 5

Synthesis Gas and Chemicals

5.1 SYNTHESIS GAS

5.1.1 Steam Reforming of Hydrocarbons

Steam reforming is an efficacious process for converting hydrocarbons catalytically into a mixture of hydrogen and carbon monoxide, which is designated as synthesis gas. Synthesis gas was abundantly produced from coal or coke during nineteenth century and even now this technology prevails. Before 1970, steam reforming was extolled and systematically coal based technology was abandoned in favour of this. Soon the boisterous oil crisis round the world gave a lease of life to this coal based technology mainly due to the local availability of coal; that decreases the dependency upon the import of oil. Now-a-days piquantly the swing is again in favour of gas based technology for the simple offset reason; the discovery of copious amount of natural gas. The enviable advantage of this process is the flexibility of feedstock, i.e. right from natural gas to heavy fractions. Though the heavier hydrocarbons do not vehemently disturb the reaction sequence, doubts arise of economic feasibility. Excessive coke formation on catalysts diminishes the activity and life span of catalyst. Inspite of cost and availability of hydrocarbon stocks, reforming is an impeccable route for the production of hydrogen or synthesis gas or substitute natural gas.

80% of world's ammonia is accounted by this process, 70% of steam reforming operations are based on natural gas. The remaining part is shared by naphtha and heavy fractions. Heavy fractions (fuel oils etc.) are innovatively and inordinately processed through an alternative route known as partial oxidation or controlled oxidation.

Oxo process produces varieties of chemical from synthesis gas mixture, which enticed the chemical industry for tenable adaption. In our country, the gas based and LSHS based units shall become the major producers of urea and ammonia. Though the synthesis gas intake for chemicals till today is marginal, still it finds its immunable position in some of the chemical industries.

Steam reforming is basically a catalytic reaction, though many catalysts contemplate to do the same function, the predicable gains cited by nickel catalyst, established its impregnable position in the industry. A nickel content of 18 to 20% on inert medium (silica-alumina), promoted by alkaline metal oxides, in pellet or cylindrical forms upto sizes of 20 to 30 mm is the usual catalyst form.

5.1.2 Nitrogen Fertilisers-Feed Mix in India

The installed capacity of nitrogen fertilisers in 1988-89 is 8.148 MMT while the production was 6.712 MMT. With a rapid change in feed stock mix and commencing additional capacity this has become possible, further the production is likely to increase to the projected demands of 14 M^2T by 1995. Natural gas/Associated gas will be the major feed stock in coming years accounting to 41% or more as shown below:

	Gas	*Naphtha*	*Fuel oil*	*Coal*	*Others*
1987-88					
Installed capacity	34.1	35.3	16.6	6.4	7.6
Actual Production	40.2	31.8	19.0	2.5	6.5
1988-89					
Installed capacity	41.0	30.5	14.4	5.6	8.6
Actual Production	43.1	31.4	15.4	1.5	8.6

It is likely naphtha and electricity may continue their share, F.O. consumption may be decreased due to higher amounts of gas availability.

5.1.3 Technologies Adapted

With the present accent on gas based units, modern technologies from multinationals have been freely imported. The major technologies include : 1. Snam Progetti, 2. Halder Topsoe, 3. ICI, 4. Kinetic Technology International, 5. Humphreys, 6. Glasgow, 7. Kellog Lummus, 8. C.F. Baraun. Energy saving is the main theme of modern units, though this may not reflect in the reduction of cost of fertiliser. With all the eight new fertiliser units based on gas, the capacity utilisation going more than 80%, the growth rate of 11% is well anticipated. With all the existing units plus (the total investment in fertiliser industry till today is about Rs. 5,000 crores which would hike to about Rs. 22,000 crores with infrastructure facilities made available, the gas based industries would raise the target to 14 million tonnes by the end of 8th plan. It may be more scientific to name steam reforming as oxygenolysis. It is a very high temperature process demanding insatiable amount of heat *i.e.* as high as 60 x 10^3 k.cals/ metre sq./hr. This large amount of heat demand calls for a special type of reactor. A series of reactions takes place over catalyst in presence of steam. These are listed in Table 5.1.

Table 5.1: Steam Reforming Reactions at T : 25° C

Reforming Reactions
1. $CH_4 + H_2O \rightarrow CO + 3H_2 - 49.3$ kcal/mole
2. $C_2H_6 + 2H_2O \rightarrow 2CO + 5H_2 - 89.11$ kcal/mole
3. $n\,C_4H_{10} + 4H_2O \rightarrow 4CO + 9H_2 - 115.7$ kcal/mole
4. $C_7H_{16} + 7H_2O \rightarrow 7CO + 15H_2 - 280.5$ kcal/mole at 500 symbol C
Subsidiary reactions
5. $CO + H_2O \rightarrow CO_2 + H_2$ 9.84 kcal/mole
6. $CH_4 + CO_2 \rightarrow 2CO + H_2 - 59.3$ kcal/mole
7. $CH_4 + 1/2\,O_2 \rightarrow CO + 2H_2 - 8.3$ kcal/mole
Carbon formation reactions
8. $2CO \rightarrow C + CO_2 + 41.2$ kcal/mole (Boudouard reaction)
9. $CH_4 \rightarrow C + 2H_2 - 17.9$ kcal/mole

Carbon elimination reactions
10. $C + O_2 \rightarrow CO_2 + 97.2$ kcal/mole
11. $C + 1/2\, O_2 \rightarrow CO + 29.2$ kcal/mole
12. $C + 2H_2O \rightarrow CO_2 + 2H_2 - 19$ kcal/mole
13. $C + H_2O \rightarrow CO + H_2 - 28.6$ kcal/mole at 1000 symbol C
Chain formation reactions
14. $CO + 2H_2 \rightarrow CH_2^- + H_2O + 39.4$ kcal/mole
15. $2CO + H_2 \rightarrow CH_2^- + CO_2 + 48.9$ kcal/mole
16. $CO + 2H_2 \rightarrow CH_3OH + 21.664$ kcal/mole

5.1.4 Reactivity of Hydrocarbons

For a given steam to hydrocarbon ratio, increase in the molecular weight of feedstock increases the reactivity rapidly. As in cracking, the aromatics are quite dormant, hence less preferred. Impressive yield of hydrogen with low molecular weight paraffins is reported. It is found that paraffins are more active however the lower molecular weight paraffins offer more resistance because of higher activation energy (Methane 110 kJ/mole, while C_5H_{12} has 50 kJ/mole). Aromatics also vehemently oppose conversion. The limitation of heavy feedstocks for reforming in tubular reactors may be more attributed to the desulphurisation problems rather than the boiling range. It is observed that a feedstock of light gas oil could be completely gasified, when the sulphur content is brought down to 0.1. ppm. Sulphur being a severe poison for nickel catalysts, due to strong chemisorption on metal surfaces easily deactivates catalyst. Hence a threshold concentration of 0.1 ppm weight of sulphur should not exceed. Even at low levels of sulphur also catalyst is likely to be affected. The sulphur equilibrium coverage on the nickel surface is a function of temperature and H_2S/H_2 ratio. A good discussion presented by R.J. Rostrup-Nielsen[1] is presented in the graphical fashion in Fig. 5.1. The figure shows the coverage of catalyst surface as the hydrocarbon travels down through the tube for different sulphur levels of feed :

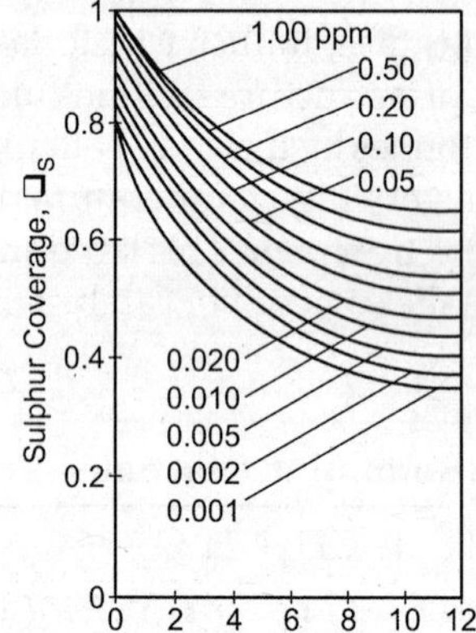

Fig. 5.1: Sulphur Coverage of Nickel Surface of Catalyst in Tubular Reformer.

$$\frac{Z}{m} = \frac{\text{Axial}}{\text{Catalyst bed}}$$

As seen in the steam crackers, where sulphur inhibits the carbon formation, it is also true in case of steam reforming; that partly covered catalyst with sulphur can give carbon free operations. As the partial pressure of steam increases, the carbon formation is usually found to be lowered. Arsenic and phosphorous are also found to be poisonous specially for alkali promoted catalysts. Chlorine although is a severe poison at higher concentrations (above 100 ppm), there is no impact on refractory catalyst at lower concentrations. A component which would block the nickel surface may be treated as pugnacious material. The maximum conversion of naphtha or natural gas is at the upper end of the tube and gradually equilibrium approach at the outlet end.

The composition of the outlet gases for naphtha and methane from the reformer is given below :

Feedstock	H_2O (*mole*) /*C-atom*	H_2	CO	CO_2	CH_4	H_2O
CH_4	3.7	39.12	5.04	6.00	5.54	rest
Naphtha	4.5	34.60	5.33	8.03	2.42	rest

The effect of temperature on equilibrium conversion and compositions at different temperatures are shown in Tables 5.2 (*a*) and (*b*).

Table 5.2. (*a*): Equilibrium Composition of Reformer Gases at different Temperatures (Methane reforming)(2)

Temp. symbol C	CH_4	H_2O	CO	H_2
500	33.3	33.3	8.33	25.0
580	21.7	21.7	14.3	42.6
635	12.5	12.5	18.8	56.2
710	5.6	5.6	21.4	67.4
835	2.7	2.7	23.7	71.4
940	0.5	0.5	24.8	73.5

Table 5.2 (*b*)

Equilibrium constants a, b, c Temperature / K	K^1 CO to $P + H_2O = CO_2 + H_2$ (*reaction* 2)	K^1 CH to $P + H_2O = CO + 3H_2$ (*reverse of reaction* 2)	K^1a 2 C to $PO = C + CO_2$ (*reaction* 8)	K^1a CH_4 to $p = C$ to $P + 2H_2$ (*reaction* 9)
273	4.555×10^5	0.066×10^{-29}	5.678×10^{24}	0.820×10^{-11}
373	3.587×10^3	0.269×10^{-19}	8.082×10^{15}	0.606×10^{-7}
473	2.299×10^2	0.487×10^{-13}	6.047×10^{10}	1.283×10^{-5}
498	1.380×10^2	0.747×10^{-12}	0.661×10^{10}	3.582×10^{-5}
523	8.737×10	0.892×10^{-11}	8.941×10^8	0.914×10^{-4}
548	5.780×10	0.858×10^{-10}	1.451×10^8	2.153×10^{-4}
573	3.973×10	0.682×10^{-9}	2.758×10^7	4.736×10^{-4}
598	2.823×10	4.592×10^{-9}	6.028×10^7	0.981×10^{-3}
623	2.066×10	2.669×10^{-8}	1.489×10^6	1.924×10^{-3}
648	1.552×10	1.361×10^{-7}	4.101×10^5	3.597×10^{-3}
673	1.192×10	0.618×10^{-6}	1.244×10^5	6.445×10^{-3}
698	9.352	2.528×10^{-6}	4.111×10^4	1.111×10^{-2}
723	7.470	0.941×10^{-5}	1.468×10^4	1.849×10^{-2}

Temp. °C	CH_4	H_2O	CO	H_2
748	6.066	3.222×10^{-5}	5.620×10^{3}	2.984×10^{-2}
773	4.999	1.021×10^{-4}	2.291×10^{3}	4.678×10^{-2}
798	4.176	3.016×10^{-4}	0.989×10^{3}	7.144×10^{-2}
823	3.530	0.837×10^{-3}	4.495×10^{2}	1.066×10^{-1}
848	3.018	2.189×10^{-3}	2.143×10^{2}	1.554×10^{-1}
873	2.606	5.429×10^{-3}	1.067×10^{2}	2.223×10^{-1}
898	2.270	1.282×10^{-2}	5.523×10	3.120×10^{-1}
923	1.995	2.968×10^{-2}	2.967×10	4.303×10^{-1}
948	1.767	6.267×10^{-2}	1.647×10	5.843×10^{-1}
973	1.541	1.305×10^{-1}	9.438	7.817×10^{-1}
998	1.415	2.623×10^{-1}	5.564	1.031
1023	1.278	5.098×10^{-1}	3.368	1.344
1048	1.160	0.961	2.090	1.731
1073	1.059	1.759	1.327	2.203
1098	9.717×10^{-1}	3.133	8.606×10^{-1}	2.775
1123	8.953×10^{-1}	5.444	5.695×10^{-1}	3.462
1148	8.284×10^{-1}	9.236	3.838×10^{-1}	4.280
1173	7.694×10^{-1}	1.533×10	2.633×10^{-1}	5.245
1198	7.172×10^{-1}	2.491×10	1.836×10^{-1}	6.374
1223	6.708×10^{-1}	3.969×10	1.300×10^{-1}	7.689
1248	6.294×10^{-1}	6.209×10	0.934×10^{-1}	9.207
1273	5.923×10^{-1}	0.955×10^{2}	6.794×10^{-2}	1.095×10
f''	1.000	97.402	0.10133	9.869

a graphite basis
b partial pressures in MPa
c K_p^1 values should be multiplied by f′ if partial pressure are in atm. $K = kp = f'\ K_p^1$.
Courtesy: J.R Rostrup-Nielsen : Catalyst Steam Reforming; Springs Verlag; Berlin; Heidelberg.

5.1.5 Steam Hydrocarbon ratio

Steam hydrocarbon ratio strictly depends upon the molecular weight of the feedstock.

$$C_nH_m + nH_2O \rightarrow (n + m/2)\,H_2 + n\,CO - Q \quad ...(1)$$

$$CO + H_2O \rightarrow CO_2 + H_2 + 41.2\ \text{kJmol} \quad ...(2)$$

$$CO + 3H_2 \rightarrow CH_4 + H_2O + 21.2\ \text{kJ/mol} \quad ...(3)$$

Except the reaction (1), all are exothermic but overall reaction may be exothermic or endothermic depending largely upon the operating conditions.

At low steam to carbon ratio, a low temperature operation ordains the overall heat of reaction to be positive. This aspect has been regarded as beneficial for adiabatic gasification process, where substitute natural gas is the main product. Such operations are conducted well below 500° C at 1.6 to 2.0 mole of steam per carbon atom. On the other side for production of hydrogen or synthesis gas, severe conditions of temperature and heat flux are exigent.

Equilibrium mixture from the reformer contains some unreacted methane (about 8% at 700° C), hence high temperature and high ratio of steam to carbon are required to reduce the methane concentration as shown in Tables 5.3 and 5.4

Table 5.3: Per 1000 m^3 of Natural Gas; the Reformed Gas Yield (with recycle of tail gas and oxygen)

(NG : Steam : CO_2 : O_2 1 : 0.57 : 0.23 : 0.53)

Feed composition	*Outlet composition*
CH_4 – 976	CO – 934.5
C_2H_6 – 3	CO_2 – 271
C_3H_8 – 1	H_2 – 1890.5
CO_2 – 236	N_2 – 25.6
N_2 – 15	H_2O – 61.4
H_2O – 570	
O_2 (98%) – 530	

Table 5.4: Product Gas Composition for Methane and Naphtha Reforming

Destination	*Ammonia*	*Hydrogen*		*SNG*	*F-T Process*
Feed	*Methane*	*Methane*	*Naphtha*	*Naphtha*	*Methane*
H_2O mol/C - atom	3.7	2.5	4.5	1.6	1.15
Partial pressure MPa	3.3	2.7	2.7	3.5	0.5
Temperature °K Yield %	1073	1123	1073	788	1223
H_2	39.12	48.59	34.6	8.25	70.92
CO	5.04	9.22	5.33	0.43	22.44
CO_2	6.00	5.24	0.03	11.18	0.90
CH_4	5.54	5.86	2.42	32.44	1.46
H_2O	Balance	Balance	Balance	Balance	Balance

It is observed(3) that with H_2 to CO ratio less than 0.5, carbon formation becomes appreciable or in other words below a certain critical steam to hydrocarbon ratio, the carbon formation dominates. Similarly unsaturated hydrocarbons increase carbon formation hence, hydrogen recycling is recommended when high olefin stocks are handled.

5.1.6 Reformer details

Quick and high heat flux necessitated by the reforming reactions are met by tubular reactors. High temperature resisting tubes made of nickel chromium alloys (20 : 25 and carbon 0.35%) known as HK-40 are found to be useful in drastic conditions of heating where a temperature of 900°C on the tube wall is prevalent. The reformer generally contains 50 to 250 tubes, the diameter of tube ranges from 9 to 15 cm, the length goes upto 80-100 times dia. The centrifugal cast tubes are bundled in harps(4)containing 40 to 50 tubes, connected with the common parallel manifolds at the top and bottom. The harp functions smoothly to absorb thermal stresses; further tubes are suspended from springs at the top, which are calibrated and generally are adjustable to assure uniform support to harp and take care of undue axial loading. The tubes are loaded with catalyst pellets usually of size 2.5 cms. Heating of the tubes is practised in many ways, like heating from sides, bottom or by heating the walls (as seen in Radiant walls type furnaces) or from the top. In the latter case, fuel gas is admitted from the top through common manifolds to the burners placed in rows along the wall, or on the sides of tubes. The flues are collected through a common parallel duct situated at the bottom of the reformer. The reformers are precisely insulated, with provisions of peep holes. The hot flues usually expand heat in the convection section or into a separate steam raising equipment.

Reformed gas is again collected from the bottom into a common manifold and led into the adjacent unit. No bends are possible in tubes hence the feed enters straight from the top and exits from the bottom.

It is worthwhile to mention the excerpts of Schillmoller(5a) *et al*; about the alloys of tubes used in reformers. Even the above named HK-40 tubes have limitations, when the service temperature exceeds by 50°C, *i.e.* at 950°C, the life of the tube is drastically reduced to one-sixth. HP modified alloys (25 : 35 Cr : Ni) containing columbian, tungsten, titanium, zirconium, cobalt, molybdenum and silicon are commissioned and successfully demonstrated a considerable improvement in combating stress rupture and in service like particularly when the temperature reaches 950°C. Manifolds and transfer lines are made of 800 H alloy (0.1% carbon, 0.6% Cr and 20 : 32 chromium and nickel combinations), instead of HK-40. This alloy has benefitted the industry in improving weldability, creep-rupture, ductility-characteristics, withstanding high thermal stresses and contributing a greater life. In operation, when hot spots are developed the life of the tube is generally cut down, which again depends upon many factors. Maturite(5b) tubes are also suitable for such high temperature operations.[33] Presently thinner tubes of 35/25 Ni + Cr alloy modified with niobium and with trace metals, titanium, zirconium are used. These can withstand 1700°C. The life of the tube is more than one lakh hours.

5.1.6.1 Catalysts[6]

To develop a catalyst to work at that high temperature with satisfactory performance is a tedious task. Most of the support materials for nickel lose the strength at high temperature (α) and (γ) alumina are found to sinter around 500°C. While ceramic supports like MgO is found unsuitable. Similarly alkali promoters may not retain the activity due to volatilisation or carry over in presence of steam and high temperature. The modern catalysts thus have to face the drastic conditions, consequently, require selective supporters for long term strength-characteristics and uniform activity over long periods with good heat transfer characteristics. The crumbling and loss in strength of catalyst naturally brings down the efficiency and overall output. Further, the choking of tubes due to crumbles, favours hot spots; high reaction rates often lead to premature ageing of catalyst. It is true that initial activity of the catalyst may not dictate the process economy, but retention of activity over long periods certainly does. Some of the catalysts in use are shown below:

Catalyst Composition[7]						
	Feedstock	NiO	Al_2O_3	MgO	SiO	Modifiers
ICI Cat	Naphtha Nat. Gas	22 12	26 78	11 -	16 0.1	20 10
BGC	Naphtha	79	20-21	-	-	3
	Primary reformer naphtha Secondary	20% on Ca, Mg, Si and aluminium oxides. 18-20% on Cal aluminates				
UCI	Natural gas	11-20				
Modifiers	Light Hydrocarbons	7-15				

Nickel on alumina support react at high temperature to give $NiAl_2O_4$. Same trend is observed with magnesium oxide support too.

Before the reforming reaction takes places, the catalyst in industrial reactors is activated by reducing agents usually with hydrogen :

$$NiO + H_2 \rightarrow Ni + H_2O + 1.2 \text{ kJmol}$$

or

$$NiAl_2O_4 + H_2 \rightarrow Ni.+Al_2O_3 + H_2O$$

The reduction of nickel oxide starts around 250°C, supported catalysts require higher temperatures. In industry this may be activated by high steam to carbon ratio, at low pressure at more than 600°C.

Topse[8] has manufactured ceramic steam reforming catalysts, a new generation catalysts (RKG-2, R-67). Some conversion at temperatures below 650-700°C is essential with NG reforming otherwise hot band formation may occur. The conversion of C_1 - C_5 hydrocarbons on high surface area catalysts (R-67) containing magnesium aluminium spinel; being active takes place at lower temperature minimising the risk of carbon formation.

5.1.7 Process Description

Nickel catalyst being prone to poisons like sulphur, arsenic etc. the desulphurisation of the feed is most important and essential. Natural gas or refinery gas may be contaminated exclusively with H_2S. In such cases the removal of H_2S is done by carbon-zinc oxide catalyst-beds operating at 400°C. Iron-soda mass is also found to be useful. It is advantageous to compress the gas to 35-40 bars before desulphurisation, so that in later stages no further compression is required. In case of naphtha the desulphurisation assumes major proportions, modern tendency is to use hydro sulphurisation unit. Hydro desulphurisation is carried out at 400°C and 35-40 atmospheres in presence of hydrogen over Cobaltmolybdenum catalysts. This ensures the complete removal of sulphur to as low as 0.1 ppm[9]. The reforming catalyst developed for certain feed may not be useful for others. Thus catalyst developed for gas reforming would lead to excessive coke formation when opted for heavy fractions, due to fastness of reaction. Hence less reactive catalyst is employed. Usually promoters like alkaline earth metals (Potassium) are incorporated in catalysts and these are supported on silica-alumina or on magnesia base.

Reforming for hydrogen is done in two stages; first stage reformer is called primary reformer. In case of natural gas, the issuing gas from desulphuriser unit is mixed with steam (2.5 mole for C-atom) and allowed to enter the convection section of the reformer. The steam hydrocarbon mixture enters the primary reactor at 500°C and the reaction proceeds around 850°C. With feed as naphtha, the steam consumption increases, but other parameters remain same. Naphtha after desulphurisation passes into a stripper to jettison the products of hydro desulphurisation. The sweet naphtha is then mixed with steam (4.5 mole) to enter the convection section for preliminary heating and finally reformer. The outlet gases from the reformer are fed into secondary reformer along with preheated air. About 10% of methane still remains at the end of first stage reforming. Hence direct oxidation to give drastic heat requirement is gracefully allowed in secondary reformer, where the temperature crosses 1000°C. The secondary reformer is a mechanism to adjust the nitrogen and hydrogen ratio suitable for direct synthesis of ammonia.

The use of secondary reformer is obviously to convert about 10% methane still unreacted in the primary reformer and it also supplies nitrogen from air for the production of ammonia. The above technique is alluring when the aim is ammonia synthesis.

Synthesis gas alone when aimed, the necessity of secondary reformer is questionable. Hence the end use of synthesis gas must be well known before hand.

During reforming naphtha, the issuing gas may contain 6% methane, which may be further converted into CO in secondary reformer or simply eliminated by separation techniques. Also CH_4 being inert does not exert any influence on the synthesis gas mixture. Accompanying carbondioxide is absorbed by any known techniques. This leaves a mixture of 56 : 43 H_2 : CO with about 1% of methane. Fig. 5.2 shows the process flow sheet of methane or natural gas reforming for the production of hydrogen suitable for ammonia. Fig. 5.3 shows the reforming

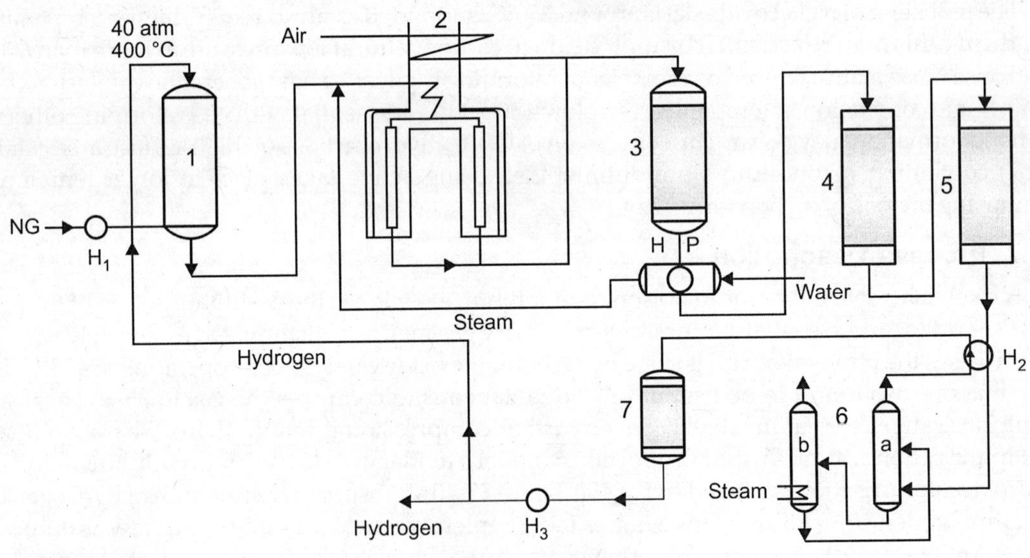

H.P. Steam Generator
1. Catalytic Desulphuriser
3. Secondary Reformer
5. L.T. Shift Converter
6. (*b*) CO_2 Desorber

$H_1 - H_3$ Heat Exchanger
2. Primary Reformer
4. H.T. Shift Converter
6. (*a*) CO_2 Removal Unit
7. Methanator

Fig. 5.2: Natural Gas—Steam Reforming

for hydrogen production only. Extremely pure hydrogen (99.99%) is obtained by steam reforming and purification by Pressure Swing Adsorption (PSA) technology.

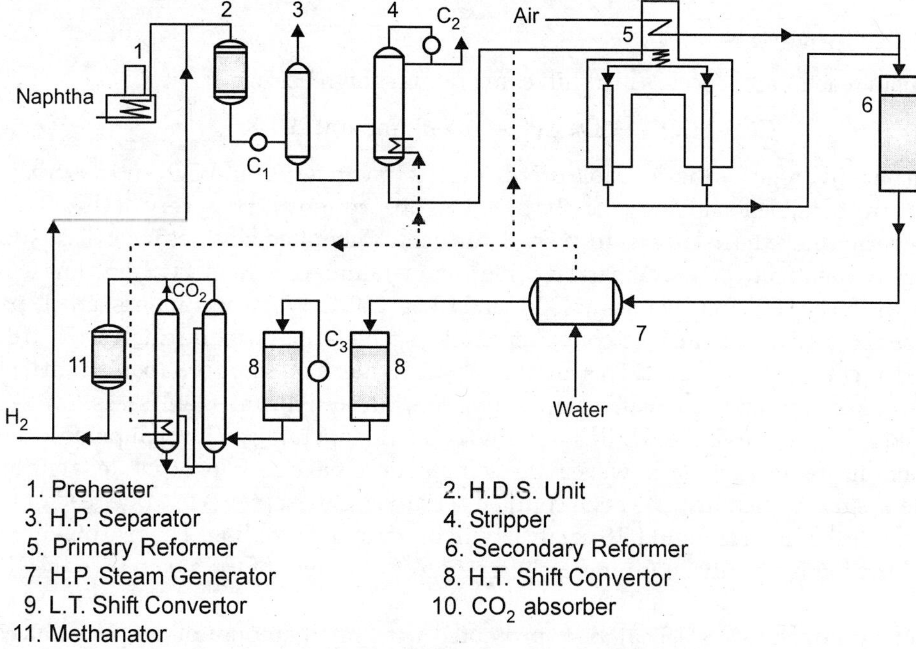

Fig. 5.3: Naphtha Steam Reforming.

5.1.8 Pressure Swing Adsorption System[10]

It is the latest technology for commercial production of hydrogen of desired degree of purity. The system operates on three or four parallel adsorbers capable of adsorbing CO, CO_2, H_2S, N_2, NH_3, C_2 to C_3 (Olefins and Paraffins), leaving behind hydrogen from the feed. The adsorption takes place in a single adsorber, cyclically, after which it is desorbed. For continuous operation three or four adsorbers are kept in parallel, when one is under adsorption, others will be desorbing. Usually desorption takes higher time than adsorption, hence, three or four units are cyclically used. The pressure is 40-50 atmospheres and the incoming gas temperature is kept below 40°C.

The adsorbers use special type of molecular sieves which are capable of amazingly reducing the impurities to 1 to 2 ppm level. The desorbed gases can be utilised, as their concentration becomes higher.

The secondary reformer is single bed reactor and uses nickel catalyst. The outlet gases from the secondary reformer go directly into HP steam generator, later cooled down to 360°C. This gas mixture is now passed into a shift converter. Precise operations require two shift converters; one operating at high temperature 450°C and the second one operating at 250°C. The reaction being exothermic, other reaction heat is sufficient to sustain reaction. The outlet gases at 450°C are cooled down to 210°C and sent into second shift converter. This converter completes the conversion of CO to CO_2. Finally a mixture containing nitrogen, hydrogen and carbon dioxide goes into absorption tower, where CO_2 is removed by ethanolamine wash. The outlet gas may still contain traces of CO, a quiet poisonous component for ammonia reactor. Hence it is converted into non-harmful methane in a methanator. Methanator again uses nickel catalyst (50 to 70% nickel oxide) on supports. In presence of hydrogen at a pressure of 35-40 bars all CO is converted into CH_4. The final gas mixture is adjusted with addition of nitrogen to bring the molal ratio of nitrogen to hydrogen in the region of 1: 3 for ammonia synthesis.

5.1.9 Shift Converters

The main reaction that occurs in a shift converter is shown below :

$$CO + H_2O \rightarrow CO_2 + H_2 + 41.8 \text{ kJ/mole at } 298°K$$

When pure hydrogen is the desirable product, all CO is converted into CO_2 and absorbed in amine solution or any suitable technology may be opted for absorption and recovery of CO_2. This reaction being exothermic no external heat is furnished, further the liberated heat has to be dissipated, which forces to use two shift converters, demarked as high temperature converters (HTS) and low temperature converter (LTS). The HTS unit operates at 430°C and LTS at 250°C. When operation is serially performed, an interstage cooler is essential. HTS converter employs tablets of iron oxide catalyst (77 to 82% Fe_2O_3) promoted by chromium oxide 9%. This reactor operates under a pressure of 30 bars. High temperature and pressure keep the activity of catalyst for a long run, but equilibrium yield is less. Only 70% of CO is converted to CO_2. Necessitated by this, a second converter operating at low temperature employing copper-zinc-chromium catalyst; however later combinations have been incorporated with alumina to provide temperature stability, has been coupled serially. Modern plants use one reactor but in two separate chests, having HTS and LTS catalyst with intermediate cooling. The rise of temperature in such combination is usually 30"C, and carbon monoxide content of the steam is reduced to below 0.2%.

Recent studies show[32] HTS catalysts promoted with small amount of copper reduces the F.T. products (like alcohols). Further, Iron based catalyst loses strength while copper promoted catalyst maintains the strength for a long period and also act as guard to poisons. So LTS catalyst is not affected easily.

Sulphur and chlorine have been recognised as major poisonous constituents for shift converters; specially for catalyst containing no alumina. Protection against such positions is widely followed by placing ZnO guards before shift converters. ZnO guard is mainly suitable for H_2S, hence modern gaskets are utilising combined sulphur-chloride guards[(11)].

The outlet stream from shift converters finally becomes rich in CO_2, H_2 and nitrogen. CO_2 is stripped off in the absorption system, employing conventional method. Of course, the equilibrium composition of the gases from absorbers show 0.1 to 0.2% of mixed oxides of carbon.

5.1.10 Methanator

Methanator[(12)] serves two purposes; firstly it increases the concentration of methane when SNG is attempted, secondly it is the effective operation for the removal of impending poisonous CO, acting like a guard to the ammonia converter.

$$CO + 3H_2 \rightarrow CH_4 + H_2O + 53.07 \text{ kcal/mole}$$

$$CO + H_2 \rightarrow CH_4 + 2H_20 + 49.3 \text{ kcal/mole}$$

The exothermic reaction generally contributes a rise in 60°C. Methanator uses a nickel catalyst (60 to 70% nickel on support). In presence of hydrogen at that high pressure all CO is converted into methane.

Schemes for production of either carbon monoxide or hydrogen are shown in Fig. 5.4.

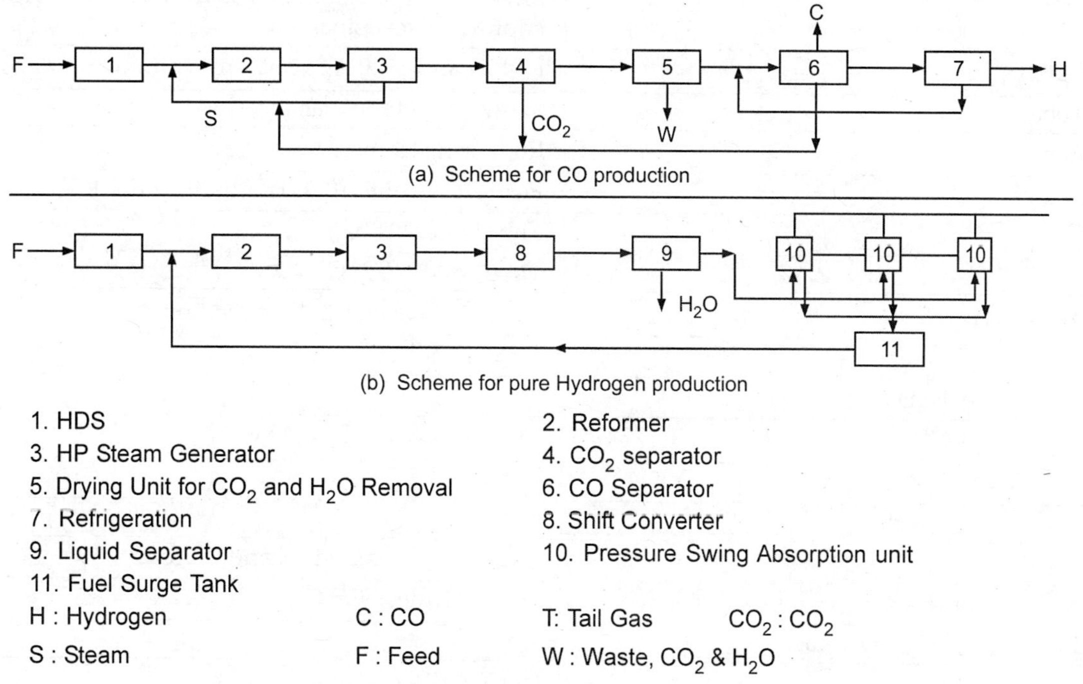

Fig. 5.4: Schemes for CO and H_2 Production

5.1.11 Production of Synthesis gas

The reforming reactions end in the secondary reformer, where the exit gas contains a mixture of carbon dioxide, carbon monoxide and hydrogen. Practically all hydrocarbon will be extinct. Hence, the section starting from shift converters are skipped to go for carbon dioxide removal unit. Thus leaving a mixture of CO and hydrogen. Usually to increase the quantity of CO, the desorbed carbon dioxide with tail gas is recycled with inlet gas streams, for a feed of natural gas or LPG. In such circumstances secondary reformer is contemplated as shown in Fig. 5.3. At the end of primary reformer, 5 to 8% methane remains unconverted, hence the tail gas for recycling contains not only carbon dioxide but hydrocarbon too.

It may be mentioned here that roughly 1 ton naphtha produces 1.7 tons ammonia and 1 ton gas, (methane) gives rise to 2 tons of ammonia. Power consumption during the last two decades has also changed. Presently approximate power involvement for different feed stocks is shown below[31] :

Reforming Natural gas	8.69 M kcal per ton of ammonia
Reforming Naphtha	9.32 M kcal per ton of ammonia
Partial Oxidation F.O.	9.57 M kcal per ton of ammonia
Gasification of coal	11.33 M kcal per ton of ammonia

A reformer operating on Kellogg design used by one of the fertilisers in India (Coromondal) shows the following highlights :

Primary Reformer	
Design	M.W. Kellogg design for a 360 MTPD plant with naphtha as feedstock
Catalyst	IC 46-1 and 46-4, Approx. 8 m^3 volume
Tubes	3 rows of 28 tubes each
Burners	Roof firing from 4 rows of burners
Temperatures	Furnace : about 970° C (Radiant section)
	Tube inlet : 421° C
	Tube outlet : 793° C
Naphtha to reformer	About 508 kg moles/hr. as carbon.
Steam to reformer	1732 kg moles/hr.
Exit gas analysis (Dry basis)	
H_2 : 64.24%	CO : 11.54% CO_2 : 15.06%
N_2 : 0.38%	CH_4 : 8.78%
Secondary reformer	
Catalysts	UCIL Catalysts C - 14-2 and C-15-1
Temperature Inlet	793°C, Outlet : 925°C
Exit gas analysis (Dry basis)	
H_2 : 53.13%	CO : 13.50% CO_2 : 11.47%
N_2 : 21.43%	CH_4 : 0.47%
Air to secondary reformer	About 535 kg moles/hr.
Steam to secondary reformer	About 58 kg moles/hr
Shift Converter	
High temperature catalyst	UCIL C-12.1; Volume about 20 m^3
Low temperature catalyst	UCIL C-18 HC; Volume about 25 m^3.

5.1.12 Substitute Natural Gas

Techonology has been now developed to gasify coals for substitute natural gas for industrial as well as domestic suppliers. SNG production from wild gasolines and naphtha (MRG Process) has been on the increase. The gasification is similar to reforming, however, the severity is restricted. Table 5.5 shows the conditions best suited for different operations.

Table 5.5

	temperature °C	*H_2O/C (mol/atom)*
SNG 99% Methane	400-500	1.6-2
Moderate calorific value gas	650-700	1.8-2.2
Lean gas	750-780	2.5-3
Methane		
Ammonia	800-810	3.5-4.0
Methanol and Hydrogen	800-850	4.0-6.5

At lower steam hydrocarbon ratio at high pressure and around 450°C, the conversion to methane is high. Hence these conditions are allowed to prevail in the reformer. Most of the gasifiers operate under adiabatic conditions. Under the chosen operating conditions, the reforming reaction tends towards exothermic, causing a temperature rise of 25 to 50°C; to prevent a temperature run away, steam hydrocarbon ratio must be closely controlled. Most of the SNG plants produce gas fully interchangeable with pipe line gas (Wobbe[13] No. 1250 to 1350). No commercial catalyst has sufficient activity below 350°C to produce almost pure methane. Therefore the feed after converting to H_2, CO and methane (approximate heating value 60 kcal/m^3) is routed through a second reactor containing a gas council catalyst[14, 15] operating around 300°C. In some designs the feed is divided into two separate stages and processed, others may employ two or more reactors in series. In another method the first reactor operates around 480°C, the rich gas from this stage is cooled and fed into another gasifier operating at 450°C.

The rich gas composition from the reactors is given below :

	Vol%
CO_2	20.3
CO	0.6
H_2	12.2
CH_4	66.9

The heating value of such gas is poor hence methanation and CO_2 removal will change the composition of the gas to approximately as given below of a calorific value of 8000 kcal/m^3.

CO_2	0.6	H_2	3.84
CO	0.06	CH_4	95.5 (9000 kcal/m^3)

MRG gasification for SNG is shown in Fig. 5.5.

Hence after desulphurisation the mixture is mixed with stream to be gasified at 400 – 450°C over very active nickel catalysts under adiabatic conditions. The issuing gas is heat exchanged and cooled to 300°C and sent into the methanator. The excess carbon dioxide is absorbed and the remaining gas is processed in a methanator.

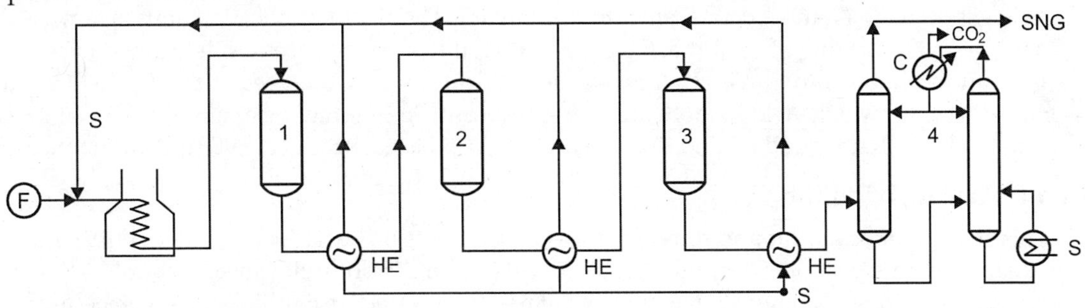

F : Desulphurised Naphtha
HE : Steam Generators
1,2. Two Stage Gasifiers (Reactors)
4. CO_2 Removal System
S : Steam
C : Coolers
3. Methanator

Fig. 5.5: SNG Production from Naphtha.

Substitute natural gas is now-a-days gaining significance and most of the development is based upon coal; the full scale utilisation of this technology has not taken place, because of the cost of production. Lurgi, Kopper-Totzek, BiGas, CO_2 acceptor process-Hygas, Synthane are some of the well known processes. Lurgi Gasification process and Kopper Totzek Gasifiers are commercially available. The processes are aimed at producing low calorific value gases only.

5.1.13 Lurgi Gasification process

Coal is gasified in a fixed bed using oxygen steam mixture at a pressure of 35 bars. Usually continuous operation is practised. The gases are then purified to remove sulphur components, followed by shift conversion.

A typical composition of gas obtained from different gasification processes is shown in Table 5.6.

Table 5.6[16]: Gases obtained in Coal Gasification

Gasification process	*Fixed bed*		*Fluidised bed*		*Entrained bed*	
	Lignite	*Hard*	*Lignite*	*Hard*	*Lignite*	*Hard*
Vol%	34	28	19	24	11.9	10.5
CO	14.4	22	38	31	55.9	5.5
H_2	36.2	38.7	40	41	29.6	32.1
CH_4	13.1	9.8	2	2	0.1	0.1
Hydrocarbons						
H_2S etc.	Rest	–	–	–	–	–

Entrained bed gasification has been developed to utilise coal and lignite at atmospheric or at 30 bar pressure. Additionally autothermal process[17] has been reported by Ruhr Chemie (RCH) and Ruhr Kohle (RAG) further developed the first generation processes by eliminating the disadvantages of fixed bed process, restricting the grades of coal. Entrained beds are operated at 80 bar pressure with less consumption of oxygen, but operating at high temperatures.

Acid gases are first removed and sent into methanator. The aim of gasification units is same as the case with steam reforming of naphtha. CRG Gas Council Research process is developed to convert straight run naphthas into SNG. High pressure gasification in two stages at 550°C, 400°C is the essential difference which yields a methane concentration of 96-97% having a calorific value of 9×10^3 kcals/m^3. Talchar, Ramagundam and Korba plants are coal based, producing 900 tpd ammonia each.

5.1.14 Partial Oxidation

This being a non-catalytic and self-sustaining process, naturally has been exalted. Further, there is virtually no limitation on selecting a feedstock. Its nature is non-relevance, it should be rich in carbon and hydrogen elements[18]. The significant reactions based on methane are given below :

$$CH_4 + O_2 \rightarrow CO + 2H_2$$

$$CH_4 + 2O_2 \rightarrow CO_2 + 2H_2O \qquad \text{...A}$$

$$CH_4 + 1.5\,O_2 \rightarrow CO_2 + 2H_2$$

This is true even with heavier fractions :

$$C_nH_m + O_2 \rightarrow CO + H_2$$
$$C_nH_m + O_2 \rightarrow CO_2 + H_2O \qquad \text{...B}$$
$$C_nH_m + O_2 \rightarrow CO_2 + H_2$$

The reactions in group (A) and (B) are oxidation reactions. The heat evolved due to complete combustion will be sufficient to induce reforming reactions:

$$CH_4 + CO_2 \rightarrow 2CO + 2H_2$$
$$2CH_4 + 2H_2O \rightarrow 2CO + 6H_2 \qquad \text{...C}$$

In the absence of catalyst, no doubt group (C) reactions proceed slowly. Further with heavier feedstocks, (Carbon to hydrogen ratio is more unlike methane), much heat is contributed due to CO_2 formation; hence it is advisable to use steam for maximisation of hydrogen in the mixture and retarding combustion reactions.

Due to exothermic reactions, the temperature in the combustion zone hits round 1500°C. Though the effect of pressure is negative, considering all the reactions, still it is preferable to operate at high pressures which will render benefits in the following operations. Natural gas or fuel oil mixtures are generally chosen raw materials for this process, hence both would be suitable for high pressure operations (40 bars). The use of air may not be fully accepted, as it is feared the reaction mixture may not acquire the temperature of 1500°C, due to dilution with nitrogen. The advantages are many if oxygen is used in gasifier. Oxygen can be obtained from liquid air separators. These liquid air fractionators supply nitrogen for ammonia synthesis and oxygen for combustion purposes, further necessary refrigeration in the process.

Multiple advantages in this process except for the fact of air liquification, have enticed the industries. Another disadvantage felt during fuel oil gasification process is high amount of carbon (0.018 to 0.03 kg/kg F.O.) produced. This has to be removed systematically or recycled, if possible. About 3.5 cubic metres of hydrogen is produced per kg of oil.

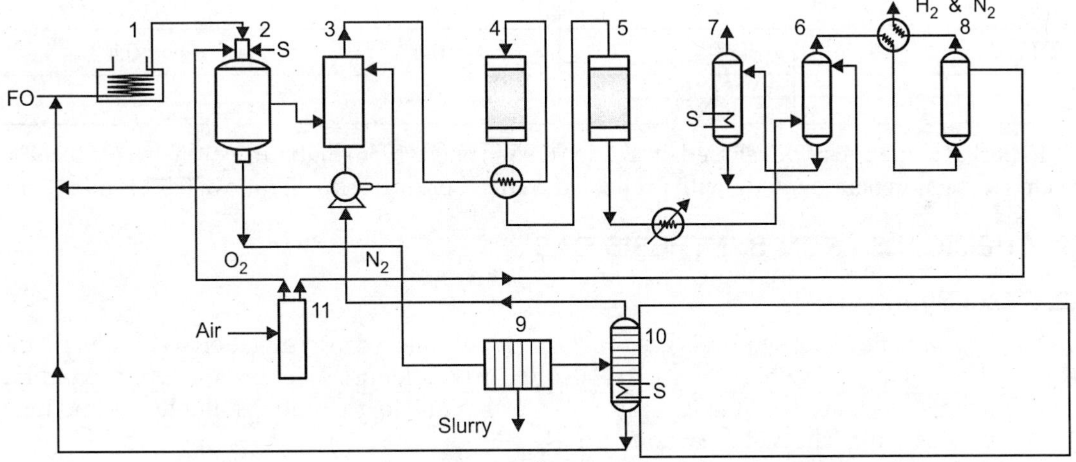

1. Fumace 2. Reactor 3. Scrubber 4. HTS, 5. LTS, 6. CO_2 H_2S SO_2 7. Gas Stripper absorber 8. Nitrogenerator 9. Filter 10. Light Oil Striper 11. Liquid Air Fractionator

Fig. 5.6: Synthesis Gas Via Partial Oxidation.

5.1.15 Process Description

As shown in Fig. 5.6 the fuel oil, mixed with recycled, heavy or light fractions is heated in a furnace (1) and sent into the reactor (2), where oxygen and steam join the stream. The reactor is vertical having a refractory lining. A high temperature over 1000°C is common. Steam is essential when the feed is heavy fuel oils. The outlet gas from the converter is scrubbed in a scrubber (3) with process fractions (light naphtha) to remove heavy condensates and other hydrocarbons. The gases after heat exchanging go into shift converters 4 and 5. The gases from shift converters are purified by absorbing acid gases (CO_2) in absorber (6) and CO_2 regenerated in stripper (7). Most usually Rectisol process is employed, which uses methanol as solvent (at 200°K) for these impurities elimination. Explicitly it differs from steam reforming process where the desulphurisation of feedstock be concluded before the first stage reformer itself, but in this process, it is attempted before ammonia reactor. If SNG is the goal, then it is done just prior to methanator. After this, gases are given a wash with liquid nitrogen, which will adjust the final ratio of nitrogen and hydrogen mix, and condensables are also easily and effectively eliminated.

Heavy tar slurry comes out of the reactor, which is filtered (9) and stripped off the lighter fractions (10). These lighter fractions are useful in scrubbing operation. The bottoms of stripper (10) are mixed with the process fuel oil and sent to reactor.

Liquid air separator (11) supplies the process oxygen and nitrogen for ammonia synthesis. Sometimes 38% oxygen containing air is also used, which automatically furnishes the required nitrogen and hydrogen ratio for ammonia synthesis.

Partial oxidation of natural gas, naphtha and fuel oil give the following composition of the gases at the outlet

	NG	*Naphtha*	*F/O*
H_2	60.9	51.6	46.1
CO	34.5	41.8	46.9
CO_2	2.8	4.8	4.3
CH_4	0.4	0.4	0.4
N_2, O_2, H_2S etc.	Rest	Rest	Rest

In partial oxidation as developed by Texaco (1957); Shell (1956) high purity oxygen (95 to 99%) is injected into gas generator along with preheated feed, where steam to oil ratio of 0.5:1 is maintained.

5.2 CHEMICALS FROM SYNTHESIS GAS

5.2.1 Oxo Synthesis

Most prominent use of synthesis gas is in Oxo reactions (Hydroformylation). The advantage of Oxo Synthesis is in getting higher primary alcohols from olefins. Alcohols are supposed to form through an intermediate step of aldehydes. It is worth while to go for lower aldehydes and higher alcohols via this route. The main reactions are :

$$RCH:CH_2 + CO + H_2 \rightarrow RCH_2.CH_2\ CHO$$

$$RCH:CH_2 + CO + 2H_2 \rightarrow RCH_2\ CH_2CH_2OH$$

Both the reactions show a gain in carbon atom. By choosing a proper ratio and reaction conditions aldehydes or alcohols may be produced.

The reaction is conducted liquid phase, catalysed by cobalt carbonyls $[Co(CO)_4]_2$ or $[CO(CO)_3]_4$. The reaction conditions are 100-150°C at 200 bars. As the temperature increases the solubility of CO decreases in carbonyls hence high pressure is used. The reaction mechanism seems to be vague, however an explanation to remote satisfaction is given below.

Cobalt carbonyls are always in equilibrium with hydrocarbonyls. These hydrocarbonyls react with olefin to transfer hydrogen atom and setting free carbonyl and carbon monoxide.

$$CO\,(CO_4)_2 + H_2 \rightarrow 2HCO(CO)_4$$

$$2HCO(CO)_4 + RCH:CH_2 \rightarrow RCH_2.CH_2CHO + CO + (CO(CO)_3)_2$$

$$2\,(CO(CO)_3)_2 \rightarrow CO_2(CO)_8 + 4CO + 2\,(CO(CO)_4)$$

It is supposed that tricarbonyl is in equilibrium with octa carbonyl. The reaction products are filtered and catalyst is recycled.

A typical case of propionaldehyde production is shown in Fig. 5.7 which is simply a representative aldehyde reaction. Synthesis gas mixed with recycled gas enters into the reactor (1). Ethylene (olefin) and cobalt catalyst are incessantly admitted into the reactor. Catalyst is usually a slurry in toluene. The ratio of $CO : H_2$ and ethylene is kept 1 : 1 : 5 : 1. The temperature is maintained at 110°C by circulating cold water and pressure is kept at 150 bars in the reactor. From the top of the reactor, the mixture of reactants and gases are withdrawn into a settler (2). The column may be sprayed with aldehyde to absorb carbonyls. The products are cooled in a condenser (3), from where condensed products are withdrawn into a separator (4) and the gases uncondensed from gas collector (5) are recycled. The product from the separator is distilled and a part of aldehyde is cycled to column settler (2). Except when ethylene is used as the starting olefin, a mixture of straight and branched chain alcohols are produced. Propionaldehyde yield is 70% while alcohol yield is around 20%. For getting alcohol, the product may be hydrogenated at 200°C under 0.1 MPa pressure.

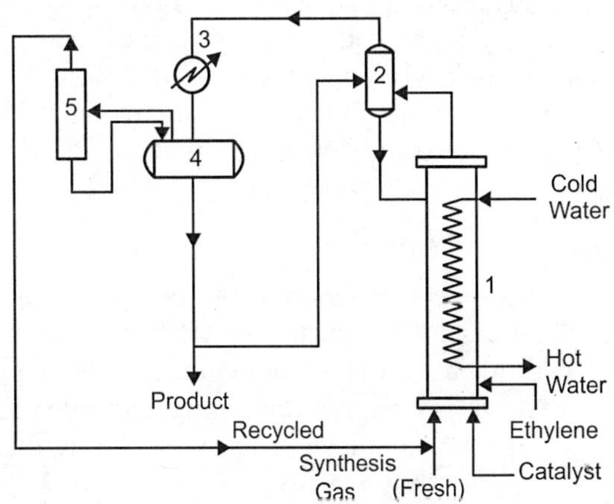

Fig. 5.7. Oxo Synthesis (Propionaldehyde Production)

By Oxo process, propylene can be converted into n and iso butyraldehydes. For increasing normal aldehyde content, the reaction is conducted at fairly lower temperature with recycle of branched compounds, however the ratio between normal to iso cannot be less than 4 to 1. Modified Oxo catalyst *i.e.* cobalt carbonyl with zinc acetyl acetonate brings about this reaction. For maximising aldehydes, substitution of rhodium for cobalt seems to be working allright.

The chemicals that are made on large scale based on this synthesis; are n-butanol, iso octanol, decanol etc. Olefins produced by cracking are best suited for primary alcohols of C_7 - C_9 range. These alcohols are converted to phthalates to be used as plasticisers.

5.2.1.1 Chemicals Based on Carbon monoxide

Carbon monoxide reacts with methyl amine to produce dimethyl formamide. CO can also combine with chlorine to give phosgene. The reaction is carried over activated charcoal at 250°C under pressure. Carbon monoxide with olefins in presence of steam reacts to form carboxylic acid.

$$RCH:CH_2 + CO + H_2O \rightarrow RCH_2CH_2\,COOH$$

Drastic operating conditions like a temperature of 350°C and a pressure of 100 MPa are required. Phosphoric acid on charcoal catalyses the reaction. Propionic acid is produced by reacting 90% CO and 8% steam with ethylene. If methyl alcohol is chosen instead of water, naturally methyl ester is formed.

$$CH_2:CH_2 + CO + CH_3OH \rightarrow CH_3CH_2COOCH_3$$

5.2.1.2 Oxo-Aldol Combined Process[19]

Propylene is the most important olefin since it is the starting materials for higher aliphatic alcohols. By direct oxidation to n-butyraldehyde followed by aldolization and hydrogenation gives 2-ethyl-l-hexanol (2EH). In the combined process propylene reacts on a special catalyst with CO and hydrogen to form 2EH. This catalyst is based on tri butyl phosphine, cobalt carbonyl complex. Similarly branched chain hexadecyl alcohol is also made by combined aldol process. To institute aldol condensation, the cobalt-phosphine combinations are treated with potassium hydroxide. These phosphines (PR_3. HCO. $(CO)_3$) yield straight chain products (about 80%) and are active by[21] 10^2 to 10^4 times the usual oxo catalysts.

5.2.2 Vinyl Acetate

Vinyl acetate monomer can be produced entirely from synthesis gas without any ethylene[20]. Methanol is first produced, which then reacts with acetic acid (incidentally acetic acid can be produced by carbonylation of methanol to acetic anhydride) to give methyl acetate. Methyl acetate again reacts with CO and H_2 to give ethylidene diacetate (EDA). Carbonylation of methyl acetate to get ethylidene diacetate is the main step. The next step would be cracking of EDA to vinyl acetate and acetic. The reactions are shown below :

$$2CH_3COOCH_3 + 2CO + H_2 \rightarrow \underset{\text{Ethylidene diacetate}}{CH_3CH(OOC.CH_3)_2} + CH_3COOH$$

$$CH_3CH\,(OOCC.CH_3)_2 \rightarrow \underset{\text{Vinyl acetate}}{CH_{2.}\,CH.\,(O.O.C.CH_3)} + CH_3COOH$$

Net reaction

$$2CH_3OH + 2CO + H_2 \rightarrow CH_2:CH.(OOC.CH_3) + H_2O$$

5.2.3 Acetic Acid

Acetic acid production through methyl alcohol has gained significance in these years. This is high pressure catalytic reaction (65 bars), carried out at a temperature of 250°C in liquid phase. Carbon monoxide is to be added to methanol. Much heat is liberated during reaction (53 Kcal/kg acid). The catalyst in the system consists of iodide and acetate of cobalt. Fig. 5.8 shows the outlines of the process as developed by BASF based on Reppe's Synthesis. Monsanto[22] process employs iodine promoted rhodium catalyst. This has been able to reduce the pressure in the reaction to about 15 bars. In this synthesis[23] a complex containing CO and methanol on catalyst is formed, which on hydrolysis separates the acid.

5.2.3.1 Reppe's Method

Another synthesis based on Reppe's chemistry is production of butanol from olefin and CO and H_2 mixture.

Carbon monoxide with olefins in presence of steam reacts to form alcohols.

$$RCH:CH_2 + 3CO + 2H_2O \rightarrow RCH_2.\,CH_2.\,CH_2OH + 2CO_2$$

When propene is the olefin the reaction proceeds in liquid phase at 100°C under a pressure of 1.5 MPa in presence of iron carbonyls; water and butanol act as solvent for the catalyst. Butyl pyrolidene is also added to the reacting mixture to enhance the activity of catalyst. The reaction is moderately exothermic (52.2 kcal/mole of butanol). The reactor-outlet gases contain CO_2 and propane-hydrogen mixture; these are separated before the olefin : CO mixture is recycled into the reactor. The catalyst can be easily decomposed, hence iron pentacarbonyl along with butanol comes as over-head product, while butyl pyrolidene obtained as bottom product for recycling. Some iron carbonate is also formed which is periodically removed from the acid column as shown in Fig. 5.8.

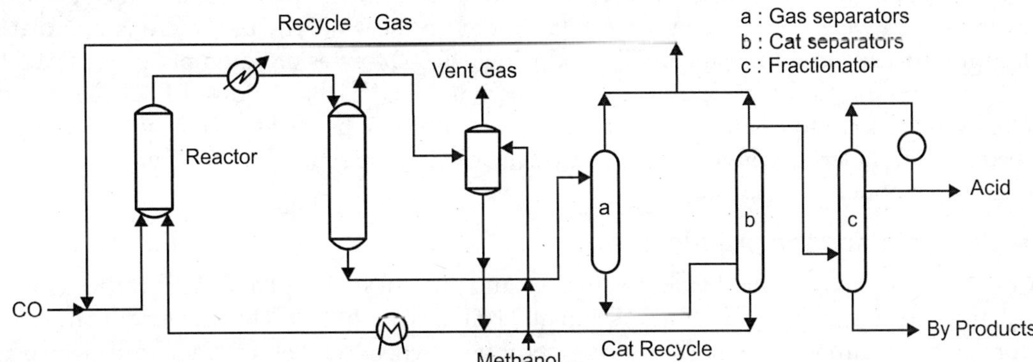

Fig. 5.8: Production of Acetic Acid.

In the figure it is shown that the recycle gas condensate separates into three phases; the top most layer is a mixture of iron carbonyl with dilute butanol, while the middle layer is highly water, and the heaviest layer being carbonyl-water mixture. The crude butanol is degassed with carbon monoxide, whereby iron carbonyl is carried away by CO for reactors (*a*) and (*b*). The bottom product is dehydrated into n and i butanol.

Hydrocarboxymethalation[(24)] of olefin to fatty alcohols has become an economic route than hydroformylation. Usually metals of group 8A catalyse the reaction giving 85% selectivity. Generally in hydroformylation reactions 50%, alcohols are produced. Thus the process has become varient to fatty alcohols from olefins. The reactions are shown below:

$$CH_2:CH_2 + CO + CH_3OH \rightarrow CH_3.CH_2.COOCH_3$$

$$CH_3.CH_2.COOCH_3H_2 \rightarrow CH_3.CH_2CH_2OH$$

5.3 FISCHER-TROPSCH SYNTHESIS

Originated in pre 1939, Gemany, Fischer-Tropsch Synthesis was planned as a unidirectional route to convert coal to hydrocarbons or mineral oil components, which were short in Germany. The work started with iron oxide catalysts with which Fischer-Tropsch converted synthesis gas mixture into olefins and paraffins as per the following reactions:

$$CO + H_2 \rightarrow (-CH_2-) + H_2O; \qquad H = -116.5\ kJ$$

$$2CO + H_2 \rightarrow (-CH_2-) + CO_{2;} \qquad H = -206kJ$$

$$CO + 3H_2 \rightarrow CH_4 + H_2O; \qquad H = -215\ kJ$$

$$CO + H_2O \rightarrow CO_2 + H_{2;} \qquad H = -415\ kJ$$

$$2CO \rightarrow CO_2 + C; \qquad H = -132\ kJ$$

$$CH_4 \rightarrow 2H_2 + C$$

Over the whole range of temperature 200-400°C formation of CH_4 and decomposition of CO are favoured thermodynamically over the desired olefin production. Carbon formation reactions are troublesome as they foul the catalyst but elimination is impossible.

Fischer-Tropsch reaction is an interesting route from coal to hydrocarbons. After oil crisis FT regained great actuality in producing chemicals. Synthesis gas when passed over an iron, nickel or cobalt catalyst at 150-300°C at near normal pressure a mixture of alkanes and alkenes with a broad range of molecular weights is formed. Hydrocarbons are predominantly of $C_5 - C_{11}$ straight chain and side products like aldehydes, acids and alcohols are also formed. With Ruthenium catalysts still higher molecular weight hydrocarbons result. To-day in the whole world, there is one plant at Sasolburg(S. Africa) operating since 1955 producing 240,000 tonnes of liquid products. Fixed bed reactors are used for producing high wax. The gloom in oil crisis that followed 1973-74 led to coal being re-examined as a source of fossil carbon. Today naphtha substitutes via F-T synthesis appears to be economical under special conditions. However, olefin production from F-T route will be more promising and economic in near future.

5.3.1 Catalysts and the products

Cobalt or iron catalysts yield mainly saturated and unsaturated aliphatic hydrocarbons ranging from methane to high melting paraffins. As the equilibrium is reached slowly a mixture of saturated or unsaturated hydrocarbons results. The stoichiometric yield when complete conversion of olefin occurs is about 208.5 gm/m^3 of synthesis gas. The maximum yield is possible only when offered ratio of CO and H_2 is equal to consumption ratio as shown below:

Olefin yield in gms per m^3 of gas[(25)]

Consumption ratio	*Offered ratio*	
	1:2	2: 1
1:2	208.5	104.3
2:1	104.3	208.5

The above observation prevails irrespective of catalyst used.

Iron, Cobalt, Nickel and Ruthmium are proved active metals is FT synthesis. These contain partially occupied d-orbitals which form active centres for adsorption of CO or hydrogen to lower the metallic state. CO is at least three times more adsorbed than hydrogen on cobalt and iron catalysts. The reaction velocity depends upon CO partial pressure

$$-\frac{d(\text{CO})}{dt} = K.\frac{p_{H_2}^{m}}{p_{C_o}^{n}}$$

Obviously the over all conversion depends upon the weakly adsorbed hydrogen on the surface. As a consequence of the reactivity of cobalt or iron, primary products that result are olefins only, hydrogenation being fast, secondary products may be formed more easily. To suppress this hydrogenation, cobalt or nickel should not be used. Other catalysts made out of iron, magnesium, chromium and vanadium shall be successful to that extent.

Alkali salts have been used as activators for Iron catalysts to shift the distribution of products; 0.2 to 0.1% of K_2CO_3 in iron catalyst doubles heat of CO absorption, thus increasing CO and reducing hydrogen on surface. Similar modifications for cobalt or nickel catalysts were not possible, hence they produce only saturated compounds; thus the suitability of these catalysts for FT synthesis is doubtful. Iron catalysts offer highest selectivity and conversion too (less methane production), however the two side reactions invincible, at temperature above 250°C are shift conversion of CO and Boudouard reaction; which results in large CO_2 formation. At present the doped iron catalysts with oxides of titanium, vanadium, molybdenum, tungsten and manganese are considered to have the best prospects for F-T synthesis. Synthesis over nickel and cobalt catalyst is conducted with CO and H_2 mixture in ratio of 1: 2 . 2 while with iron catalysts it is done with a ratio 2:1 or 1:1. Carbon selectivity of 31.3% for ethylene[27] and 22.2% for propylene was obtained at 320°C over a catalyst containing 100 Fe, 100 Mn, 10 ZnO and 4 K_2O_3.

Precipitated iron oxides on silica, promoted by alkali are wax making catalysts. Synthol reactors (fluidised bed/slurry reactors) use fused iron oxide catalysts. During fusion of oxide promoters are added. As an example fused iron catalyst used in F-T reactions has a composition of Fe_3O_4 94.5%, MgO 4.0%, Cr_2O_3 0.6%, SiO_2 0.6%, K_2O 0.52% and Mn_3O_4 0.03%.

5.3.1.1 Reactor Models

Basically FT synthesis is acccomplished in three types of reactors. Fluidised bed (Entrained bed), Oil-circulating (fixed bed) and slurry processes are the widely prevailing methods used for FT reactions.

5.3.1.2 Oil Circulating Process (Fixed Bed)

In this technology carbon monoxide, hydrogen and circulating oil are allowed to pass over the iron oxide catalyst. The product consists of largely unconverted gases along with methane and upto high molecular weight components (solid waxes). The circulating oil carries away the products and disposes heat effectively (*i.e.* maintaining temperature in the reactor). With fused iron catalyst and hydrogen to carbon monoxide ratio 1:1 the gasoline yield (C_3 – 204 °C) is about 56% wax. High pressure in the system (about 25 bars) increases high boiling components. Increase in temperature increases gasoline and wax. To permit the range of operating conditions at high levels, massive[26] iron catalysts were developed. Iron in the form of 0.8 to 1.6 mm dia. steel shots or turnings of commercially available pure iron and carbon steel were used. Iron shots, turnings were dried with steam (maximum 20% for turnings and 8% for shots) and then impregnated with 2% potassium carbonate solution and reduced with hydrogen and stored in an inert atmosphere. Such catalysts were capable of producing high amount of gasoline and less amount of wax.

5.3.1.3 Slurry Process

It differs from the previous process by suspending the finally divided catalyst in oil, and is allowed to circulate through reactor along with feed gases. Promoted iron catalysts are found to be the most useful ones in the process. The catalysts are composed of iron, copper and potassium oxide in the ratio of 100 : 10 : 1 respectively. The product separation is similar to the previous case.

5.3.1.4 Entrained Bed Reactor

Sasol plant is having this type of reactor. These reactors are efficient in heat disposal and bringing about the reactions; in fact rates of reactions are very high. These reactors work like fluidised bed reactors, hence thorough mixing is there. The product selectivities in fixed bed and entrained bed are shown in Table 5.7.

Product	Fixed bed 220°C	Entrained bed 325°C
CH_4	2.0	10
C_2H_4	0.1	4
C_2H_6	1.8	4
C_3H_6	2.7	12
C_3H_8	1.7	2
C_4H_8	3.1	9
C_4H_{10}	1.9	2
$C_5 C_{12}$	18.0	40
$C_{12}-C_{18}$	14.0	7
$C_{19}-C_{23}$	7	–
$C_{24}-C_{35}$	20.0	4
C_{39}	25	–
Non acids	3	5
Acids	0.2	1

5.3.1.5 Reaction Mechanism

Though no convincing mechanism is available, out of the two existing mechanisms namely Pichler and Anderson-Emmett; the latter is found to be more appropriate. Recently Shutz and Flory explained the mechanism on the basis of general polymerisation. However, Anderson-Emmett mechanism[25] is able to explain diferent prdoucts from a single intermediate. The reaction sequence of Anderson-Emmett mechanism begin with the adsorption of CO on metal ion, subsequently it changes into enolic form by reacting with hydrogen.

$$M + CO \rightarrow M = C = O$$

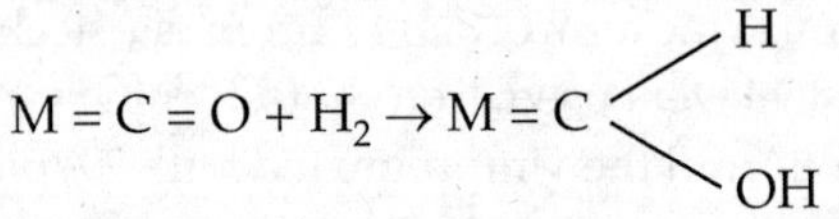

$$M = C = O + H_2 \rightarrow M = C \begin{matrix} \diagup H \\ \diagdown OH \end{matrix}$$

Two enolic complexes can join during which elimination of water molecule takes place.

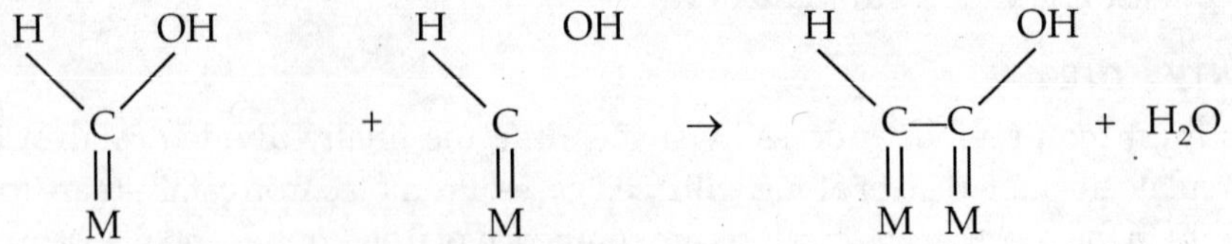

$$\begin{matrix} H & & OH \\ & \diagdown \; \diagup & \\ & C & \\ & \| & \\ & M & \end{matrix} + \begin{matrix} H & & OH \\ & \diagdown & \\ & C & \\ & \| & \\ & M & \end{matrix} \rightarrow \begin{matrix} H & & & OH \\ & \diagdown & \diagup & \\ & C - & C & \\ & \| & \| & \\ & M & M & \end{matrix} + H_2O$$

This double metal ion enolic complex is hydrogenated to give a methyl group in chain accompanying successive elimination of metal ion

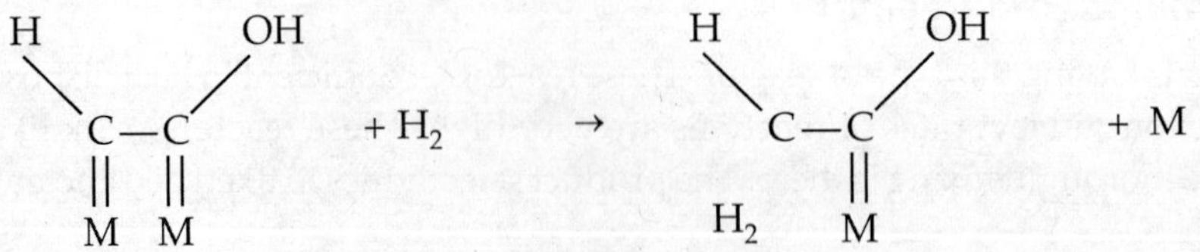

$$\begin{matrix} H & & & OH \\ & \diagdown & \diagup & \\ & C - & C & \\ & \| & \| & \\ & M & M & \end{matrix} + H_2 \rightarrow \begin{matrix} H & & & OH \\ & \diagdown & \diagup & \\ & C - & C & \\ & & \| & \\ & H_2 & M & \end{matrix} + M$$

This inital chain formation is grown up suitably with inital enolic complex with elimination of water.

$$
\begin{array}{c} OH \\ | \\ H_3C-C \\ \| \\ M \end{array}
\;+\;
\begin{array}{c} H \quad\; OH \\ \diagdown \;\; \diagup \\ C \\ \| \\ M \end{array}
\;\rightarrow\;
\begin{array}{c} \qquad\quad OH \\ \qquad\quad \diagup \\ H_3C-C—C \\ \quad\;\; \| \quad \| \\ \quad\;\; M \;\; M \end{array}
\;+\; H_2O
$$

Sequence of successive hydrogenations and elimination of metal ions ultimately leads to bigger chain length.

$$
\begin{array}{ccc} CH_3 & & OH \\ | & & | \\ C & — & C \\ \| & & \| \\ M & & M \end{array}
\;+ H_2 \;\rightarrow\;
\begin{array}{c} \qquad\qquad OH \\ \qquad\qquad | \\ CH_3-CH_3-C+M \\ \qquad\qquad \| \\ \qquad\qquad M \end{array}
$$

This chain attached with the metal ion is supposed to be the origin for various hydrocarbons to be formed as per the schemes illustrated below :

$$
\begin{array}{c} RCH_2\,CH_2OH \\ +M \end{array}
\leftarrow
\left|\begin{array}{cccc} R & & CH_3 & OH \\ \diagdown & & | & | \\ CH_2 & & C & C \\ + & & \| & \| \\ H_2 & & M & M \end{array}\right|
\rightarrow
\begin{array}{c} RCH_2\,.\,CHO \\ +M \end{array}
$$

$$\downarrow$$

$$
\begin{array}{c} R=CH_2 \\ \diagdown\; H_2 \\ RCH_3 \end{array}
\leftarrow
\begin{array}{c} H \quad\; OH \\ \diagdown \;\; \diagup \\ C \\ \| \\ M \end{array}
\rightarrow HCHO+M
$$

Anderson-Emmett mechanism can be readily proved by experimental work, as it has favourable energy formation, either aldehyde or alcohol or a paraffin can be produced.

Syngas, thus can be upgraded to alcohols or other products, because of positive thermodynamics and low yield of by-products.

Earlier (Chapter 2) methanol production has been presented in detail. A competitive process for the production of C_1-C_5 alcohol mixture, suitable as motor fuel blend has already begun.

Zinc, chromium-potassium based catalysts give high conversion per pass but the product consists mainly of 10-20% (wt %) C_2 and alcohols in methanol. This can be replaced by high yielding copper based methanol synthesis catalysts. These catalysts can be used at high temperature (380°C) and pressures upto 15 MPa; these conditions could yield 20-30% heavy alcohols. I.F.P[(29)] has taken lead in this regard, based on a homogeneous mixed oxide system containing copper and cobalt. These metals are associated with the other metals like aluminium, iron, cerium, lanthanum etc. While producing heavy alcohols simultaneously water production (30) in the reactions also increases. The separation of water and concentration of alcohol are achieved by extractive distillation, using solvents like diethylene glycol. The composition of alcohols appears to be :

CH_3OH	50 to 70
C_2H_5OH	16 to 20
C_3H_7OH	8-14
C_4H_9OH	4 to 7
$C_5H_{11}OH$	2 to 3
C_6^+	1 to 3

REFERENCES

1. JR. Rostrup-Nielsen, Catalysis, Science and Technology, Vol. 5, Springs Verlag, Berlin.
2. P. Belov; Fundamentals of Petroleum Chemical Technology, Mir Publications. Moscow, 1070, 253.
3. David C. Gardner; CH. Bartholomew; I.E.C. Proc. Res. Dev Vol 20, No. 1, 1981, 81.
4. J.A. Finneran; L.J. Buividas and N. Walen; HCP, April, 72, 129.
5. (*a*) C.M. Schillmoller and W. Vanden bruck, HCP, Dec. 84, 56.
   (*b*) HCP, January 74, 13.
6. J.R. Rostrup-Nielsen; Steam Reforming Catalysts 1974, (C) Hador Topsoe Copenhagen.
7. Lewis, F. Hatch and C. Matar; From Hydrocarbons to Petrochemical; GPC, Houston.
8. Peter, B. Tottrup; HCP, March 82, 89.
9. F. Marschner and H.J. Renner; HCP, April 72, 176.
10. HCP, Nov. 73, 137, Sept. 72, 221.
11. John, F. Lombard; HCP Aug. 69, 115.
12. J.P. Kapoor; Indian Chemical Engineer, April-June 1977, Vol XLX No. 2,1 and II.
13. George Long HCP, Aug. 72,91.
14. DP. Thornton, D.J. Ward and R.A Erickson; HCP, Aug. 72,81.
15. Gilbert E. Klingman and Rober P. Schaff; 101; HCP, April 72, 96.
16. Hermann F. Staege, HCP March 82, 95.
17. W. Konkol, P. Ruprecht, B. Cornils, R. Durrfeld and J. Longhoff, HCP, March 82, 97.
18. Martin Hess, HCP, Nov. 76, 99.
19. Edward J. Wickson and Herbert. P. Dengler, HCP, Nov. 72, 73.
20. J.L. Ehrler and B. Juran, HCP, Feb. 82, 111.
21. Kurt Christian Traneker and Rita Payen; HCP, June 75, 83.
22. H.D. Grove, HCP, Nov. 72, 77 and D.N. Freeman and R.J. Voorhoe, HCP, March 83, 59.
23. A.L. Waddams; Chemicals from Petroleum, GPC; 244.
24. Peter Hoffmann, Kurt Kosswing and Wolfgang Scharefer, Ind Chem. Pro. Res. Dev. 1980. No. 3, 19, 330.
25. B. Bussemeier, C.D. Frohning and B. Cornils; HCP, Nov. 76, 106.
26. Michael J. Baird, Richard R. Schehl, William P. Haynes and James, T. Cobb. IEC. Pr. Res. Div. Vol. 19, Nov. 1980, 177.
27. Tuhin K. Roy and S.K Gupta, Chemical Age of India Vol. 35, Nov. 3, March 84, 149.
28. Dry. M.E. Chem. Tech Dec. 82, 744 and HCP, Aug. 82, 121.
29. Ph. Courty, J.P. Arlie; A. Corivers, P. Mikitenks and A. Sugier; HCP, Nov. 84, 105.
30. Wolf Dieter Deckwer and Yalcin Serpremen IEC; Process Design and Milos Ralek et. al. Dev. Vol. 21, No. 2, 1982 Feb. 231 and 223.
31. I.K Suri C E W, Nov. 89,37.
32. J.H. Cartensen, J. Bogil Hansen and P.S. Pedersen, HCP, March, 1990, 57.
33. D.L. King, C.E. Bochow; HCP 2000, May, 39.

CHAPTER 6

Polymers of Olefins

6.1 POLYMER STRUCTURE

In all polymers, natural or synthetic, like organic compounds, the covalent bonds have the same length and angle; although every polymer may differ from the polymer of the same origin in its degree of repeating units known as degree of polymerisation(DP). Thermosets are three dimensional networks of primary bonds as polymerisation proceeds in all directions. Therefore, they are relatively rigid. Rather they become harder due to completion of any left-over polymerisation due to heat. Long chain polymers can be sub-divided into three types :

(*i*) Plastics (*ii*) Fibres; (*iii*) Elastomers.

Plastics play an invincible role in modern life; not only in household but also in other key sectors of economy. Main applications are in the field of agriculture, water management, automobile and transportation, electronics and communication, construction, defence, etc. The consumption of the plastics have grown up at a yearly rate of 13%. A consumption of 2.1 lakh tonnes in 1984-85 reached 5.4 lakh tonnes in 1989-90 and still expected to race with the same growth rate. Some of the demand projections are given below.

	Demand 1989-90	*Projections 1994-95*	*2000 AD (Thousand tonnes)*
LDPE/LLDPE	229	482	710
HDPE	178	378	560
PP	72	286	420
PVC	232	485	713
PS	38	82	120
Total	750	1,713	2,523

Long chains are randomly oriented in plastic form, therefore, the plastic has no directional property unless mechanically worked to align the chains. Fibres have chains aligned in lengthwise. Elastomers also have long chain molecules exhibiting unique rubbery behaviour. Usually polymers display a molecular weight 10,000 to over millions.

Physical and chemical properties of polymers are governed not only by molecular weight but by the structural configurations too. Generally three forms of structures are prominent, and they are isotactic, syndiotactic and atactic.

While polyethylene is a crystalline polymer of linear structure, polypropylene is a stereo regular polymer. Stereo regular polymer has a definite regular arrangement of the substituent groups in all the macromolecular units with respect to the back bone of the chain. Polymerisation of an olefin or monomer with substituent group ($CH_2 : CHR$) may result in the above mentioned structures as shown below:

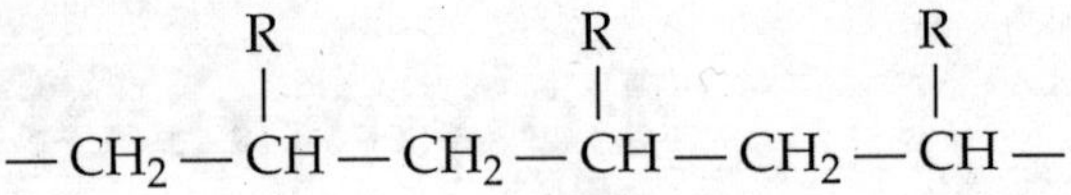

$$\begin{array}{cccccc} & R & & R & & R \\ & | & & | & & | \\ -CH_2- & CH & -CH_2- & CH & -CH_2- & CH- \end{array}$$

Isotactic

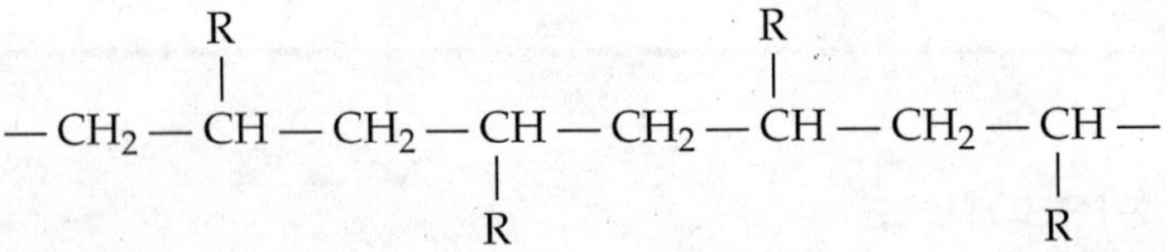

$$\begin{array}{cccccccc} & R & & & & R & & \\ & | & & & & | & & \\ -CH_2- & CH & -CH_2- & CH & -CH_2- & CH & -CH_2- & CH- \\ & & & | & & & & | \\ & & & R & & & & R \end{array}$$

Syndiotactic

$$\begin{array}{cccccccc} & R & & & & & & R \\ & | & & & & & & | \\ -CH_2- & CH & -CH_2- & CH & -CH_2- & CH & -CH_2- & CH- \\ & & & & & | & & \\ & & & & & R & & \end{array}$$

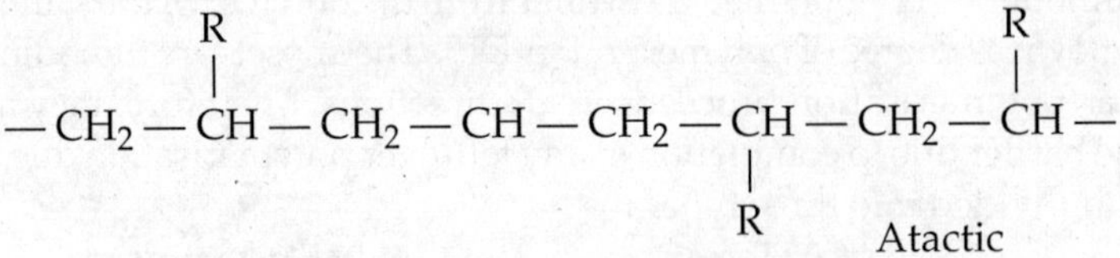

Atactic

Polymer Crystallinity

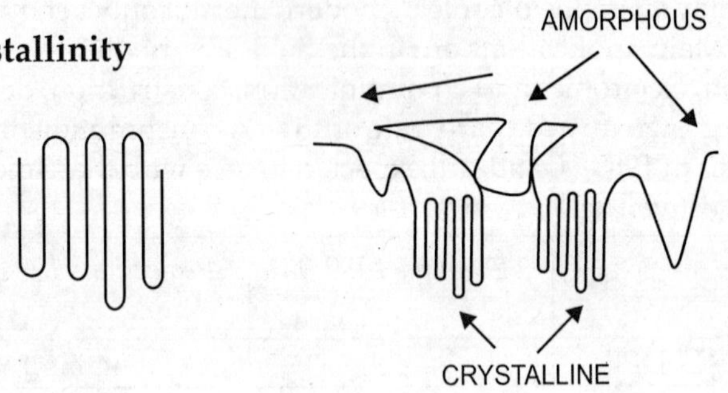

AMORPHOUS

CRYSTALLINE

- Most polymers are folded, not straight
- For polyethylene, the length the chains will stretch before they fold is about 100 angstroms.
- Straight polymers-ultra high molecular weight polyethylene and Kevlar
- Polymers have crystalline and amorphous regions

In the above structure, R is the substituent group, which is oriented on one side of the back bone (isotactic), or regularly arranged on either side of the back bone (syndiotactic)or without any orderly arrangement(atactic). These look to be branches; if the monomer happens to have a substituent group in its back bone such as vinyl chloride or acetate, 'R' becomes chloride or acetate, it cannot be called branching, but if a polymer from a monomer, like ethylene exhibits such monomer group, ($R = C_2H_5$) then it is termed as branching. Stereoisomers differ in their properties. Stereo regular polymers form crystallites even if the substituent group is large; such polymers have high mechanical strength, melting point and a lower creep than atactic ones. The properties of stereo-blocks depend upon length to width ratio; smaller lengths lower the crystallinity, molecular weight and melting point, thus a change in block size contributes to different properties. The valuable property of stereoblock polymers is high elasticity in oriented direction.

In a linear polymer, carbon atoms are linked in a continuous sequence and sometimes these carbon atoms are linked periodically or non-periodically to the other groups or carbon atoms, thus giving branches.

If the original monomer contains pendant groups like acetate, chlorine or stearate, but carbon atoms are only linked in back bone chain, then also it is called linear polymer. When the carbon atoms link in a cross-wise fashion, the resultant polymer will have space networks and these are called network polymers.

Irrespective of branching, polymers are softened by heat, thus assuming the name of thermoplastics. The polymers hardened by heat, and cannot be remelted are known as thermosets. In addition to the above mentioned structures, often we come across, head to head and head to tail configurations which depend entirely on monomer configuration. As an illustration head to head configuration of 2, 3 dichlorobutane.

```
     H   Cl  Cl  H
     |   |   |   |
 H — C — C — C — C — H        Head to head
     |   |   |   |            2, 3 dichlorobutane
     H   H   H   H
```

Head to tail configuration differs from the above one as

```
     H   Cl  H   Cl
     |   |   |   |
 H — C — C — C — C — H        1, 3 dichlorobutane
     |   |   |   |
     H   H   H   H
```

Polymerisation of 1, 4 disubstituted 1, 3 butadiene would give a polymer structure of such polymer which is known as tritactic(1).

```
— CH — CH = CH — CH — CH — CH — CH = CH — CH
  |              |    |              |    |
  R              R    R              R    R
```

Random configuration has no systematic arrangement of substituent group is often visible in high molecular compounds.

6.1.1 Methods of Polymerisation

There are in general three methods of polymerisation namely, mass, solution, emulsion/slurry. Mass or block polymerisation takes place usually in the state of monomer. In mass polymerisation, monomer and initiator are involved; the reaction being fast (going by free radicals) and the end product with a high molecular weight results instantaneously, hence the heat of dissipation becomes a major problem, only in case of few monomers like ethylene, methacrylates, VCM, it has been used successfully. In solution polymerisation a diluent, which can dissolve catalyst, polymer and monomer is used. This process is used for commercial poly olefins and butyl rubber production. In suspension (slurry) and emulsion polymerisation, the monomer and catalyst are suspended in medium; which is either stabilised in aqueous suspension or as an emulsion. In suspension polymerisation, the catalyst is not usually soluble, but in emulsion polymerisation water soluble catalysts are used (Ex-potassium persulphate). Highest molecular weight is achieved in emulsion polymerisation. It is used for vinyl products, SBR, acrylonitrile products.

The information on polymerisation process is condensed and given below :

Process	State of Monomer	Product form
Bulk	Monomer in its natural form	Melt, in solution or in slurry
Bulk in solution	Monomer in solvent	Liquid (usually for adhesives) slurry
Suspension	Monomer in water (Ingredients added for suspension)	Beads in slurry
Emulsion	Monomer in water (Ingredients added for emulsion)	Latex

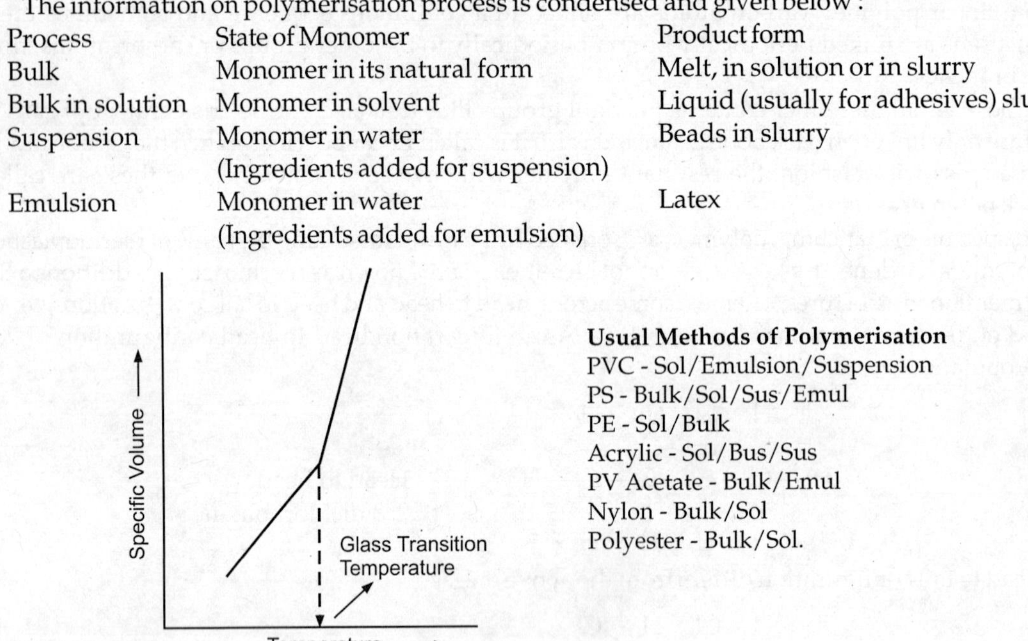

Usual Methods of Polymerisation

PVC - Sol/Emulsion/Suspension
PS - Bulk/Sol/Sus/Emul
PE - Sol/Bulk
Acrylic - Sol/Bus/Sus
PV Acetate - Bulk/Emul
Nylon - Bulk/Sol
Polyester - Bulk/Sol.

Fig. 6.1: Glass Transition Temperature of Polymers.

6.1.2. Rheology and Polymer Structure

Most of the low molecular weight organic compounds have sharp melting points. But in case of polymers of low molecular weight, where the change occurs from viscous liquid to glass state is important, which is called glass transition temperature. As shown in[2] Fig. 6.1, an abrupt change in density of the polymer defines the glass transition temperature. Different polymers and their glass transition temperatures are given in Table 6.1. Below glass transition temperature, the polymer becomes hard and brittle like glass. Plasticisers are added to decrease the glass transition temperature.

Table 6.1: Properties of Plastics

Plastic	*Glass Transition Temp.°C*	*Solubility*
Polyethylene	–70 to –20	Diphenyl ether
Polyvinyl chloride	81	Cyclohexanone, ketones
Polyvinyl acetate	28	Benzene, chloroform
Poly acrylonitrile	107	DMF, DMSO
Polytetrafluoro-ethylene (Tefflon)	+113	Insoluble in almost all solvent
Polymethyl methacrylate	105	Acetone/ CCl_4 Soluble only in fluorinated kerosene
Polystyrene (isotactic)	+ 105	Decalin, Chlorinated solvents

Nylon 6,6	+47	
Polybutadiene	–108	
Polyethylene terephthalate	+ 69	
Polypropylene	–16	Soluble in aromatics above 80°C
Polyisoprene cis Trans	–73	
Polyisobutylene	–73	
Polychloroprene	–50	
Polyethylene adipate	–70	

6.1.3 Degree of Crystallinity

Degree of crystallinity is defined by the following expression

$$c = \frac{d_c}{d}\,\frac{d-d'}{d_c-d'}$$

Where d_c - density of high density stereo regular polymer
d' - density of amorphous polymer
d - density of test sample

Depending upon the crystallinity of physical properties like the density, rigidity softening point vary as shown below (For Polyethylene)

Crystallinity	*Density (10^3) kg/m^3*	*Relative Rigidity*	*Softening point°C*	*Elongation, %*
65	0.92	1	104	500
75	0.93	2	118	300
84	0.94 –0.95	3	146	100
95	0.95 –0.96	4	149	20

Depending upon branching, the polymer exhibits variable properties, for polyethylene melting point crystallinity and density are furnished with respect to CH_3 groups in branching.

CH_3(3) groups/ 1000°C-atoms	*Density × 10^3 kg / m^3*	*Crystallinity %*	*MP°C*
87			105
83	0.9100		
60		21	
48	0.9170		
46	0.9250		
35		50	115
26	0.926		
16	0.9269		
10	0.9420	77	124
7	0.9504	88	143
2	0.9600	95	148

6.1.4 Chemistry of Polymerisation

Olefins and their derivatives like vinyl chloride, vinyl acetate undergo a process of self-addition under the influence of heat or catalysts yielding a very high molecular weight compounds called polymers. When polymer is formed from one reacting species called monomer, it is known as homopolymer, if two different monomers are used, it is known as co-polymer. Addition polymerisation starts with initiation by radicals or by ions called initiators (catalysts). Many organic peroxides, persulphates, oxygen, azo compounds are used as common initiators. This process of addition takes place due to initiators which split into radicals, or ions, while radicals are free of charge, and acquire the activity because of spin orientation (unpaired electrons) in an atom, ions are capable of chemically reacting, through positive or negative charges, known as cationic or anionic polymerisation.

6.1.5 Thermodynamic Considerations

The feasibility or reaction under a given set of conditions depends upon the sign and magnitude of Gibbs-Free energy change. During polymerisation in most cases a decrease in entropy results, as a consequence the polymer formation progresses specially at low temperatures. As an example ethylene free energy at 227°C is -5.45, at 1227°C it is +26.71, which shows that polymerisation is impossible at that high temperature. Even at 227°C, the feasibility of polymer formation is there, with uncertainty of its existence because of decomposition. Also if the rate of formation of polymer is more than the rate of decomposition, then there is always surplus polymer.

6.1.5.1 Molecular weights of polymers and significance

In general, properties like physical, chemical and electrical are essentially dependent upon molecular weight. In case of amorphous polymers, melt viscosity, softening point, tensile and impact strength depend upon the length of chain. The properties readily increase with chain length or molecular weight initially and level off. For all commercial applications, polymer must be above the threshold value of (minimum length) molecular weight.

6.1.5.2 Mechanism of Stereoregulation

In general terms stereo regular polymer formation by ionic mechanism depends on the effective separation of growing ion and counter ion, so that the growing monomer of transitions state complex is available for a large number of transition steps. Further a large number of molecules of monomer must enter the transition complex. The fact that some catalysts generate stereo regular or atactic polymer depends on how stereo specific sites on the catalyst surface exist.

6.1.5.3 Kinetics in polymerisation

In most of the cases, except for low monomer concentration, first order kinetics apply for homogeneous and heterogeneous systems, second order kinetics have been observed for the polymerisation of ethylene with ethyl zinc/$TiCl_4$ and butyl lithium/$TiCl_4$. Crystalline $TiCl_4$/ metal alkyl catalysts have shown rates proportional to surface area of the catalyst. With precipitated catalysts like $TiCl_4$, the polymerisation rate passes through a maximum, sometimes initial adjustment periods were also observed.

6.1.6 Cationic polymerisation

In general all proton donating and carbonium ion forming compounds like Friedel-crafts (Lewis Acids) are good catalysts. In most of the cases, they require co-catalysts for carbonium ion contribution.

The initiator or catalyst splits into ions, giving rise to positive or carbonium ions to activate the monomer. Compounds like BF_3, $AlCl_3$, H_3PO_4 are some of the cationic catalysts industrially used. These are effective in presence of water or other co-catalysts. Of all the acids sulphuric acid seems to be the best, but produces only low molecular weight polymers. Just like free radical polymerisation, here also , initiation, propagation and chain transfer are the sequences; in polar- bond mechanism, proton donation to monomer and proton donation to counter ion followed by hydride ion abstraction and isomerisetion results in the polymer. The mechanism of carbonium ion initiation for polymerisation is shown below in various catalysts.

$$BF_3 + H_2O \rightarrow BF_3\,OH_2\,(BF_3OH^- + H^+)$$

$$(CH_3)_2C=CH_2 + BF_3OH_2 \rightarrow (CH_3)_3C^+ + BF_3\,OH^-$$

$$[(CH_3)_3\,C]^+ + (CH_3)_2\,C=CH_2 \rightarrow [(CH_3)_3.C.CH_2\,C^+(CH_3)_2]$$

$$(CH_3)_3\,C\,CH_2\,C^+\,(CH_3)_2 + BF_3\,OH^- \rightarrow (CH_3)_3\,C{-}CH_2{-}CH_2\,CH=CH_2 + BF_3\,OH_2$$

Reaction with $AlCl_3$

$$AlCl_3 + H_2O \rightarrow AlCl_3\,(OH)^- + H^+$$

$$CH_2=C\,(CH_3)_2 + H^+ \rightarrow CH_3\,C^+\,(CH_3)_2 \text{ and so on}$$

With sulphuric acid

$$H_2SO_4 \rightarrow HSO_4 + H^+$$

$$\begin{array}{l} CH_3\diagdown \\ \qquad\quad C=CH_2+H^+ \\ CH_3\diagup \end{array} \rightarrow \begin{array}{l} CH_3\diagdown \\ \qquad\quad C^+.CH_3 \\ CH_3\diagup \end{array}$$

$$\begin{array}{l} CH_3\diagdown \\ \qquad\quad C^+.CH_3\ + \\ CH_3\diagup \end{array} \begin{array}{l} CH_3\diagdown \\ \qquad\quad C.=CH_2 \\ CH_3\diagup \end{array} \rightarrow \begin{array}{l} CH_3\diagdown \qquad\qquad\qquad\quad CH_3 \\ \qquad\quad CH-CH_2\,C^+ = \qquad CH_2 \\ CH_3\diagup \qquad\qquad\qquad\quad CH_3 \end{array}$$

$$\rightarrow \begin{array}{l} CH_3\diagdown \\ \qquad\quad C-CH \\ CH_3\diagup \end{array} \hspace{-3em}\begin{array}{c} CH_3 \\ | \\ \\ \end{array} = \begin{array}{c} CH_3 \\ | \\ C-H \\ | \\ CH_3 \end{array}$$

or in a single step

$$\begin{array}{c} CH_3 \\ | \\ C \\ | \\ CH_3 \end{array} = CH_2 \quad \begin{array}{c} HO \\ | \\ +\,SO_2 \\ | \\ HO \end{array} \rightarrow \begin{array}{c} CH_3 \quad CH_3 \\ \diagdown\ \diagup \\ C- \\ | \\ CH_3 \end{array} O\,SO_2\,OH + \begin{array}{c} CH_3 \\ \diagup \\ C \\ \diagdown \\ CH_3 \end{array} = CH_2$$

$$\rightarrow (CH_3)_3\,.\,C.\ CH{:}\ C(CH_3)_2 + H_2SO_4$$

The example of sulphuric acid shows the intermediate salt formation, evidently addition goes through a number of steps, as suggested by Butlerov. This is known as stepwise polymerisation and each step requires additional energy. Such polymerisation is suitable for low molecular weight polymers, usually a monomer repeats 10 to 15 times and the ultimate product is liquid polymer. This process is also known as oligomerisation, as the number of molecules of monomer are clearly known. Concentrated phosphoric acid at 300°C and 35 bars can give butane, isobutane and liquid hydrocarbons from ethylene. Liquid hydrocarbons are also produced from ethylene using hydrogen flouride catalyst, at moderate temperatures and pressures. Synthetic lubricants were produced from ethylene and butene using hydrogen flouride as catalyst, at moderate temperature and pressures. Synthetic lubricants were produced using anhydrous aluminium chloride catalyst at 100- 150°C and 20-40 bars pressure, monomer being ethylene. These are all the examples of step-wise polymerisation. The energy required in each case depends upon the bond scission and the polymer gradually builds up unlike chain polymerisation. The energy for bond breaking is different for different monomers as given below:

Ethylene	93.87	kJ/mole
i-butene	53.76	kJ/mole
Styrene	69.30	kJ/mole
i-propene	75.18	kJ/mole
Butadiene	72.66	kJ/mole
Chloroprene	68.04	kJ/mole
Vinyl acetate	89.46	kJ/mole
Methyl methacrylate	54.60	kJ/mole

It can be inferred, that unifunctional molecules (like acetic acid) can not be polymerised, bifunctional (olefins), trifunctionals (phenol-formaldehyde) can be easily polymerised. Again energy required for initiation is different in each case as given below :

C–C	294	kJ/mole
C–N	205	kJ/mole Unifunctional
C–S	226.8	kJ/mole does not polymerise
C–O	294	kJ/mole
C–H	411.4	kJ/mole
C=C	420	kJ/mole Bi-functional
C=N	394.8	kJ/mole Polymerise
$C=F_2$	432.6	kJ/mole
C=O	625.8	kJ/mole High energy requirement
C≡N	630	kJ/mole for initiation
C≡C	516.6	kJ/mole Trifunctional

6.1.7 Anionic Polymerisation

Like cationic polymersation anionic (Carbanion) polymerisation is also complex and versatile. Anionic polymerisation has been widely used in dienes polymerisation using lithium or sodium as initiators. Isoprene or butadiene polymerised with lithium gives a high molecular weight (million) polymer, almost immediately, that too in trans 1, 4 structural configuration identifying with natural rubber. Other catalysts would yield 1, 2 or 3, 4 structures of isoprene.

Some monomers like styrene, are polymerised in solvents like tetrahydrofuran. Alkyl lithium initiated styrene polymerisation gives termination free polymer distinguishing as living 4 polymer; as shown by the following reaction.

$$R\,Li + Styrene \rightarrow R\,St\,Li$$
$$R\,St\,Li + n\,St \rightarrow R\,Stn\,St\,Li$$

Most of the liquid rubbers, in the form of living polymers can be produced, with Li-alkyl catalysts, even with introduction of functional groups.

6.1.7.1 Chemistry of Anionic Polymerisation

The initiation of reaction is caused by electropositive metallic compounds and cations become counter ions. The rate of initiation, propagation are depending upon the electropositivity of the initiators, electronegativity of substituents in the monomer; and dielectric constant of the medium and the stability of the carbanion ion. Molecular weight of the polymer inversely depends upon the activity of the initiator (organolithium). The reactivity decreases from alkyl to the triphenylmethyl via benzyl and vinyl, *i.e.* styrene produced with the tri-phenylmethyl, phenyl and t-butyl[(5)] lithium gives under the identical conditions a mol wt. of 66,000, 12,000 and 3,200 respectively.

Highly electronegative substances like acrylonitrile, vinyl chloride may be polymerised by weak electropositive initiators. Monomers, with less electronegative substituents (such as styrene), require more electropositive initiators such as n-butyl lithium. The following reaction shows the anionic polymerisation of acrylonitrile catalysed by amine.

$$K\,NH_2 \rightarrow NH_2^- + K_{2+}$$

Initiation

$$NH_2^- + H_2C = \underset{\displaystyle CN}{\underset{|}{CH}} \rightarrow NH_2 - CH_2 - \underset{\displaystyle CN}{\underset{|}{HC^-}}$$

Propagation

$$NH_2 - CH_2 - \underset{\displaystyle CN}{\underset{|}{\overset{\displaystyle H}{\overset{|}{C}}}} + nH_2\,C = \underset{\displaystyle CN}{\underset{|}{CH}} \rightarrow NH_2\,(CH_2\,CHCN)n - H_2C - \underset{\displaystyle CN}{\underset{|}{\overset{\displaystyle H}{\overset{|}{C}}}}$$

Termination

$$H_2N\,(H_2C\,.\,CHCN)n - \underset{\displaystyle H}{\underset{|}{\overset{\displaystyle H}{\overset{|}{C}}}} - \underset{\displaystyle CN}{\underset{|}{\overset{\displaystyle H}{\overset{|}{C}}}} \longrightarrow (CH_2CHCN)n\,CH_2CH_2CN + NH_2^-$$

6.1.7.2 Alfin Catalysts

Alfin[(6)] catalysts developed by Morton and co-workers and they are special combinations of sodium salts, amylsodium, sodium isopropylate and alkyl sodium. These catalysts produce butadiene polymers of trans 1, 4 structure having a high molecular weight. Stereo regularity is an offspring of such catalysts. Alfin catalysts (Alcohol-Olefin) are pyrophoric, can ignite if spilled in air. The method of preparation involves an inert atmosphere and a low temperature (–10°C). Amylchloride is slowly added to sodium till it dissolved completely, to which isopropanal is added. Propylene is slowly bubbled into the reaction mixture. The resulting mixture is known as Alfin catalyst. The reactions are shown below:

$$C_5H_{11}Cl + 2Na \rightarrow C_5H_{11}Na + NaCl$$
$$C_5H_{11}Na + iC_3H_7OH + CH_2{:}CH.CH_3 \rightarrow CH_2 = CH\,CH_2\,Na + Na\,Cl + iC_3H_7ONa$$

6.1.7.3 Coordination Polymerisation

In this, anions play the role of initiation. Ziegler found that ethylene could be polymerised by certain heterogeneous catalysts in an inert hydrocarbon solvent at low pressure and temperature, incidentally it happened to be the real foundation for modern polymer industry. In this system a composite catalyst consisting mainly alkyl aluminium with transition metal halide brings about the initiation; still it is confounding to explain the exact mechanism. Phillips process, Standard Oil use chromium oxide supported on silica-alumina and 8% molybdena on alumina support respectively; again heterogeneous catalysts of different nature but work on the same lines. Comprehensively either a reducible compound of transition metal and reducing organometallic compound, as in Ziegler's system or metal oxide catalyst with high electron accepting surface area of proper valency state is a prerequisite for all these polymer processes. This in general predicts that monomers containing electron withdrawing groups more easily polymerise anionically, however, the reactive centre and the metal are appreciably covalent, and the process is called coordination polymerisation; even though there is no anion. Anionic polymerisation usually fails in the case of co-polymers preparation, as the difference in stability of organic ions is higher than the corresponding radicals. The initiation step in anionic polymerisation is the production of an anion from a monomer. This is shown in the following equations:

$$CH_2:CH_2 + Al(C_2H_5)_3 \rightarrow (C_2H_5)_3 . CH_2CH_2^- . Al^+$$

or with amides of potassium

$$CH_2:CH_2 + NH_2 . K \rightarrow NH_2 . CH_2 . CH_2^- \, K^+$$

6.1.7.4 Ziegler-Natta Catalysts

Ziegler-Natta catalysts are the most important and widely used coordination catalysts. These are complexes of a transition metal of IVth to VIIIth group and an organometallic compound of a metal from 1st to IIIrd group elements. Usually transition metal halides are known as catalysts and organometallic compounds of I to IIIrd group are known as co-catalysts. Catalysts like zirconium, cobalt and iron are also found useful while co-catalysts like hydrides, alkyl, aryls of Al, Li, Zn, Sn, Cd, Be and Mg are prevalent in usage. Most useful catalysts in polymerisation of lower olefins is composed of ethyl (alkyl) aluminium and titanium tetra (tri) chloride. Though the polymerisation is not clearly understood with these catalysts, yet these have gained significance in production of stereo regular polymers. Before polymerisation with an olefin is set in, various types of halides, and alkyl metal halides are produced by the reacting catalyst as shown below.

Organic titanium compounds are believed to undergo reduction via homolytic bond cleavage.

$$3\,TiCl_4 + Al\,R_3 \rightarrow 3R\;TiCl_3 + Al\,Cl_3$$
$$TiCl_4 + Al\;R_3 \rightarrow R\,Ti\,Cl_3 + R_2\,AlCl$$
$$AlR_2\,Cl + TiCl_4 \rightarrow AlR\,Cl_2 + Ti\,R\,Cl_3$$
$$R\,TiCl_3 \rightarrow TiCl_3 + R$$
$$Ti\,R\,Cl_2 \rightarrow TiCl_2 + R$$
$$Ti\,R\,Cl_3 \rightarrow TiCl_3 + R$$
$$R + TiCl_4 \rightarrow TiCl_3 + R\,Cl$$
$$TiCl_4 + TiCl_2 \Leftrightarrow 2\,TiCl_3$$

The reaction continues till $TiCl_3$ is formed. Soluble complexes of Ti and Al have tetrahydral configurations, with localised active sites on catalyst surface. Organo metallic compounds are believed to activate the sites by alkylation of transition metal atom surface. The crystalline complex of this nature is shown below.

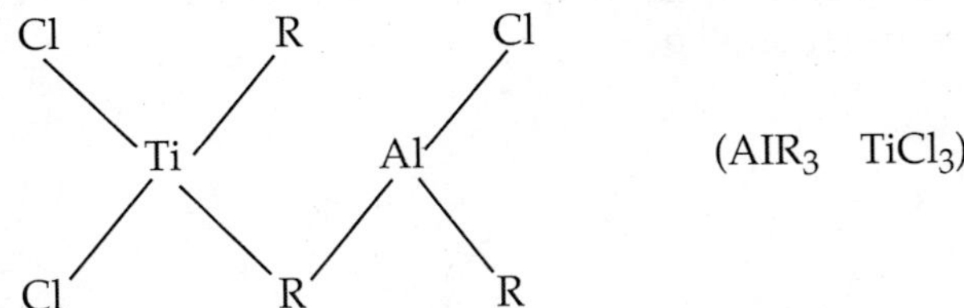

Active site theory, bimetallic mono[7] metallic theories are in literature to explain the polymerisation pattern though not quite clear.

$TiCl_3$ is usually present in (α), (β), (γ), (δ) forms; except (β) variety all others given rise to high degree of stereo regularity. The following extracts show the stereo regularity obtained in propylene polymerisation.

Catalyst	*Stereo regularity % (Triethyl aluminium)*
$TiCl_4$	48
$TiBr_4$	42
$TiCl_3$ (α,ϒ, δ)	80–90
$TiCl_3$ (β)	50
$ZrCl_7$	55
$CrCl_3$	36
VCl_4	48

Also the change in alkyl group of co -catalyst with $TiCl_3$ and $TiCl_4$ system show

Alkyl radical	*Stereo regularity percentage*	
	$TiCl_3$	*$TiCl_4$*
Ethyl	80	48
n-propyl	72	51
i-butyl	75	30
Hexyl	65	26
And		
R_3 Al + (α) $TiCl_3$	84	
(β)	45	
R_3 Al +Na F		97
R_3 Al (Halogen)	90-99	
R_3 Al + Ti(ethyl) Cl_2	85	

6.1.7.5 Mechanism of Catalyst[8]

The active site on catalyst must contain at least one P electron, so that locking of monomer takes place, which forms the propagation step

$$\begin{array}{c}\text{Cl} \quad\quad \text{R} \\ \diagdown \quad \diagup \quad \diagdown \\ \text{Ti} \quad\quad\quad \text{Al} \\ \diagup \quad \diagdown \quad \diagup \\ \text{Cl} \quad\quad \text{R}\end{array} + CH_2{:}CH\,R' \rightarrow \begin{array}{c} \quad\quad R' \\ \quad\quad | \\ CH_2\text{—}CH \\ \diagup \quad\quad \diagdown \\ \text{Ti} \quad\quad\quad\quad R \\ \diagup \diagdown \quad\quad \diagup \\ R\text{—}Al \end{array}$$

Further addition of $CH_2{:}CH\,R'$

$$\begin{array}{c} CH_2\,CH\,R' \quad\quad CH_2\,CH_2\,R \\ \diagdown \quad \diagup \quad\quad\quad \diagup \\ \text{Ti} \quad\quad\quad\quad \text{Al} \\ \diagup \quad \diagdown \quad\quad \diagup \quad \diagdown \\ R \end{array}$$

Thus, initiation shows

$$\text{Cat } R' - CH_2{:}CH\,R' \rightarrow \text{Cat } CH_2\,CH\,R\,(R')$$

Propagation

$$\text{Cat} - CH_2\,\underset{\displaystyle R}{\underset{|}{CH}}\,R' + n\,CH_2{:}CH\,R \rightarrow \text{Cat } CH_2\,\underset{\displaystyle R}{\underset{|}{CH}} - (CH_2 - \underset{\displaystyle R}{\underset{|}{CH}})_n - R'$$

Chain Transfer with Al R_3

$$\text{Cat}.CH_2\,.\,\overset{\displaystyle X}{\overset{|}{CH}} - (CH_2\,.\,CH\,X)_n\,R + Al\,R_3 \rightarrow$$

$$\rightarrow \text{Cat } R + R_2\,Al\,.\,CH_2\,.\,\overset{\displaystyle X}{\overset{|}{CH}} - CH_2 - \overset{\displaystyle X}{\overset{|}{CH}} + R$$

Internal hydride transfer

$$\text{Cat } CH_2 - \overset{\displaystyle X}{\overset{|}{CH}} - (\overset{\displaystyle X}{\overset{|}{CH_2}}\,CH_2) - R + H_2 \rightarrow \text{Cat } H + CH_3 - \overset{\displaystyle X}{\overset{|}{CH}} - CH_2\,.\,\overset{\displaystyle X}{\overset{|}{CH}}\,R$$

The last step may also be treated as termination step. Unless catalyst hydride bond is first alkylated chain propagation cannot continue. Natta suggested that the termination results form spontaneous disproportionation of growing chain. The rate of polymerisation is proportional to concentration of monomer catalyst. Heterogenous systems usually result in stereo regularity. Further, because of the solvating power of the medium different rates of propagation are likely to occur in different media. Similarly poisons like nickel metal can prevent due to nickel adsorption. But still, it offers a more difficult conclusion for termination thus differing from free radical polymeriastion.

Termination may result, due to proton transfer from solvent or monomer , or weak acid.

6.1.8 Mechanism of Free Radical Polymerisation

Homopolymerisation of ethylene is governed by free radicals or monomer, or weak acid like initiation, propagation and termination.

6.1.8.1 *Initiation*

Initiation of free radicals is caused by energy, usually thermal or photo chemical.

$$O_2 \rightleftharpoons \dot{O} + \dot{O} \quad \text{Oxygen Free Radicals}$$

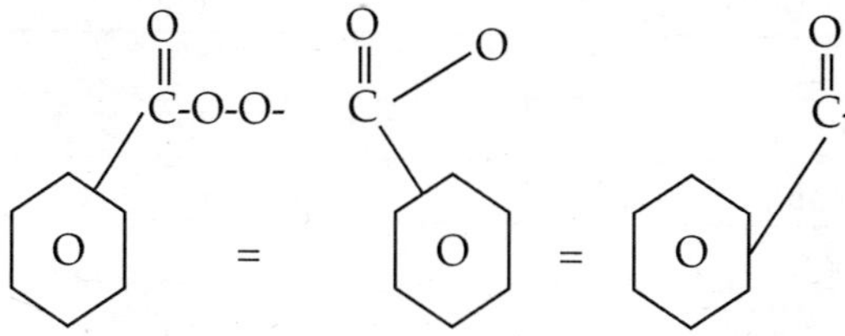

$$C_6H_5-\overset{O}{\overset{\|}{C}}-O-O-\overset{O}{\overset{\|}{C}}-C_6H_5 \rightleftharpoons C_6H_5-\overset{O}{\overset{\|}{C}}-O \quad \text{Benzyl Peroxide Radicals}$$

$$Cl_2 \rightleftharpoons \dot{Cl} + \ddot{Cl} \quad \text{(Photo Chemical)}$$

Sometimes azo compounds, persulphates also contribute free radicals. Polyamines, per sulphates, sulphates are used as activators, which increase the rate of initiation at lower temperature. Activators often work as oxidation reduction systems, thus help in intensive formation of free radicals. Ethylene polymerisation depends upon free radicals which have a half-life of a minute or less at 150- 200°C.

6.1.8.2 *Propagation*

Free radicals being highly reactive induce chain growth by creating radicals in ethylene molecule.

$$O + O_2 \quad \dot{O} + \dot{O} + \dot{O}$$

$$CH_2:CH_2 + O \quad CH_2:CH_2 + O$$

$$CH_2:CH_2 + CH_2:CH_2. \quad -(CH_2.CH_2)CH_2.\dot{C}H_2$$

$$(CH_2.CH_2)_1.CH_2.CH_2 + CH_2:CH_2 \quad (CH_2CH_2)_2CH_2CH_2$$

and so on,....... $(CH_2.CH_2)_n.CH_2.CH_2$

6.1.8.3 *Termination*

Chain growth terminated by coupling or disproportionation

$$2O \rightarrow O_2$$

$$(CH_2.\dot{C}H_2)(CH_2.\dot{C}H_2) \rightarrow -CH_2.CH_2.CH_2^-$$

$$2R\,CH_2.CH_2 \rightarrow R\,CH_2.CH_3 + R\,CH:\dot{C}H_2$$

$$2\,CH_2.CH_2 + O \rightarrow \underbrace{CH_2.CH_2 + CH_2.\dot{C}H_2}_{O}$$

Branching reactions also often show their presence (hydrogen transfer)

$$CH_2.\dot{C}H_2 + (CH_2.CH_2)_n \rightarrow (.CH_2.CH_2)_{n-1} + \underset{\displaystyle |}{\overset{CH:CH_3}{}}CH_2CH_3$$

Polymer radicals can be terminated by chain transfer with solvents as shown below :

$$R\,CH_2.\ \dot{C}H_2\ . + SH \rightarrow R\,CH_2.\ CH_3 + \dot{S}$$

$$\dot{S} + CH_2 : CH_2 \rightarrow S\,CH_2\,CH_2$$

Propagation and termination constants for some free radicals is shown below.

	At 50°C	
	K_p *lit/mol sec.*	K_t *lit mol sec.* $\times 10^{-6}$
Styrene	123	65
Vinylchloride	11.000	2,100
Ethylene	242	540
Acrylonitrile	32,500	4,400
Butadiene	100 at 60°C	100 at 60°C
Methylacrylate	2100	0.5
Isoprene	5.0	

6.2 Polyethylene

Oligomers of ethylene were common in earlier days. During second world war synthetic lubes were obtained by polymeristion of ethylene. It was only an accidental discovery of polyethylene (1933) by I.C.I. Scientists, Fawcett and Gibson, during their studies on behaviour of ethylene under high pressure that led to the birth of this polymer.

Butlerov, Lebdev, Gurwitch, Namyatkin were other scientists who served a lot to the polymer industry in general. Since the discovery, it has become the biggest commercial triumph in the petro-chemical industry. With Zieglor's development of complex catalyst system, another landmark was set up not only in polyethylene industry but in all polymerisation industries. Present day, the importance of polyethylene is undescribable, in a single statement in may be difficult to locate a field impregnable to its influence. An invincible chemical, as it is, leaves in doubt to name a country, which does not produce it.

Vinyl Polymers

Monomer Polymer

$H_2C{=}CH_2$ → $-(CH_2-CH_2)_n-$

Polyethylene

$H_2C{=}CH(OCOCH_3)$ → $-(CH_2-CH(COCH_3))_n-$

Poly(methyl acrylate)

$H_2C{=}CHCl$ → $-(CH_2-CHCl)_n-$

Poly(vinyl chloride)

$H_2C{=}CH(C_6H_5)$ → $-(CH_2-CH(C_6H_5))_n-$

Polystyrene

$H_2C{=}CH(CH_3)$ → $-(CH_2-CH(CH_3))_n-$

Polypropylene

$F_2C{=}CF_2$ → $-(CF_2-CF_2)_n-$

Teflon

$O_2 \rightleftharpoons \dot{O} + \dot{O}$ Oxygen Free Radicals

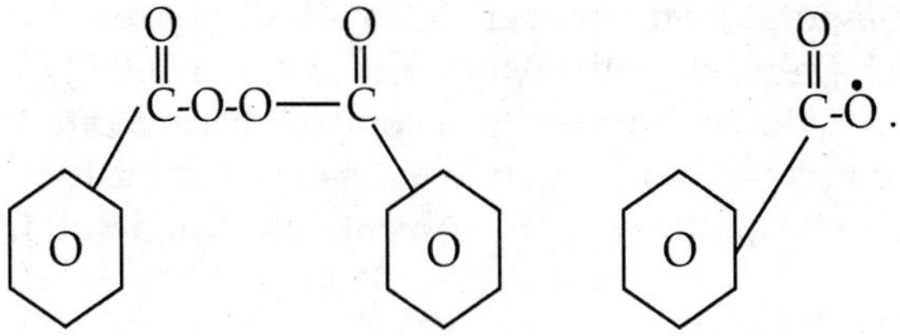

$C_6H_5-C(=O)-O-O-C(=O)-C_6H_5$ $\quad$ $C_6H_5-C(=O)-\dot{O}$. Benzyl Peroxide Radicals

$Cl_2 \rightleftharpoons \dot{C}l + \ddot{C}l$ (Photo Chemical)

Polythylene

Linear polyethylene (high density- HDPE) mw 200,000 – 500,000

Ultra high molecular weight UHMWPE mw – 3,000,000 – 6,000,000 stronger than Kevlar Branched polyethylene (low density - LDPE)

Industrially polyethylene is manufactured by three distinct processes, namely high, medium and low pressure. The products of these processes exhibit different properties as shown in Table 6.2. It may be appended that polymerisation can be conducted in gas phase or liquid phase (Solution/ emulsion/slurry).

Table 6.2: Properties of Polyethylene produced by different methods .

	High pressure	*Low pressure*	*medium pressure*
Crystallinity%	54-68	80-90	95
Density kg/M^3	915–925	940–965	950–965
Melting point°C	108–110	120–140	140–143
Brittle point °C	–70 and power	–70 (lower)	
Tensile strength kg/cm^2	120–16	220–450	400–450
Elongation% (cross wise)	400–700	100–1000	20
Dielectric strength	460–700	450–500	
Kv/mm	45–60	45–60	
Organic solvents affecting	Paraffins, Aromatics dissolve ; serviceable upto (50°C)	Paraffins Aromatics; service able upto (80°C)	
Branching and unsaturation per 1.000 C-atomes (as methyl groups)	20 –30	7–10	2
Melt index	0.2 to 0.3	0.to 4.0	
Resistant to heat	Upto 98°C	Upto 110°C	

6.2.1 High pressure polyethylene (Low Density polyethylene)

The Process owes its birth to ICI. The polymerisation is carried out under a very high pressure (pressures ranging upto 350 MP$_a$) in a temperature range of 200°C to 240°C. The reaction is credited to free radical mechanism. This being a chain reaction the exothermic heat of 100 kJ/mole has to be successfully dissipated. Otherwise the reaction mixture goes uncontrollable at a rate of temperature rise of 10°C to 12°C per 1% ethylene converted. Chain reaction to be perpetual, ethylene of onerous purity is essential. Such purity obviously is costly hence the solvent/ emulsion polymerisation has been opted.

6.2.1.1 Process Description of High Pressure Polyethylene

E. Ethylene
O. Oxygen
W. Waxes and Liquid Polymers
PE. Polyethylene
IIW. Hot Water Circulation
HP. High Pressure Reliever
LP. Low Pressure Reliever
C_1 & C_2 compressors

1. Reactor (Tubular)
2. Separator & Cyclone (High pressure)
3. Separator (Low pressure)

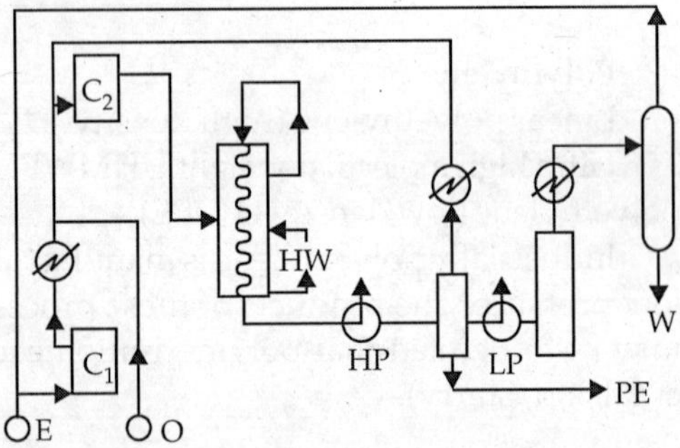

Fig. 6.2: Polyethylene (LD) Production.

Fig. 6.2 represents the process flow sheet. The main steps involved in the process are:

(*a*) Purification and compression of ethylene in two stages (50 bars and 3000 bars).
(*b*) Introduction of free radical initiators.
(*c*) Raising the reaction mixture to a temperature of 200°C and maintaining around 250°C-270"C.
(*d*) Release of pressure in a number of stages.
(*e*) Separation of polymer from unreacted ethylene which is recycled.

Polymer grade ethylene (99.9%) mixed with 600-800 ppm oxygen is compressed mainly in two stages c^2 and c^1 upto 350 MPa. The compressed gas is passed into a tubular reactor (1), (the reactor is tube-in-tube type), where heat dissipation is conducively carried out by circulating dowtherm or hot water. Another method of heat removal is to heat the incoming gas by out going mixture. The temperature of the reactants in the entering zone is around 150°C, but at the end of the coil, the temperature reaches 240°C. The temperature in the reactor outlet should never drop below 200°C, as the crystallisation of polyethylene begins, giving obstruction to flow. The reaction products are directed to pass into high pressure separator (2). The pressure of the gases, obtained from this stage is still above 35 MPa, hence, this after purification can go directly to second stage compression (c_1). For the high pressure separator, the melt goes into a second stage separator (low pressure) (3) where the remaining ethylene is expelled and routed to 1st stage (c_2) compressor, through purification system. In fact, after first high pressure separator, there may be a series of separators to catch polymer dust. The gases are rich with ethylene and oxygen but contain some carbon dioxide and formaldehyde. Hence purification is essential before the gas enters 2nd stage compression. A safety system is desirable while handling such high pressure gas, to actuate exit automatically, if disturbance in heat balance or flow obstruction appears in the reactor. It is always desirable to blow the whole system with nitrogen before the polymer gas enters the reactor. As high as 80% ethylene is recycled due to inklingly low conversions per pass. LDPE formed is collected from low pressure and high pressure separators. The gases leaving out of the system pass through a separator where some liquid polymers gets condensed.

LDPE can also be produced by solution polymerisation, where the pressure and temperature are aminably low order, 100 MPa, and temperature 190°C. The solvent constitutes a equivolume mixture of benzene and water. Oxygen as in the former case acts as initiator which is introduced into ethylene stream. After reaction, the products are separated, and unconverted ethylene goes back. This process gives a high conversion rate upto 20% per pass.

6.2.2 Medium Pressure Polyethylene (Phillips-Standard Oil of Indiana)

Phillips process is first commercialised in 1957. The polymerisation is conducted at a pressure of 5 MPa in solvent medium; either in a slurry fashion or in a solution. Phillips solution catalyst is prepared by muffling chromium-impregnated silica alumina-supports at selected conditions. Usually hexavalent chromium oxide on any inert base (silica-alumina) in a poorer solvent for polyethylene; will serve as polymer catalyst. Some active ingredients like calcium, barium, tungsten oxides are also used. Sometimes halogenated chromium oxide is prevalent in the industry to increase the molecular mass of the polymer. The polyethylene obtained from this system is linear and highly crystalline. There are other catalysts like nickel, cobalt, molybdenum oxides. With different catalysts, the conditions of the reactor are likely to vary. Medium pressure process was developed by Phillips based on chromium oxide catalyst. Standard oil of Indiana pioneered with molybdenum catalyst on (δ)-alumina. Though these are predated to Ziegler's process, industrial exploitation was rather slow. These processes are capable of yielding high molecular weight polymers than even the reputed Ziegler's, hence inevitably require chain stoppers. The swapping of these processes for making polypropylene failed considerably, perhaps due to the nonavailability of a sound mechanism. *A* flow sheet on this technology is presented in Fig. 6.3. Medium pressure process employs a pressure of 35 MPa and a temperature of 130 – 170°C. 0.1% Cr_2O_3 treated with HF gives a molecular mass of 45,000, otherwise the polymer mass does not exceed 20,000. Phillips uses cyclohexane as solvent and the concentration of ethylene in solution is kept around 5%. The solvent serves the dual role as a solvent to ethylene and polymer, and permit the chain length to grow irrespective of the presence of chain breakers. Viscosity, heat transfer problems and control of ethylene consumption are also attended successfully by solvents.

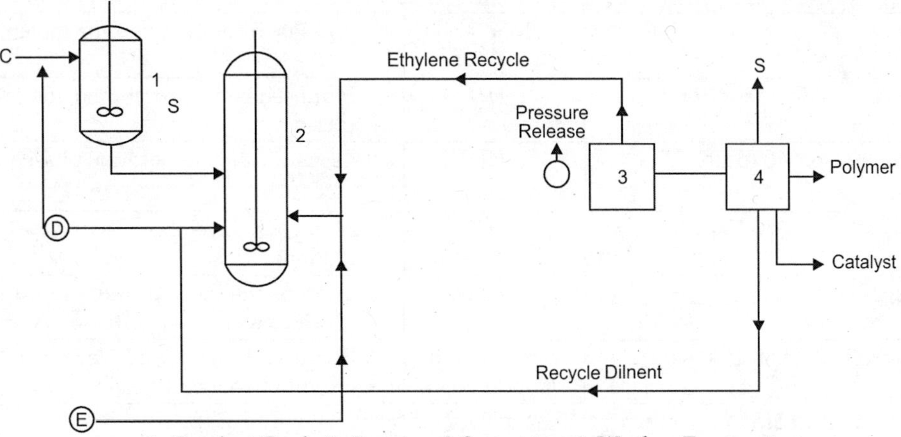

1. Catalyst Tank, 2. Reactor, 3 Separator, 4. Washer-Evaporator, C: Catalyst, D : Diluent, E : Ethylene, S : Solvent

Fig. 6.3. Polyethylene Production (Medium pressure process).

6.2.2.1 Phillips Particle Form Process[9] (PF)

In this process a double loop reactor is employed and polymer is continuously drawn off from the bottom of the reactor in slurry form. The process employs the catalyst similar to that of Phillips solution process, except the support is pure silica. Because of high efficiency of catalyst, de-ashing is not required.

6.2.3 Low Pressure Polyethylene (Ziegler's Process)

In recent versions, solution process gives high solids (40%) content with no-ashing problem. In the slurry process, higher yield of product per given reactor is an advantage. Further, because of poor solvency nature, the polymer or copolymer separates from the diluent in fine particles. Thus the viscosity of solution does not increase as seen in solution process. Of course slurry process takes a longer time (2-4 hours). Further all slurry processes are credited with yielding very high mole weights, perhaps of no commercial value, hence chain stoppers are inevitably sent into the system.

Karl Ziegler developed a composite catalyst system for polymerising ethylene at low pressure and low temperature. The conditions required are almost ambient conditions, viz., atmospheric pressure and a temperature of 60°C, though it was a profounding discovery the manufacturers were reluctant to replace the older technology. Providentially, Giulio Natta extended the complex catalyst to get polypropylene of high molecular weight, a linear polymer, of course stereo-regular. This prompted other to work on these catalysts and ultimately all types of polymers have become a reality. The discovery thus paved the way to polymerise different monomers, where free radical polymerisation failed. Table 6.3 shows the different technologies for the production of HDPE.

6.2.3.1 The Catalyst and Mechanism

The composite catalyst is formed by reacting tri-alkyl aluminium and titanium tetrachloride. In general alkyl aluminium (ethyl, propyl, butyl) with titanium tri or tetrachlorides is found to be active. It is likely that the active sites generated owing to the electron deficient complex, favour the bond formation with a monomer. With the result, propylene (or olefin) inserts itself in between the titanium and alkyl group. With the loosely bound olefin, the vacancy of electron deficiency exists favouring unending chain formation in no time.

Table 6.3: HDPE Process-Catalysts and Reaction Conditions

Stamicarbon	HD and MD	Catalyst directly to reactor solvent medium. d : 0.93 to 0.97
Unioncarbide, 20 Bars 85 to 100°C	HD	Fluidised bed (chromium oxide cat) gas phase
Montecatini Edison, 50 to 100°C 1 to 15 atm	HD	Organic catalysts Solution phase
Phillips petroleum 85 to 150°C;7 to 35 bars	HD	Slurry with cyclohexane Cr_2O_3.(0.5%) on alumina.
Snam progetti	HD	Titanium halide and Aluminum alkyl hydride (Slurry process) aliphatic solvent.
Solvay and CIE 55 to 85 C 30 bars	(Homo and Copolymers)	Supported catalyst Ethylene, hydrogen com-onomer in solvent hexane.
Friedrich Uhde GMBH 10 bars 80 to 90°C	HD	
Mitsubhishi Chemical Industries Ltd.	HD	Al and Titanium complex copolymer

30 to 90°C 5 to 15 bars		(slurry), solvent
Montedison SPA. 50 to 100°C loser 15 bars	HD	Supported solid cat. of Ti+alkyl aluminium
Union Carbide 80 to 10°C 21 bars	HD	Cr_2O_3 fluidised bed
BASF	HD	Ziegler type
B-P Chimie	HD	Ziegler-super active catalysts.

The complex is obtained in a number of steps. Firstly, direct reaction of aluminium and ethylene to yield triethyl aluminium :

$$Al + 3\,C_2H_4 + 3/2\,H_2 \rightarrow Al\,(C_2H_5)_3 + 344\text{ kJ}$$

This synthesis takes place at a temperature of 110-120°C under a pressure of 2.5 MPa. Alternatively, it can be produced by reacting aluminium with ethyl chloride, afterwards treating aluminium ethyl chloride with sodium to expel chlorine atom. The ratio between ethyl aluminium and titanium tetra chloride is significant in determining the size of polymer. The ratio is usually kept in between 1 : 1 to 1 : 1.2. This delivers a molecular mass of 70,000 to more than a lakh. By increasing the ratio of alkyl aluminium to halide a polymer of higher molecular weights *i.e.* a million are likely. Even diethyl aluminium is suitable to make polymers of lower masses. Polymerisation can be done either in continuous or in batch fashion.

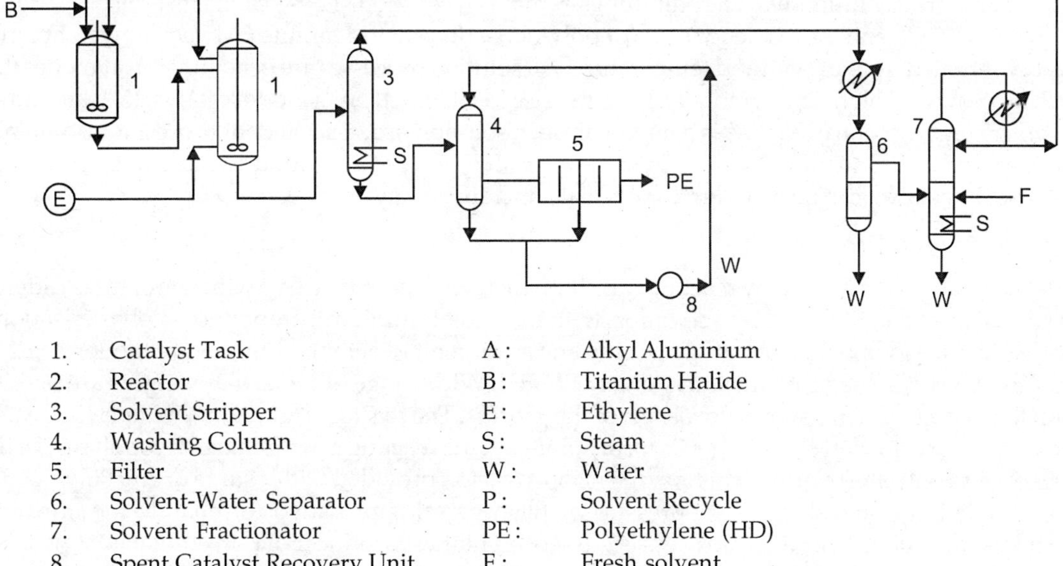

1.	Catalyst Task	A :	Alkyl Aluminium
2.	Reactor	B :	Titanium Halide
3.	Solvent Stripper	E :	Ethylene
4.	Washing Column	S :	Steam
5.	Filter	W :	Water
6.	Solvent-Water Separator	P :	Solvent Recycle
7.	Solvent Fractionator	PE :	Polyethylene (HD)
8.	Spent Catalyst Recovery Unit	F :	Fresh solvent

Fig. 6.4: Polyethylene Production – Ziegler Process.

6.2.3.2 Production of HDPE

Fig. 6.4 shows the outlines of the process. In the catalyst vessel (1) a fresh and dilute solution of composite catalyst (max. 0.1 gm/litre) in pentane or hexane is made and this acts as a reservoir too. The solution is fed continuously into reactor (2) where solvent and ethylene streams enter. A stirrer in the reactor is helpful in homogenising the mixture. The polymer mixture after reaction, is allowed to enter a flasher (3), where unreacted monomer and inerts are vented out. The solid material is given a thorough wash (4) with water to decompose the catalyst. The strippers (6 and 7) separate the organic solvent from water - hydrocarbon azeotrope and solvent is recycled. The polymer material is washed many times, either with water or with methanol, after which it is filtered (5) and dried. Catalyst is obtained in a slurry and is recovered (8) for usage. Additives for preserving the quality of polymer are added at the stage of granulation. It is found that by alcoholic wash, the ash content of the polymer decreases very much. Before start of the operation all units are submitted to nitrogen breathing, and is essentially important to see nowhere water or air seeps in. Fig. 6.5 shows typical reactor conditions encountered in various HDPE process.

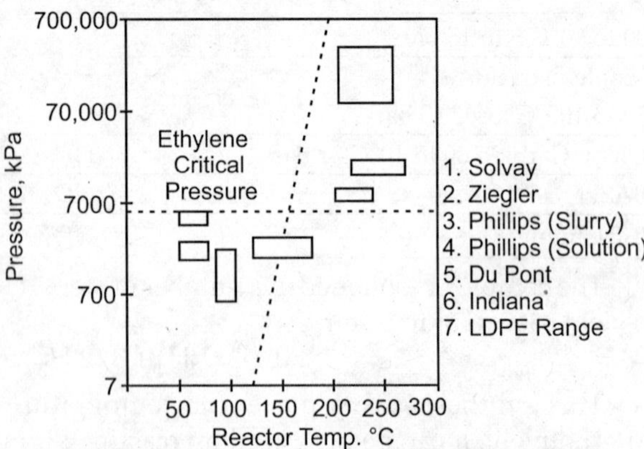

Fig. 6.5: Temperature-Pressure Ranges in HDPE Processes.

6.2.3.4 Du Pont Process

This is a solution process. Du Pont uses soluble Ziegler catalysts, although it differs from Phillips process in a true solution way, *i.e.*, all things remain in solution. These catalysts are comprised of $TiCl_4$, VCl_5, $VOCl_3$ and the residence time for polymerisation is 3 to 4 minutes. Molecular weight can be easily controlled by hydrogen and temperature. All solution processes are generally run at moderately high pressure and high temperatures. This gives rise to increased efficiency of catalyst (500 kg/gm). A polymer concentration of 35-40% is common in such solution process. Du Pont process is also suitable for co-polymer production. These polymers from this process are found to have less ash content because the catalyst can be dislodged very easily and effectively.

6.2.3.5 Solvay Process

The catalysts are formed by reacting transition metal compound ($TiCl_4$) with one of three different supports at a time, followed by activation with organo-aluminium compounds. Polymerisation is carried out in a loop type reactor using n-hexane as diluent. Reactor conditions are moderate; 25-35 bars and 50-90°C. The resultant slurry is centrifuged. Wide range of molecular weights are possible with this process. Fig. 6.6 shows flow sheet of the process. The reactor is tubular provided with a jacket for circulating cold water. Hydrogen is incorporated in the reacting mixture, so that copolymerisation is smoothly controlled. Further, this is a closed loop reactor, provided with a surge drum. Polymer fluff is available in hexane slurry. A depressuriser flash vessel is provided to separate the unreacted reactants for circulation. Solvay process is the second highest producer of polyethylene.

6.2.4 Revolution in Polyethylene

Polyethylene, being a major plastic of commerce, received a wide attention of researchers and producers. Incessant quests have been resulting in ameliorating performance of the catalysts and simplified techniques in processing. Thus, meeting exigency of the industry. The fruits of the searches could be seen in the appearance of polymers of widely different structures due to copolymerisation and cross linking.

On catalyst's side the following developments have been widely acclaimed.

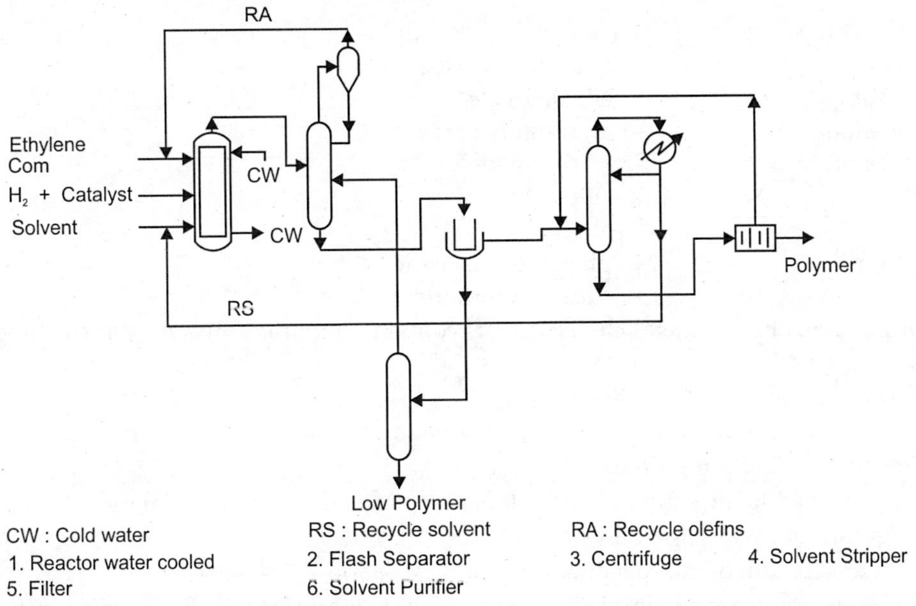

CW : Cold water
1. Reactor water cooled
5. Filter
RS : Recycle solvent
2. Flash Separator
6. Solvent Purifier
RA : Recycle olefins
3. Centrifuge
4. Solvent Stripper

Fig. 6.6: Solvay Process for HDPE

6.2.4.1 Improvement In Catalyst

In the production of HDPE, conventional Ziegler's catalysts are found to have certain disadvantages like low efficiency (< 1500 gm polymer per[10] gm of transition metal) and inability to complete removal of catalyst residues from polymer. Further, the polymer obtained from these catalysts exhibit inflexibility for a wide range of applications and any modification in molecular weight has to be carried out in a separate step. These first generation catalysts of Ziegler-Natta complexes have been gradually on the wane, with the second generation supported or super active catalysts taking their place. Perhaps, these may be quoted as the most important category of ethylene polymerisation catalysts, from the stand point of recent developments. These active supported catalysts are gifted with incredible yields of polymer (6000 kg/gm of transition metal). Many new catalysts are capable of producing 100 to 500 kg of polymer per gm of catalyst[11]. It is further stated that the ash content of polymers with these catalysts is surprisingly low (2 to 10 ppm) against conventional catalysts that give 1700 ppm.

Stami carbon process for poly ethylene requires about 1025 kg ethylene, solvent about 36 kg, steam about 1150 kg. This process can produce a polymer of Melt Index 0.2 to 30 and the density can be varied from 0.93 to 0.97. Union Carbide using fluidised bed reactor, in gas phase using comonomer to 0.02% can produce ashless product because of high efficiency of catalyst *i.e.* 300 tonnes of polymer can be produced per lb of chromium.

6.2.4.2 Catalyst Compositions

Many producers have been specifying their catalysts and elate performance. However, these are basically separated into two kinds, namely based on metal-oxide systems and reducible metal halide-metal alkyl catalysts or supported catalysts.

Phillips and Indiana processes pioneered in using metal oxide catalysts. But now non-deashing processes using metal oxide based catalysts are in the forefront. As an example, organochromium compounds on silica supports as developed by the Union Carbide is achieving a rapid commercialization.

Chiefly, catalysts are composed of metal oxide pairs as shown below :

Chromium - aluminium oxide
Chromium - silicon oxide
Molybdenum - aluminium oxide
Molybdenum - silicon oxide
Chromium - silica-aluminium
Titanium - aluminium oxide
Vanadium - phosphorous oxide
Titanium - silicon-chromium oxide etc.

Reducible metal halide-metal alkyl catalysts contain titanium compounds as the fundamental ingredients such as:

- Fischer Catalyst (Al, $AlCl_3$, $TiCl_4$)
- Silicon-aluminium-titanium halide compounds
- Ether containing compounds
- Titanium halide-aluminium halide ether complex and magnesium halide
- Magnesium containing catalysts
- $Mg(OH)Cl_2$ with titanium halide or Vanadium halide
- Similar formulation on nickel, tin, vanadium, zinc, etc. were described vividly by Sittig[12].

Supported catalysts usually contain a reaction product of an inorganic compound of polyvalent metal having at least a valency of three, and a solid bivalent metal hydroxylated inorganic phosphate or hydroxy chloride giving hydroxylated phosphate support with a metal, a hydride or an organometallic compounds of groups IV-b, V-B, and a metal halide of group Ill-b or V-B.

$$Mg(OH)_2 + TiCl_4 \rightarrow MgOTiCl_3 + HCl$$

$$2MgCl_2 + TiCl_4 \rightarrow MgCl_2\,Mg(TiCl_6)$$

$MgCl_2$ Mg ($TiCl_6$) acts as a catalyst while Mg forms support.

On the process side two exciting developments are on the scene :

Firstly, the fluidized bed reactor developed by Union Carbide, using gas-phase polymerization.

Fluidized bed reactor developed by Union Carbide [13,14] uses gas phase with a diluent like hydrogen. Catalyst in dry powder form is introduced into the reactor. Co-monomer can also be used. Because of small amount of catalyst usually it may not be necessary to dislodge from the polymer. DSM version[15] of Zieglers route departs from usual practice in that it employs unsupported dissolved catalyst that remains with polymer giving 100 kg polymer per gm. catalyst. Polymer from such gas phase process (fluidized) is in powder or granules-form. Molecular weight can be controlled by diluent. Monomer to co-monomer ratio and operating conditions basically dictate isotacticity and melt index. Fig. 6.7 shows the outlines of the process. Pressure and temperature in the reactor are 20 bars and less than 100°C. Gas phase fluidization is credited with reducing fixed and operating costs by more than 10%.

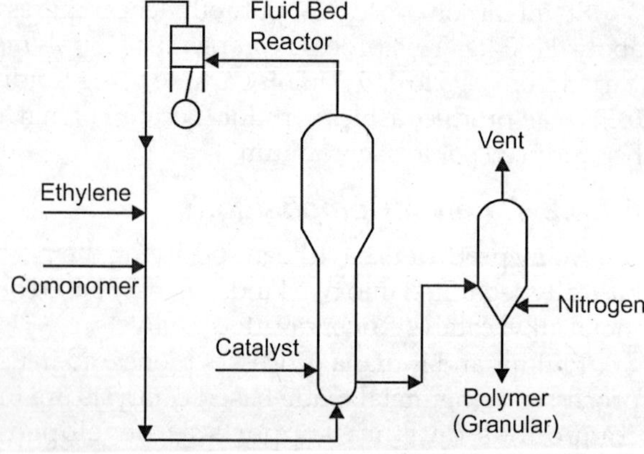

Fig. 6.7: Polyethylene production (fluidised bed).

A highly active dissolved[16] catalyst in solution starts HDPE production and goes by the name Compact Process, using a special reactor, ethylene is compressed, absorbed with steam and an (α)-olefin as a comonomer, passes into an aliphatic solvent. The solvent along with the contents is cooled to – 40°C before the mixture enters the reactor. Under adiabatic conditions, the polymerization proceeds. The product is said to have a density range of 0.93 - 0.97 and melt index 0.2 to 0.35 gm/10 min.

6.2.5 Developments in LDPE Polymers (LLDPE)

LDPE is a linear branched homopolymer of ethylene produced by high pressure, free radical polymerisation. Union Carbide had done extensive work in developing a fluid bed reactor for HDPE, under trade name Unipol. Lately (1980) LLDPE, a copolymer of ethylene, in which branches are provided by butene-1, hexene-1, or octene-1, was introduced by Union Carbide. It is a random polymer, having more short chains with no further branching, differing from LDPE which has fewer but longer and widely branched chains. Its physical properties lie between LDPE and HDPE. LLDPE is a copolymer of ethylene and butene-1 produced by high pressure process, while butene-1, pentene-1 or hexene-1 subscribed to side chains in slurry process. Ziegler type catalyst is used. A new variety Ziegler type catalyst (Uno-transition metal catalyst), comprises of Cr_2CO_3, tetra-isopropyl titanate and ammonium silica fluoride $(NH_4)_2 SiF_6$ on silica support has been developed for LLDPE, which yielded a low density as 0.917 gm/cm^3. Another type catalyst titanium trichloride ($TiCl_3$) and magnesium chloride activated by tetra hydrofuran consists of (THF) on triethyl aluminium. THF with magnesium titanium complexes on silica support are actively used in vapour phase process. LLDPE is consumed in a major proportion as a blend 30 : 70 or 50 : 50 with HDPE.

6.2.5.1 *High Pressure Process for LLDPE*

In this process reaction takes place under super critical conditions, whereby the homogeneous gas phase behaves like a liquid. A reactor of 1500 metres long with 50 mm dia is used. The reaction conditions are : temperature range 130 to 180°C, 1500-3500 bars pressure. The reaction time lasts 60-300 seconds.

LLDPE can also be produced either by the low pressure or the high pressure processes, and also in solvent process as developed by Du Pont, DSM and others, or can be produced using slurry technique developed by Phillips and licensors. End product appears in wide range of resins.

Fluidized bed[7] process developed by B.P. Chemie used superactive Ziegler catalysts tailored for the process. By this technique a remarkably large range of products are obtained just by changing the operating conditions. *A* variation in comonomer to ethylene ratio can change the density of polymer from 0.965 to 0.919. Diluent introduction brings a large change in melt index.

6.2.5.2 *Ultra high molecular weight polyethylene*

Ultra high molecular weight polyethylene (UHMWPE), also known as high modulus polyethylene (HMPE) or high performance polyethylene (HPPE), is a *thermoplastic*. It has extremely long chains, with molecular weight numbering in the millions, usually between 3.1 and 5.67 million. The high molecular weight results a very good packing of the chains into the crystal structure. This results in a very tough material, with the highest *impact strength* of any thermoplastic presently made. It is highly resistant to corrosive chemicals, with exception of oxidizing acids. It has extremely low moisture absorption, very low coefficient of friction, is self lubricating and is highly resistant to *abrasion* (10 times more resistant to abrasion than Carbon Steel). Its coefficient of friction is significantly better than nylon and acetal, and is comparable to teflon, but UHMWPE has better abrasion resistance than teflon. It is odorless, tasteless, and nontoxic, UHMWPE finds use in high *modulus* fibres (for example,

Spectra or *Dyneema*) for bulletproof vests. Due to its low friction and wear resistance it is used in industrial impact, wear, and sliding applications in both normal and corrosive environment. It is also used in orthopaedic implants (artifical hips, knees). UHMWPE is a type of olefin and, despite relatively weak Van der Waals bonds between its molecules, derives ample strength from the length of each individual modecule. It is made up of extremely long chains of *polyethylene*, which all align in the same direction. Each chain is bonded to the other with so many Van der Waals bond that the whole can support great *tensile* loads.

When formed to fibers, the polymer chains can attain a parallel orientation greater than 95% and a level of crystallinity of up to 85%. In contrast, *Kevlar* derives its strength from strong bonding between relatively short molecules. Its melting point is around 144 or 152 degrees *Celsius*. It becomes *brittle* at temperatures below -155° C. The simple structure of the molecule also gives rise to surface and chemical properties that are rare in high-performance polymers. For example, the *polar* groups in most polymers easily bond to water. Because olefins have no such groups, UHMWPE does not absorb water readily, but it also does not get wet easily, which makes bonding it to other polymers difficult. Similarly, *aromatic* polymers are often susceptible to aromatic solvents due to *aromatic stacking interactions*, and effect *aliphatic* polymers like Dyneema are also immune to, it is very resistant to water, moisture, most chemicals, UV radiation, and micro-organisms.

6.2.5.3 Production

These are molecules of *polyethylene* which are several *orders of magnitude* longer than familiar, *high density polyethylene* due to a synthesis process based on *metallocene catalysts*. HDPE molecules generally have between 7000 and 1,800 monomer units per molecule, while UHMWPE molecules tend to have 100,000 to 250,000 monomers each. The material is also known as high modulus polyethylene (HMPE) or high performance polyethylene (HPPE). To make fibers like *Dyneema*, they are dissolved and drawn into fibers as the solvent evaporates, causing the polymer chains to orient in the direction of the fiber.

6.3 POLY PROPYLENE

Poly propylene, though a late comer (1960) into the field of polymers, it has occupied the commercial scene at an outstandingly rapid pace. Unlike polyethylene which can be produced by free radical process, this cannot be produced, because of the reactivity of allylic-hydrogen atom, hence only a process analogous to that of high density polyethylene is indispensible.

Earlier use of propylene was restricted up to tetramers. The oligomerisation proceeds with considerable ease in presence of Friedel-Crafts catalysts. The reaction product usually consists of many isomers of tetramers. Hence it was used as good commercial process for the production of polymer gasoline, further the tetramer (Dodecene) has become a commercially important component for detergent manufacture.

Polypropylene is commercially produced by only one method that is anionic polymerisation. Phillips process failed in this case because of low crystallinity (20%). The process is accomplished in hydrocarbon solvents with the same Ziegler type composite catalyst, but with a difference, titanium trichloride is opted over tetrachloride. Polypropylene produced by tetrachloride contains less isotactic polymer (35-40%) hence trichloride has been recommended; which yields about 80-85% stereo regular polymer. $TiCl_3$ exists as porous crystals exhibiting active sites uniformly. $TiCl_3$ diethyl aluminium chloride yields minor atactic form only; 2:1 mole ratio of DEAC to $TiCl_3$ is used industrially to give good amount of stereo regular polymer. Higher amount of $TiCl_3$ spoils the active sites of catalyst. Production based on traditional catalysts involved after polymerisation, removal of catalyst dust and separation of isotactic polymer from atactic polymer. During deashing the unreacted monomer and solvent get poisoned, hence extensive purification is appended. The new catalyst developed by Montedison[18] and Mitsui Petrochemicals, allow polymer structure, yield selectivity and rheology to be controlled; resulting in simplification of the process.

Mitsui Petrochemicals catalyst system result from the reaction of three major components titanium supported on magnesium halide, organometallic compound and a Lewis base.

Biaxially oriented polypropylene film (BOPP) is produced by orienting long chain molecules in two directions. BOPP has a configuration ABA or BAB where A is PP and B is PP copolymer. This exhibits a density of 0.905 gm/cc. Because of its excellent clarity and high barrier for moisture, it has become an excellent packing medium. It is manufactured by using stereo specific catalysts during polymerisation to give pre-programmed arrangement of isotactic form. The following concerns are manufacturing this film :

BOPP Film

M.M. Rubber, Madras	800	tonnes/year	1981
Cosmos Films, Aurangabad	30,000	tonnes/year	1982
United Polypropylene Ltd.	800	tonnes/year	1983
Gujarat Propack	800	tonnes/year	1987
Industrial Development Corp. Haryana	2000	tonnes/year	1989
Punjab State Industries			
Devp. Corp. Punjab	2000	tonnes/year	1989

Usually 10 to 40 kg of polymer results under varying conditions of pressure 7 to 15 bars and time of polymerisation at 70°C, though liquid monomer gives highest yield. With the above catalyst systems, isotacticity goes upto 90 to 95% with a narrow particle size distribution of 300 to 400μ. Presently three technologies are available :

1. Diluent slurry 2. Gas phase 3. Bulk (Boiling reactor/tubular loop reactor).

Each of these have relative merits and demerits.

6.3.1 Diluent Slurry Process

Based on traditional catalysts, yields and isotacticity are low. Operating conditions are not severe. Range of copolymer production is narrow and the heat exchange problems do not permit big reactors.

6.3.2 Gas Phase Process

Dry polymer is obtained with reactor volumes increasing for the same amount of polymer production, as the efficiency of catalyst regarded as constant, it increases additional burden from to that of bulk reactors. However deashing problem is eliminated.

6.3.3 Boiling Reactor

In this, heat of polymerisation is absorbed by boiling monomer. Agitation may become difficult due to uneven phases namely polymer and monomer.

6.3.4 Catalyst Preparation

Titanium trichloride is highly deliquescent hence, its contact with air would render it fribble. Titanium is ground to fines in an atmosphere of inert nature. The second component, triethylaluminium or diethylaluminium chloride is used to complete the composite catalyst system. The system must be out of contact with oxygen and water. In addition to that Ziegler-Natta[19] catalysts of 1st generation have also inherent disadvantages in their consumption pattern. In case of stereo regular isotactic polypropylene, titanium tetrachloride reduced with metallic aluminium activated with O-dichloro benzene ether or phosphorous compounds with stereo regulators like methylene diamine is opted. The yield with this complex is many times more than with the 1st generation catalysts with 96% stereo regularity[20]. $TiCl_4 - D_1\ MgCl_2 - AlR_3 - D_2$ (where D_1 and D_2 are[21] centers as stereo additives ; can yield about 300 to 800 kg PP per gm. titanium with a high degree of stereo regularity).

6.3.5 Process Description

The reaction is carried out at 10-20 bars in the temperature range of 65-70°C. Polymerisation is done in a medium for which a solvent is chosen to act either as solvent or medium for slurry. Though conventionally slurry process is more pronouncive, solvents like heptane or gasoline with no unsaturates is preferred. Propylene of very high grade purity (99.9%) is essential, propane-propylene (35%) mixture is also used, at times. Polymerisation is carried out in batch or continuous wise. Desirable aluminium to chloride ratio is 3 : 1 to 2 : 1 in continuous process. Propylene concentration in solvent is kept very high. The heat of polymerisation which is only 60% of polymer heat of polyethylene makes safe for easy handling, in fact no external heat dissipation is required. One way of extraction of heat is by allowing propane to evaporate, when the feed gas is propane-propylene.

Fig. 6.8 illustrates the process. A 10% solution composite catalyst is made in a solvent usually heptane and kept in tank (1). The solution is proportionally charged into reactor (2), where olefin and some more solvent are admitted. The reactor is kept under required pressure and temperature for about 3 to 4 hours to complete the reaction. Feed gas and catalyst thoroughly mix due to agitation. After the reaction, products are transferred into a stabiliser (3), where pressure is decreased, to let off unreacted gas. The catalyst is decomposed by means of alcohol. All the polymer product is centrifuged (4) and the solvent-alcohol mixer is separated by steam stripping in a stripper (5). Recovered solvent is distilled for purification; to begin circulation. A separate stripper-purifier section is essential for this. The heavy polymer with adhering solvent is stripped in another stripper (6) by means of steam, and the recovered solvent mixture is processed for alcohol and solvent in section (7). Polypropylene collected from the bottom of the stripper (6), is further centrifuged, mixed with stabilisers and sent to storage. Polypropylene requires about 1.1 tonnes of monomer per each tonne.

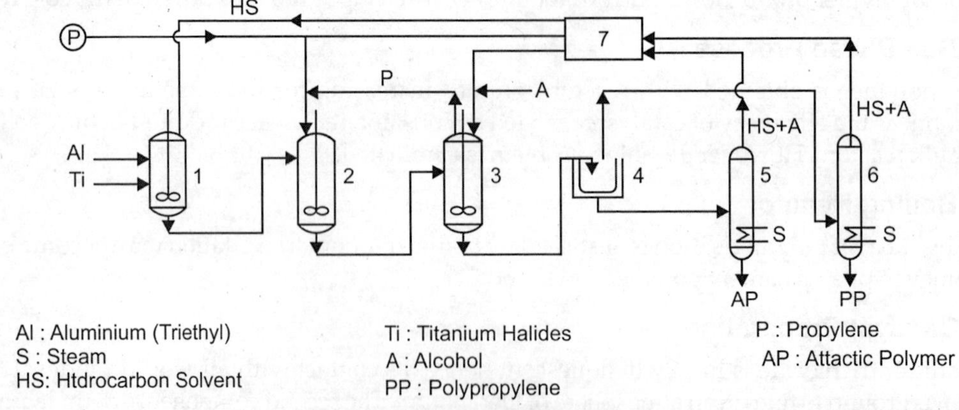

Fig. 6.8: Production of Polypropylene.

Polypropylene being less denser than polyethylene (0.91 to 0.92), it has become an attractive thing for making water-floating objects. Ropes made out of polypropylene exhibit the same strength as steel wires, still floating and non-corroded, hence appealed for water floating ropes, tugs and vessels. Fibres of polypropylene can be knitted easily, all these advantages have made it as an impregnable rival to polyethylene.

LLDPE and high impact polypropylene can be produced by gas phase fluidised bed process. Unipol and instituted 5.5 M^2 T plants on this technology CdF Chimie has installed plants of 2 M^2T capacity.

Poly propylene Friedrich Uhede Process (FWW) uses Ti Cl_3 and organic aluminium as catalyst. Hercules Process uses Ti tetra chloride with alkyl aluminium in the temperature range of 50-100°C Montedison process uses Ti-tri chloride with alkyl aluminium at high pressure.

6.3.6 More Information about Polypropylene

Polypropylene or polypropene (PP Chemical name 1 methyl ethylene) *a thermoplastic polymer*, used in a wide variety of applications, including food packaging, *textiles*, laboratory equipment, loudspeakers, automotive components, and *polymer banknotes*. An *addition polymer* made from the monomer *propylene*, it is unusually resistant to many chemical solvents, bases and acids.

Chemical & Physical properties

Most commercial polypropylene has a level of crystallinity intermediate between that of low density polyethylene (LDPE) and high density polyethylene (HDPE); its Young's modulus is also intermediate. Although it is less tough than LDPE, it is much less brittle than HDPE. This allows polypropylene to be used as a replacement for engineering plastics, such as ABS. Polypropylene has very good resistance to fatigue, so that most plastic living hinges, such as those on flip-top bottles, are made from this material. Very thin sheets of polypropylene are used as a dielectric within certain high performance pulse and low loss RF capacitors the density of amorphous PP is 850 kg/cu.m and crystalline 950 kg/cu.m.

Polypropylene has a melting point of 160 degrees Celsius and degradation is at 286°C. Food containers made from it will not melt in the dishwasher, and do not melt during industrial hot filling processes. For this reason, most plastic tubs for dairy products are polypropylene sealed with aluminium foil (both heat-resistant materials). After the product has cooled, the tubs are often given lids of a cheaper (and less heat-resistant) material, such as LDPE or polystyrene. Such containers provide a good hands-on example of the difference in modulus, since the rubbery feeling of LDPE with respect to PP of the same thickness is readily apparent.

A rubbery PP can also be made by a specialized synthesis process. Unlike traditional rubber, it can be melted and recycled, making it a thermoplastic elastomer. To produce a rubbery polypropylene, a catalyst can be made which yields isotactic polypropylene, but with the organic groups that influence tracticity held in place by a relatively weak bond. After the catalyst has produced a short length of polymer which is capable of crystallization, light of the proper frequency is used to break this weak bond, and remove the selectivity of the catalyst so that the remaining length of the chain is atactic. The result is a mostly amorphous material with small crystals embedded in it. Since each chain has one end in a crystal but most of its length in the soft, amorphous bulk, the crystalline regions serve the same purpose as vulcanization. An important concept in understanding the link between the structure of polypropylene and its properties is tacticity. The relative orientation of each methyl group (CH_3 in the figure at left) relative to the methyl groups on neighboring monomers has a strong effect on the finished polymer's ability to form crystals, because each methyl group takes up space and constrains backbone bending.

Unlike most other vinyl polymers, useful polypropylene can not be made by radical polymerization. The material that results from such a process has methyl groups arranged randomly, and so is called atactic. The lack of long-range order prevents any crystallinity in such a material, giving an amorphous material with very little strength and few redeeming qualities.

A Ziegler-Natta catalyst seems to be able to limit incoming monomers to a specific orientation, only adding them to the polymer chain if they face the right direction. Most commercially available polypropylene is made with titanium chloride catalysts, which produce mostly *isotactic* polypropylene. With the methyl group consistently on one side, such molecules tend to coil into a helical shape; these helices then line up next to one another to form the crystals that give commercial polypropylene its strength.

More precisely-engineered *Kaminsky catalysts* have been made, which offer a much greater level of control. Based on metallocene molecules, these catalysts use organic groups to control the monomers being added, so that a proper choice of catalyst can produce isotactic, *syndiotactic*, or atactic polypropylene, or even a combination of these. Aside from this qualitative control, they allow better quantitiative control, with a much greater ratio of the desired tacticity than provious Ziegler-Natta techniques. They also produce higher molecular weight than traditional catalysts, which can further improve properties.

6.4 POLYVINYL CHLORIDE

This is one of the most widely used large tonnage thermoplastic credited with diverse applications. Two types of homopolymers are produced one being flexible and the other a rigid type. Flexible polymers have a good elasticity possessing a density of 1150 to 1600 kg/m^3. Rigid type polymers exhibit a density of 1150 to 1800 kg/m^3 with a good tensile strength 42-60 MPa. Rigid polymers are suitable for making articles of hard type sheets, papers, gears etc. Both the types show low crystallinity due to random orientation of chlorine atom. Polyvinyl chloride is obtained by radical polymerisation leading to head to tail configuration as shown here. PVC is molded into phonograph records and is good for coatings and it is also used in the form of sheets.

$$-CH_2-\underset{\displaystyle Cl}{\underset{|}{CH}}-CH_2-\underset{\displaystyle Cl}{\underset{|}{CH}}-CH_2-$$

Properties of VCM:

Boiling point	=	– 13.8°C
Flash point	=	– 42°C
Melting point	=	– 153.5°C
Sp. gr.	=	0.98
Explosive limits in air (U.)	=	22%
Volume of air (L.)	=	4%

PVC is hard material which curtails its utility, though it can provide an array of products when proper plasticization is applied. Different plasticizers with varying amounts can yield stiff or flexible materials. Di-octyl phthalate and other phthalate esters are generally used in the much formulations. Epoxide oils are secondary type plasticisers and their action is more towards stabilising rather plasticising ability.

It can be used with or without plasticisers. It is white amorphous solid, soluble in chlorinated solvents, further it is very easily welded at 200°C, with rods made of PVC. However, it suffers from low ultimate temperature (65°C) at which articles made of it can be used. To overcome this difficulty vinyl chloride is copolymerised with vinylidene chloride or acrylonitrile. Vinylidene chloride is a comonomer upto 20%, in copolymer with vinyl chloride, and the product is known as Saran. These copolymers soften around 185-200°C and they are not easily soluble in organic solvents. Processing of vinylchloride copolymers is much easy, and are suitable for making fibers such as saran, or floorings etc.

PVC can also be further chlorinated to manufacture anti-corrosive lacquers and enamels. A special fibre is produced by wet spinning of PVC in acetone. This fibre is found to get charged easily when in contact with skin, which permitted its use in special applications. PVC with 10% acrylonitrile copolymer dissolves in acetone, which when spun, yields flame resisting fibres.

6.4.1 Polymerisation of Vinyl chloride

PVC can be produced by bulk polymerisation initiated by sunlight. However due to the handling difficulties of the physical form of polymer, and high heat disposal rates, the conversion never exceeds 60% and the resultant polymers are of low molecular mass. Hence prepolymer technology is employed(22).

Gel effect is noticed with increase in polymerisation, up to 50% conversion the rate of polymerisation increases but soon decreases, this necessitated prepolymer technique. Though PVC is supposed to be linear polymer still branching is noticed.

In this method, first pre-polymerisation is carried out in essentially liquid medium (autoclave) to 7% conversion. The next operation consists of polymerising this prepolymer by dilution with monomer and 'beads' grow to high molecular product. By this way polymers of any desired type are produced. Presently this technology is not used anywhere.

6.4.1.2 Suspension Polymerisation (25)

Superficially it resembles emulsion polymerisation. Monomer is dispersed as fine droplets (50 to 250 (μm) in the aqueous phase. Agitation is required to keep the monomer in a finely dispersed state, and to protect colloidal formations. Catalyst is insoluble in aqueous phase. Peroxides of benzyl, laural, p-chlorobenzoyl or isopropyl carbonate or azo compounds initiate the polymerisation. Suspension agents are water soluble agents, like polyvinyl alcohol, starch, gums, gelatins etc. and are found useful. Polymer is not soluble in monomer, while monomer is soluble in polymer. The reaction mechanism and kinetics in all suspension polymerisations is similar to bulk polymerisation.

6.4.1.3 Emulsion Polymerisation(24)

Monomer is dispersed into water, which is emulsified with surface active agents like ammonium salts of higher fatty acids, sulphonates, palmitates and sterates. A solution of the following type is used. Water soluble initiator, like persulphate is also added into the aqueous phase along with a buffer to maintain P_H around 4.

Water	100 parts
Emulsifying agent	0.5—1
Buffer	0.05—0.1
Ammonium persulphate	0.25
Sodium hydrosulphite	0.2
VCM	55—90

Aqueous phase consisting of emulsifying agent with soluble initiator and buffers is charged into an autoclave. VCM is discharged into the emulsion in droplets (1 mm) at 200 - 1000 kN/m^2 pressure and the temperature kept as low as possible. The reaction goes on for 6 hours. Unpolymerised material and latex containing 35-40% polymer are later discharged into coagulator, where electrolytes coagulate the polymer from latex. The solvent is stripped off the polymer, washing, filtering, drying of the polymer are finishing operations. It is preferred to keep monomer in liquid state. Fig. 6.9 shows the production of PVC by emulsion polymerisation. The polymerisation is done in autoclaves of capacity 40 to 100 cubic metres. Autoclaves are cooled by water to remove the polymer heat. After the reaction is over the contents are degassed, unreacted monomer is purified and recycled. Polymer is separated by centrifugation. Lower temperatures give rise to high molecular weight polymer.

Carbon tetra chloride is used as modifier to prevent branching.

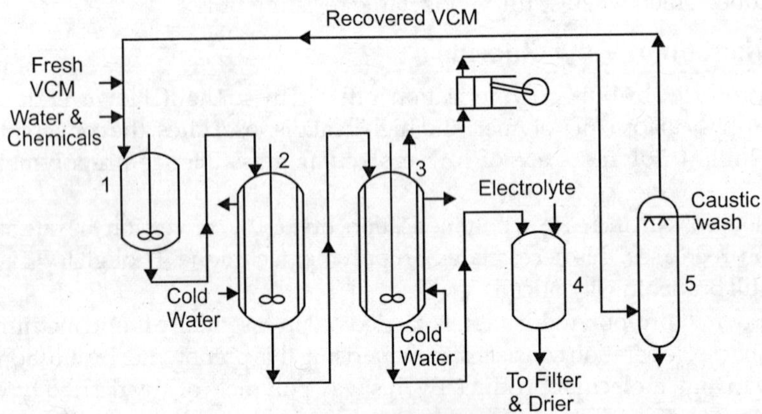

1. Tank for Emulsion Preparation 2. Polymerisation Reactor 3. Degassing Operation
4. Coagulation Tank 5. Impurities-Eliminator 6. Compression System

Fig. 6.9: Production of PVC.

6.4.2 Polyvinyl Acetate

Polyvinyl acetate is mainly used in adhesives, water based paints and for the production of polyvinyl alcohol.

At the time of start of PVA bulk polymerisation was used, gradually solution polymerisation replaced it. Though bulk polymerisation avoids solvent recovery and polymer drying, still the control of reaction is difficult, thus limiting the polymer to low molecular weight. The solution polymerisation is carried out in ethyl-acetate methanol benzene medium at 70-80°C. Initiators like benzoyl peroxides and other peroxides are found useful in polymerisation. Acetaldehyde is used as a regulator with benzene solvent; a high molecular polymer results.

6.4.3 CIS-Polybutadiene

1, 4 polybutadiene is produced by solution polymerisation from butadiene. The polymer is non-pigmented, oil extended rubber. The process resembles the conventional polymer process. Butadiene (99.3% purity) is mixed with recycle stream of butadiene and solvent and sent into the reactor. Catalyst is added at a controlled rate to the reaction mixture before it is fed into the reactor. Solvent consists of a mixture of two types, one the conventional one, for polymerisation and the other part consists of a modifier. Moderate pressure and temperature are regarded sufficient for polymerisation. Solvent to monomer ratio is carefully chosen so that the polymer solution viscosity does not give any trouble in handling. Usually 10-12% polymer solution is trouble free. Any conventional autoclave reactors are sufficient and three to four reactors are arranged in series for polymerisation. After the polymerisation is over, a suitable short stop and anti-oxidant are added to the polymer solution. Even oil and carbon black are added to the polymer solution for oil extended elastomers. Unreacted butadiene, solvent are removed after coagulation of rubber crumb. The resultant crumb slurry is passed through a series of strippers, where the remaining hydrocarbon solvent is removed. The overhead solvent is purified and sent back to the reactors. The polymer is washed with water to remove catalyst residues and short stops, after this the polymer is dried. Fig. 6.10 shows the process outlines based on Good Rich Chemical Co.

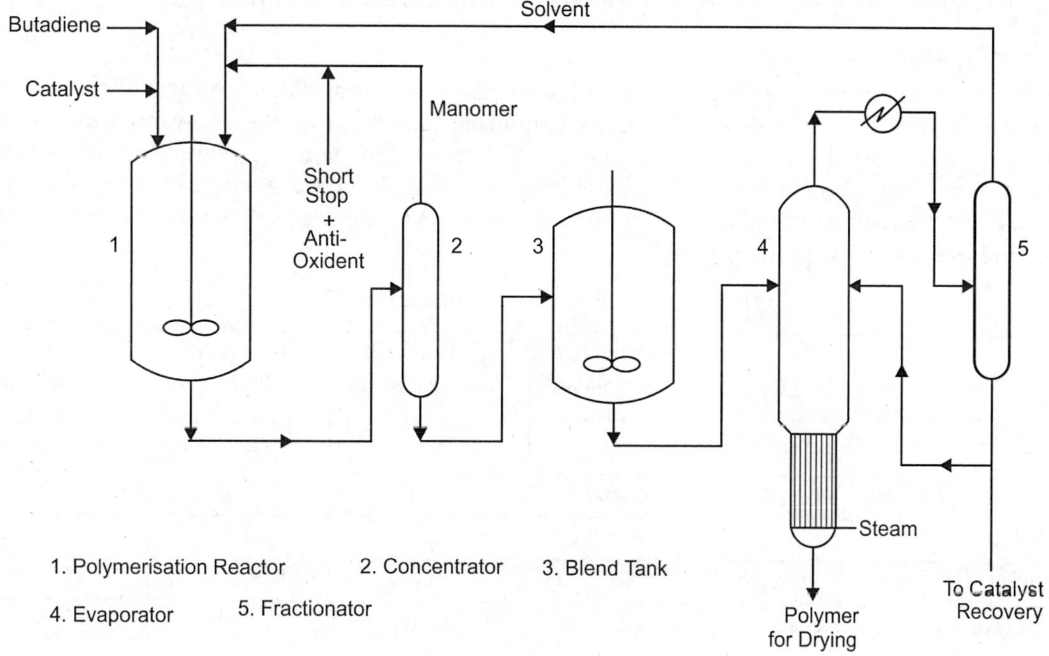

Fig. 6.10: Polymerisation of Butadiene.

6.4.4 Polyacrylonitrile

Polyacrylonitrile is produced by free radical initiators like peroxides, diazo compounds. Polymerisation can be conducted in toluene-solvent or water emulsion process. It is obtained as a white hard powder and is insoluble in many solvents. Orlon fibre is made out of this acrylonitrile, which is suitable for blending with woollen.

6.4.4.1 Properties

- *Optical Clarity* : Excellent light transmission. Will not yell after prolonged sun exposure.
- *Weather Resistance* : Despite heat, cold, sunlight, acrylic maintains its original appearance and color.
- *Safety* : Shatter-resistant, earthquake safe, and burglar-resistant. Increase safety with windows glazed of acrylic.
- *Light Weight* : Even with its strength and durability, it is half the weight of glass.
- *UV Light Resistance* : Clear acrylic sheet resists ultraviolet degradation. Each sheet has a ten-year-limited warranty against yellowing and loss of light transmission.
- *Chemical Resistance*: Acrylic sheet resists many chemical including ammonia, dilute acids, and aliphatic hydrocarbons.

6.4.5 Polytetra Flouroethylenes

Polytetra flouroethylene is a heat resistant polymer, degrades rapidly above 350°C evolving flourine. It is produced under a moderate pressure of 50 bars at 60°C in emulsion. The reaction is initiated by persulphates, perodoxides.

Different plastics are available, the properties of which are shown in Table 6.4.

6.4.6 Polybutenes

Isobutene polymerises with poor yields giving rise to low molecular products like lubricants. However, high molecular masses above 50,000 exhibit elastomeric properties. Low molecular weight polymers serve as good additives for lube oil viscosity improvement. Being a saturated compound it cannot be vulcanised. Polybutene rubbers exhibit high thermal plasticity, hence cannot be used directly as rubber but used with natural or other synthetic rubbers. Polyisobutene blended with waxes, resins result in high grade insulating putties.

Table 6.4: Typical Properties of Fluoro-plastics

	Chloro tri-flouroethylene (CTFE)	*Poly Tetra fluoroethylene (TFE)*	*Ftuorinated ethylene propylene (FEP)*	*Vinylidine flouride (VF2)*
Density kg/m^3(10^3)	2.15	2.18	2.14	1.76
TS (MP_a)	36	22	20	48
Impact Strength ft. 1b/in notch	2.6	3.0	no break	3.5
Max. Temperature°C to heat resistance	86	288	204	149
Dielectric constant	2.58	2.1	2.1	8.4
Effect of acids/alkalies	nil	nil	nil	H_2SO_4 may attack
Organic solvents	nil	nil	nil	nil

6.4.6.1 Production

High polymers can be produced by Friedel-Crafts catalysts, BF_3 or $AlCl_3$. In presence of these catalysts, the polymerisation is instantaneous, releasing a large amount of heat. Polymerisation can be carried out at ordinary temperature or at low temperature (–70°C). Polymerisation, even at moderate temperature hinders heat removal hence degradation of polymer results. To obtain high molecular weight polymer, the reactants are cooled to low temperature. Polyisobutene formed at - 80°C has a mole weight over 100,000. When $AlCl_3$ is the catalyst HCl is added as a regulator in small quantities (less than 1%).

Commercially high molecular mass isobutene polymer is produced by batch, semicontinuous or continuous process. In continuous process, the polymerisation is carried out in liquid isobutene-butane mixer. The reaction heat is absorbed by boiling butane. After the polymerisation the reaction mixer is allowed to settle, whereby the solvent can be removed, while the polymer goes for degassing operation. The catalyst is washed from the polymer with water and the polymer mass is stripped or distilled to separate low molecular mass.

6.4.7 Rubbery Poly Isobutenes

There are many technologies available for the production of rubbery poly isobutenes. One such technique is known as conveyor polymerisation (Fig. 6.11).

Liquid ethylene at –40°C under 16 bars pressure is fed to a chiller to liquefy butene. The temperature is kept at –104°C due to evaporation. Liquid ethylene then goes to isobutene cooler, cooling it to –85°C. The polymerisation is carried out on an endless steel belt 35 cm wide and 18 metre long. A mixer of liquid ethylene butene fraction is continuously delivered on one end of the belt. Catalyst in ethylene is delivered on to the mixer at this point. Polymerisation is complete in a few seconds, *i.e.* before the belt reaches the other end. Reaction heat is removed by ethylene. Since the belt moves slowly and the reaction would be soon over before the belt reaches the other end, from where the rubbery sheet is delivered. The whole conveyor belt is assembled in an air tight chamber. Evaporated ethylene goes through an outlet to purification section and further recycles. Fig. 6.11 shows the outlines of the rubbery poly isobutene.

1. Ethylene Liquifier
2. Liquid Ethylene Tank
3. i-Butene Tank
4. Cooler for i-Butene (- 85°C)
5. Catalyst Sol. BF_3 in liquid Ethylene
6. Belt Polymeriser
7. Ethylene Purifier
8. Compression-cooling system for Ethylene Liquifaction.

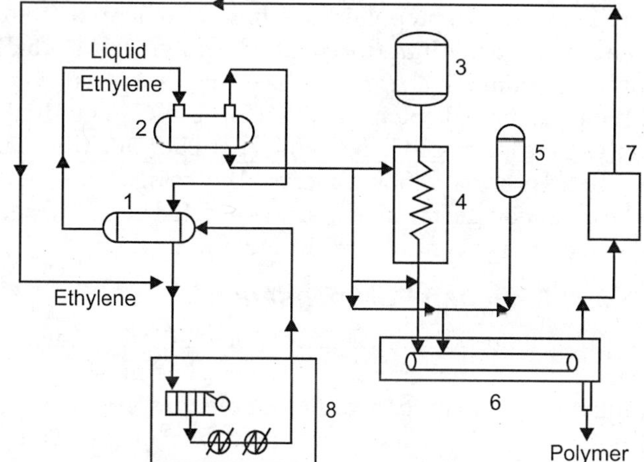

Fig. 6.11. Polyisobutene production

6.4.8 Polystyrene

Next to polyethylene and polypropylene, polystyrene is the widely used plastic. Though the first commercial production was in 1925, by Naugatuck Chemical Company, real commercialisation started with Dow Chemical Company in 1937 under a trade name Styron.

Polystyrene can be made in a number of ways to suit the end use. As the molecular weight increases, it slowly turns from liquid to solid state. High strength polystyrene is usually having a molecular weight more than 100,000; these are difficult to mold; the intermediate range polystyrene 70,000-100,000 is easy to mold. As the molecular weight drops below 70,000 ; the tensile strength, impact strength and elongation fall of rapidly; however, these are classified as general purpose polystyrenes. Polystyrene has got many favourable properties, like ease of fabrication high thermal stability, low specific gravity (900 kg/cu m) etc. Homopolymers of polystryene are extremely brittle and rigid hence copolymers are opted.

6.4.8.1 Methods of Polymerisation

Polystyrene production is carried out by free radical initiation or by co-ordinated catalyst. Bulk, suspension and emulsion polymerisations are in use. The polymer produced by initiators is atactic, while Ziegler's polymers are isotactic. Polystyrene is a linear polymer, though the phenyl groups often prevent ordered arrangement of macromolecules and crystalline structure. At 80°C and above amorphous polystyrene passes into rubbery state till it reaches 150°C, whereby it becomes plastic. Further heating above 300°C results in decomposition. It dissolves in most of the aromatic solvents. Isotactic polymer melts at 220°C. Its brittleness decreases with the addition of plasticisers, thus lowering its glass transition temperature.

6.4.8.2 Mass Polymerisation[25]

In making impact polystyrene, rubber is cut and dissolved in styrene, filtered to remove gels and fed to stirred autoclaves. After 35-40% polymer content is reached the material is placed into a press for further polymerisation beginning at 90°C and going up to 125°C for 12 to 14 hours. The material from this pressblock is removed for purification and drying.

6.4.8.3 Impact Polymerisation[26]

This is a sort of prepolymer technique, where copolymerisation takes place. 5 to 10% of synthetic or natural rubber is introduced into the polymer system. Polymerisation is carried out in autoclaves. Rubber monomer is charged in the form of suspension (water) into the reactors. The temperature of polymerisation is kept around 90°C to 130°C in gradual steps for about 10 hours. Sometimes two catalysts are introduced, like benzyl peroxide and ditertiary butyl perbenzoate. A slight deviation in the impact process is known as crystal process, where prepolymerisation is not called in. In suspension polymerisation the raw material is converted into polymer unlike continuous mass polymerisation. It gives a thorough control over the molecular weight.

6.4.8.4 Emulsion Polymerisation

This is more popular version of polymerisation. Latex or suspension polymerisation can be affected to produce different granular polymers. In emulsion polymerisation initiators are presulphates and emulsifiers are soaps or sodium petroleum sulphonates. Fig. 6.12 shows the outlines of the process. Polymerisation is accomplished in 3 to 5 m^3 enamelled reactors fitted with a water jacket stirrer and reflux condenser. The monomer is suspended in aqueous phase using a stabiliser (1). Sodium sulphate is added to control PH. A thorough agitation keeps the monomer suspended in medium. The aqueous phase emulsified and mixed with monomer (2). The emulsion is sent into reactor (3) which is heated and kept at 60°C. There may be number of reactors in series, and the polymerisation is carried out in nitrogen atmosphere. Catalysts and chain transfer agents are added into the reactor. The reaction takes place for 3 to 6 hours, after which it is sent into coagulator (4). The polymer mass is coagulated by heating to 120°C later allowed to settle. The polymer from the latex is separated by centrifugation (5). The polymer is washed and sent to dryer.

6.4.8.5 Modified Styrenes

Modified styrene also can be polymerised in a similar way. Allyl styrenes like α-methyl styrene can be polymerised with difficulty, although copolymerisation is relatively easier. O, m, p methyl styrenes are used in polymer processes. Ethyl toluene derivatives with styrene give on polymerisation ion-exchanged resins. Divinyl benzene (0.5 to 15%) is used for production cationic exchange resins. Similarly halogenated styrenes, like dichlorostyrenes are polymerised to give improved heat resistance and fire resistance polymer. Other low molecular weight resins are produced from low grade styrene and phenol or polyesters. Copolymers with styrene are plenty, like methyl methacrylate-styrene, poly phenylene-styrene, acrylonitrile-styrene, acrylic acid-butadiene-styrene graft polymer etc.

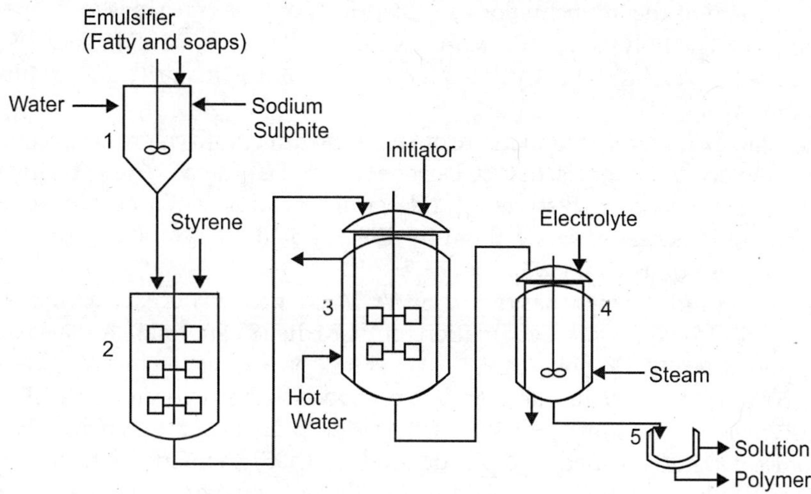

Fig. 6.12: Polymerisation of styrene

6.4.8.6 Environmental Facts About Polystyrene VCM - ACN

Polystyrene itself is made from styrene, a petroleum by-product, through a chemical reaction process known as polymerization. Polymers made from styrene are also used extensively today in the manufacture of automobile parts, electronic components, personal computer housings, boats, recreational vehicles, and synthetic rubber products.

While polystyrene can be manufactured using many different processes, the two most common forms of food service polystyrene are oriented polystyrene sheet (OPS) and foamed polystyrene. OPS is made from polystyrene resin pellets that are extruded into a solid, clear, "glass-like" sheet without using a blowing agent. The sheet is reheated and transformed into various shapes and colours, depending upon the size of the mold used and the dye added. Generally, the OPS process is used in the food service industry to make cold drink cups and clear salad and carry-out containers.

Foamed polystyrene is made using one of the two methods: an extrusion process or an expandable bead ("steam chest molding") process. The former requires that polystyrene resin pellets heated and mixed with a blowing agent to Create a flat sheet of plastic with a cellular structure. The resultant foamed extruded sheet is over 95 percent air and can be reheated and formed into a variety of shapes. This process is used to manufacture plates, trays, bowls and hinged-lid containers, including hamburger "clamshell" containers.

Dart foam cups are made using expandable bead polystyrene that is steam chest molded. This process originates when polystyrene resin granules are impregnated with a blowing agent. After undergoing a steamed "pre-expansion" process that produces foamed beads, the material is placed in a separate cup mold and heated with steam. The polystyrene is expanded further until the beads fuse together, forming the finished product, which is 95-98 percent air. Foaming also gives the material its white colour.

The most common forms of blowing agents today are carbon dioxide and pentane. Carbon dioxide is non-toxic, non-flammable, and does not contribute to low-level smog, nor does it deplete the stratospheric ozone layer. Its use in this manufacturing process results in no net addition of carbon dioxide to the atmosphere. Pentane also has no effect on the upper ozone layer, albeit this gas, if not recaptured, contributes to low level smog. Where smog formation is a large concern, many manufacturers use state-of-the-art technology to recapture pentane emissions.

See, for example : The Polystyrene Packaging Council, *Polystyrene and Its Raw Material Styrene: Manufacture and Use* (November 1993; Updated January 1996). The majority of Dart plants recapture and reuse pentane as fuel.

Polystyrene foam food service products are not manufactured with chlorofluorocarbons (CFCs) or any other ozone-depleting chemicals. In fact, Dart never used CFCs in manufacturing molded foam cups. Those manufacturers of polystyrene food service products that employed CFCs in their manufacturing operations ceased using them by 1990. Judd H. Alexander, In *Defense of Garbage* (Westport, CT : Praeger Publishers, 1993): 55.

Since 1974, federal and state law has regulated VCM as a proven human carcinogen. OSHA was the first federal agency to promulgate a regulation for VCM. Its 1974 permanent standard for VCM set the allowable exposure to VCM to not exceed 1 part per million averaged over an eight-hour workday, or not to exceed 5 ppm for any 15 minutes. Since exposure to VCM can occur in VCM manufacturing plants, PVC polymerization plants and PVC fabrication plants (plants where raw PVC resin or compound is fabricated into finished PVC plastic products), OSHA intended its standard to apply to all three of these workplace environments. EPA soon followed with its own regulation, which focused on atmospheric exposure to communities surrounding VCM and PVC facilities. Other agencies have regulated VCM since 1974.

VCM as a multi-potential carcinogen for almost thirty years and it is recognized as a potent cause of a wide variety of cancers arising in various organs of the body, both in humans and in animals. These cancers are not at all limited to vinyl chloride's "signal neoplasm," angiosarcoma of the liver, whose very existence in a given population strongly implies a larger occupational cancer risk caused by vinyl chloride. The most commonly recognized vinyl chloride related cancers include primary non-angiosarcoma liver cancer, primary brain cancer, lung cancer, lymphoma, and cancers of the blood and blood-forming organs. However, these non-malignant diseases are probably often obscured by their non-specific nature (Raynauds syndrome, scleroderma, acroostelysis, *etc.) with* alternate etiologies, including such common ailments as autoimmune disease, arthritis, dermatitis, idiopathic sclerosis, and even carpel-tunnel syndrome. However the limited studies that are available, for example a study of former workers at a "modern" Vinatex PVC plant in Chesterfield England that began operations only in 1969 have shown an enormous prevalence of non-malignant diseases (diseases that the industry long ago declared to be all but extinct) even in workers whose first exposures occurred after 1974.

6.4.8.7 ACN toxity

Concentrations in the air near a factory producing or using acrylonitrile average less than 1 part per billion (ppb). Extremely small amounts of acrylonitrile may be found in water near some factories that make or use it, but acrylonitrile rapidly breaks down and disappears from water. Plastic food containers that are made from acrylonitrile are regulated by the Food and Drug Administration (FDA) such that only 0.17 ppb can enter food; therefore, acrylonitrile intake from food packaging would be extremely low. Acrylonitrile can pass through skin, food, water. Inside the body, acrylonitrile is broken down into other chemicals, including cyanide. Most of these breakdown products are removed from the body in the urine. Small children are more likely to be affected than adults. In several cases, children died following exposures that adults found only mildly irritating. It should be noted that specific levels of acrylonitrile causing death were not reported. Exposure to large amounts of acrylonitrile for a short period of time, as might occur in the case of an industrial accident, results mainly in effects on the nervous system.

The U.S. Department of Health and Human Services has determined that acrylonitrile may reasonably be anticipated to be a carcinogen. Long-term exposure to acrylonitrile in air or water may increase your chances of getting cancer. Birth defects have been seen in animals exposed to high concentrations of acrylonitrile in the air or drinking water. Reproductive effects have been seen in animals given acrylonitrile in drinking water for three generations.

In humans, breathing acrylonitrile at a concentration of 16 parts of acrylonitrile per million parts of air (ppm) causes headaches, nausea, and disorientation. This concentration is close to that at which acrylonitrile can be smelled in air (about 21 ppm).

Regulations

The U.S. Environmental Protection Agency (EPA) recommends that acrylonitrile levels in water not exceed 0.058 ppb. Any release to the environment of more than 100 lbs must be reported to the federal government. The Occupational Safety and Health Administration (OSHA) has established a legally enforceable maximum limit of 2 ppm.

6.5 COPOLYMERS OF ETHYLENE-PROPYLENE

It is very common to produce co-polymers with propylene. Ethylene-propylene polymers are in great demand due to characteristic physical and mechanical properties with high elasticity. Only consideration reckoned being the ratio of ethylene to propylene, all other conditions remains same. Usually 5 to 10% propylene is used up in the co-polymer production, remaining being ethylene. Co-polymers can also be processed into thin films like polyethylene, they have high melting points and excellent rigidity too than either of these monomers.

Co-polymers can be made by using the same Ziegler's composite catalyst, although compositions of trihexyl aluminium and vanadium tri and tetrachlorides are also in service. Of late gas phase polymerisation of propylene is also being ventured, still no commercialisation has taken place.

Ethylene-propylene polymers (EPDM) are recently attempted polymers. The advantage of these co-polymers is that they can be readily processed into rubber. The rubbers produced with 40 to 50% propylene as co-monomer behave like saturated ones and possess good elongation and heat resisting properties. Co-polymer production is shown in Fig. 6.13.

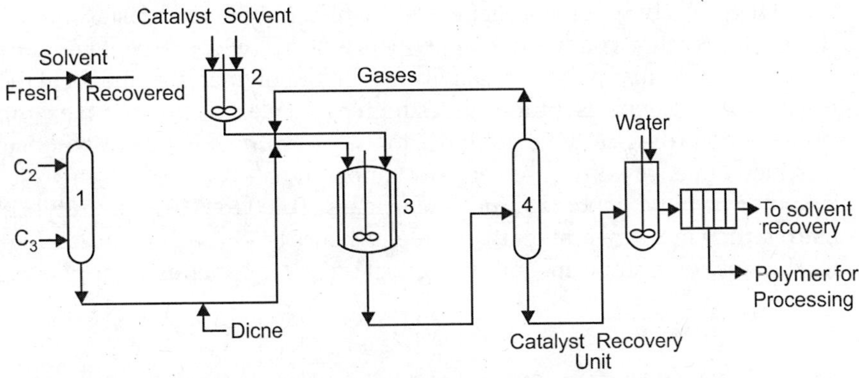

1. Absorber, 2. Catalyst Tank, 3. Reactors (2 ro 3 in series),
4. Flasher. 5. Catalyst Decomposer, 6. Filter (to remove solvent from Polymer)

Fig . 6.13: Copolymers (EPDM) Production.

6.5.1 Ethylene-butadlene Copolymer[27]

The catalyst system is crucial in ethylene-butadiene co-polymer synthesis. It has to be active in the presence of low moderate concentrations of butadiene and give good homogeneous polymer. This homogeneity gives good crosslinking efficiency. Further the catalyst must preferentially lead configurations in 1, 4 positions. Polymerisation is conducted at 50-120°C under a pressure of 50 bars. The polymer must not be soluble in hydrocarbon medium. Molecular weight regulation is done by hydrogen. Fig. 6.14 is based on Snam Progetti Technology; alcohol is used to regulate the chain formation. The recovered slurry is water washed to remove catalyst particles from polymer. Vanadium based complexes serve as the best catalyst in this system.

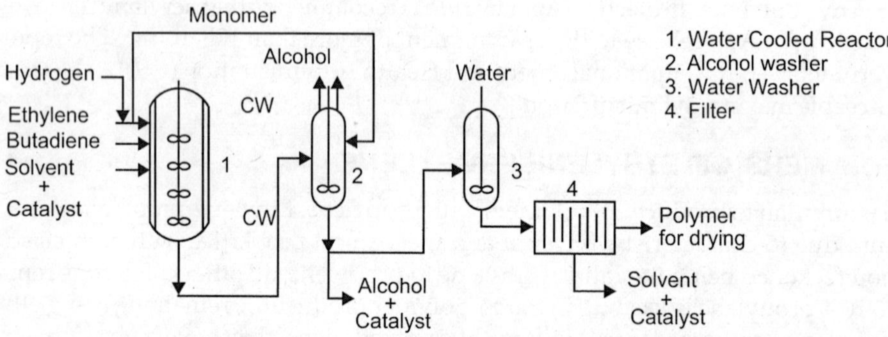

Fig. 6.14: Ethylene- Butadiene Copolymer Production.

6.5.1.1 Polymer properties

Co-polymers with butadiene concentration of 0.5-3 mole% and melt index 0.5-20 have been found to be useful for vulcanization process above 170°C. Vulcanization increases molecular weight and subsequent gel formation lead to greater impact resistance.

6.5.2 Isobutene Co-polymers

Isobutene forms co-polymers with other unsaturated hydrocarbons. The difficulty in getting co-polymers is due to high reactivity of isobutene itself, which polymerises faster than other olefins. However, by altering the reaction conditions co-polymers with propylene or pentenes can be obtained. These polymers are low molecular polymers usually meant for polymer gasolines or at the maximum, lube oil categories. With diolefins, isobutene gives high molecular masses, as an example isoprene reacts with isobutene according to the equation 2 to 3% of isoprene is mixed with isobutene and polymerised in an inert diluent medium at a very low temperature (– 80 to – 90°C), in presence of $AlCl_3$. The product is known as butyl rubber, of molecular mass 60,000 to 80,000. In butyl rubber, olefin serves as the main starting monomer rather than diolefin, hence less unsaturation. It is highly resistant to oxygen attack; because of its impermeability to gas, it is used for pneumatic appliances.

$$n\,\underset{\substack{|\\CH_3}}{\overset{\substack{CH_3\\|}}{C}}=CH_2 + n\,CH_2=\overset{\substack{CH_3\\|}}{C}-CH=CH_2 \rightarrow \left[-\underset{\substack{|\\CH_3}}{\overset{\substack{CH_3\\|}}{C}}-CH_2-\right]\left[CH_2-\overset{\substack{CH_3\\|}}{C}=CH-CH_2-\right]_n$$

6.5.3 Ethylene/Ethylacrylate and Ethylene/Vinylacetate Co-polymers

These two types of polymers have the same type of behaviour. Both possess side groups of polar nature on the polyethylene chains and these decrease the crystallinity of polyethylene, imparting plasticising effect. As the amount of acrylate or acetate is increased, the density progressively increases at 40% of these groups the polymer becomes amorphous. Further the solubility of these polymers in aromatic solvents increases or at high amounts of acetates the polymer is completely soluble in alcohol also. By incorporating these monomers, the tensile strength of polymer can be reduced to one-third of the original value of LDPE. Commercially important polymers have variable melt-index in the range of 2 to 150. Melt indices of about 20 are suitable for extrusion and injection moulding. These polymers have excellent properties suitable for use at low temperatures and are good adhesives too.

6.5.4 Ionomers

These are ethylene based co-polymers containing carboxyl groups. These groups almost eliminate the crystallinity of polyethylene. Ionomers are usually transparent, with high tensile strength equivalent to HDPE and tough and resilient. lonomers such as ethylenemethacrylic acid polymer known as (surlyn) can be produced by high pressure polymerisation process. Sometimes, a portion of carboxylic is converted into sodium salt to reduce the affinity for water so that they can be used for coatings or insulations.

6.6. LIQUID CRYSTALS

Mixtures of different polymers/copolymers whose composition is not independent of molecular size, cannot be characterized by conventional GLCs[1] but on line viscometers can do. Some polymers are homogeneous with respect to overall composition but vary long enough in branch frequency of comonomer spacing, exam. Ethylene-α olefin in LLDPE. The relative frequency of comonomer placements is reflected in changes in frequency, this influence processing behaviour of the polymer. These are inhomogeneous polymers, Aromatic polyamides/polyesters/are stiff chain polymers, p-phenylene terephthalamide (Kevalar) can be made by reactions of HMP hosphamide with n-methyl pyrrolidene.

Polymers that display structures that are intermediate between 3-dimensional ordered crystalline state and the disorderly fluid state has been investigated over the years. These polymers can be considered for isotropic amorphous polymer models where semi ordered regions have been proposed to exist. Liquid crystalline phases have been known to exist in low molecular weight systems. Lehman introduced the term Liquid Crystals. With increasing concentration a transition from non ordered isotropic solution to an ordered isotropic solution can take place. Allotropic liquid crystals, thermo tropic behaviour on heating they become isotropic. Mono tropic liquid crystalline, phases are stable only in super cooled state. General description of order of LQ structures has to include physical structure on molecular scale which is characteristic by position as well as orientation order of neighbouring molecules and super molecular arrangement of assemblies of molecules/parts of molecules. Mole structure of liquid crystalline phase is known to be intermediate between 3 dimensional crystalline order and disorder of isotropic fluid state. The crystalline order is characterised by long range positional order. The thermal, mechanical properties are strongly dependent on the existence of long range order. For nonspherical particle, in addition to positional order, the orientation order has to be characterised. It has been known that[2] not all molecular systems completely lose their order at melting point. There are systems that make transition from crystalline state to disordered state in more than one step. In some systems the oriental state is lost at phase transition, still position order is maintained *i.e.* above transition temperature. Texture of liquid crystalline phases determine the optical properties of these materials to a great extent, that can be brought about by mechanical, thermal, electrical, magnetic forces.

Anisotropic geometrical shape and stiffness of a particle are fundamental requirements for forming liquid crystal phases.

Polymers like[3] PF, PS, Polymethamethacrylate, polycarbonates are considered as flexible chain polymers. Liquid crystalline order is exhibited by side groups where as the order of main chain is less defined. Cholestric liquid crystals reflect the incident light in a specific wave length range with circular polarization. Linear and non linear optical properties have been explored. There are basically 3 devices (*a*) based on molecular reorientation with applied field (response time is millisecond) (*b*). functional moieties as a part of the LC structures may respond to light forces (*c*). LCs have been explored for passive device applications in which no switching is involved. Vitrified Liquid Crystals (VLCs) have emerged as novel class of mesogenic materials for the preparation of well aligned LC solid film to facilitate a diversity of optical and optical electronic applications as shown in Fig. 6.15 (*a*).

Liquid Crystals

- Flow like liquids but have optical properties of solid
- Polarized light
- molecules are lined up like those in solids

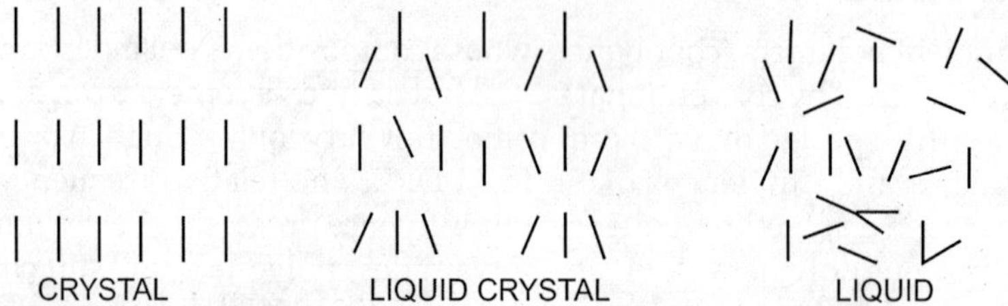

The shapes of polymers determine their properties

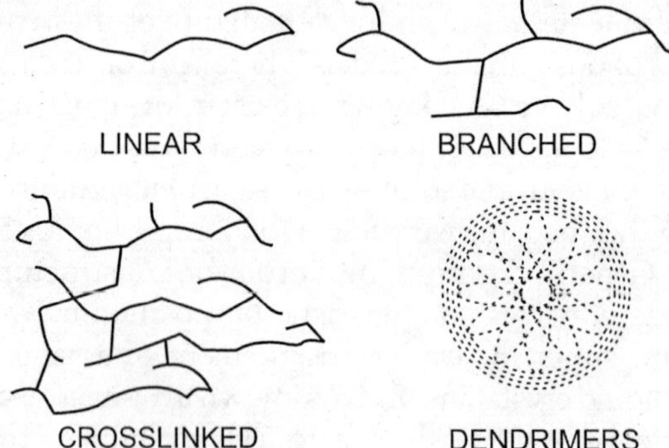

6.6.1 Introduction to Polymer-Dispersed Liquid Crystals

Polymer-dispersed liquid crystals (PDLCs) are a relatively new class of materials that hold promise for many applications ranging from switchable windows to projection displays. These materials, which are simply a combined application of polymers and liquid crystals, are the focus of extensive research in the display industry. PDLCs consist of liquid crystal droplets that are dispersed in a solid polymer matrix. The resulting material is a sort of "swiss cheese" polymer with liquid crystal droplets filling in the holes. These tiny droplets (a few microns across for practical applications) are responsible for the unique behaviour of the material. By changing the orientation of the liquid crystal molecules with an electric field, it is possible to vary the intensity of transmitted light.

6.6.1.1 Preparation of Polymer-Dispersed Liquid Crystals

Polymer-dispersed liquid crystals have been prepared in several different ways including : encapsulation (emulsification) and phase separation; the latter process has become the primary method of manufacture. Each method produces PDLCs with different properties and characteristics. Among the factors influencing the properties of the PDLC material are the size and morphology (shape) of the droplets, the types of polymer and liquid crystal used, and cooling and heating rates in production. The relationship between the method of production and these factors is explained below. The work discussed here has nematic liquid crystals but studies of droplet dispersions containing chiral nematics have also been reported.

6.6.1.2 Encapsulation

Early attempts to produce PDLCs were made with a technique known as microencapsulation. In this method, a liquid crystal is mixed with a polymer dissolved in water When the water is evaporated, the liquid crystal is surrounded by a layer of polymers. Thousands of these tiny "capsules" are produced and distributed through the bulk polymer. Droplets produced with this method tend to be non-uniform in size and can even be interconnected with each other. Materials manufactured by encapsulation are referred to as NCAP or nematic curvilinear aligned phase.

6.6.1.3 Phase Separation

In order to obtain PDLCs by phase separation, a homogeneous mixture of polymer (or prepolymer) and liquid crystal is first produced. The liquid crystal droplets are then formed by the separation of the two phases. The separation can take place in one of the following three ways :

6.6.1.4 Polymerization-Induced Phase Separation

Polymerization-induced phase separation, or PIPS, occurs when a liquid crystal is mixed with a solution that has not yet undergone polymerization (a prepolymer). Once a homogeneous solution is formed, the polymerization reaction is initiated. As the reaction progresses, the liquid crystal molecules come out of the solution and begin to form droplets. The droplets grow until the polymer binder becomes solid enough that the molecules are trapped and can no longer move easily.

The two main factors that influence the size of liquid droplets in PIPS are the cure temperature and the type and proportions of materials used. The cure temperature affects the speed of the polymerization as well as the diffusion rate and solubility of the liquid crystal in the polymer. These factors can greatly influence the size of the liquid crystal droplets which translates into different macroscopic optical properties.

6.6.1.5 Thermally-Induced Phase Separation

TIPS, can be used when the polymer binder has a melting temperature below its decomposition temperature. In this method, a homogeneous mixture of liquid crystal and a melted polymer is formed. The solution is cooled at a specific rate to induce phase separation. Liquid crystal droplets begin to form as the polymer hardens. The droplets continue to grow until the glass transition temperature of the polymer is crossed.

Droplet size is affected the most by the cooling rate of the polymer melt/liquid crystal solution.

Fast cooling rates tend to produce small droplets because there is not sufficient time for large particles to form.

6.6.1.6 Solvent-Induced Phase Separation

The third common type of phase separation is called solvent-induced phased separation, or SIPS. This process requires both the liquid crystal and polymer to be dissolved in a solvent. The solvent is then removed typically by evaporation at a controlled rate to begin the phase separation. Droplets start growing as the polymer and liquid crystal come out of solution and stop when all of the solvent has been removed. Droplet size in SIPS is influenced by the rate of solvent removal.

6.6.2 Imaging of PDLCs

Infrared (IR) spectroscopy can be used to chemically characterize polymeric systems. It provides information on the relative concentration of various species in a sample and their spatial distribution. It allows the difference in the chemical state and interactions at the interface from either phase bulk to be studied. These features are illustrated in the following figure.

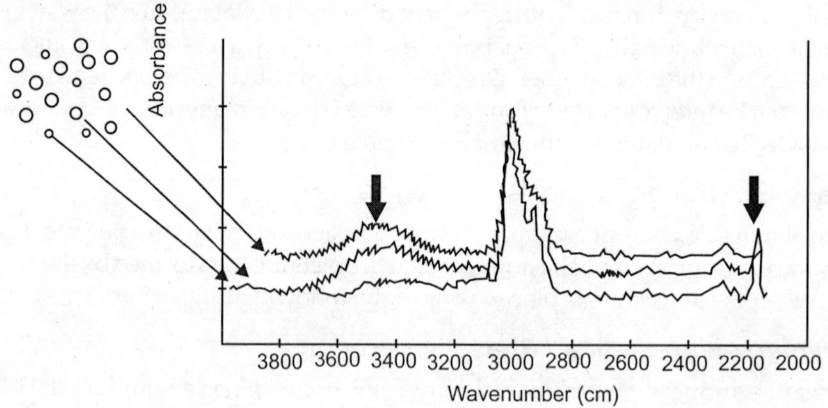

Use of spectroscopy in determining the distribution of relative concentration is illustrated by the red and blue curves which correspond, respectively, to the matrix and the liquid crystal droplets. Here the absorption at the wavenumber (frequency divided by the velocity of the radiation) of the characteristic OH peak is chosen to represent polymer concentration with blue representing the least concentration (in droplets) and red the greatest concentration (matrix). The green curve shows its use in exploring interfaces through relative absorbance. At the interface, intensity of transmitted IR radiation is lowered because of scattering and refraction. This leads to an apparent absorbance that can be used as a measure of concentration. The loss in intensity depends on the refractive index ratio and refractive index varies depending on the frequency of radiation.

Because the IR images depend on refractive indices, they can also be used to study the effect of an applied electric field. First, the IR images below were taken at non-absorbing regions. This means that neither component, liquid crystal nor the matrix, normally absorb at this wavenumber (2600 cm^{-1}). As shown in figure (*a*) the interface shows absorbance at this wavenumber, which is due to the refractive index mismatch between the liquid crystal and the matrix that causes the incident radiation to be refracted. The high absorbing regions within some of the liquid crystal droplets is due to coalescence (liquid crystal molecules are still coming together to form larger droplets). Refractive indices are matched due to the applied field as seen in figure (*b*), since there is no longer absorbance. The interface still causes a small amount of scattering, however. This is partly because of imperfect matching between the liquid crystal and matrix indices, which could be due to a predominance of one of the components of the liquid crystal at the boundary. Higher potential difference decreases the amount of scattering but does not totally eliminate it as shown in figure (*c*). This residual scattering could be due to liquid crystal molecules near the boundary not switching with the bulk.

6.6.3.1 Droplet Configuration

The configuration of the liquid crystal droplets in a polymer matrix is the focus of much current research. Many different configurations have been observed and they depend on factors such as droplet size and shape, surface anchoring, and applied fields.

The radial configuration occurs when the liquid crystal molecules are anchored with their long axes perpendicular to the droplet walls. The axial configuration of the liquid crystal droplets also occurs when the molecules are oriented perpendicular to the droplet wall, but only when there is weak surface anchoring. When an electric field is applied to a radial droplet, the molecules adopt the axial configuration. The radial configuration is returned when the field is removed. Crystals exhibit special symmetries when they slide in certain directions or rotate through certain angles. These symmetries can be compared to those encountered when walking in a straight line through empty space. Regardless of the direction or distance of each step, the view remains the same, as there are no landmarks by which to measure.

6.6.3.2 Applications of PDLCs

Polymer-dispersed liquid crystals hold potential for a variety of electro-optic applications ranging from displays to light shutters. PDLC windows are based on the ability of the nematic director of the liquid crystal droplets to align under an electric field as discussed in the previous section. In a typical application, a thin PDLC film (about 25 microns thick) is deposited between clear plastic covers. The plastic substrates are coated with a very thin layer of a conducting material known as indium tin oxide (ITO).

Transmission of light through a PDLC window depends primarily on scattering which in turn depends on the difference in refractive index between droplets and their environment. In the case of high droplet density, the environment consists mainly of other droplets, which makes the relative orientation of their directors an important factor. The droplets are anisotropic with the index of refraction parallel to the director different from that perpendicular to it.

6.6.3.3 PDLC Display Materials

The area of PDLC materials is a recent technology that has been rapidly picked up, improved upon, and developed for display application by Japanese scientists (Doane, 1991). The physical concepts behind this technology have origins in English patent literature, but materials and processes to bring it about largely began in the United States. The team found that nearly all display companies in Japan had an interest in and maintained a research and development program on these materials. In displays, these materials offer improved brightness because they do not require polarizers and they are relatively simple to fabricate. They are principally interesting for use in projection television, but many companies foresee their use in direct-view displays. Since PDLC materials require the active matrix for high resolution, most research programs focus on efforts to lower the drive voltage and increase resistivities required for the AM TFT. Two companies have made significant strides in this direction: Asahi Glass reported a full-color video projection prototype using PDLC materials, and Dainippon Ink has developed a PDLC material and is now working with other display companies to develop display products. Both companies show materials that can be used on AM TFT substrates.

There are variations in the amount and type of polymer used in PDLC materials; the amount generally varies from 20% to 70% polymer by weight. Recent materials using gel polymers contain approx. 2% polymer. Both aqueous and nonaqueous polymers have been used. For high resistivities, both the polymer and liquid crystal material must be of high purity. Hysteresis can be a problem. Nematic materials most desired are those with large delta e and delta n. Dainippon Ink reports the use of fluorinated materials to achieve high-purity nematic materials. An example is the fluorinated tolans, which also exhibit a high delta *n* :

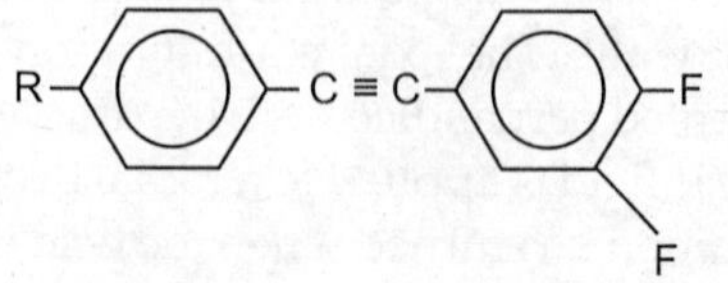

R–C₆H₄–C≡C–C₆H₃(F)–F

GLOSSARY

Discotic liquid crystal: The component disc-shaped molecules self-assembles in a way that resembles stacks of coins. The discs are stacked on top of each other to form columns, which in turn are packed on a two-dimensional, usually hexagonal, lattice.

Macroscopic molecular orientation: The property that allows the orientation of liquid crystals to be seen only on a large scale and cannot be seen if a single layer is compared to any other single layer.

Main-chain polymer liquid crystals: Polymer liquid crystals in which the mesogenic cores are a part of the main chain or backbone of the polymer.

Melting transition temperature: The temperature at which the substance loses its translational and orientational order, changing from a solid phase to a liquid phase.

Mesogen: Rigid rodlike or disclike molecules which are components of liquid crystalline materials.

Mesogenic cores: The mesogenic core is the primitive structural unit of a polymer having the requisite anisotropic shape and attractive interactions to establish long range intermolecular order in its liquid phase. That is, it is what gives a polymer liquid crystal its liqsquid crystal properties.

Mesomorphic substance: Another term for a liquid crystal material.

Mesophase: Equilibrium liquid crystalline phases formed with order less than three dimensional (like crystals) and mobility less than that of an isotropic liquid. Parallel orientation of the longitudinal molecular axes is common to all mesophases (long-range orientational order). See the Liquid Crystal Phases section.

Micelle: A spherical formation caused by an amphiphilic substance in a solution. The lyophilic end of the molecule tends to orient itself toward the outside of the sphere while the lyophobic end tends to orient itself toward the inside of the sphere.

Chiral molecule: A molecule that is not identical to its mirror image. This gives a chiral substance its characteristic twisted shape, due to the fact that its molecules do not line up when combined.

Cholesteric liquid crystals: Also known as Chiral Nematic. Similar to the nematic phase, however, in the cholesteric phase, molecules in the different layers orient at a slight angle relative to each other (rather than parallel as in the nematic). Each consecutive molecule is rotated slightly relative to the one before it. Therefore, instead of the constant director of the nematic, the cholesteric director rotates helically throughout the sample. Many cholesterol esters exhibit this phase, hence the name cholesteric.

Cholesteric mesophase: Nematic liquid crystals with chiral centers form in two dimensional nematic-like layers with directors in each layer twisted with respect to those above and below so that the directors form a continuous helix about the layer normal. Many cholesteric esters exhibit this phase, hence the name cholesteric. This mesophase exhibits circular dichroism and is optically active.

Nematic mesophase: Liquid crystals are characterized by long-range orientational order and the random disposition of the centres of gravity in individual molecules. Nematics may be characterized as the simplest spontaneously anisotropic liquids. Nematic phases are composed of rod-shaped molecular aggregates that are arranged with parallel but not lateral order.

Liquid crystals can be divided into thermotropic, lyotropic and metallotropic phases. Thermotropic and lyotropic LCs consist of organic molecules. Thermotropic LCs exhibit a phase transition into the LC phase as temperature is changed. Lyotropic LCs exhibit phase transitions as a function of both temperature and concentration of the LC molecules in a solvent (typically water) Metallotropic LCs are composed of both organic and inorganic molecules; their LC transition depends not only on temperature and concentration, but also on the inorganic-organic composition ratio.

Examples of liquid crystals can be found both on the natural world and in technological applications. Most modern electronic displays are liquid crystal based. Lyotropic liquid-crystalline phases are abundant in living systems. For example, many proteins and cell membranes are LCs. Other well-known LC examples are solutions of soap and various related detergents, as well as tobacoo mosoic virus.

Basic Scientific Description

Normally, we consider matter to have three distinct states: solid, liquid, and gas. However, there are states of matter which do not meet the necessary requirements of any of these three categories. For example, a substance such as mayonnaise is somewhere between a liquid and a solid.

Nematic Liquid Crystals. Nematic are polarizable rod-like organic molecules on the order of 20 Angstroms in length. Because of their tendency to organize themselves in a parallel fashion, they demonstrate interesting and useful optical properties; the digital watch you used to wear back in the 80's functioned using nematic liquid crystals. Today, many more useful and interesting properties of nematics are known and explored.

Smectic Liquid crystals are different from nematics in that they have one more degree of orientational order than do the nematics. Smectics generally form layers within there is a loss of positional order, while orientational order is still preserved. There are several different molecules align perpendicular to the layer planes, and Smectic C, where the alignment of the molecules is at some arbitrary angle to the normal.

Basic Scientific Description

Normally, we consider matter to have three distict states : solid, liquid and gas. However, there are states of matter which do not meet the necessary requirements of any of these three categories. For example, a substance such as mayonnaise is somewhere between a liquid and a solid.

Liquid crystals are also not quite liquid and not quite solid. Physically, they are observed to flow like liquids, but they have some properties of crystalline solids. Liquid crystals can be considered to be crystals which have lost some or all of their positional order, while maintaining full orientational order. For example, imagine a large number of toothpicks put into a rectangular box and shaken. When you open the box, the toothpicks will be facing in about the same direction, but will have no definite spatial organisation. They are free to move, but like to line up in about the same direction. This is a primitive model for **nematic** liquid crystals. Ematics are polarizable rod-like organic molecules on the order of 20 Angstroms in length. Because of their tendency to organize themselves in a parallel fashion, they demonstrate interesting and useful optical properties.; the digital watch you used to wear back in the 80's functioned using nematic liquid crystals. Today, many more useful and interesting properties of nematics are known and explored.

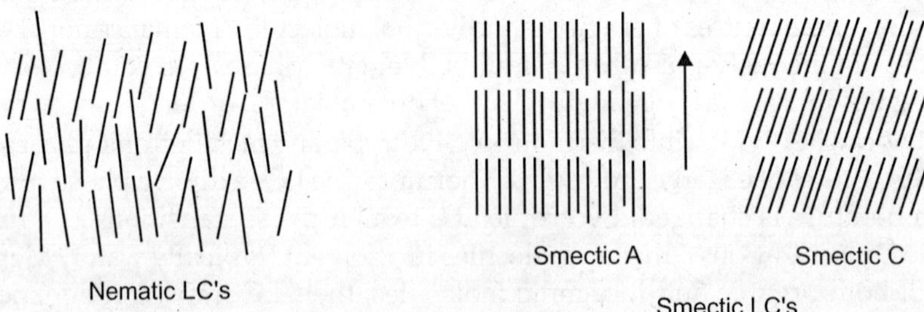

Nematic LC's

Smectic LC's

Smectic Liquid crystals are different from nematics in that they have one more degree of orientational order than do the nematics. Smectics generally form layers within which there is a loss of positional order, while orientational order is still preserved There are several different categories to describe smectics. The two best known of these are Smectic A, in which the molecules align perpendicular to the layer planes, and Smectic C, where the alignment of the molecules is at some arbitrary angle to the normal.

REFERENCES

1. Malcolum P. Stevens, Polymer Chemistry, Addison Wesley Pub. Co., Massachusetts, 1975, 55-180.
2. Reymond B. Seymour, Introduction to Polymer Chemistry, McGraw Hill, 1971.
3. J. Barandrup and CH. Immergut, Polymer Hand Book, Inter Science (NY), 1966. p. VI-41.
4. Joseph P. Kennedy and Erik G.M. Torngvist, Polymer Chemistry of Synthetic Elastomers Part II, Interscience, 848.
5. Stanely R. Sandler and Wolf Karo, Polymer Synthesis, Academic Press, NY. 1974, 33.
6. L. Reich and A. Schindler, Polymerisation by Organometallic Compounds, Wiley Interscience. N.Y., 1966, Chapter V.
7. AD. Ketly, The Stereo Chemistry of Macromolecules, Marcel Dekker 1967, Vol. 1, Chapter 2 and 3.
8. J. Parker Foreman, HCP, Nov. 1972, p. 131-136.
9. S. Sivaram. Ind. Engg. Chem. Pr. Res. and Dev., Vol. 16, 2, 1977, p. 121 and 20.
10. Akiokageyama and Atsushi Hirotani; HCP, Nov. 1972, p. 97.
11. Marshall Sittig; Poly Olefin Production Processes, Latest Developments, NDC, New Jersy, 1976, p. 113, 230.
12. Donald, M. Rasassen, Chem. Engg., Sept. 1, 192, p. 105
13. Fredrick J. Kard; Chem. Tech., April 1983, p. 222.
14. John C. Davis, Chem. Engg., Oct. 16, 1972, p. 62-64.
15. S.D. de Bree, Chem. Engg., Dec. 11, 1972, p. 73.
16. C.R. Ranfast, HCP, May 84, p. 105.

17. Glovanni, Di Drusco and Roberto Rinaldi, HCP, Nov. 84, p. 113-118.
18. Lawrence I. Chriswell, CEP, April 83, p. 84.
19. G.D. Bukatov, S.A. Sergeev and V.A. Zaharo, Advances in Catalysis Science and Technology, Edited by T.S.R. Prasadaa Rao, Wiley Eastern Ltd. 1985, 525.
20. K.V.C. Rao, Jayadev Sreedhar, S.K. Gupta and P.K. Traama, Proced of the Symposium on Industrial Polymers and Radiation, Sardar Patel University, Feb. 12-14, 174, 2000.

CHAPTER

7 Petroleum Aromatics

7.1 OCCURRENCE

Aromatics are the most important intermediates for plastics, fibres, resins, dyestuffs, detergents, and many other things of daily requirements. The chemical industry highly depends upon these materials. Decades back coaltars were the principal suppliers of aromatics of all ranges. Since 1950 the thinking was changed in favour of petroleum feeds as the source of aromatics. Initial draw-back of technology hunt was successfully met and the industry adopted the highly efficient technology for the production of aromatics namely catalytic reforming process. This technology was available in petroleum industry purely as a producer of quality gasoline. The demand for aromatics is growing at a very fast rate as shown by the Table 7.1. The production of light aromatics does not end with reforming process alone, because of the demands alternative processes were also pressed into service. Some crudes are enriched with aromatics, these can be effectively extracted; an example Bombay High Crude (1) contains 12% light aromatics, which are extracted.

Present day refineries are equipped not only with catalytic reformers, but also with the polymerisation, alkylation and other cracking units which have the capacity to match the production of aromatics in tune with the growing demands. Jet fuels, kerosene and lube oils are predominantly apathetic towards aromatics, thus a simple extraction of aromatics from these fractions and other high boiling stocks would be able to supply huge amount of stocks to the chemical industry, which the coal industry can never match, as the progress of iron industry is not that fast, to absorb coke.

Aromatics in our country are primarily coal based, as our oil industry is still growing. However, as cited earlier, Bombay high crude itself shall be able to supply 1.5 to 1.6 MMT of aromatics just by the simple extraction process; for which Bharat Petroleum has commissioned one such unit with the capacity of 34,000 tonnes per year for the production of BTX. Anticipated growth of petroleum refining capacity in our country is placed at 90 MMT from present 52 MMT by 2000 A.D. This would be able to fulfil the demand of petrochemical industry.

The highly useful aromatics are low boiling stocks namely, benzene, toluene and xylene and the consumption and demands are shown below :

Table 7.1: Aromatics Production Consumption (tonnes)

	1988-89	*1989-90 capacity*	*Demands (90)*	*Demands 1997-98*	*Installed capacity*	*Demands 2001-2002*
Benzene	1,76,000	2,30,000	4,00,000	618.000	561,000	1,037,000
Toluene	32,900	45,000	52,000	139,000	215,000	203,000
O-xylene	25,500	28,000	67,000	155,000	51,000	230,000
P-xylene	64,000	46,000	2,30,000	539,000	324,000	913,000

Phenol	43,000	52,000	48,640			
LAB	1,19,200	1,00,000				
DMT		1,20,000	1,25,000			
PTA		1,35,000	1,45,000			

7.1.1 Production of BTX

Primarily BTX production is done through catalytic reforming. With the advent of I.C. engines of high compression ratio, thc quality of fuel had to undergo changes. Catalytic reforming is the process by which low obtain gasoline is upgraded to high octane value with the help of a catalyst. This is the major route for the production of BTX. The following reactions explain the formation of BTX from highly paraffin stocks of low boiling range (80–180°C).

(*a*) *Dehydrogenation*: Napthenes are converted to aromatics

$$\text{(cyclohexane)} \rightarrow \text{(benzene)} + 3H_2$$

(*b*) *Isomerisation :* This yields the highest octane number because of isomerisation of n-paraffins to i-paraffins. Isomerisation in aromatics side chains also takes place

$$CH_2.\,CH_2.\,CH_3 \text{ (on benzene ring)} \rightarrow \text{(ring with } C_2H_5 \text{ and } CH_3\text{)} \text{ or } \text{(ring with } CH_3, CH_3, CH_3\text{) etc.}$$

(*c*) *Cracking of paraffins:* In this larger molecules of paraffins crack to smaller molecules. Long side chains in aromatics also follow the same pattern

$$CH_2.\,CH_2.\,CH_2.\,CH_3 \text{ (on benzene ring)} \rightarrow CH_3 \text{ (on benzene ring)} + C_3H_6$$

(*d*) *Hydrogenation:* Hydrogenation of unsaturates to saturates takes place because of hydrogen

$$C_6H_{12} \rightarrow C_6H_{14}$$

$$CH_3 \text{ (on benzene ring)} + 3H_2 \rightarrow CH_3 \text{ (on ring)}$$

(*e*) *Naphthenes Isomerisation and dehydrogenation:* This reaction also produces aromatics

$$\text{(methylcyclopentane, } CH_3\text{)} \rightarrow \text{(cyclohexane)} \rightarrow \text{(benzene)} + 3H_2$$

(*f*) *Dehyrocyclisation followed by dehydrogenation:* A large amount of aromatics is obtained by this route. Intermediate naphthene formation takes place

$$-C-C-C-C-C-C-C \longrightarrow \text{(cyclohexane ring with } CH_3\text{)} + H_2 \longrightarrow \text{(benzene ring with } CH_3\text{)} + 3H_2$$

The other reactions are elimination reactions like denitrogenation, desulphurisation etc.

It is seen that dehydrogenation and isomerisation reactions are fast. Cyclisation requires severe conditions. Hydrocracking and other elimination reactions are slow in nature. Pressure, temperature and space velocity are the decisive factors but catalyst composition is also equally important and discussed below :

7.1.1.1 Effect of Pressure

Low pressure increases the rate of dehydrogenation, while isomerisation is unaffected by pressure. Coking significantly increases at low pressures. Dealkylation is negated by high pressure. Further hydrocracking and elimination reactions are directly related to partial pressure of hydrogen.

7.1.1.2 Effect of temperature

Except hydrogenation reactions which are exothermic, other reactions are favoured by increasing temperature. The increase of temperature invariably results in carbon formation and degradation of product. Hinderance to coke deposition and maintenance of catalyst activity over a long run period and prevention of unnecessary formation of olefins can be adequately assured by hydrogen circulation. An economic operation, pleads for overall adjustment of reaction conditions as shown below :

Temperature	:	450-500°C
Pressure	:	12-50 bars
Space velocity	:	2
H_2 recycle ratio	:	1-10 (vol)

7.1.1.3 Catalysts

In reforming many reactions occur simultaneously, hence the catalyst activity towards all these reactions is highly important. Platinum formulations are most widely used. These catalysts are usually graded by the amount of platinum deposited on bases like silica or silica-alumina. Low platinum catalysts are non-regenerative type and they usually contain less than 0.3 platinum. High platinum catalysts differ from the former by increased metal content upto 0.5% and constitute regenerative type.

Bimetallic formulations like rhenium-platinum are good in resisting fouling and can be operated even below a pressure of 20 bars. Further they can be sustained with lower space velocities and low hydrogen recycle ratios. Englehard Magno Forming process details a catalyst containing platinum on bimetallic formulations. The rugged nature of bimetallic catalyst permits operation at much severe conditions and tolerates sulphur comfortably upto[(2)] 20 ppm; although most of the feedstocks are desulphurised to 2-5 ppm sulphur content(3). Platinum catalysts are susceptible to sulphur and nitrogen and these elements are kept preferably below 5 and 3 ppm respectively; while most frequently 0.5 to 3 ppm sulphur is permitted in feedstocks.

Non-noble catalysts are also in operation. Selecto forming uses such catalysts based on zeolites. These are dual function catalysts and permit only straight chain compounds to react. Activators like potassium, enhance the capacity of the catalyst. Other non-noble formulations are based on chromium, molybdenum, cobalt etc. Of course, the use of these formulations is no match to platinum combinations.

7.1.1.4 Processes

The advancement of catalytic reforming has given birth to a number of reforming processes like Catalytic reforming, (Institute duPetrole; Englehard Minerals and Chemicals Corp.) Houdri forming, (Houdry Div.) Platforming, (UOP); Power forming (ESSO Research); Rhein-forming, (Cheveron Research Co.) etc. operation of all these processes is more or less similar, the difference occurs in number of reactors and arrangement. Fig. 7.1 shows the process outlines of catalytic reforming.

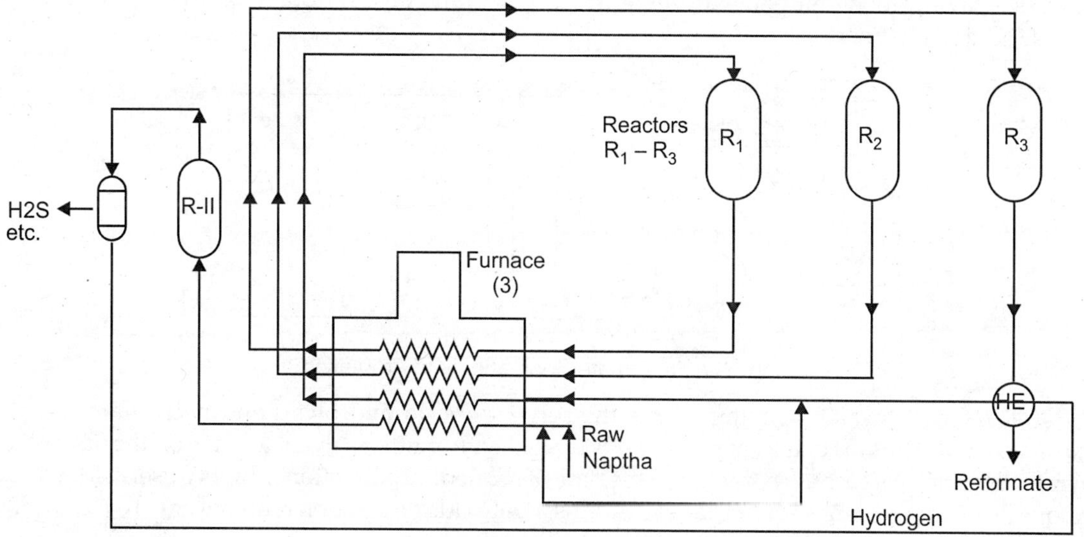

Fig. 7.1: Catalytic Reforming.

Raw naphtha is mixed with hydrogen and sent through a furnace (3). The mixture is hydrodesulphurised by hydrotreating in Reactor R-II. This eliminates sulphur in the form of H_2S etc. and the gases are purged out in a separator (2) by releasing pressure. The sulphur free naphtha is mixed with recycled hydrogen and passed through a process heat exchanger. The feed is further heated to around 500°C in furnace (3) and the vaporised feed is sent into reactors $R_1 - R_3$ serially with intermediate heating. Fourth reactor is called a swing reactor (not shown in Fig. 7.1) takes the place of any of the reactors when regeneration is to be done. Pressure in the reactors is maintained around 15-75 bars. Desired cyclisation reactions being slow, are carried out at severe conditions in the last reactor where the temperature is higher by 20-30°C, than the previous reactor. Catalyst is usually in 3 mm pellet form. Most usually one furnace maintains all the coils for different reactors. Sometimes intermediate furnaces are provided.

The reaction products are cooled by exchanging heat with cold water. The gaseous product, usually hydrogen, is obtained through a high pressure releaser (1). Fig. 7.2 shows the recovery of aromatics from reformate. The liquid product is stabilised (at a pressure of 5-7 bars) to drive off the low boiling components as well as dissolved gases (2). The aromatic rich stock is then supplied to solvent extraction system. Many solvents like DMF (4), DMSO (5), ethylene glycols (6) are good for extraction of light aromatics.

Extractor (3) consists of packed or plate column, where the feed is introduced at the bottom, and the solvent is fed countercurrent to feed. The temperature of the extractor is kept around 40-50°C. In the present case extract is the desirable product. Extract and raffinate are settled in a settling column (4).

Most of the raffinate is routed to extractor as reflux. The rich extract from the bottom of the column (4) goes to a stripper, where solvent and aromatics are separated. Aromatics, thus freed from extract phase are contaminated with solvent. This extract is then given water wash in column (6). BTX is then obtained as top product from the washer. Raffinate being deficient can be usually purified by washing with water alone (7). Non-aromatic raffinate may be further processed by mixing with incoming naphtha or can be sent to crackers or steam reformers. The solvent washings from all units are collected and distilled to get solvent in pure form. Udex process (Ethylene glycol+water solvent) is one of such processes having good reputation throughout the world, and this is adopted in Aromatic complex of Baroda (Gujarat Refinery).

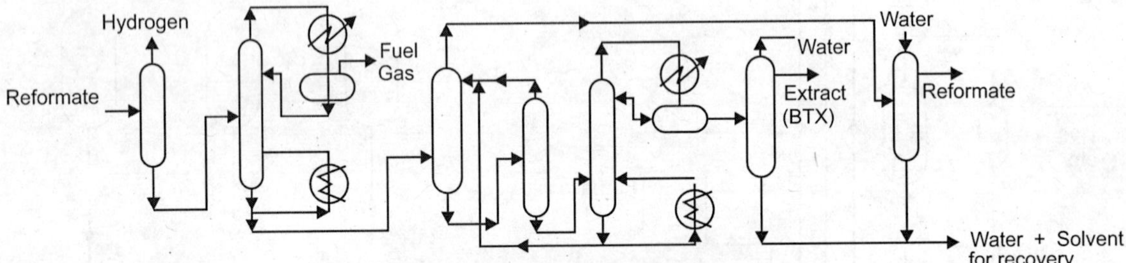

Fig. 7.2: Reformate Separation into Aromatics.

In some cases water washing affects the solvent due to hydrolysis and necessitates another extraction unit. Thus the extract of reformate is treated with another solvent to leave off the first solvent in raffinate phase of second extraction. This type of sequential extraction is highly useful for DMSO. Above extraction scheme requires three columns, while Udex operation requires only two columns.

After extraction, separation of BTX into individual components follows: The aromatic mixture containing BTX is fractionated in column 1, 2, 3 as shown in Fig. 7.3. The separation is purely guided by relative volatility. Benzene and toluene can be obtained as pure components by fractionation. C_8 aromatics consist of ethylbenzene, o, m, p-xylenes and are taken out as overhead product of column (3). The heavier fraction C_9^+ is obtained as bottom product. The separation of xylenes involve an elaborate procedure and was discussed in the Chapter 1.

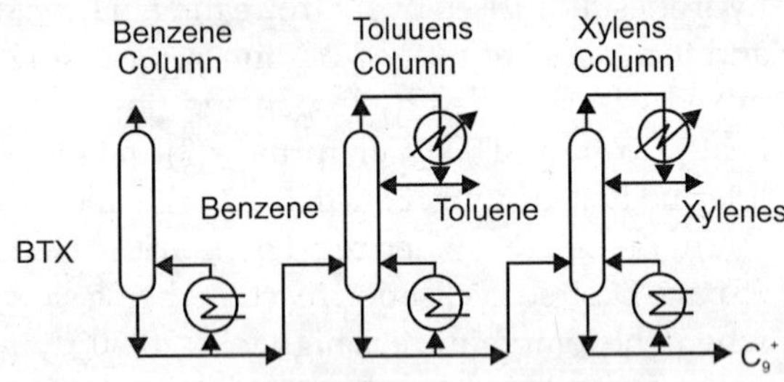

Fig. 7.3. Separation of BTX.

Further purification of benzene and toluene can be done by fractionation or selective treatments depending upon the purity requirements.

7.1.1.5 *Aromatics from polymer gasoline* [7]

Pyrolysis gasoline is obtained as a by-product in a number of thermal cracking operations. Sometimes 50-60% of BTX is present in pyrolysis gasoline. Because of the presence of free radicals in thermal gasoline, the gasoline is extremely active and ready to polymerise, hence storing over a long period is not advisable. However an initial hydrogenation step helps in eliminating this difficulty. By hydrotreatment most of the unsaturates, usually diolefins would be converted to saturates leaving aside aromatics. By subjecting the hydrogenated gasoline to further hydrotreatment other impurities such as S, N, O can be eliminated.

Thus two stages of hydrogenation are required; initial heating of stock to eliminate impurities would facilitate gum formation hence initially hydrogenation under mild conditions followed by severe conditions, would enhance the purity of BTX by elimination of S, N and O. The two stage hydrotreating of pyrolysis gasoline is shown in Fig. 7.4. Raw gasoline mixed with H_2 is heated in a furnace F_1 and sent into packed bed reactor R_1. The vapors from the reactor are heat exchanged and cooled and then sent to pressure relieving system (S_1). The separator gas is purified and hydrogen is recycled back. The liquid fraction, containing light and heavy fractions are normally rejected by distillation (d_1) and the $C_6 - C_8$ fraction is fed into the second stage hydrogenator (R_2). By this way the load on second hydrogenator which operates under more severe conditions than the first reactor, can be kept down. The product from R_2 is cooled as vent gases are separated (S_2). The product is finally distilled (D_2) to separate hydrotreated stock from lights.

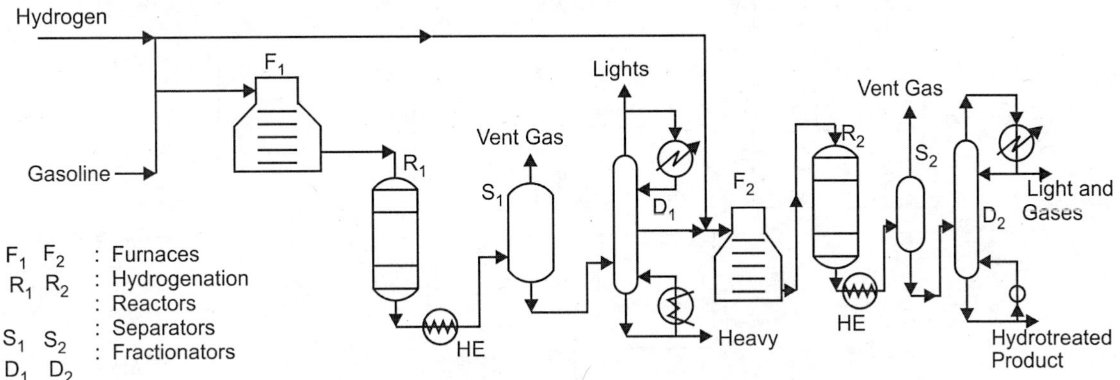

Fig. 7.4: Pyrolysis Gasoline Hydrogenation.

Many technologies, as listed below account for the bulk of hydrogenation capacity of the world's pyrolysis gasoline.

B.P. selective hydrogenation - British Petroleum
HPN Process - Englehard Industries
HPG Process - Houdry
DPG Hydrotreating - Lumus
IFP Hydrogenation Process - IFP

7.1.1.6 Benzene: Purification by Extractive Distillation

Extractive distillation is normally applicable for recovering a single aromatic component from a predistilled fraction that is enriched with a high concentration of required aromatic. Therefore, extractive distillation has proven beneficiable for recovery of benzene from pyrolysis gasoline and xylenes from reformates. More recently extractive distillation has been used to recover benzene and toluene in a single step. Though good quality benzene is obtained by this method, same is not true with toluene.

7.1.1.7 BTX Fractionation

Crude BTX from reformers may not contain much light and heavy fractions, but polymer and alkylated gasolines are usually mixed up with higher fractions. Hence crude BTX has to be prefractionated to depentanise to remove lighter components than benzene. The tendency is to separate benzene and toluene by distillation, while xylenes stream is fed to extraction unit. High purity benzene and toluene are obtained by distillation scheme as shown in Fig. 7.5.

The feed BTX is first stabilised in column (1) to remove fractions upto C_5. The bottom product of this column is sent to distillation column (2), whereby $C_6 - C_7$ fraction is obtained as over-head product. If this $C_6 - C_7$ fraction contains other than aromatics, it is sent to an extraction unit (3), where benzene and toluene can be removed by the solvent, and the solvent recovery is done in column (4) as described earlier. The bottoms of the column (2) contain C_8^+ fraction, which may be passed through an adsorbent (7) to remove any resinous material, before it is distilled into two cuts C_8 and C_9^+ in column (6). The benzene, toluene fraction collected is distilled in a precise fractionator (5), to obtain benzene as an overhead product and toluene as bottom product. The column uses 50-60 plates and operates under a reflux ratio of 3.5 to 4.5.

Depending upon the demands of benzene and xylene, the toluene fraction can be dealkylated or disproportionated as described below.

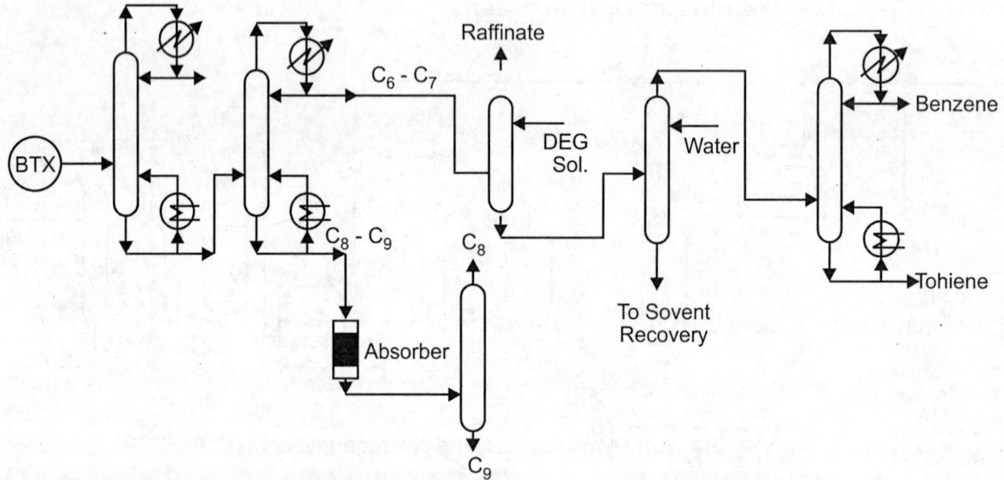

Fig. 7.5: BTX Separation from Crude BTX.

7.1.1.8 Benzene by dealkylation of Toluene

Surplus toluene can be converted into benzene by hydrodealkylation in presence of a catalyst.

$$C_6H_5CH_3 + H_2 \rightarrow C_6H_6 + 504\ \text{kJ/mole} + CH_4$$

The above reaction shows dealkylation of toluene in presence of hydrogen and steam, over a catalyst. The catalyst may be metal oxide type of metals on silica alumina or alumina. It is more or less a thermal reaction, however aided by catalyst. The temperature ranges between 600 to 700°C and pressures vary from 40 to 60 bars. Silica alumina catalysts are good enough to give a conversion of 60-90%. Unconverted toluene is recycled. Molybdenum compounds are also used as catalysts.

In fact hydrodealkylation and pyrolysis gasoline would in future account for 50% benzene available for industry. The gradual decline of reforming process for aromatics is already an indicative for recovering of BTX from pyrolysis gasoline and dealkylation routes.

7.1.1.9 Disproportionation of Toluene to Benzene and Xylenes[8,9]

This is the modern method to produce simultaneously relatively more useful products like benzene and xylenes. Transalkylation, known by the other name saves hydrocarbon wastage and results in all xylenes, though p-xylene is less pronouncive. The reaction is given below:

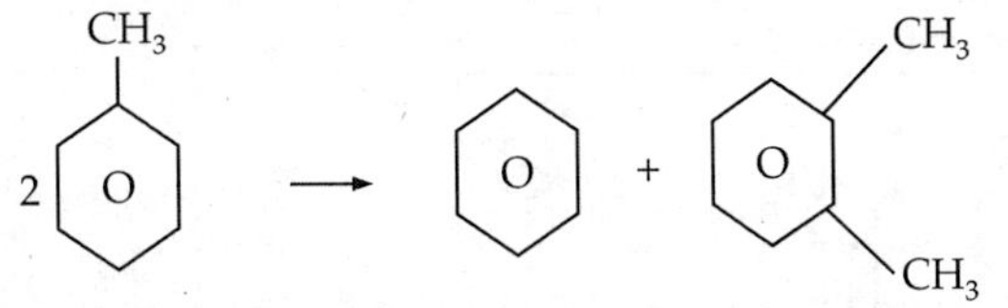

All the xylenes and other isomeric products are common. Diphenyls also result from condensation reactions prompting less yield of benzene

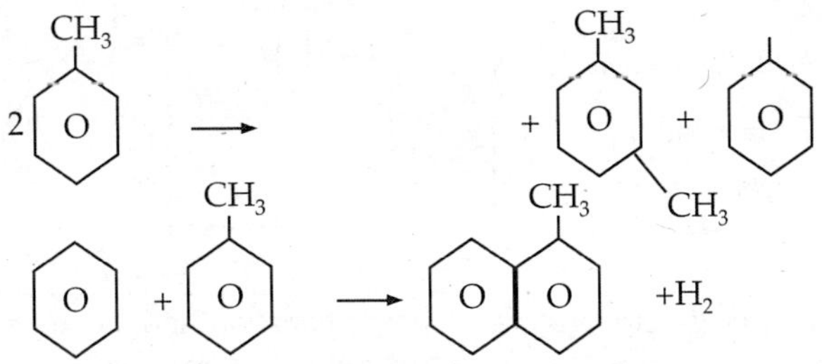

These are reversible reactions and are sensitive to the concentration of reactants and products go up to even 2 to 3% mole.

Without hydrogen, toluene can be disproportionated at low temperature, *i.e.* in liquid phase, Mobil's process employs zeolite catalyst treated with rare earths to give rise the disproportionation reaction. The temperature of disproportionation is around 300-350°C and toluene is kept under liquid phase due to pressure (50-60 bars). Ethyl benzenes, polymethyl benzene and diphenyls also form as impurities; specially ethyl benzene elimination is not desired at all when isomerisation step is likely to be followed. Usually product from disproportionation reaction shows 20 to 40% xylenes, where the equilibrium concentration of xylenes show 24% p, 54% m and 22% othoxylenes. Mordanites treated with cerium are good catalysts specially at a temperature of 400°C-450°C.

Fig. 7.6 shows the low temperature disproportionation[10] (LTD) developed by Mobil. By this process, xylenes obtained from reforming unit is supplemented by xylenes from toluene. As shown in Fig. 7.6 fresh toulene and recycled toluene are heated and sent into Low temperature Disproportionation Reactor (LTD, R-l). The reactor is packed with catalyst. The outlet stream is fractionated to separate non-aromatics and benzene in a benzene column (1). The bottoms from the column are again fractionated in column (2) to separate toluene, which goes for recycling. The heavy fraction, mixed with C_8^+ aromatics obtained from the reformate (after separation of ethyl benzene by distillation in column 5) is fed into xylene splitter 3, where *p*-xylene is obtained as a top product. The heavy product from column 3, is fed into column to separate o-xylene from heavy aromatics (C_9^+). Major part of over head from column 3, is mixed with toluene and sent into a Low Temperature Isomerisation Reactor (LTI, R-2). The feed in presence of toluene is easily isomerised to a product of 95-98% p-xylene in the stream. The reaction is carried out in liquid phase at 300-350°C. The product from the reactor is distilled to separate the light fractions and sent through the aromatic separation units for getting *p*-xylene.

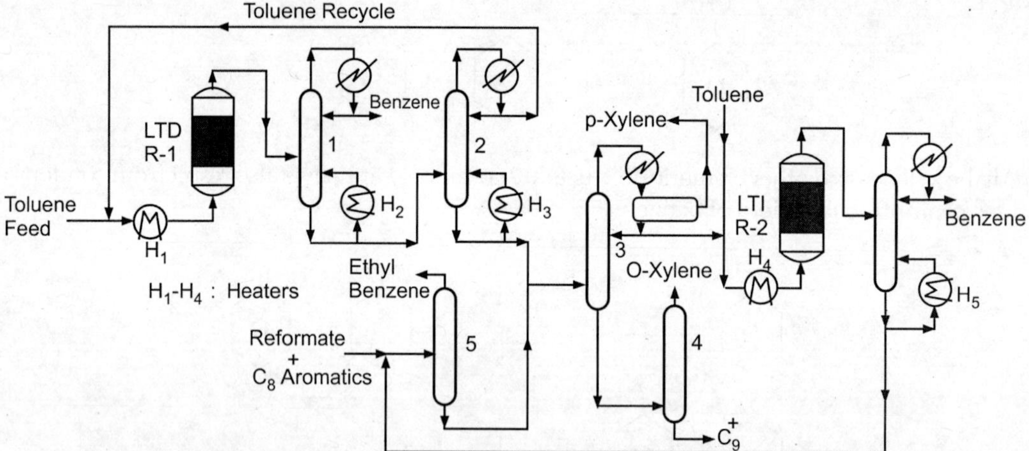

Fig. 7.6: Low Temperature Disproportionation and Isomerisation Process for Maximum p-Xylene.

7.1.1.10 Isomerisation of Xylenes[11]

Isomerisation of xylenes is carried out to alter the composition of reformates of dispropor-tionated xylenes. The most important xylene being, *p*-xylene other xylenes are successfully converted into *p*-xylene by isomerisation. ICI process employs fixed bed, non-noble metal catalyst at near atmospheric pressure and 400-500°C to isomerise mixed xylenes to maximum *p*-xylene. Fig. 7.7 shows the process. Mixed-xylenes stream after separation of lighter fractions in a distillation column (1); fed into xylene splitter (2). The overhead xylene stream is cooled, crystallised in a number of crystallisers to separate p-xylene. The mother liquor is then passed through the isomerisation reactor (Reactor 1). The outlet stream is heat exchanged and hydrogen gas is separated. The product stream is taken into the fractionator (1) to separate the lighter products, if any.

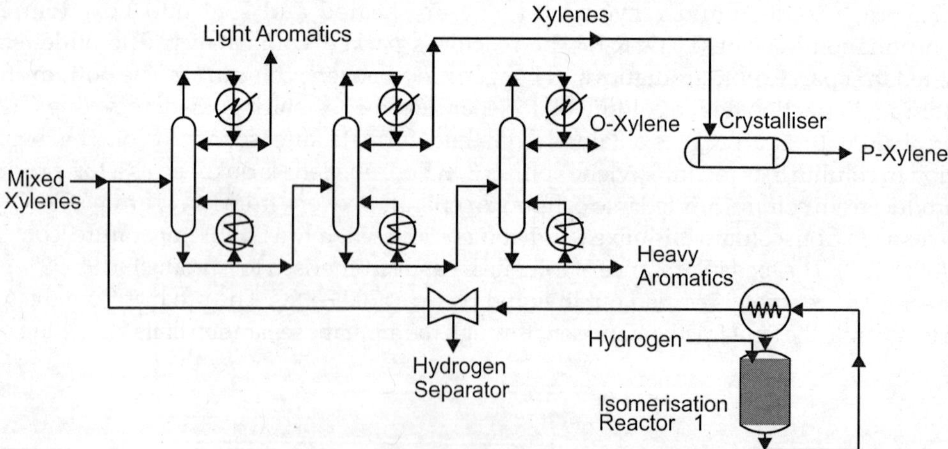

Fig. 7.7: Isomerisation of Xylenes to p-Xylene.

Isomerisation is done by using a number of catalysts, like Friedel-Crafts as well as noble metals and silica-alumina catalysts.

In the recent method developed[18] by Japan Gas Chemical, xylenes are separated by HF – BF_3 extractive solvent. This solvent is found to be very effective for removing m-xylene. Once m-Xylene is removed, p and o-xylenes can be easily separated by fractionation and crystallisation.

Alternatively antimony trichloride for complete separation of *p*-xylene from *m*-xylene has also been suggested, *m*-xylene can be eliminated by careful sulphonation from other xylenes. By far distillation and partial crystallisation combination is adopted worldwide. All the distillation columns 1,2,3 as shown in the figure contain 150-300 plates. Ethyl benzene column has got 150 plates, while xylene splitters will have 250-300 plates.

7.2 BENZENE DERIVATIVES

The delocalised electrons associated with benzene ring impart special properties to aromatic hydrocarbons. They exhibit the properties of saturates and unsaturates and some other properties of their own. They react like paraffins by substitution under mild conditions and addition compounds result under severe conditions, as an example:

$$C_6H_6 + 2NO_2 \rightarrow C6H_5NO_2 + HNO_2$$
$$C_6H_6 + 3Cl_2 \rightarrow C_6Cl_6$$

7.2.1 Derivatives with Halogens

Monochlorobenzene is a well known substituted compound of benzene. It can be produced by batch and continuous processes. The position of chlorine in benzene ring can be at ortho or para position and usually more than a single substituted compound results depending upon mole ratio of benzene to chlorine and time, temperature and catalyst empolyed :

Reactants		Product		Name
O + Cl_2	→	O–Cl	+ HCl	Monochloro Benzene
O + 2 Cl_2	→	O (Cl, Cl ortho)	+ 2HCl	Ortho-dichloro Benzene
O + Cl_2	→	Cl–O–Cl (para)	+ 2HCl	Para-Dichloro Benzene

The reaction conditions are :

Temperature	80-100°C
Pressure	normal
Catalyst	Chlorides of Al, Fe, Tin etc.

80% monochlorobenzene is produced in the above substitution reaction, and the remaining dichlorobenzenes share the rest. By continuous chlorination higher amounts of monochlorobenzene can be produced. Monochlorobenzene can be produced in vapour phase by oxychlorination process. Hydrogenchloride is oxidised by oxygen *in situ* to generate chlorine, which can easily react with benzene in presence of catalyst like copper oxide on silica. The conversion mainly depends upon the availability of chlorine, the temperature in the reaction is around 250°C. Conversion can be increased with higher temperature and pressures.

The main use of monochlorobenzene is found in the production of phenol and aniline. Other uses are for the production of DDT, polychlorobenzenes and nitrobenzenes; all these uses are collectively insignificant when aniline and phenol are compared.

7.2.2 Aniline[12]

Aniline is one of the building blocks of organic chemicals notably dyestuffs. It is produced from two chemicals namely phenol or nitrobenzene.

7.2.2.1 Production of Aniline from Phenol

Phenol is catalytically reacted with ammonia to yield aniline. As shown in Fig. 7.8, a mixture of ammonia and phenol is preheated and sent into a reactor (1). The reactor effluent is partially cooled, unconverted ammonia is stripped of from the products in a column (2); later purified and recycled. The product containing phenol, aniline and water is distilled and water is distilled in two columns (3 and 4) in series to drive of water; aniline is collected as overhead product from the last column. Halcon process employs copper salts as catalyst although silica-alumina would serve. The reaction is conducted at 200-250°C under a pressure of 60–80 bars. Ammonia in gaseous form or in an aqueous solution of 20-30% concertration can be used.

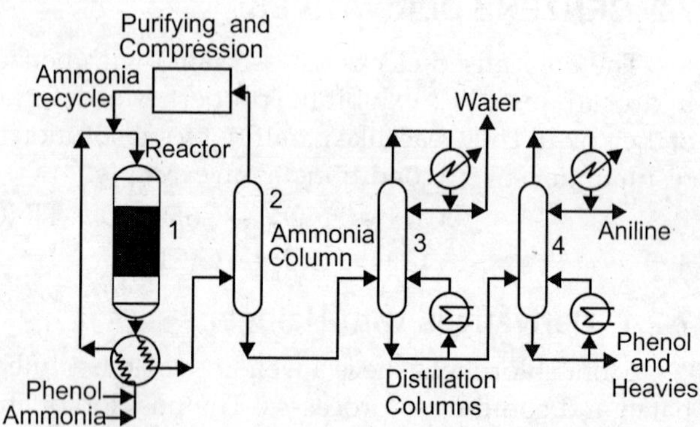

Fig. 7.8. Production in Aniline.

This is a much simplified technology, specially when low grade phenols in large amounts are available.

7.2.2 Aniline through Nitrobenzene

The other method of production of aniline is from nitrobenzene. In the first step benzene is nitrated using a conventional mixer of nitric acid and sulphuric acid at 60–80°C. The reaction is fast even at lower temperature. The crude nitrobenzene from the product mixture is stripped of by steam, further purification may be resorted by simple distillation. The nitrobenzene in the second stage is hydrogenated to produce aniline. This is done catalytically in vapour phase using Raney nickel as the catalyst at 270–300 °C under moderate pressure. The reactions are shown below:

In presence of sulphuric acid

$$\text{(O)} + HNO_3 \xrightarrow[H_2SO_4]{+} \text{(O)}\text{-}NO_2 + H_2O\ (H_2SO_4) + 110\ \text{kJ}$$

$$NO_2\text{-(O)} + 2H_2 \longrightarrow NH_2\text{-(O)} + H_2O$$

or

$$NO_2\text{-(O)} + Fe + H_2O \quad NH_2\text{-(O)} \longrightarrow + Fe_3O_4 + 534\ \text{kJ}$$

An alternative process employs chlorobenzene as the starting material. Chlorobenzene is reacted with ammonia at 220°C and 60–70 bars pressure using copper salts (Cu_2Cl_2) as catalysts.

7.2.3 Production of phenol

Phenol can be produced through many processes. Already a process has been described on the technology of cumene (Isopropyl benzene) oxidation to phenol and acetone. Other processes are based on benzene. Benzene can be converted into compounds like benzene sulphonic acid or chloro benzene; these two compounds on hydrolysis yield phenol. However, industrially important processes are cumene oxidation and chlorobenzene hydrolysis. It can also be produced from toluene, the reactions involved in these technologies are shown below:

$$C_6H_5Cl + H_2O \xrightarrow[\text{Over Silica}]{300°C} C_6H_5OH + HCl$$

From Sulphonic Acid

$$C_6H_6 + H_2SO_4(SO_3) \longrightarrow C_6H_5SO_3H + H_2SO_4$$

$$C_6H_5SO_3H + NaOH \xrightarrow{\text{Fusion}} C_6H_5OH + Na_2SO_3$$

$$C_6H_5SO_3Na + H_2O \longrightarrow C_6H_5OH + Na_2SO_3$$

From Toluene

$$C_6H_5CH_3 \xrightarrow{\text{Oxidation}} C_6H_5COOH$$

$$C_6H_5COOH \longrightarrow C_6H_5OH + CO_2$$

From Cumene

$$C_6H_5CH(CH_3)_2 \xrightarrow{\text{Oxidation}} C_6H_5C(CH_3)_2OOH \longrightarrow C_6H_5OH + (CH_3)_2CO$$

7.2.3.1 *Phenol from Chlorobenzene*

In this process, benzene is first chlorinated to chlorobenzene by *in situ* oxychlorination. Outlines on this technology is presented in Fig. 7.9.

$$C_6H_6 + HCl + \frac{1}{2} O_2 \longrightarrow C_6H_5Cl + H_2O$$

Benzene vapours enter the oxychlorinator through a heater, which brings it to reaction temperature. Aqueous hydrochloric acid vapours from hydrolysis unit (4) and benzene vapours are mixed with air and sent into reactor (1). The reaction takes place over copper chloride and some per cent of di and tri chlorobenzenes also form. The reaction mixture goes into a column (2), where un-reacted benzene is distilled off and chlorobenzene is separated in column (3). The remaining mixture is hydrolysed (4) to yield phenol and hydrochloric vapours. Hydrochloric vapours are routed to reactor while phenol is purified by distillations. Two distillation columns are necessary to distill the product into dichlorobenzene and phenol. Dichlorobenzene, benzene are routed back to reactor.

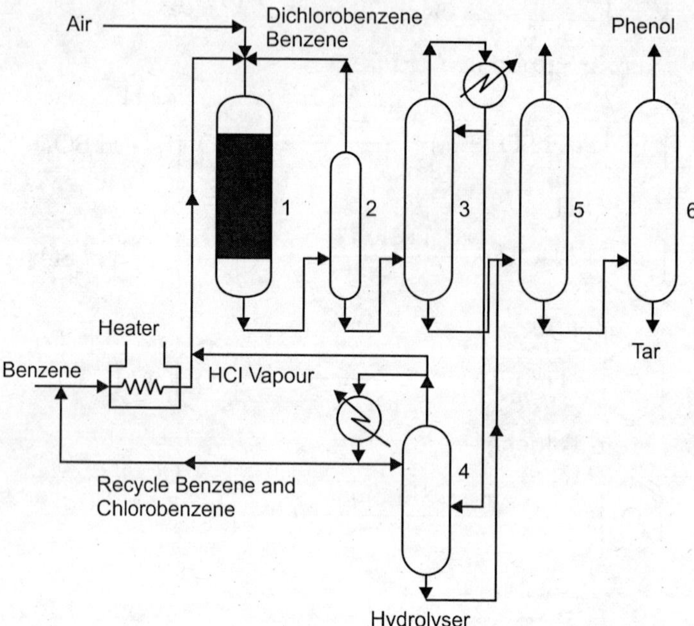

Fig. 7.9: Phenol production from benzene (via Chlorobenzene).

7.2.4 Alkylation of Benzene

Alkylation of benzene with propylene has been described in Chapter 4. In this chapter alkylation of benzene with ethylene is presented.

Ethyl benzene is the starting material for styrene production. In alkylation reactions with aluminium chloride, the difficulty is located in feeding dry aluminium chloride, hence a new method of feeding aluminium chloride as a composite catalyst with ethyl chloride has been initiated. Ethyl chloride reacts with benzene giving rise to ethyl benzene and hydrogen chloride. This hydrogen chloride is utilised as a promoter for aluminium chloride and lowers its consumption. The reaction proceeds as shown below[13]:

$$CH_2 : CH + HCl \longrightarrow CH_3CH_2Cl$$

$$CH_3CH_2Cl + AlCl_3 \longrightarrow (CH_3CH_2)^+ + AlCl_4^-$$

$$(CH_3CH_2)^+ + AlCl^-{}_2 + \text{(O)} \longrightarrow \text{(O)}\text{-}C_2H_5 + HAlCl_4$$

$$HAlCl_4 \longrightarrow HCl + AlCl_3$$

The reaction is exothermic generating 110 kJ/mole. Hence reaction heat has to be removed. In view of this, the reactor is divided into four compartments. Each compartment is provided with a jacket for circulating steam or cold water. During startup, the reactor is heated and cooled during normal operation. The reactor is ceramic or glass lined, normally packed with ceramic material.

Ethylene and catalyst enter from the bottom of the reactor and the products from the top pass through a reflux condenser where the benzene condenses and gravitates to reactor but exhaust gases may be allowed to escape after a water wash. The alkylate from the top section of the reactor enters a settling tank to permit settling of catalyst; overflow from this tank is given a thorough wash with water and alkali to remove the acid tinge. The light organic phase containing alkylate and benzene is distilled to separate benzene and polyalkyl benzenes, which are formed in small amount. These are obtained from the distillation column, where ethyl benzene appears as overhead product. If diethyl benzene is also used for composite catalyst production then this is separated by distillation and recycled to the reactor. Ethyl benzene can also be produced using catalysts alumina silicate and phosphoric acid.

7.2.4.1 Dehydrogenation of Ethyl benzene for Styrene

Styrene is produced by dehydrogenation of ethyl benzene. The reaction proceeds over a catalyst containing Fe_2O_3 with other metallic oxides or on zinc oxide at 600–630°C. All these catalysts are mixed with a small amount of potassium carbonate, this facilitates the removal of carbon deposits on the catalyst surface prolonging the activity. For styrene production : 0.5 to 1.0% chromium oxide(14) is mixed with the support, *i.e.* either iron oxide or zinc oxide. The reaction temperature is kept in the range of 560-600°C. Tubular reactors are used for dehydrogenation and they look exactly the reactors of butane dehydrogenators. In the dehydrogenator, the partial pressure of hydrocarbon is kept around 0.1 bar by using a high ratio of steam to hydrocarbon. Styrene concentration in the product goes upto 35%. The difficulty in the end product (styrene) separation from ethyl benzene has to be enocountered in a proper way. Styrene is distilled under a high vacuum (30 mm Hg), where the boiling point of styrene is 54°C.

The method of extraction of styrene from other isomers has been described in Chapter 2. Fig. 7.10 shows the combined alkylation and dehydrogenation.

7.2.4.2 Cumene(14)

By alkylation with propylene, benzene yields I-propyl benzene which has been described in chapter 4. Cumene is mainly used for the production of phenol and (α) methyl styrene. The alkylation reaction is carried out with phosphoric acid deposited on kiesulguhr or pumice. Aluminium chloride is a useful catalyst for the production of ethyl benzene and propyl benzene. However phosphoric acid is the most convenient. Sulphuric acid is also used for alkylation however the formation of polymers with high strength of acid cannot be ignored. Hence 90% strength acid at 20-30 is used for the reaction.

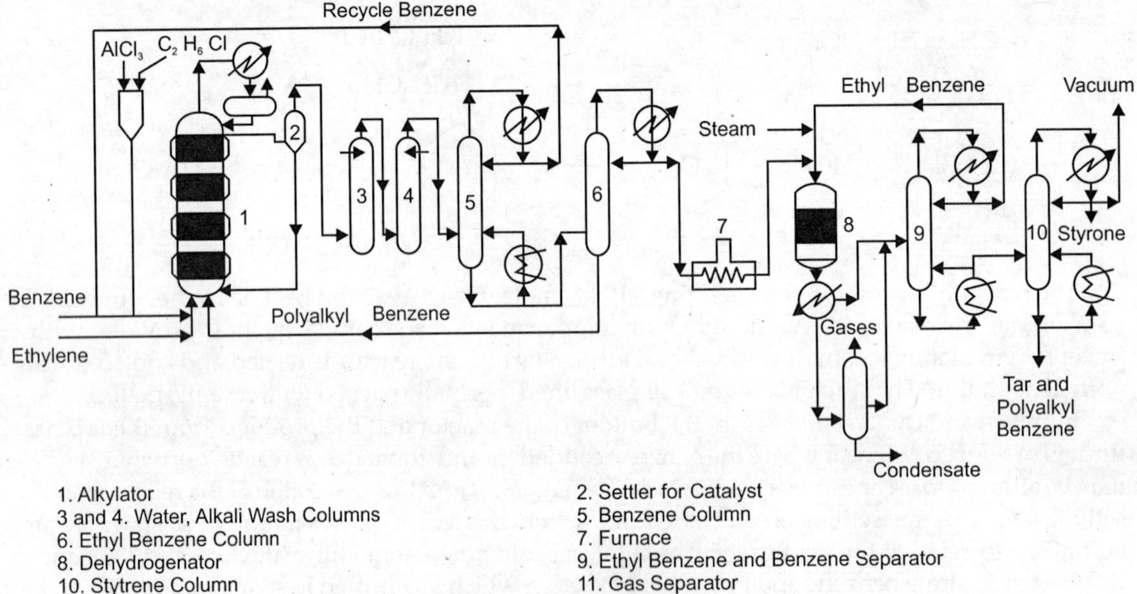

Fig. 7.10: Styrene from Benzene.

7.2.4.3 Phenol from Cumene

This seems to be the most economical route for the production of phenol; because of the important side product, acetone. The process has been described in detail in Chapter-4.

7.2.4.4 Oxidation of Benzene

Oxidation of benzene to phenol without ring opening is of scientific interest only.

$$C_6H_6 + 1/2\,O_2 \rightarrow C_6H_5OH$$

Catalysts like V_2O_5, WO_3, MoO_3 and platinum are used for oxidation. It is carried out at 300–400°C under a high pressure of 150-200 bars using air or oxygen. The reaction has no industrial significance. But, the ring opening reaction leading to maleic anhydride is quite alluring. In previous chapters oxidation routes of butanes, butenes have been described; and from benzene the description follows :

7.2.4.5 Maleic Anhydride

Benzene oxidation for maleic anhydride with air is the oldest process and is still viable.

$$2 \; \bigcirc\!\!\!\!\!\text{O} \; + \; 9O_2 \; \rightarrow \; 2 \; \begin{matrix} CHC=O \\ \| \qquad \quad \; \rangle O \\ CHC=O \end{matrix} + 4\,CO_2 + 4H_2O$$

This reaction is carried out in a fixed bed reactor, using oxides of vanadium and molybdenum. The reaction temperature is maintained around 380°C and the pressure is kept at normal conditions. The yield is about 65-70%. Flow sheet in Fig. 7.11 is based up on this technology.

A stream of benzene is sprayed (1) into hot air kept at 400–450°C, this gives a mixture of 1 to 30 benzene and air ratio. The hot mixture is passed into a reactor (2), which is provided with bundle of tubes containing catalyst pellets. In the initial stages heating may be done by circulating dowtherm into shell space of reactor; during reaction exothermic heat can be removed by the same salt. The vapour mixture is cooled in a series of coolers, whereby maleic anhydride is condensed and separated in the separator (3). The gases from the separator are treated with water in a scrubber. This gives a lean solution of acid, and waste gases escape from the top of the scrubber. The lean solution of acid and crude MA are mixed together and evaporated in spray type evaporator (5) under vacuum. Maleic anhydride in concentrated form is transferred from the dehydrator to distillation column (6), where maleic anhydride is obtained as an overhead product. Further purification leaves a product of 99.7% purity.

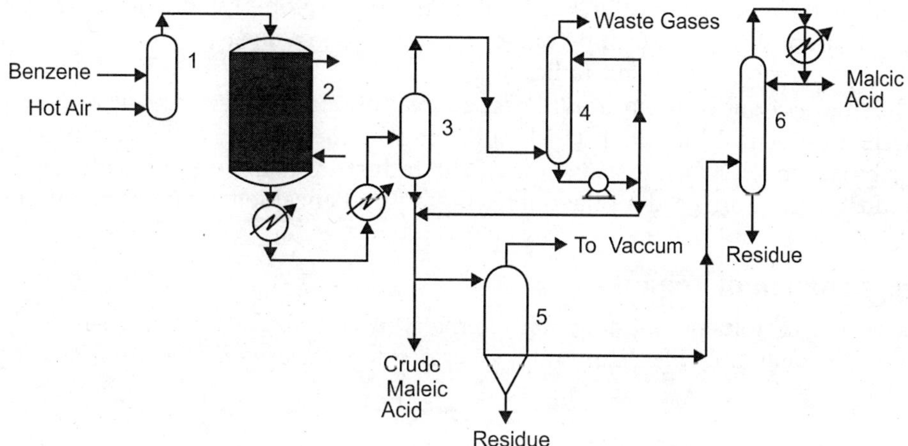

Fig. 7.11: Maleic acid from Benzene – Air Oxidation.

7.3 PRODUCTS FROM TOLUENE

Next to benzene, toluene forms important aromatic compound. Alkyl aromatic hydrocarbons behave in a different manner upon chlorination. The reaction proceeds by two routes : (a) Nucleus chlorination is an ionic reaction favoured by low temperature and chlorine carriers, like aluminium and iron chlorides, (b) Substitution of hydrogen by chlorine in side chains follow free radical mechanism. This reaction proceeds at slightly higher temperature above 100°C and radical initiators are required. Hence nuclear substituted or side chain substituted products are obtained by halogenation at different conditions. The following discussion pertains to nuclear substitution.

7.3.1 Chloro-Toluenes

Ferric chloride influences profoundly in ring substitutions, three different isomers, ortho, meta para are obtained. Ferric chloride is good for ortho substitution like other Lewis acids. Metal salts and metal sulphides are good in yielding para substituted products.

The chlorination can be carried out either in a continuous fashion or batch wise. Higher chloro toluenes are obtained by further chlorination of mono compounds. Antimony trichloride, zirconium tetrachloride are helpful in chlorination of *p*-chloro toluene. Ferric chloride produces 2, 3, 6 isomers to considerable extent, while chlorinating *o*-chlorotoluene. Similarly metal sulphides also act as good catalysts to produce 2, 3, 4, 5 tetra chloro toluenes. Exhaustive chlorination in an organic solvent results in penta chloro toluene.

7.3.1.1 Uses of Cholorotoluenes

O-chlorotoluene is used as solvent, p-chlorotoluene is used as an intermediate in herbicide production, p-chloro benzaldehyde is used in dyestuff industry. O-chloro toluene is converted to o-chloro benzaldehyde that has great application in dyestuff industry. Most of the other higher chlorotoluenes are also used for similar purposes.

7.3.1.2 Side Chain Substitution

Photochemical chlorination of toluene at about 100-140°C yields benzylchloride, benzalchloride and benzo trichlorides. Photochemical reaction is brought out by means of 40 W. fluorescent lamp fitted centrally in the reactor. Simultaneously all the hydrogen atoms can be substituted by chlorine. Sometimes phosphorus trichloride or peroxides can be used as catalysts, these catalysts significantly accelerate the rate of formation of free radicals.

In addition to side substitution simultaneous substitution in benzene ring also takes place.

Benzylchloride, benzalchloride and trichloride are colourless liquids with pungent odour. Benzyl alcohol, or butylbenzylphthalate are some of the industrial compounds produced from these chlorides. Most of these are used for flavour or fragrances. Benzaldehyde is another useful product from these.

7.3.2 Sulphonatlon of Toluene

Oleum is used for toluene sulphonation; alternatively chlorosulphonic acid can also be used for sulphonation. Major amount of sodium sulphonate of toluene is used for the production of cresol.

$$C_6H_5CH_3 + SO_3 \longrightarrow CH_3C_6H_4SO_3H \quad \text{(a)}$$

$$C_6H_5CH_3 \quad SO_3 + NaOH \longrightarrow CH_3C_6H_4ONa + NaSO_2 + 2H_2O \quad \text{(b)}$$

$$CH_3C_6H_4ONa + SO_2 + H_2O \longrightarrow CH_3C_6H_4-OH + Na_2SO_3 \quad \text{(c)}$$

The reaction (*a*) is easy and can be accomplished either in liquid phase or in vapour phase. Theoretically SO_3 is the best sulphonating agent as there is no side product formation, as seen in the case of sulphuric acid or chlorosulphonic acid. The resultant product is entirely p-sulphonic acid, however, with sulphuric acid all the three isomers appear, particularly at higher temperature, higher reaction times, metasulphonic acid dominates.

If the sulphonation with SO_2 is carried out in presence of sodium sulphite, tolyl sulphones can be greatly reduced. Usually the reaction is carried out in three reactors kept in series, first one operating at O°C while the other two operate at 30°C. Usually excess toluene is distilled of for recycling. The mixed sulphonic acids are converted into o and p cresols. The main isomer of industrial significance is p-sulphonic acid, which is used as catalyst in organic syntheses. p-toluene sulphonylchloride is obtained by treatment of sulphonic acid with PCl_5; it is used in organic syntheses, and in disinfectant formulations.

7.3.3 O-cresols

Cresols are produced for *p* and *m* sulphonic acid isomers.. Crude *p* and *m* sulphonic acid is melted and added to molten caustic soda containing 4 to 5 mole per cent caustic potash at 350°C. Then sodium cresylates are formed. These cresylates are extracted with water and then decomposed with SO_2 in presence of steam. Crude cresols are finally distillied for separation.

The reactions are shown below.

$$CH_3.\,C_6H_4.\,SO_3 + 3N_aOH \rightarrow CH_3.\,C_6H_4.\,ON_a + Na_2SO_3 + 2H_2O$$

$$2CH_3C_6H_4ONa + SO_2 + H_2O \rightarrow 2CH_3C_6H_4OH + Na_2SO_3$$

It has been reported that much difficulty is experienced in agitating the mass, if sodium hydroxide alone is used in fusion; hence a small amount of caustic potash is also used. In the reaction product 40% sodium cresylate is likely to be present after fusion is over. The fused mass is then leached with water to extract sodium cresylate and the resulting slurry is centrifuged to get the crystals of sodium sulphite. The remaining solution is then blown with sulphur dioxide. The crude cresols of separated and distilled for further purification.

Cresolscan also be produced by a number of processes; like phenol produced in Dow process, chlorotoluene hydrolysis yields cresol. Considerable amount of isomers are expected. The optimum conditions for hydrolysis is chlorotoluene are listed as : reaction time 3 to 4 hours, composition of mixture of 3.5 : 1 mole ratio of NaOH and chlorotoluene and temperature 370-380°C. Cresol yields from different chlorotoluenes are given below:

O-chlorotoluene	Cresols	57%
	Tolyl ethers	12%
	Tolyl cresols	6%
p-chlorotoluene	Cresols	48%
	Tolyl ethers	13%
	Tolyl cresols	10%

The cresols isomer distribution 0-32% m-43.5% and p-24.5% is the usual pattern available in cresols obtained from hydrolysis product of chlorotoluene. Tolyl ether may be recycled to improve the yield of cresols. But the hydroperoxidation of cymenes is similar to the cymenes hydroperoxidation process for the synthesis of phenol; cymene hydroperoxidation may be used for the production of cresols. Though complications are mandatory during alkylation of toluene with propylene because of difficulty in separating the isomers of cymene, Alkylation may be done by using aluminium chloride, titanium tetrachloride or phosphoric acid on kiesulguhr. Lewis acid catalysts like $AlCl_3$, BF_3 are used at temperature 75-100°C while $TiCl_4$ and $AlCl_3$ mixture can be used at 70-80°C. However with phosphoric acid the temperature is about 300°C. *o*, *m* and *p* cymenes are present 40%, 31% and 29% in a mixture from phosphoric acid alkylation.

The next step involves in hydroperoxidation of p-cymene . p-cymene is converted easily into hydroperoxide at 85°C in presence of 2% sodium bicarbonate containing benzol peroxide initiator[(8)].

The cleavage of *p*-cymene hydroperoxide yields p-cresol and acetone of about 80%; cleavage is done with sulphuric acid.

7.3.3.1 Through Methylcyclohexane Oxidation

Cyclohexane is utilised for the production of phenol, while methyl cyclohexane can be used for the production of cresol. This process however does not appear to be attractive. Hydrogenation of toluene, and oxidation of methylcyclohexane using boric acid catalyst in sequence gives methyl cyclohexanol which on dehydrogenation gives rise to cresol isomers of equal concentration.

Most of these cresols are used as plasticisers, a chemical for phenolic resins. As antioxidants they are used in petroleum industry very much.

7.3.4 Nitration of Toluene

Toluene nitration is precursor to many variety of chemicals like dyestuffs, isocynates etc. Toluene nitration is an electrophilic substitution. It is carried out by a mixture of nitric and sulphuric acids. Mononitro toluene is obtained in three isomeric forms, when the nitration temperature is low (30 to 40°C) and the concentration of NO_2^+ ions is less. Usually 2 and 4 nitrotoluenes will be in large quantities while 3 nitrotoluene is formed upto 5% only.

The b.p. and m.p.s. of these isomers along with their densities are cited in Table 7.1.

Table 7.1: Isomers of some Aromatics

Xylenes	*B.P.*	*M.P.*	*Sp.Gr. at 20°C*
o-xylene	144.41	– 25.17	0.880
m-xylene	139.10	– 47.40	0.864
p-xylene	138.35	13.26	0.861
Ethyl benzene	136.18	– 94.97	0.867
Cumenes			
Pseudo cumene	169.4	– 43.8	0.8802
Mesitylene	164.7	– 44.7	0.8696
Hemi mellitene	176.1	– 25.4	0.8987
i-propyl benzene	152.5	– 96.9	0.862

7.3.4.1 Dinitrotoluenes

Dinitrotoluenes are formed under forcing conditions. Mixed mononitrotoluenes are utilised for producing dinitrotoluenes. These are used as good starting materials for tolylene diamine and tolylene disocynates as shown in the following way :

Furter Nitration

Nitration

Maximum product Traces Less

The main product from dinitrotoluene is tolylenediamine. Dinitrotoluene is hydrogenated in presence of alcohol and nickel catalyst at moderate pressures.

$$CH_3C_6H_3(NO_2)_2 + H_2 \longrightarrow CH_3C_6H_3(NH_2)_2 + 1100\ kJ$$

Continuous process employs a slurry catalyst. After complete reduction the liquid stream from the reactors is stripped of solvent and any toluidines. Tolylenediamine reacts with phosgene to give rise to toluene diisocynate.

7.3.4.2 Toluene Diisocynates (TDI)

Main product of commerce is available as a mixture of 2:4 and 2:6 diisocynates. It is a colourless liquid boiling at 251°C. Most of it goes for polycarbonates preparation and fungicides, disinfectants etc. Toluene diisocynate is currently manufactured by reacting toluene-diamine with phosgene in a solvent like o-dichlorobenzene at a temperature of 100°C under a pressure of 10 bars. The solvent chosen should be inert to all reactants and products. The product is highly toxic and affects the respiratory system. Though the reactions are simple, the chemistry involved is very complicated.

Another route suggested for the production of TDI is by carboxylation of DNT. The reactions are shown below :

$$C_6H_5CH_3 \xrightarrow{HNO_3} CH_3C_6H_3(NO_2)_2 \xrightarrow{6CO_2} CH_3C_6H_3(CON)_2$$

Palladium chloride is used as carboxylation catalyst. The reaction proceeds at 190°C and high pressure 300 bars. This route is not promising due to the short life of catalyst. The process outlines are shown in Fig. 7.12.

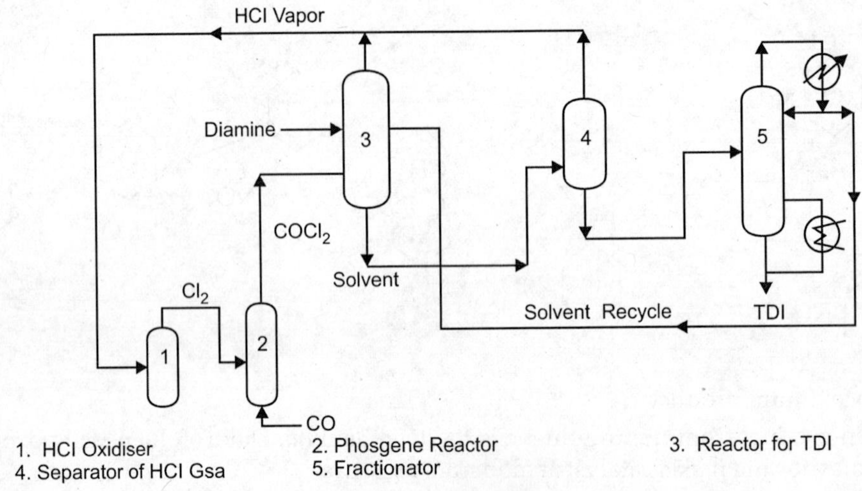

Fig. 7.12: Production of TDI.

7.4 OXIDATION PRODUCTS OF TOLUENE[15]

By oxidation of toluene, benzyl alcohol benzaldehyde and benzoic acid are obtained.

7.4.1 Benzaldehyde

It is produced by oxidation of toluene. This is low yield process because the oxidation goes through an intermediate stage of alcohol.

$$C_6H_5CH_3 + O_2 \longrightarrow C_6H_5CH_2OH \xrightarrow{O_2} C_6H_5CHO \longrightarrow C_6H_5COOH$$

Successive oxidations are more rapid and create a problem for stopping the reaction short of acid. This is tackled by using air oxidation for a short time with selective catalyst. Typical reaction conditions are :

Temperature	500°C
Catalyst	UO_2 or MnO_2
Conversion	10-20%

Many chemical oxidising agents like potassium dichromate in presence of sulphuric acid, manganese dioxide and sulphuric acid, lead peroxide and sulphuric acid, and sulphur dioxide are reported for the oxidation to toluene. However, industrially oxidation is carried out in liquid phase or in vapour phase using air. In liquid phase oxidation cobalt acetate, naphthenate etc. are used as catalysts. Vapour phase oxidation uses zinc oxidechromium catalyst or uranium molybdate, copper molybdate on inert support. Liquid phase reaction takes place at 150-160°C under a pressure of 10 bars. Air to toluene ratio is maintained around 2 M^3 per kg toluene. Catalyst concentration of 0.02 to

0.05% of toluene in feed is allowed. Under these conditions, yields of over 90%, at a conversion per pass of 30-35% are reported. The reaction proceeds *via* free mechanism. The main product is benzoic acid, while benzaldehyde and benzyl benzoate are side products.

$$\overset{CH_3}{\bigcirc} + O_2 \longrightarrow \overset{CHO}{\bigcirc} + H_2O$$

$$2\overset{CH_3}{\bigcirc} + 3O_2 \longrightarrow 2\overset{COOH}{\bigcirc} + H_2O$$

Fig. 7.13: shows the liquid phase oxidation of toluene to benzoic acid.

Toluene and catalyst (cobalt acetate dissolved in water) are charged into the reactors, kept at 140–145°C under a pressure of 10 bars. Air is fed continuously into this system. The heat of the reaction is absorbed in external heat exchanger. The product average 35% benzoic acid and has to be stripped of toluene in a steam stripping column. Toluene and others collected are recycled back. Benzoic acid is obtained as a side cut to be further purified in a finishing column. Catalyst is collected from the bottom of the stripper and recycled after scrubbing. Vapour phase oxidation of toluene is carried out with air 400–500°C over zinc oxide-chromium catalyst. Vanadium pentoxide on inert material has been also cited.

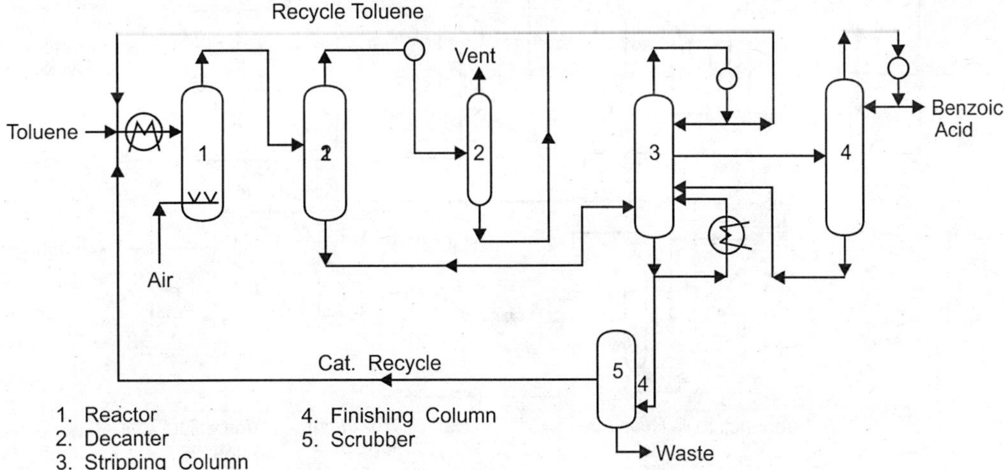

Fig. 7.13: Benzoic Acid Liquid Phase Oxidation of Toluene.

7.4.1.1 Benzoic Acid Products(16)

Phenol can be produced from benzoic acid by subjecting benzoic acid to catalytic oxidation. The reaction is shown below :

$$2\overset{COOH}{\bigcirc} + O_2 \longrightarrow 2\overset{OH}{\bigcirc} + 2CO_2$$

The reaction conditions are :

Temperature	220–240°C
Pressure	Atmospheric
Catalyst	Magnesium, cupric benzoates (initiated by magnesium benzoate)

Conversion usually exceeds 50% in presence of steam and air.

Molten benzoic acid and soluble copper catalyst along with promoters is fed through a heater to the oxidiser. The vapours from the reactor are passed into an azeotropic column where water is removed as an azeotrope with toluene. From the azeotropic column, the side-cut is passed into a stripping column to remove the low volatiles. The bottoms from this column are finally fractionated in another column to get phenol as overhead product. The bottoms from azeotropic column are routed to the oxidiser.

7.4.2 Caprolactum

Benzoic acid is the starting material for the production of caprolactum. Benzoic acid is hydrogenated over conventional catalysts like palladium at 170°C under a pressure of 15-20 bars. Two or three stirred tank reactors (1) in series may be required. Conversion is about 85%. The product being cyclohexane carboxylic acid, this when (2) dehydrated yields ketone as shown in the reactions.

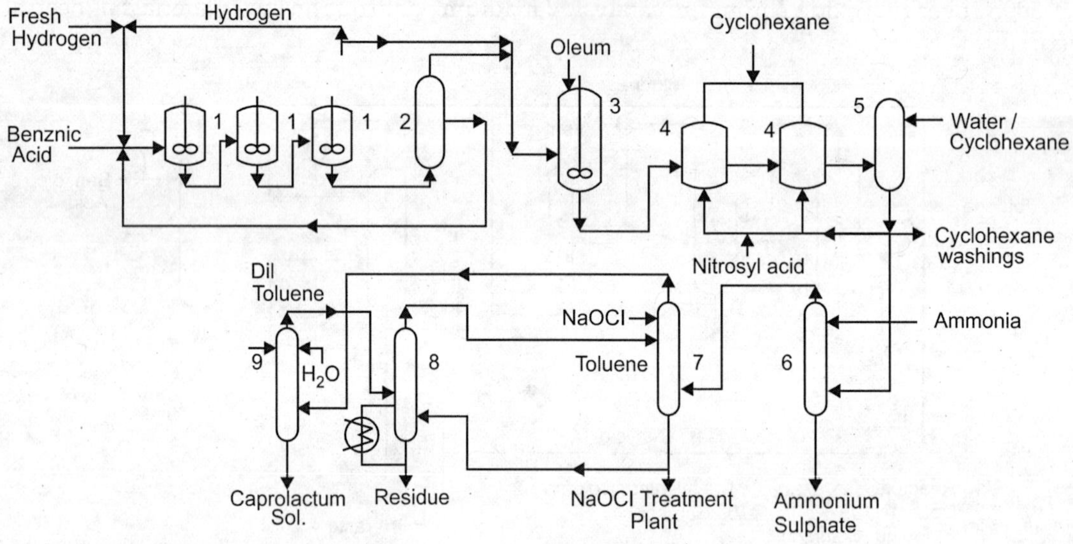

1. Hydrogenators
2. Separators
3. Oleum Treater
4. Nitrosation Reactors
5. Water/Cyclohexane Scrubber
6. Ammonia Treater
7. Hypochlorite - Toluene Extractor
8. Toluene Regenerator
9. Water Extractor for Caprolactum

Fig. 7.14: Production of Caprolactum

COOH COOH

O $+H_2$ ⟶ O $\xrightarrow{-H_2O}$ O

O

NOH C=O

⟶ O ⟶ $(CH_2)_6$

$+NOHSO_4$ NH

Solution of ketone is mixed with oleum (3) and sent into nitrosation reactor (4) nitrosyl ion attacks ketone forming oxime. The oxime arranges into caprolactum due to Beckman rearrangement.

Nitrosyl sulphuric acid is produced by absorbing oxides of nitrogen in oleum. The effluent after the nitrosyl reaction is diluted with water (5) and unreacted cyclohexane carboxylic acid is extracted with cyclohexane (5) solvent. The solution of caprolactum in sulphuric acid is first neutralised with ammonia (6) which results in two layers, upper layer being organic contains crude caprolactum. Caprolactum can be purified by extraction with toluene and washed with sodium hypochloric solution. The solution of toluene containing caprolactum is again extracted with water and the dilute solution is evaporated in a falling film evaporator to get crystalline form of caprolactum.

7.4.3 Terephthalic Acid[17]

Though terephthalic acid is produced conventionally by oxidation of p-xylene, it can also be obtained from toluene. Toluene dissolved in acetic acid is reacted with formaldehyde and hydrochloric acid and the product is oxidised in presence of a metal oxide catalyst by oxygen. Chloromethylation gives rise to para and ortho chloromethyl toluene almost in equal proportions. Next stage is conversion of P and O chloromethyl toluene to phthalic and terephthalic acid by oxidation.

COOK COOK

2 O ⟶ O + O

COOK

COOK COOK COOK COOK

O + O ⟶ O + O

COOK COOK COOK COOK

CO

COOK COOK COOK

O + O O O ⟶ O + O

COOK CO COOK COOK

COOK COOK

O ⇌ O

COOK COOK

In Henkel's process potassium benzoate is disproportionated to terephthalic acid and benzene by reacting with CO_2 over cadmium catalyst, at 400°C under a pressure of 50 bars.

7.5 CHEMICALS FROM XYLENE

Most of the xylenes are fetched from reformates. At one time coal- tar products convincingly supplied a large amount of armonatics including xylenes. There are three isomeric xylenes and these have to be separated into individual cuts for economic benefits before chemical manufacture is attempted. Most captivating of these xylenes is *p*-xylenes and the separation of which is also regarded as difficult. Description of the chemicals obtained from different xylenes follow.

7.5.1 Xylene

The separation of the mixed xylenes into individual components has been described earlier (chapter 2). Xylenes are present in reformates to the extent of 20% and separation becomes difficult specially when ethylbenzene is present due to the closeness of boiling point (2.9°C difference only). Usually ethylbenzene is extracted from the reformate cuts or it can be separated by using columns of 300 plates with a very high reflux. It is desirable that all xylenes are isomerised to p-xylene, the most important of all the ones. When *o*-xylene is wanted, it is obtained by precise fractionation from *p*-xylene.

Most important chemical produced from o-xylene is phthalic anhydride o-xylene on oxidation in presence of a catalyst like vanadium pentaoxide at 550°C yields phthalic anhydride. Though phthalic anhydride may be produced by oxidation of naphthalene, this forms the major method of consumption of *o*-xylene.

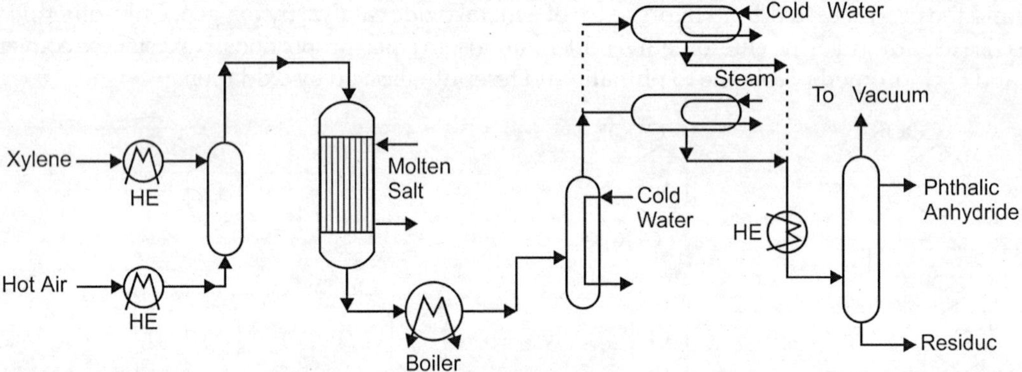

Fig. 7.15: Phthalic Anhydride from o-xylene.

The production of phthalic anhydride by air oxidation is outlined in Fig. 7.15. Preheated *o*-xylene is sent into a hot stream of air which is kept at 140-150°C and under a pressure of 2 bars. Air is enriched with vapours of *o*-xylene and care should be taken that the mixture is outside the combustion limits. The air-vapour mixture enters the fixed bed reactor, where the catalyst is packed in tubes. Exothermic heat is absorbed by circulating molten salt in the space between tubes and shell. The hot vapours are allowed to exchange heat in waste heat boiler. By cooling further phthalic anhydride is sublimed and it deposits on the surface of switch condenser. The deposited film of phthalic anhydride is melted by means of circulation of steam, usually two parallel condensers are kept in the circuit, while one acts as a cooler another acts as a melter. The molten anhydride is again distilled under vacuum of purification. Alongwith phthalic anhydride a small amount of benzoic acid and maleic anhydride also accompany.

Maleic anhydride can be recovered, if required. Phathalic anhydride is mainly converted to phthalic acid, which forms the mother substance for phthalates used in polyesters and alkyl resins. Phthalic anhydride is converted into phthalic acid through potassium phthalates. Potassium phthalates are first isomerised to terephthalates in presence of zinc-cadmium catalyst. By hydrolysing potassium terephthate phthalic acid can be obtained.

A new process developed by Alusuise Italia shows the following advantages over conventional processes:

(*a*) Increase in catalyst activity by more than 40%

(*b*) A reduction in air-oxylene ratio from 20 : 1 to 9.5 : 1

(*c*) Reduction in capital cost and other investment charges.

Usually about 50 kgs. of meleic anhydride accompanies one ton of phthalic anhydride.

7.5.1.1 Phthalonitrile

O-xylene is oxidised by indirect method, using a metal oxide catalyst in presence of ammonia in a fluidised bed reactor. The reactions are shown below :

$C_6H_4(CH_3)_2$ + $NH_3(O_2)$ → $C_6H_4(CN)_2$ | $C_6H_4(CH_3)(CN)$; $C_6H_4(CONH_2)_2$; $C_6H_4(CO)_2N$; CO + CO_2 + H_2O ; N_2

Major products obtained by oxidation are o-toluonitrile and phthalonitrile. Side products like phthalumide/phthalmide also accompany.

7.6 M-XYLENE

m-xylene is the least preferred one of all the xylenes. Most of this xylene is isomerised to get p-xylene. The only useful derivative so far known from m-xylene is isophthalic acid. As in the oxidation of o-xylene, this can be oxidised for the production of iso acid. The oxidation is carried out by ammonium sulphite in liquid phase :

$$C_6H_4(CH_3)_2 + 2(NH_4)_2SO_3 \longrightarrow C_6H_4(COOH)_2 + 2H_2S + 4NH_3 + 2H_2O$$

Like in other oxidation reactions first methyl group is rapidly oxidised while the second one exhibits adamancy. Isophthalic acid is used in the manufacture of polyesters alkyd resins. Isophthalonitrile can aso be produced by ammonium sulphite oxidation.

7.7 P-XYLENE[18]

Virtually *p*-xylene is the only versatile xylene used in chemical industry. Its main use is as raw material for terephthalic acid and dimethyl terephthalate DMT. (B.P. 282°C ; M.P. 140.62) *p*-xylene demand had gone beyond 10 lakh ton by 1995 from a demand of 4.5 lakh ton in 1984. It is planned to reduce C_6 fraction in gasolines and increase C_7 to C_9 which possibly justifies the supply of xylene demands. This will in fact demand more oxygenates in gasoline resulting in MTBE consumption to cross 20 MT by 1997.

Liquid phase oxidation, in presence of cobalt acetate catalyst gives a high yield of terephthalic acid. Cobalt acetate solution is some usually activated with methyl ethylketone. Oxidation is conducted at a temperature of 200°C under a pressure of 15-30 bars, using air. Mobil process uses a temperature of 130°C, oxidant used is oxygen instead of air. Cobalt acetate in dissolved form as acetic acid solution is used.

The process outlines are shown in Fig. 7.16 *p*-xylene cobalt catalyst with activator (MEK) are introduced into a reactor (1). Oxygen is blown into this solution. The gaseous products are separated in a separator (2). The liquid products are washed with acetic acid and the washings are sent for acid and catalyst recovery. The filtrate from the washer (3) is sent to recover acetic acid and catalyst. Activator MEK is also converted to acetic acid, which is recovered alongwith other acids. Crude TPA product is many times washed with acetic acid and heated. The solution is finally evaporated and crystallised (4) or centrifuged. The crystals are melted and sent into esterification unit (5) alongwith methanol vapour, unconsumed methanol vapour is recycled back. The esterified product is distilled to separate water and methanol vapour. Ester is obtained as the bottom product of the column (6).

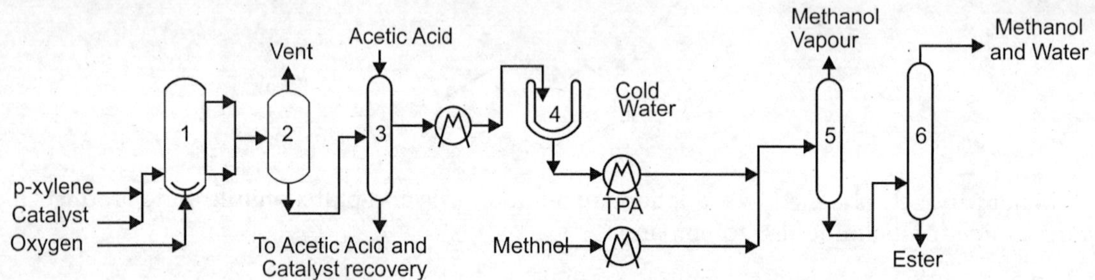

Fig. 7.16: Dimethyl Terephthalate from P-Xylene.

7.7.1 Xylenes Isomerisation

Any non-equilibrium mixture of C_8 aromatics can be efficiently isomerised towards equilibrium. From the product mix, the desired xylene is separated and the remaining is recycled to extinction.

Charge stock may be reformate or hydrotreated pyrolysis gasoline. A lot of ethylbenzene is likely to present in pyrolysis gasoline, which is no detriment to operation. Isomerisation consumes relatively less energy and is conducted in presence of hydrogen to arrest coke formation on catalyst. Purity of the product essentially depends upon the separation procedures involved. Temperature and pressure are moderate any catalyst of available acid sites is useful. Silica-alumina, platinum and palladium are supposed to be the best catalysts. The product from any non-equilibrium feed shows 2 : 1 ratio of para to orthoxylenes.

7.8 HIGHER AROMATICS

Next to xylenes, cumenes are important. The alkylation of benzene may lead upto hexa methyl benzene; but the discussion is limited upto trimethyl benzenes. Trimethyl benzenes are usually isolated from reformer or cracker streams. The isolated stream is conveniently isomerised.

7.8.1 Poscudocumene[20]

Like other aromatics, it gives on oxidation trimellitic anhydride (TMA). Liquid phase air oxidation in presence of manganese catalyst results in this acid. TMA is basically used as plasticiser and a curing agent for epoxy resin. It is also used for polyester-imides and in the production of vitamins. The chemicals of importance are shown by the reactions.

CH_3 CH_3 O CH_3 — Oxidation → $O = C - O$, $C = O$, O, COOH — Trimellitic Acid (Anhydride)

Pseudocumene

Nitrogen / Reduction → NH_2 CH_3 O CH_3 CH_3 — Pseudo Curmidine

H_3C CH_3 O CH_3 — Nitrogen / Reduction → NH_2 CH_3 CH_3 O CH_3 — Mesitidine

Mesitylene

Oxidation → COOH COOH COOH O — Trimesic Acid

7.8.2 Mesitylene

It is obtained by aromatic extraction of C_9 streams of reformate or naphtha cracker. It is also obtained by isomerisation of other C_9 fractions. It is used as a raw material for trimesic acid and aminoderivatives (anti oxidants). Its nitration product gives trimethyl aniline useful for dyes.

7.8.3 Cumene

1-Methylethyl Benzene 2 - phenylpropane, o-propylbenzene-cumene, $C_9 H_{12}$, are all isomeric.

It is normally liquid, substituted aromatic compound in the benzene, toluene, ethylbenzene series. It is the major intermediate chemical used in the worldwide production of Phenol, acetone and 1-methylstyrene all of which are components for plastic resins. High purity cumene is normally manufactured from propylene and benzene although it is a minor constituent of most gasolines.

Properties : Freezing point – 96°C. Boiling point 152.39°C

Density 0.87 g/cm^3 at 30°C : viscosity 0.791 Cp at 20°C.

Manufacture

Cumene as a pure chemical is manufactured exclusively from propylene and benzene utilising as acidic catalyst. Many different catalyst systems have been proposed, including both crystalline and noncrystalline silicas, boronfluoride, aluminiumchloride and phosphoric acid but current plants employ solid phosphoric acid catalyst.

The process utilises an excess of benzene in the reactor charge to minimise side reactions to increase product purity and yield. The propylene feed must be essentially free of ethylene and butylene but can contain substantial quantities of propane such as that occurring in a refinery stream from fluid catalytic cracking operations. A small portion of the propylene charge oligomerizes to form materials boiling with cumene.

Specifications of Cumene:

1. Specific gravity	0.864 to 0.867
2. Sulphur compounds	2 PPm (max)
3. Cumene wt%	99.9 (min)
4. Butyl benzene	500 PPm
5. n-propyl benzene	500 PPm
6. Ethyl benzene	500 PPm

7.9 NAPHTHALENE[21]

Usually higher aromatics are obtained from coal tar. Naphthalene can be produced by hydro dealkylation of methyl, dimethyl naphthalene of reformer bottoms. Purity of naphthalene obtained from such streams is of 80% order. Naphthalene is mainly utilised for the production of phthalic anhydride by air oxidation using V_2O_5 as catalyst. Naphthol is the second major compound. It is produced by first sulphonation of naphthalene at high temperature and the sulphonated product is fused with caustic. (β)-naphthol is used for naphthalysamine production, anti-oxidant, α-naphthol is used as insecticide "Sevin". Sulphonation preceded by alkylation can give rise to detergents. The reaction shows :

O O $\xrightarrow[SO_3]{H_2SO_4}$ O O (SO_3H, SO_3H, SO_3H) $\xrightarrow[\text{Hydrogenation}]{HNO_3}$ O O (NH_2, HO_3S, SO_3H, SO_3)

O O (NH_2, HO_3S, HSO_3, HO_3S) $\xrightarrow[\text{With Na(OH)}]{\text{Fusion}}$ O (NH_2, OH, SO_3Na, SO_3Na)

Higher aromatics other than naphthalene are generally obtained from coal tar distillation only. However the reactions of benzene and chlorine with xylene and toluene are usually reflected in the case of higher aromatics.

REFERENCES

1. Z. Quadir, B.K.B Rao & S. Chand; Urja, (INDIA) Aug. 85, 93.
2. B. K. B. Rao: Modern Petroleum Refining Processes; Oxford-IBH; 1984.
3. HCP; April 72, 131.
4. K. Venkateswarulu, B. K. B Rao, & T. S. Banerjee, Chemical Industry Development; Aug 72, 21.
5. B. K. B. Rao, HCP Nov, 76; 237.
6. B. K. B. Rao, HCP Dec. 78,105.
7. EH. Hamcock, Toluene, Xylenes and their Industrial Derivatives, 225.
8. K. M. Wang and J. H. Lansford; J. Cat. 69, 1981, 392.
9. M. K. Barooah, Shrichand, B. K. B. Rao, Advances in Catalysis; Science and Technology, Wiley Eastern Ltd. N. Delhi, 365.
10. D. Grandio, F.H. Schneider, A.B. Schwartz & J.J. Wise : HCP Aug. 72, 85.
11. HCP, Aug. 69, 109
12. Manfred Gans; HCP, Nov. 76,148.
13. D.M. Samuel; Industrial Chemistry Organic, Royal Institute of Chemistry 2nd, 1972.
14. N.E. Ockerbloom, HCP, Dec. 71, 101.
15. James E. Lyons, HCP; Nov. 1980, 108.
16. J.B. Saunby & B.W. Kiff; HCP Nov. 1976, 251.
17. Ichikawa & Y. Takeuchi. HCP, Nov. 72, 104.
18. N.E. Ockerbloom, HCP, Jan. 72,93 and HCP, April 1972,115.
19. Eliosantacesaria, Massino Mobidetli, IEC, Process Design & Dev. Vol. 21, B, 1982, 440.
20. Albert V. Hahn. The Petrochemical Industry, McGraw Hill Book Co. 1970.
21. Luigi Verde and Amleto Neri; HCP, Nov. 1984.

CHAPTER 8

Synthetic Fibres

8.1 INTRODUCTION

Man habitually used different types of fibres since the dawn of civilization.

All the fibres can be classified into three types namely, natural, man-made and inorganic fibres like silicon, glass and carbon. Natural fibres are obtained from plants and animals (Ex. cotton, jute, wool, silk etc.). Man-made fibres are divided into two categories, cellulosic and synthetic. All types of fabrics produced together account for 29,000 million metres (97-98). Cellulosic fibres are reshaped or rearranged fibres from the available cellulose, like rayons and acetates. Other synthetic fibres strictly derived from petroleum based monomers. The consumption of synthetic fibres is increasing in every country and it is no exception to India. The pattern of production and consumption of synthetic fibres also change and the change is rapid. During sixties nylon was holding a dominant role accounting 60% of all the synthetic fibres, prior to which rayons occupied that position. After 1970 the production of nylons witnessed a steep fall to about 30% while polyesters accounted about 50% and the acrylics and other fibres barely constituted 20%. By 2000 A.D. the synthetic fibre production round the world may be about 20 million tonnes, constituting 48 to 50% of the total man-made fibres. In India the fibre consumption is not much, *i.e.* only 15 metres per capita of which the synthetic fibres constitute only 15%. However, the growth rate of synthetic fibres may be around 2% in the coming years. Synthetic fibres produced by various concerns are shown in Table 8.1.

Table 8.1: Producers of Synthetic Fibres

Nylon Filament Yarn	*Licensed Capacity*	*Installed Capacity (Tons)*
Baroda Rayon Corporation Ltd. Bombay	2436	2436
Garware Nylons Ltd. Bombay	4830	4830
Century Enka Ltd. Bombay	3640	3340
J.K. Synthetics Ltd. kanpur	5376	5400
Gujarat Industrial Development Corporation	2100	2100
Modi Pon Ltd. U.P.	4760	4760
Nirlon Synthetics & Fibres, Chemicals Ltd., Bombay	5308	3840
Shree Synthetics Ltd., Ujjain	3452	2175
Stretch Fibres, Bombay	4200	540
Taplon Ltd. Punjab	2000	—
Nylon Staple Fibre		

J.K. Synthetics Ltd. Kota	1000	1000
Nirlon Synthetics, Fibres & Chemicals	2190	2190
Shree Ram Fibres, Delhi	3000	3000
Asthan Import, Export Co.	—	300
Nylon Tyre Cord		
Baroda Rayon Corp. Bombay	2000	2000
J.K. Synthetics Ltd. Kanpur	3000	3000
National Rayon Corp. Ltd. Bombay	4627	3300
Nirlon Synthetic Fibres & Chemicals	4000	2190
Shree Ram Fibres Ltd.	6500	5000
Acrylic Staple Fibre		
IPCL	12,000	12000
J.K. Synthetics Ltd. Kanpur, Kota	4000	4000
Polyester Staple Fibre		
Ahmedabad Manufacturing & Calico	12,000	7800
Bongaigon Refinery & Petrochemicals	30,000	—
Chemicals & Fibres India Ltd. Bombay	10,000	10,000
J.K. Synthetics, Kanpur & Kota	12,000	6000
Indian Organic Chemicals, Madras	10,000	9000
Swadeshi Polytex, Ghazibad	8000	6100
Punjab State Industrial Development Corpn.	—	6000
Chandigarh		
Polyster Filament Yarn		
Baroda Rayon Corpn. Udhana	1800	1000
Century Enka Ltd. Bombay	1500	4200
Garwara Nylons Ltd. Bombay	1700	1710
Modi Pon Ltd. U.P.	1700	1700
Nirlon Synthetics & Fibre Ltd. Bombay	2655	1752
Orkay Silk Mills, Bombay	6000	6000
Reliance Textiles, Haldia	—	2500
Titagraph paper Mills, Calcutta	—	2500
Shree Synthetics Ltd., Kubba, Ujjaini	10,000	10,000
Stretch Fibres, Nagpur	500	1097
Petro Foils Co-operative Ltd. Baroda	7000	7000
National Rayon Corpn. Ltd.	7000	—
Kesoram Rayon, Tribeni, Hoogly	—	2000
Polyropylene Staple Fibre		
Neomer Ltd. Vadodadra	6000	4000
Indo Rama	25,000	

8.2 PRODUCTION TECHNIQUES OF SYNTHETIC FIBRES

A synthetic fibre is supposed to be at least 100 times longer than its width of a monomer constituting the polymer, and the molecular weight lies in the range of 20,000-50,000. The manufacture of synthetic fibres basically involves the physical change of same or different linear polymers from a relatively disordered state into a continuous filament, where the chain of molecules orient in a longitudinal axis of the polymer. The suitability of a polymer for filament production is best judged by the following properties :

1. High melting point, about 90°C or above.
2. Linear and symmetrical structure and high molecular weight.
3. Chemical resistant.
4. Durability, easy processing and dyeing etc.
5. Good mechanical properties.
6. Reclaimability (may not be always possible.)

Generally three major processes are used for production of fibres from polymers; namely melt spinning, dry spinning and solution (wet) spinning.

8.3 MELT SPINNING

This is used for polymers like polyesters, nylons, where the polymer can be melted without decomposition. The melt is extruded through spinnerets into a continuous filament. The filament is in the disoriented state and possesses less tensile strength. Hence the filament is drawn or stretched by sets of rolls. The rolls operate with differential speeds, when the fibre passes through each set of rolls, the drawing takes place resulting in increase of strength, crystallinity and orientation of polymer.

8.4 SOLUTION SPINNING

The polymer is dissolved in a suitable solvent. The solvent evaporates into air. Acrylic fibre in dimethyl formamide as solvent is one such example. This process is called as dry spinning. Wet spinning takes place due to coagulation of polymer material in a bath containing an electrolyte. Nylon(1) 66 is wet spun from 90% formic acid solution resulting in porous fibres, a slight alteration in bath conditions result in hollow fibres. In all these operations, the filament is washed thoroughly and stretched mechanically to improve the mechanical properties, at times heat treatment also improve properties. With technical advancements in fibre production, bulk and texture are improved by thermal or mechanical texturing so that artificial twist is thrusted upon. Further development constitute combination of different monomers for good lustre and feel.

The most important synthetic fibres are grouped as polyesters, polyamides, polyacrylics and others.

8.5 POLYESTERS

Polyester is a category of polymers, or, more specifically condensation polymers, which contain the ester functional group in the main chain. Although polyesters do exist in nature, polyester generally refers to the large family of synthetic polyesters (plastics) which includes polycarbonate and above all polyethylene terephthalate (PET). Polyester clothing is considered to have a "less natural" feeling; compared to natural fibres. Quite frequently, polyester fibres are spun together with fibres of cotton, producing a cloth with some of the better properties of each. Photographic film (after cellulose triacetate, polyester is the most important substrate). A common matrix for glass- reinforced plastic (commonly called "fiberglass") and other composite materials. Liquid crystal displays Holograms and Filters. Liquid crystalline polyesters are the first industrially used liquid crystalline polymers. In general they have extremely good mechanical properties and are extremely heat resistant. Thermosetting polyester resins are commonly used as casting materials, fiberglass laminating resins, and non-metallic auto-body fillers. In such applications, polymerization and cross-linking are initiated through an exothermic reaction involving an organic peroxide, such as methyl ethyl ketoneperoxide or benzoyl peroxide.

Polyester fibres are very important group of synthetic fibres and their production is increasing at an unexpected pace. Though the original synthesis was credited to Carothers (1930), based on aliphatic polyesters; but real boost appeared after 1950, when aliphatic polyesters were abandoned in favour of aromatic polyesters, because of the improved qualities. Polyethylene-terephthalate is the most prominent constituent of all polyesters. Aliphatic polyesters and aromatic polyesters formed due to reaction between acids and glycols of varying number of carbon atoms. They have different melting points as shown in Table 8.2. Polyesters are defined as synthetic fibres containing atleast 80% of long chain polymers composed of an ester of dihydric alcohol and TPA.

Table 8.2: Meplting Points of Polyesters(2) (°C) (Aliphatic)

Carbon atoms in glycol	*Carbon 2*	*atoms 4*	*in acid 6*
2	172	108	50
3	69	52	46
10	80	70	77

(Aromatic)

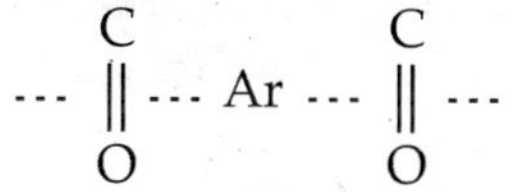

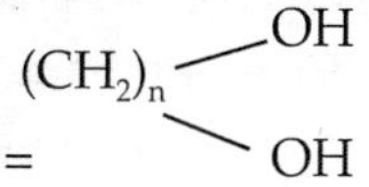

$$\cdots \overset{C}{\underset{O}{\|}} \cdots Ar \cdots \overset{C}{\underset{O}{\|}} \cdots \qquad (CH_2)_n \begin{matrix} OH \\ OH \end{matrix}$$

n =

	2	3	4	5
tere 1, 4	265	190	226	150
iso 1, 3	143			

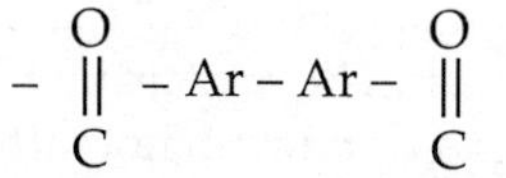

$$- \overset{O}{\underset{C}{\|}} - Ar - Ar - \overset{O}{\underset{C}{\|}}$$

2 Ar 4, 4 355

2 Ar 2, 2 132

$$- \overset{O}{\underset{C}{\|}} - Ar\ (CH_2)_4 - Ar - \overset{\overset{O}{\|}}{C} - \rightarrow 220$$

(4, 4)

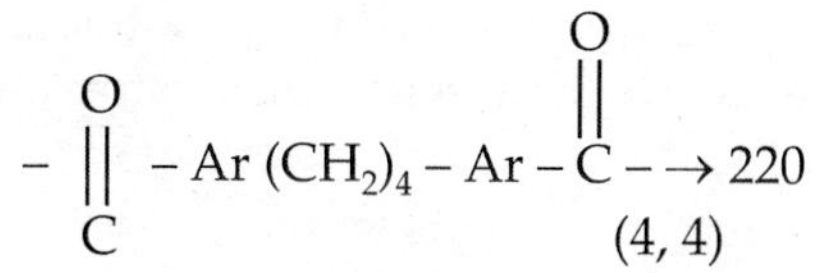

8.5.1 Production of Polyesters

Polyesters result when condensation of a diol takes place with a dicarboxylic acid. It is shown by the following reaction :

Azeotrope esterification: In this classical method an alcohol and a carboxylic acid react to form a carboxylic ester. To assemble a polymer, the water formed by the reaction must be continually removed by azeotrope distillation.

Acylation (HCl method) The acid begins as an acid chloride, and thus the polycondensation proceeds with emission of hydrochloric acid (HCl) instead of water. This method can be carried out in solution or as an enamel.

Silyl method

In this variant of the HCl method, the carboxylic acid chloride is converted with the trimethyl silyl ether of the alcohol component; trimethyl silyl chloride is produced.

$$HO - \overset{\overset{O}{\parallel}}{C} - R_n - \overset{\overset{O}{\parallel}}{C} - OH + HOCH_2\, R'_n - CH_2OH \rightarrow$$

$$- \overset{\overset{O}{\parallel}}{C} - R_n - \overset{\overset{O}{\parallel}}{C} - OCH_2 - R'_n - CH_2\, O + H_2O$$

The reaction may be repeated with dichloride instead of an acid in which case HCl will have to be eliminated instead of water. The linear polyesters suffered from lack of strength and other properties suitable for fibre, hence aromatic polyesters have been formulated.

8.6 PRODUCTION OF POLYETHYLENE TEREPHTHALATE (PET)

There are three steps *p*-xylene in polyester production. Starting material for production of PET is *p*-xylene. *p*-xylene is oxidised to terephthalic acid by means of either nitric acid or by a catalyst like U_2O_5 using air or oxygen, of course these methods are obsolete now. In Chapter 7, the production of terephthalic acid from toluene has been described. Fibre grade terephthalic acid production from *p*-xylene has been presented in Fig. 8.1. The liquid phase oxidation of p-xylene is carried out in a solvent like acetic acid. Cobalt bromide or acetate is used as a catalyst and acetaldehyde is used as an activator. The temperature and pressure are moderate, about 200°C and 15 bars respectively, p-xylene, acetic acid and catalyst are sent into a reactor (1). About 2% activator is added to the contents, and hot air is blown into the reactor. The off gases from the effluent are separated after cooling (2). The whole mass is then filtered (3). The filtrate contains catalyst, *i.e.* acetic acid, water and other oxidates. The crude TPA is again washed (4) with acetic acid to dislodge any catalyst mass. The prepurified TPA is again dissolved in acetic acid (5), filtered (6), crystallised (7), separated and dried with hot nitrogen to final fibre grade product. All the acid washings and filtrates are stripped (8) to recover acid and catalyst. Terephthalic acid has a high melting point (300°C) and is slightly soluble in ethylene glycol and is difficult to obtain in a pure form. However, its dimethyl ester is easy for transesterification, hence industrially used. Recent literature shows that p-xylene can be oxidised with p-toluic acid in presence of water as solvent catalysed by cobalt or manganese salts in the absence of activators and yields more than 90 mol % are reported (3).

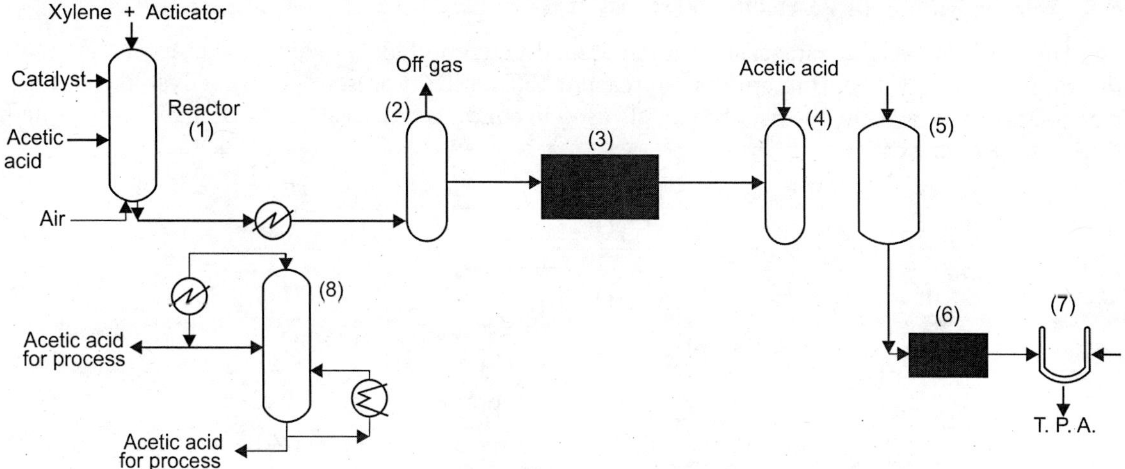

Fig. 8.1: Terephthalic Acid from p-Xylene

8.7 METHYLATION OF TEREPHTHALIC ACID

The second step in the production of polyesters is the esterification of phthalic acid with methanol. This can be done in one stage or two stages depending upon the reaction sequence. As seen in Hercules-Written Process, the air oxidation of *p*-xylene gives rise to p-toluic acid. This acid is converted into methyl p-toluate, which is further oxidised to give monomethyl terephthalate and the latter one is completely esterified with methanol to yield dimethyl terephthalate, zinc acetate, lithium oxide or sodium alchoholate work as catalysts at 190°C reaction temperature.

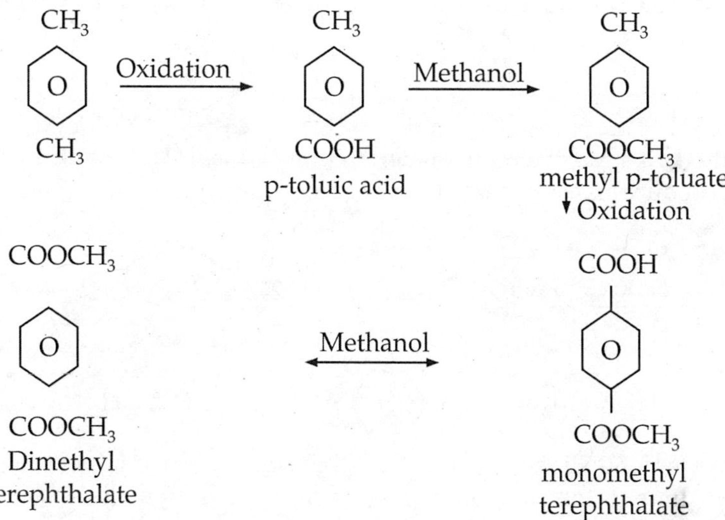

Another method is first production of terephthalic acid and esterification with methanol in a single stage.

8.8 TRANS-ESTERIFICATION WITH DIOLS

The third stage is the trans-esterification of methyl terephthalate to diglycol terephthalate with glycol. Ethylene glycol and terephthalate react at 150-200°C in presence of a catalyst like sodium alcoholate or zinc acetate; the usual catalysts used in condensation reactions. Methanol is liberated during transesterifications.

$$C_6H_4(COOCH_3)_2 + 2\,\begin{matrix} CH_2OH \\ | \\ CH_2OH \end{matrix} \longrightarrow C_6H_4(COO(CH_2)_2OH)_2 + 2CH_3OH$$

Dimethyl terephthalate → Monomethyl terephthalate

To increase the yield of glycol terephthalate, the produced water has to be removed by using a suitable catalyst.

8.9 POLYMERISATION OF THE GLYCOL TEREPHTHALATE

The last step in the production of polyester is polymerisation of diglycol terephthalate and degree of polymerisation is monitored by the melt index of polymer. The polymerisation is conducted in the batch reactors. The reaction is self-catalysed polycondensation, hence suitable terminators are added to the polymer mass. When high tenacity industrial yarns are required high molecular weight polymer are produced by continuous polymerisation. The reaction of polymerisation is presented below :

$$HO\,(CH_2)_2O-\overset{O}{\overset{\|}{C}}-C_6H_4-\overset{O}{\overset{\|}{C}}-O\,(CH_2)_2OH \rightarrow -\overset{O}{\overset{\|}{C}}-C_6H_4-\overset{O}{\overset{\|}{C}}-OCH_2-$$

$$-CH_2-O + \begin{matrix} CH_2OH \\ | \\ CH_2OH \end{matrix}$$

In recent methods, instead of terephthalate, terephthalic acid (TPA) of fibre grade is reacted with ethylene oxide to produce bis terephthalate.

$$HOOC-C_6H_4-COOH \;+\; 2CH_2\overset{O}{\frown}CH_2 \longrightarrow$$

$$HO(CH_2)_2\;O\;\overset{O}{\overset{\|}{C}}-C_6H_4-\overset{O}{\overset{\|}{C}}\;O\;(CH_2)\,OH$$

Other routes of terephthalic acid production from benzoates and toluene were discussed in the previous chapter. Lummus method comprises of two sections. In the first section *p*-xylene and *p*-tolunitrile (obtained from the process) react with ammonia in a fluidised bed reactor in presence of metal oxide catalyst. Metal oxides supply the oxygen required for the reaction. The reaction is supposed to proceed at high selectivity to terephthalnitrile (TPN). In the latter section the purified TPN is hydrolysed and steam stripped to form monoammonium terephthalate (MAT). This is removed and decomposed to form TPA. Mother liquor and ammonia are recycled. The catalyst is reoxidised to be used. The reactions are shown below.

Bis terephthalate in the next step is polymerised at 270°C-300°C under low pressure in presence of a catalyst; may be an amine or quaternary ammonium compound and the polymerisation goes in steps, i.e. first dimer formed, which reacts with monomer to form trimer and so on.

$$CH_3-C_6H_4-CH_3 + NH_3(O_2) \longrightarrow CN-C_6H_4-CN \longrightarrow COON-C_6H_4-COON$$

$$CH_3-C_6H_4-CN \quad \text{(p-toluonitrile)}$$

$$COON-C_6H_4-COON + H_2O \longrightarrow COONH_3-C_6H_4-COOH \quad \text{(MAT)}$$

$$COONH_3-C_6H_4-COOH \qquad COON-C_6H_4-COON$$

$$2HO(CH_2)_2\,O\overset{O}{\overset{\|}{C}}-C_6H_4-\overset{O}{\overset{\|}{C}}\,O\,(CH_2)_2OH \longrightarrow$$

$$HO(CH_2)_2\,O\overset{O}{\overset{\|}{C}}-C_6H_4-\overset{O}{\overset{\|}{C}}\,O\,(CH_2)_2\,OH\,O\overset{O}{\overset{\|}{C}}-C_6H_4-\overset{O}{\overset{\|}{C}}\,O\,(CH_2)_2OH + \begin{matrix} CH_2OH \\ | \\ CH_2OH \end{matrix}$$

The advantage is easy purification of bis terephthalate, even relatively impure terephthalic acid may be used. The catalyst used for polymerisation may be of organometallic (Ziegler's type).

Fig. 8.2 presents outlines of the batch-wise process of PET. DMT and ethyleneglycol in a ratio of 1 : 2 with catalyst are charged into autoclave (1), where the contents are heated upto 150°C. The whole mass is transferred into another autoclave (2) under nitrogen pressure, where the transesterification concludes. Simultaneously methanol liberated is distilled off to collect in the condensation system (3). After all the methanol is driven off, the product is again transferred under nitrogen pressure to autoclave (4), where polycondensation occurs at about[4] 230°C in presence of catalyst like antimony trioxide under low pressure (about 0.5 -1 mm Hg). Excess and liberated glycol are distilled off and condensed (5). Distillation is necessary to shift the equilibrium to forward direction and ultimately a molecular mass of 25,000 to 35,000 results. The melt is forced out by nitrogen pressure in the form of a ribbon on to a drum and the solidified ribbon to finally machines for chopping. Crystalline polyester having a molecular weight of 10,000–25,000 is white opaque substance melting between 250°C–265°C and is insoluble in ordinary organic solvents. The polymer is fabricated into Lavsan (Russian name for terelene) by melt spinning, after which the fibres are stretched and given twist and allowed to set at a temperature of 120°C.

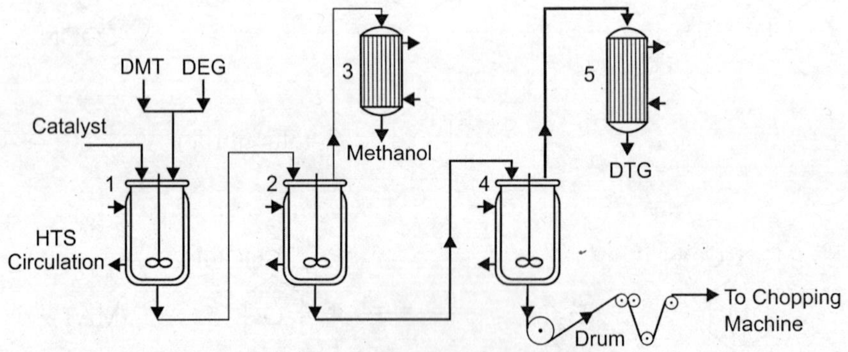

Fig. 8.2: Production of Polyester Ribbon.

A continuous process[5] offered by Vickers-Zimmer closely follows the above pattern and the advantages like no methanol recovery less terephthalic but pure acid are seen.

8.10 POLYCARBONATES

Polycarbonates are a new kind of polyesters which are classified as thermoplastics and serve as engineering materials. Usually these are transparent, heat and break resistant and defined as linear low crystalline high molecular weight (18,000) compounds, where linking elements are carbonate radicals. These are however easily attacked by alkalies, acids and many solvents. Because of the thermal stability these can be molded even at a high temperature of 500°C. The growth rate of polycarbonates is high about 8%.

8.10.1 Reaction

Polycarbonates are polyesters of carbonic acid $HO-\overset{\overset{O}{||}}{C}-OH$ and are produced by condensation polymerisation of sodium salt of bisphenol in the presence of an organic solvent.

$$n\,NaO\quad Ar\quad \underset{\underset{CH_3}{|}}{\overset{\overset{CH_3}{|}}{C}}\quad Ar\quad ONa + n\,Cl\quad \overset{\overset{O}{||}}{C}\quad Cl \longrightarrow$$

$$\left[O-Ar-\underset{\underset{CH_3}{|}}{\overset{\overset{CH_3}{|}}{C}}-Ar-O-\overset{\overset{O}{||}}{C}- \right]_n + 2n\,NaCl$$

Instead of sodium salt of bisphenol, simply bisphenol can be used.

Another method of production by transesterification of carbonic acid is shown below :

$$n\,HO-Ar-\underset{CH_3}{\overset{CH_3}{\underset{|}{\overset{|}{C}}}}-Ar-OH-n\,OR-\overset{O}{\overset{||}{C}}-CR \rightarrow \left[-O-Ar-\underset{CH_3}{\overset{CH_3}{\underset{|}{\overset{|}{C}}}}-Ar-\overset{O}{\overset{||}{C}}-O\right]_n + n\,ROH$$

Diphenylol propane[6] (bisphenol) can be condensed with phosgene in presence of pyridene and a solvent at a temperature of 4°C as the reaction time lasts about 4 hours.

$$(OH)-Ar-\underset{CH_3}{\overset{CH_3}{\underset{|}{\overset{|}{C}}}}-Ar\,(OH) + Cl-\overset{O}{\overset{||}{C}}-Cl \rightarrow \left[O-Ar-\underset{CH_3}{\overset{CH_3}{\underset{|}{\overset{|}{C}}}}-Ar-O-\overset{O}{\overset{||}{C}}\right] + HCl$$

When carbonic acid esters of aromatic groups are used, phenol (C_6H_5OH) is liberated and it is removed with great difficulty. Hence the chance for side reactions to proceed always exists.

Polycarbonates have good impact strength and good electric properties. They can be used to make gears, valves etc. They can be coloured easily and used for varnishes and cements.

These can be produced in two ways :

1. Ester exchange between carbonate diesters and a dihydroxy aromatic compound.
2. Phosgenation of dihydroxy aromatic.

8.11 POLYAMIDES (NYLONS)

Polyamides are high molecular compounds containing - NH - C - group in their back bone chain. They are formed by the reaction between dicarboxylic acid and diamines (Nylons 6, 6, 6, 10 etc). These are named as Nylons by Dupont, exhibit a molecular weight of 7000. Nylon 6 (Capron) and enant are formed with slight difference in structural changes. Nylon 6, 6 was the first synthetic fibre appeared in the market (around 1940) owing to the endless devotion of Carother. Nylon 6 was developed in Germany by I.G. Farben.

8.11.1 Nylon 6,6

Nylon 6,6 has been obtained by reacting hexamethylene diamine with adipic acid and this usually has a molecular weight of 15,000— 20,000. Adipic acid may be treated as the starting material for nylon as hexamethylene diamine can be produced from it.

Adipic acid production had been practised through different routes, the starting material may be benzene/cyclohexane/phenol. The different reactions are almost interlinked as shown below :

To elaborate further, phenol is hydrogenated over a nickel catalyst at 140-150°C under a pressure of 15-20 bars. A packed bed reactor having catalyst in tubes is used. The heating is done initially by suitable means (by circulating dowtherm into the space between shell and tubes). The reaction being exothermic liberates about 1900 kJ per kg phenol consumed, hence the temperature of the reactor should be carefully controlled. Almost all phenol is consumed giving rise to cyclohexanol. The crude cyclohexanol is distilled and then sent into a dehydrogenator. Cyclohexanol is dehydrogenated over copper-zinc catalyst at about 450-500°C. The reaction is conducted in a packed bed reactor, in vapour phase. Along with cyclohexanone, cyclohexanol is obtained as a by-product which is separated by simple distillation. Cyclohexanone is oxidised by nitric acid to yield adipic acid.

FROM BENZENE

Benzene $\xrightarrow{\text{Hydrogenation}}$ Cyclohexane $\xrightarrow{\text{Oxidation}}$ $(CH_2)_4(COOH)_2$

FROM CYCLOHEXANE

Cyclohexane $\xrightarrow{\text{Oxidation}}$ Cyclohexanone $\xrightarrow{\text{Oxidation}}$ $(CH_2)_4(COOH)_2$

FROM PHENOL

Phenol (OH) $\xrightarrow{\text{Hydrogenation}}$ Cyclohexanol (OH) $\xrightarrow{\text{Dehydrogenation}}$ Cyclohexanone (O)

Cyclohexanone (O) $\xrightarrow{\text{Oxidation}}$ $(CH_2)_4(COOH)_2$

8.11.2 Adipic Acid from Cyclohexane[7]

Fig. 8.3 shows the outlines of adipic acid production from cyclohexane. Cyclohexane is oxidised by air in presence of a catalyst like cobalt naphthenate at a temperature of 140-165°C under a pressure of 10-12 bars, as shown in reactor (1). The reactor product is washed (2) with caustic and water to remove acids. Unconverted cyclohexane is separated by distillation in a column (3). Heavy organics are removed from the bottom of distillation column. The product from the side cut contains cyclohexanol and hexanone both are fed into oxidiser (4), where the oxidation is carried out in presence of a catalyst composing copper-vanadium complexes, by the oxidant, 60% nitric acid. The off gases from the reactor are absorbed in an absorber (6) to get dilute nitric acid which along with the dilute process acid, is used for recirculation. From the bottom of oxidiser the adipic acid is taken out into an evaporator and crystalliser (5) system. The crystals are separated by filtration from the mother liquor and purified further by dissolving in any suitable organic solvent and crystallising.

In USSR, the oxidation of cyclohexane is carried out by air in presence of boric acid at 155-180°C. The boric acid esterifies with cyclohexanol and can be easily hydrolysed to obtain pure cyclohexanol. Similarly the mixture of hexanone and hexanol can be oxidised by air at 150-160°C in presence of acetic acid (which acts as solvent for copper manganese catalyst). While air oxidation or nitric acid oxidation can be used for cyclohexane to cyclohexanol but cyclohexanol to adipic acid is done by nitric acid oxidation alone.

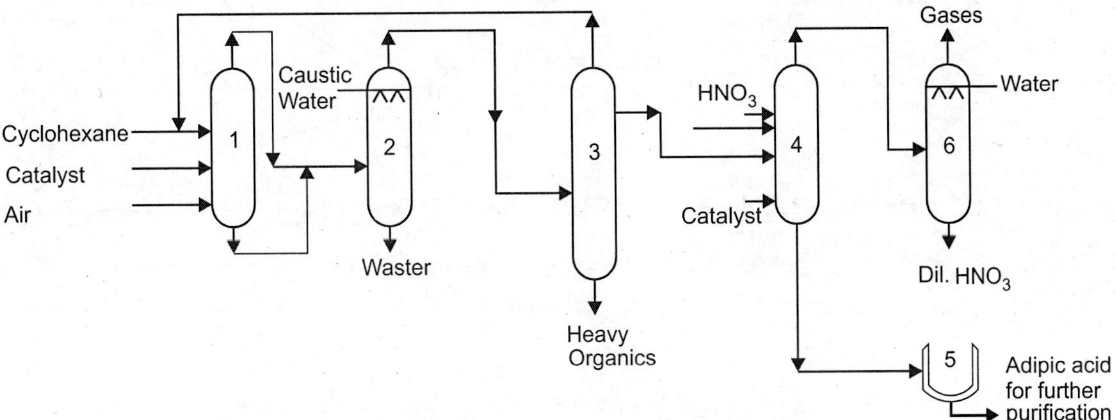

Fig. 8.3: Adipic Acid Production from Cyclohexane.

The next step is to convert adipic acid to adiponitrile. This is achieved by first ammoniation and then hydrogenation of adiponitrile.

$$(CH_2)_4\begin{matrix}COOH\\ COOH\end{matrix} + NH_3 \rightarrow (CH_2)_4\begin{matrix}COONH_4\\ COONH_4\end{matrix}$$

$$(CH_2)_4\begin{matrix}COONH_4\\ COONH_4\end{matrix} \rightarrow (CH_2)_4\begin{matrix}CN\\ CN\end{matrix} + 4\,H_2O$$

$$(CH_2)_4\begin{matrix}CN\\ CN\end{matrix} + 4\,H_2 \rightarrow (CH_2)_6\begin{matrix}NH_2\\ NH_2\end{matrix}$$

Ammoniation and dehydration are simultaneously carried out by using MoO_3 or WO_3 catalyst at 250°C. The resulting adiponitrile is later hydrogenated at 80-100°C under a hydrogen pressure of 50-100 bars over nickel or cobalt catalyst to produce hexamethylene diamine. Hexamethylene diamine can also be produced through other routes, one being butadiene. Butadiene is reacted with hydrogen cyanide to yield dicyanide, which when hydrogenated yields hexamethylene diamine through an intermediate stage of adiponitrile. Other method is hydrodimerization of acrylonitrile.

$$2CH_2 = CH \,.\, CN \xrightarrow{H_2} \begin{array}{l} CH_2\,CH_2\,CN \\ | \\ CH_2\,CH_2\,CN \end{array} \longrightarrow (CH_2)_6 \begin{array}{l} NH_2 \\ \\ NH_2 \end{array}$$

$$CH_2 : CH - CH : CH_2 + HCN \longrightarrow CNH_2\,C\,.\,CH = CH\,.\,CH_2\,CN \xrightarrow{\text{Hydrogenation}} (CH_2)_6 \begin{array}{l} NH_2 \\ \\ NH_2 \end{array}$$

Hexamethylene diamine and adipic acid are reacted to yield polycondensate product AH salt. Equimolecular mixture of these two compounds is dissolved in methanol, on cooling AH salt precipitates out. This salt is dissolved in steam condensate, a small amount of acetic acid is added as a stabiliser. The mixture is then heated in an autoclave and held at 250°C under a pressure of 15 bars for six to eight hours during which prepolymerisation sets in. The water produced is removed by distillation (by lowering the pressure) and subsequent condensation (by lowering the pressure) continues to yield polyamide melt. Water is again removed by lowering the pressure, the molten polymer is forced out by using nitrogen through spinnerets. The glassy jets from the spinnerets are solidified in a water bath.

8.11.3 Production of Nylon 6

Nylon 6, has six carbon atoms in one monomer molecule, unlike polyamide where adipic and hexamethylene diamine contribute six carbon atoms each. Caprolactum when polymerised yields Nylon 6, *i.e.* polycapromide. The production of caprolactum has been described in the previous chapter. Pure caprolactum is polymerised in presence of water.

Water opens the ring of caprolactum to give amino acid.

$$\underbrace{CH_2\,(CH_2)_4 - C = O}_{NH} + H_2O \rightarrow HO - \overset{\overset{\large O}{||}}{C} - (CH_2)_5 - NH_2$$

Caprolactum, water and slight amount of acetic acid are charged into a reactor. The reactor is heated to 270°C and maintained under a pressure of 20 bars. During this process, water is driven off, but the heating continued in presence of nitrogen. Polymerisation takes place for about 8 hours, the molten polyamide is forced under pressure through spinnerets to form filaments.

$$(CH_2)_4 \begin{array}{l} C = O \\ \quad \diagdown \\ \text{———} NH_2 \\ \quad / \\ CH_2 \end{array} \longrightarrow [(CH_2)_5 \cdots \overset{\overset{\large O}{||}}{C} - NH -]_n$$

Nylon 6

As described above any dibasic acid and any diamine can react to give nylons of various designation as shown below.[8]

$(CH_2)n\,(NH_2)_{n-2}$	$(CH_2)_{n-2}\,(COOH)_2$	*Designation Nylon*	M.P. °C	
			Salt	*Polyamide*
$n = 4$	$n = 4$	4, 4	204	278
4	5	4, 5	138	233
4	6	4, 6	—	278
4	7	4, 7	175	233
6	4	6, 4	183	250
6	6	6, 6	—	165
6	7	6, 7	—	202
6	8	6, 8	170	209
6	10	6, 10	—	215
10	4	10, 4	142	230

8.11.4. Nylon 6, 10 is obtained by reacting sebasic acid $(CH_2)_8\,(COOH)_2$ and diamine. Other types of nylons include 4, 11 and 12.

Nylon 12 (Polylaurylamide)[9] is produced from butadiene. In the first stage butadiene is trimerised. The trimer is epoxidised with peracetic acid. The resulting epoxide is hydrogenated and the saturated epoxide is rearranged to the ketone with catalyst magnesium iodide at 100°C. The oxime of this ketone is then rearranged to laurolactum.

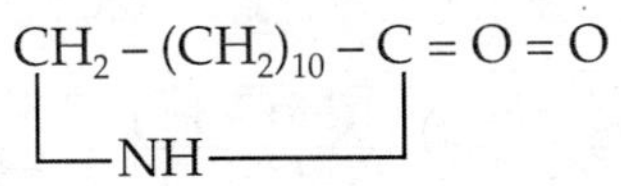

$$\underbrace{CH_2 - (CH_2)_{10} - C}_{NH} = O = O$$

8.11.5. Nylon 11 is produced commercially from 11-amino decanoic acid $(H_2N\text{-}(CH_2)_{10}\text{-}COOH)$. The polymerisation starts by heating the monomer in water suspension. Water is driven off during polycondensation. Nylon 11 is used extensively in coatings. Nylon 4 (poly butyramide) is produced from the monomer in a solvent 2-pyrolidone through several routes. One is acrylonitrile route. Acrylonitrile and HCN react to give an additive product $CN(CH_2)_2CN$. This compound leads to production of 2-pyrolidone of the structure

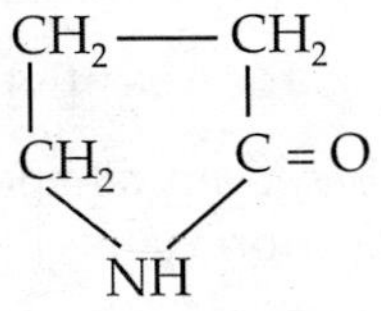

$$\begin{array}{ccc} CH_2 & - & CH_2 \\ | & & | \\ CH_2 & & C = O \\ & \diagdown \; \diagup & \\ & NH & \end{array}$$

which on polymerisation yields a Nylon of structure.

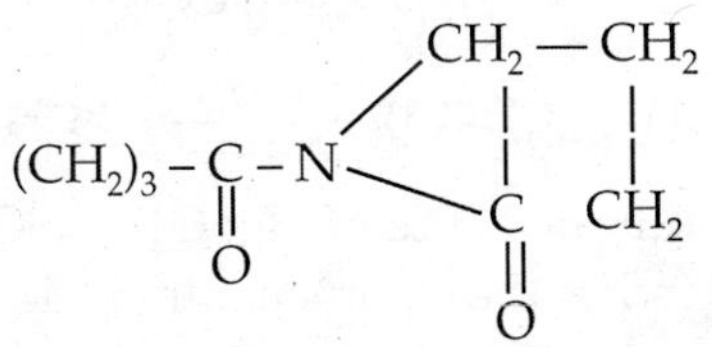

$$\begin{array}{ccccc} & & CH_2 - & CH_2 & \\ & \diagup & | & | & \\ (CH_2)_3 - \underset{\underset{O}{\|}}{C} - N & & | & | & \\ & \diagdown & \underset{\underset{O}{\|}}{C} & CH_2 & \end{array}$$

It may be essential to mention that lactums may be polymerised under acidic or basic conditions. Base catalysed polymerisation employs alkalimetal, sodium or lithium (0.5 to 1%); and the polymerisation takes place when the lactum is heated to 250°C. Caprolactum can be polymerised with acidic or basic catalysts. Acid catalysation is conducted by using HCl or acetic acid. Nylons in general can be modified by the addition of certain plasticisers, reinforcements and fillings. Depending upon the hydrocarbon character of the macromolecule, moisture absorption capacity varies. Nylon 4 is almost similar to cotton in water absorption properties. Nylons are used in tyre cords, conveyor belt manufacture because of toughness, strength and abrasive resistance. Polyamides in general are soluble in phenols and concentrated mineral acids. They are soluble in acetic, formic acid and benzyl alcohol. Polymers having odd carbon atoms of methylene groups have lower melting points than the polymers of even carbon numbers. The melting points can be increased by inducting phenyl or cyclohexyl groups in the back bone. When the number of carbon atoms in both the diamine and dicarboxylic acid are even, there is intermolecular hydrogen bonding[10]. The melting point of polyamide decreases because of irregular structure produced. Crystallinity is hindered when pendant groups are present in the chain.

Table 8.3: Properties of Different Fibres

	Polyesters	*Nylons 6, 6*	PVC (Saran)	HDPE	*Acrylics*	*P. P.*
Specific Gravity kg/litre	1.38	1.14	1.14	0.96	1.14	0.905-0.91
Melting point °C	265	260	235	135	235	172
Elongation % (Break up point)	20	30	25	10	25	25
Tenacity gm/dimer	5.0	5.0	2.8	8.0	2.0	9.0
Ultimate tensile strength						
G/denier	3.4-5.8	upto 2.8	3.5-7.0	5-5.8		
T.S.(kg/cm²)	5624	5624	3515	5624	2812	5624

8.11.6 Enant Fibres (Nylon 7)

Polyenanthoadmide is the starting material for the production of enant fibres. It is obtained by polycondensation of aminoenanthic acid.

$$n\,NH_2-(CH_2)_6-COOH \longrightarrow [NH\,(CH_2)_6-\overset{\overset{\displaystyle O}{\|}}{C}-]_n + n\,H_2O$$

Synthesis of aminoenanthic acid is based on telomerisation reaction, (the polymerisation of ethylene in presence of carbon tetrachloride to form tetra chloroalkanes). The telomers are subsequently converted into amino acids as shown below :

$$(CH_2=CH_2)_n \xrightarrow{Cl} CCl_3(CH_2)_6CH_2Cl$$

$$CCl_3(CH_2)_6(CH_2Cl) \xrightarrow[HgSO_3]{H_2O} Cl\,CH_2(CH_2)_6COOH$$

Chlorotelemer is hydrolysed in presence of $HgSO_4$ (H_2SO_4) giving rise to chloroenathic acid. Under the action of aqueous ammonia the chloroenanthic acid turns into amino enanthic acid

$$ClCH_2\,(CH_2)_6\,COOH \rightarrow NH_2–(CH_2)_6–COOH + NH_4Cl + CH_3OH$$

Polycondensation of aminoenanthic acid is carried out exactly like polycondensation of AH salt. Alternatively hydrolysis of pimelic dinitrile yields ω-aminoenanthic acid. Nylon 9, is produced by the amidation and reduction of azelaic acid.

8.12 ACRYLIC FIBRES (ORLON)

These are third major class of synthetic fibres. Dupont and Union Carbide developed these fibres under trade names Orlon and Dynel respectively. Most of the acrylics contain a major amount of acrylonitrile, the other monomers such as methacrylate, vinylacetate, vinylchloride and acrylamide present upto 15% only. Solution spinning is only used for acrylonitriles because of the heat sensitive nature. Dimethyl formamide is used as solvent for wet spinning. Modacrylics are spun from acetone solution.

8.12.1 Manufacture

Dynel fiber is manufactured by copolymerising acrylonitrile with vinylchloride. Fig. 8.4 outlines the process for the production of polyacrylonitrile (PACN). Acrylonitrile, comonomer and catalyst are sent into an autoclave (1). The polymerisation is carried out in presence of acid catalyst such as ammonium persulphate in aqueous medium at low temperature. The polymer mass after discharging, is mixed with a chain stopper in a reactor (2). The mass is then washed with water to remove the unreacted monomers. The polymer mass is then dried by hot air. The polymer powder is packed and transferred to wet spinning section. All these polymers possess, soft touch, resistance to creasing and quick drying. They look like wools and replace wool. Though these fibres also exhibit high hydrophobic nature, the new fibre developed by Bayer A.G. can absorb as high as 30% moisture, matching the properties of wool.

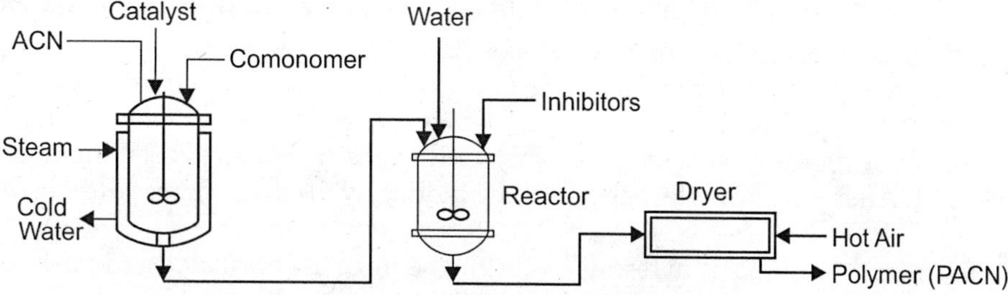

Fig. 8.4: Production of Polyacrylonitrile (Fiber grade).

Dynel polymer incorporates vinylchloride as the second monomer, because of the presence of chlorine, the fibre is less inflammable. Ethylene/ethylacrylate copolymers are used for flexible hose, bathing suits etc. Ethylene-vinylacetate resins are also used for the same purpose. These fibres are elastic with toughness.

8.13 POLYPROPYLENE

Though these are produced on large scale, still only a small fraction is utilised as synthetic fibres. The production of polypropylene has been described in Chapter 6. Isotactic polypropylene is having crystallinity 60-85% is processed into filament by using melt spinning technology. The major disadvantages with polypropylene fibres are low melting point (171°C), low resilience and critical dyeing problems and high hydrophobic nature. Fibres, including ropes and film yarn[11] accounted for one-third of the polypropylene consumption. Monofilament are oven into canvas and twisted into yarn and ropes of various sizes. Monofilament is produced in a similar manner as done in high density polyethylene films (extrusion) and its uses are similar. The meterage of polymer produced per kg is higher for polypropylene because of its low density. The tenacity of these drawn filaments is slightly higher than that of HDPE. Usually the fibres are stronger and do not easily rot.

8.14 SYNTHETIC PAPER

Like synthetic fibres, synthetic papers may play a vital role in modern civilization in future. The advantages of synthetic paper over conventional paper are many. These do not permeate air and moisture, and have very good mechanical properties like toughness high tear strength and are not weathered by ordinary chemicals or ambient conditions. Synthetic papers are made out of plastics that can be extruded into thin films. Plastics like HDPE/PP and other polymers after extrusion are given treatments like coatings with clay, filling with $CaCO_3$, SiO_2 or TiO_2 etc., in order to achieve paper resemblance in its rigidity, opacity and printability. The main disadvantage of the synthetic papers is non-absorbent nature, that gives poor ink receptivity, hence special internal methods are developed, like foaming, leaching, stretching, compounding and other surface treatments, either by coats or by reinforcing with swelling agents. Sometimes the films are treated with coagulating agents. Most of the synthetic fibres are made by extruding polymers of HDPE/PP/acrylonitrile polystyrene/vinylchloride/vinylacetate copolymers into this films and subjected post extrusion treatments. Sometimes natural pulp or rayons are added with the polymer pulp and sent to paper marking machines. Rayon/polypropylene gives rise to tissue papers for tea bags etc. Similarly straw pulp (30%) and thermoplast 70% (15% vinylchloride) can be made into tissue paper, tea bags and adhesive paper etc.

Like paper making, synthetic papers require pulp. The pulp is made by a continuous emulsion, or solution polymerisation process, from monomers, like ethylene/propylene. Solvay and Cie have developed a continuous process for synthetic pulp making. Polymerisation at low pressure (5 to 20 bars) using supported catalyst at a temperature of 250°C in case of emulsion process. Dislodging of the catalyst from polymer pulp is not required here, as the catalysts are very active, further they are not going to deteriorate the quality of paper produced. However, monomer and solvent can be separated and recycled. Dilution by addition of water is done, some resins are added to get the aqueous slurry of fibrous substance. After pulp is made, paper making is done by the usual procedure, *i.e.* Four-drinier screen[(12)].

Mono axial orientation of PP film results in the film being easily fibrillated and can be chopped to derived length to obtain opaque fibres. Pigments like kaolin, etc. are dispersable and yield free floating slurries at high solids in hydrocarbon systems. Dispersability of the pigments in hydrocarbons and polyolefins is credited due to the reduction in surface energy by kaolin. The paper developed by Honshu Paper Company, Japan is a mixture of 65% of $CaSO_4$ and 35% Polystyrene-Polymer Corporation Ltd., Canada developed PP – $CaCO_3$ film with a density of 1.3 gm/cc.

Polyart, Zexan[(13)], Aratheme are some trade names of these papers and papers are usually composed of filled and modified HDPE films. Q-kote, etc. are clay coated polystyrene papers.

Solvent treated polystyrene, PVC and filled poly allomer films are too available in the market. Though the present position shows not much enthusiasm, future is expected to be bright.

8.15 OPTICAL FIBERS

Thus, polymer fibers are strong materials with excellent tensile strength, which makes them very useful as textiles. In fact, optical fibers have polymers in their composition.

However, the term *optical fiber* doesn't tell us anything about polymers, but about two transparent, dielectric tubes or cylinders, one surrounding the other. However, the usefulness of light guidance would not be completely appreciated until 1950, when many scientists began to think in its potential applications. They were smart enough to understand that such applications ranged from medicine, enabling the visualization of inaccessible regions of the human body, to communication networks, in replace of metal wires.

But there were a number of drawbacks that arose in those first applications : light couldn't be carried too far, and the transmitted images came pretty poor and distorted.

That was because the fibers had no coating at all surrounding them to increase the *difference* in refractive index. For that reason, if such fibers got wet, for example, the glass-air interface didn't remain fixed and the way the light reflected inside was changed.

Only in the mid 50s, Dutch Abraham van Heel designed a covering layer surrounding the fiber with another cylinder of glass, of lower refractive index than that in the middle. This way, the total reflection was not affected by water, dirt or other contamination. Finally, a third layer was added, in order to protect the glass fiber from damage, and to make it easier to handle.

- An inner cylinder with high refractive index, called the *core.*
- A middle cylinder with a lower refractive index, called the *cladding.*
- An outer protective polymer layer (usually **polyurethane** or **PVC**) called the *jacket.*

For glass optical fibers, the diameter of the core ranges between 10-600 microns, the cladding thickness is between 125-630 microns, and that of the jacket varies between 250-1040 microns. For POF all diameters range between 750-2000 microns. As can be seen, one of the main differences between glass and plastic optical fibers is their diameter. This makes POF easier to handle.

It was just reflection, *total internal reflection*, as the walls of the thin fiber acted like mirrors in which the incident light bounced back and forth.

The material used for currently commercialized fibers (core and cladding) include pure glass (SiO_2), plastic, or a combination of both. The use of one or the other material will be determined by such factors as quality and economics. Plastic optical fibers (POF) have the advantage of being made of cheaper materials than glass and to operate in the visible range of the spectrum. However, they show a high loss, and for that reason their applications are confined to short distance transmission. In spite of this, POF is widely used for medical and industrial instruments, and currently research is carried out about using POF as a replacement of copper wiring for data transmission in automobiles. If you use silica glass for the core, it must be of high purity in order to allow the light to be transmitted along the core with minimal loss.

If a bit of germanium (as a germanium tetrachloride gas) is added to pure silica glass, which has 18 more electrons than silicon, acts as a dopant. Consequently, the refractive index of the core glass is increased, although the attenuation is not affected. Likewise, you can add a bit of boron or fluroine to reduce the refractive index of the cladding glass.

There are several types of fibers based both on the transmission modes and the refractive index profile. We won't discuss them in detail, but it is important to known that the design of an optical fiber depends on the needs it is supposed to meet. Key parameters like *attenuation, bandwidth, dispersion,* and *tensile strength* are the most considered.

Also the protection of the fiber from external factors like humidity, heat, cold, and water is contemplated. That's why plastic sheaths where the fiber is enclosed in, are used. The whole material comprising the single fiber or bundle of fibers, the sheaths, and the jacket, is usually referred as a *fiber cable.*

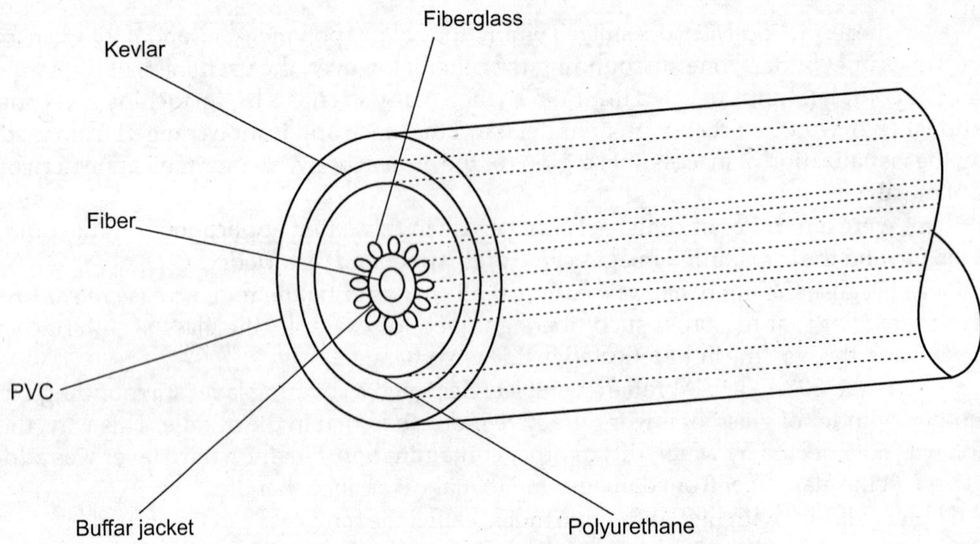

Fig. 8.5: Cross section of fiber.

Obviously these cables also must satisfy requirements such as high flexibility, resistance to kinks and crushing, and light weight. A buffer tube is usually a layer of **silicon** or **epoxyresin,** softer than the external jacket, and has no optical function. It keeps the fiber from "microbends" due to physical contact with the other components of the cable. As relatively fragile materials, fibers need some mechanical reinforcement. Many materials in form of strands or filaments can play this role. One of them is fiberglass, and in the example, we can see a bunch of fiberglass strands surrounding the buffer tube. Of course, since fiberglass is a stiff material, another layer of polyurethane is added, to provide cushioning. Finally, fiber cable may be placed in a number of different environments : in the air, into the water, or under ground. Therefore an external protective jacket becomes essential. **PVC** and **polyurethane** are the most used materials for that purpose : PVC is a better material than polyurethane when considering its resistance to: water, flame, acids, alkalis, hydrocarbons and alcohol.

Conversely, polyurethane has certain advantages over PVC when dealing with its resistance to : abrasion, nuclear radiation and low temperatures. See the Fig. 8.6 and 8.7.

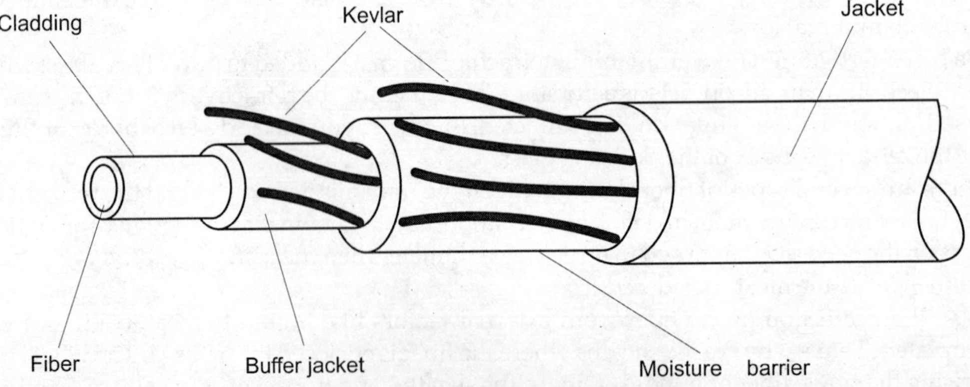

Fig. 8.6: Arrangement of fiber.

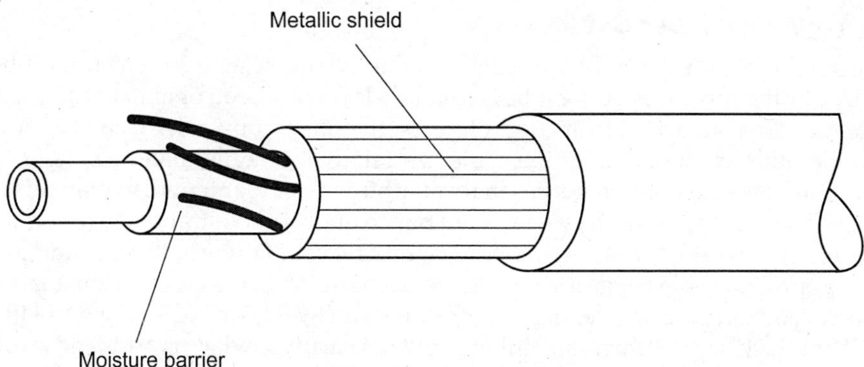

Fig. 8.7: Finished cable.

In the second example on the left (redrawn from "Fibre Optics"/S. Ungar, John Wiley & Sons, 1990) an exterior fiber cable with a buffer jacket which can be constituted by two sheaths: one of silicon, and the other of Hytrel®, extruded on silicon. Hytrel® is a neat polyether-ester block copolymer, a thermoplastic elastomer with optimum water resistance, sold by Du Pont. Between two layers of Kevlar® filaments, a moisture barrier is inserted. This moisture barrier can be made of plastic (often polyethylene), metal (particularly aluminium), or both. Lastly, a PVC coating ensures the utility of this fiber cable to be installed in free air.

The third example on the right (redrawn from "Fibre Optics"/S. Ungar, John Wiley & Sons, 1990) is a slight modification of the second. Only this time an underground fiber cable is shown. Since it requires a further protection against humidity, two sheaths have been inserted between a Kevlar® layer : a moisture barrier and a metallic shield.

The enormous advantages of the optical fiber telecommunications compared to metal wire or coaxial transmission may be summarized as follows :

- High bandwidth, completely independent of the cable size.
- Low attenuation, *i.e.*, optical loss is reduced to a minimum.
- Extremely low electromagnetic induction, induced noise and crosstalk.
- Availability of long distance light-weight cables.
- Lower cost in installation and maintenance.

Many Optical Fibers in the Market

There are several types of fibers based both on the transmission modes and the refractive index profile. We won't discuss them in detail, but it is important to know that the design of an optical fiber depends on the needs it is supposed to meet. Key parameters like *attenuation, bandwidth, dispersion, and tensile strength* are the most considered.

Also the protection of the fiber from external factors like humidity, heat, cold, and water is contemplated. That's why plastic sheaths where the fiber is enclosed in, are used. The whole material comprising the single fiber or bundle of fibers, the sheaths, and the jacket, is usually referred as a *fiber cable.*

Obviously these cables also must satisfy requirements such as high flexibility, resistance to kinks and crushing and light weight.

If you're eager to know what and how many polymers are involved in the manufacture of fiber cables, take a look at these examples :

Designing Your Own Fiber Cables

In the first example on the right we have a multi-layered system in which the fiber is firstly surrounded by a buffer tube. This buffer tube is usually a layer of **silicon** or **epoxy resin,** softer than the external jacket, and has no optical function. It keeps the fiber from "microbends" due to physical contact with the other components of the cable. As relatively fragile materials, fibers need some mechanical reinforcement. Many materials in form of strands or filaments can play this role. One of them is fiberglass, and in the example, we can see a bunch of fiberglass strands surrounding the buffer tube. Of course, since fiberglass is a stiff material, another layer of polyurethane is added, to provide cushioning. If your fiber cable needs a good tensile strength as well as electrical insulation, which is highly desirable, you can place a layer of Kevlar®. Have you ever heard of Kevlar® and its wonderful properties ? If not, click here to learn something about it. Usually Kevlar® is arranged into the cable in the form of filaments. Finally, you must think that your fiber cable may be placed in a number of different environments: in the air, into the water, or under ground. Therefore an external protective jacket becomes essential. **PVC** and **polyurethane** are the most used materials for that purpose. But be careful with the choice: PVC is a better material than polyurethane when considering its resistance to : water, flame, acids, alkalis, hydrocarbons and alcohol. Conversely, polyurethane has certain advantages over PVC when dealing with its resistance to : abrasion, nuclear radiation and low temperatures.

In the second example on the left (redrawn from "Fibre Optics"/S. Ungar, John Wiley & Sons, 1990) we have an exterior fiber cable with a buffer jacket which can be constituted by two sheaths: one of silicon, and the other of Hytrel®, extruded on silicon. Hytrel® is a neat polyether-ester block **copolymer,** a **thermoplastic elastomer** with optimum water resistance, sold by Du Pont. Between two layers of Kevlar® filaments, a moisture barrier is inserted. This moisture barrier can be made of plastic (often **polyethylene**), metal (particularly aluminium), or both. Lastly, a PVC coating ensures the utility of this fiber cable to be installed in free air.

The third example on the right (redrawn from "Fibre Optics"/S. Ungar, John Wiley & Sons, 1990) is a slight modification of the second. Only this time an underground fiber cable is shown. Since it requires a further protection against humidity, two sheaths have been inserted between a Kevlar® layer : a moisture barrier and a metallic shield.

REFERENCES

1. A. Hancock, J.E. Spruiell & J.L. White, J. Appl. Sci., 1977, 21, 1227-45.
2. Malcohm P. Stevens, Polymer Chemistry, Addison Wesley Pub. Comp. Massachusetts 1975, 65.
3. Milan Hronec & Janilavsky, IEC. Prod. Res. Dev., 1982, 21, 455.
4. Encyclopaedia of Polymer Science & Technology, Inter Science (Pub.) Supplement to Vol., 1, 448.
5. HCP, January, 1972, 98.
6. Encyclopaedia of Chemical & Process Technology (Douglas M. Considine Editor) McGraw-Hill, 894.
7. HCP, Nov., 71,124.
8. Stanley, R. Sandler & Wolf Karo, Polymer Synthesis, Vol. 1, Academic Press, N.Y. 1974, 91.
9. L.F. Hatch & Samir Matar, From Hydrocarbons to Petrochemicals, GPC, 202.
10. R. Seymour, Introduction to Polymer Chemistry, Inter Student Edition, Kogo Kusha, 106.
11. P. D. Ritchie, Vinyl and Allied Polymers, ILIFFE Books Ltd. Vol. 1, 215.
12. M.G. Halpern, Synthetic Paper from Fibres and Films, NDC, 1975,178.
13. Vladimir M. Wolpert, Synthetic Polymers and the Paper Industry, Miller Freeman Publication, 1971, 118.

CHAPTER 9

Synthetic Rubber

9.1 TYPE OF RUBBER

9.1.1 Natural Rubber Position

Prior to Second World War, natural rubber was accounted 100 per cent of the rubber goods. India, being a tropical country was a good producer of natural rubber in those days. Since World War II, like in other countries, natural rubber could not meet the demands of industrial pace and many synthetic rubbers have been introduced. Styrene butadiene rubber is produced in largest quantities. However, the special applications like, superior building tack, better processing, high resistance, retention of strength at elevated temperature, low hysteresis are some abilities of natural rubber which made to continue its existence, even in the days of competition.

The consumption and demands of natural and synthetic rubbers in India are shown in Tables 9.1 (*a*) and (*b*)[1].

Table 9.1 (*a*): Consumption of Rubber (in tonnes)

Year	*Natural rubber*	*SBR*	*Other rubbers*	*Reclaimed rubber*
1980	170,800	25,133	21,017	22,241
1981	181,915	22,860	2,090	25,385
1982	197,035	23,725	25,980	29,282
1983	205,395	23,485	29,540	29,815
1984	212,540	26,125	31,155	33,310
1990	265,000	30,000	22,440	38,000

Table 9.2 (*b*): Consumption of Rubber (in tonnes)

Year	*Natural*	*EPDM*	*Butyl*	*Chloroprene*	*Total SR*	*Others*	*Reclaimed*
1990-1	364310		27115	2941	104735	1548	52500
1993-4	450480	1457	31580	3187	113395	1500	63110
1998-9	628110	5965	44040	4680	156395	3050	63095
2001-2	638210		50100	5210	191700		47427
2002-3	695045			5410	203100	87305	52160
2004-5	696385				233000	98205	

Table 9.2. (*a*): Production of Rubber (in tonnes)

Year	*Natural*	*SBR*	*Nitrile*	*PBR*	*Reclaimed*
1980	155,280	15,847	247	6,222	23,151
1981	150,655	18,131	425	10,618	24,926
1982	165,920	13,514	615	14,536	28,290
1983	165,025	17,886	691	12,987	30,436
1984	183,925	20,817	480	16,295	33,340
1985	186,450	20,826	386	15,457	
1989-90	2,65,000	34,000	2,000	18,000	50,000 (Butyl) 40,000 + others
1994-95	—	86,000	—	—	—
2000 AD	6,00,000	1,00,000	—	30,000	60,000"

Table 9.2. (*b*)

Year	*Natural*	*SBR*	*Nitrile*	*PBR*	*Reclaimed*	*Total Synthetic rubber*
1990-1	366745					57293
1993-4	471815					49633
1998-9	622265					67590
2001-2	655215	16500				69653
2002-3	713620	18300				94700
2004-5	610660	18600				96200

While Eastern and Latin American countries are good producers of natural rubber, India trails far behind. Synthetic rubbers are mainly the property of developed countries, see Table 9.3 (*a*) and (*b*).

Table 9.3. (*a*): Natural Rubber Production in Some Countries (in tonnes)

Year	*Malayasia*	*Indonesia*	*Thailand*	*Srilanka*	*China*	*Latin American countries*
1980	1,529,994	1,020,000	501,108	133,151	113,000	19,000
1983	11,561,992	1,005,000	585,000	126,230	150,000	22,000

Table 9.3 (*b*): Synthetic Rubber Production

Year	*USA*	*Canada*	*USSR*	*FDR*
1980	2,214,839	252,792	2,040,000	3,89,908
1982	1,828,949	181,701	2,950,000	3,85,850

9.1.2 Synthetic Rubber

From the above tables it is seen that India is always in short supply of natural rubber by 20-30,000 tonnes per annum. The anticipated[(2)] demands during seventh plan for natural and synthetic rubbers is tremendous, *i.e.*, 2.8 lakh tonnes and 60,000 tonnes respectively.

9.2 STRUCTURE OF NATURAL RUBBER

The natural rubber is mainly composed of high molecular weight hydrocarbon made up of isoprene units (established by Harries in 1900) linked together to form chain like structure, which is quite flexible at room temperature. Rubber should be sufficiently reactive to allow desirable reactions such as cross-linking, but not react during service. Natural rubber and many synthetic rubbers exhibit this quality. The methyl side group in isoprene repeat unit increase the range of possible structures like cis 1, 4; trans 1, 4, 3, 4 and 1, 2. In natural rubber isoprene is linked in cis 1, 4 configuration (regular polymer) as shown below linking in 1, 4 positions gives rise to increased elasticity and strength. Guttaperch is a stereo isomer of natural rubber having trans structure, but exhibits very poor strength; hence scant commercial importance. A good rubber withstands repeated bendings, stretchings or compressions, rare qualities like elasticity, strength (tensile and impact) insulation properties, make rubber a unique engineering material of construction; especially in pneumatic appliances, chemical industries, electrical insulation etc. However, natural weathering spoils the qualities of rubber. To prevent aging (oxidation) inhibitors like amines, cresols etc. are added to rubber.

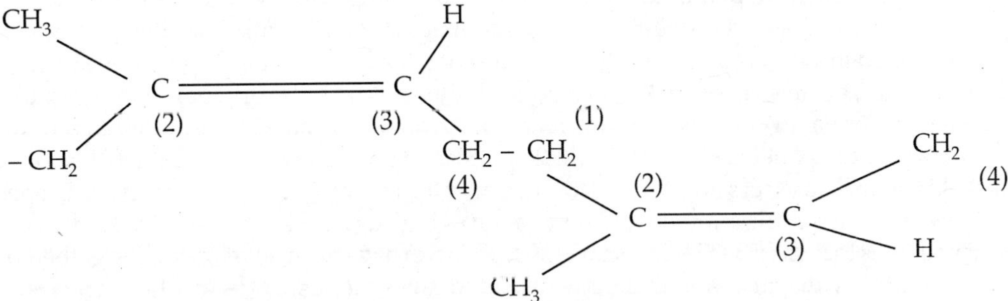

By vulcanising natural rubber, many defects can be eliminated. Vulcanisation is a process of treatment of rubber with sulphur at 140 - 150°C under an elevated pressure. Sulphur cross-links with the macromolecules giving strength, elasticity and resistance to heat and abrasion. During vulcanisation other ingredients like carbon black or kaolin as fillers, antioxidants, colour are added to rubber mixture. Plasticisers like fatty acids, petroleum tars, rosine etc. are added to soften the mixture to increase plasticity. Oil extended rubbers contain substantial amounts of petroleum oils as fillers. Polystyrene oil extended rubbers show high elasticity, strength and resistance to abrasion.

9.3 NATURAL RUBBER LATEX

Rubber latex is the milky fluid collected from the Hevea tree, and it contains 32-35% and 5% non-rubber constituents suspended in aqueous medium. Latex is first concentrated to 60% rubber content by centrifuging at the site itself. Concentrated latex is preserved by adding ammonia or other preservatives like boric acid or formaldehyde. The dried latex in sheets is later transported to rubber factories. This latex is vulcanised after mixing with various compounding ingredients for strength for elasticity and other properties like resistance chemicals and weathering.

Natural rubber possesses many disadvantages, the primary one concerned with irregular supply or short supplies at short notices and at peak time demands. This has led to the world wide quest for synthetic rubbers since 1926. S. Lebdev was the first to synthesize divinyl rubber in 1928. Soon in Germany synthetic rubbers were produced (1937). USA and other countries followed to meet the war demands. Polyisobutene was discovered by Thomas and Sparks (Standard Oil Development Co. in 1937). They found that isobutene polymerised with dienes could be easily vulcanised with sulphur resulting in good quality synthetic rubber. First commercial butyl rubber plant commenced production in USA in 1943. In fact in countries where there is scarcity or no natural rubber plantation, synthetic rubber industry advanced in leaps and bounds.

In India, the beginning of synthetic rubber industry started with alcohol based plant at Bareilly. Since that time, not much development took place except for the contribution of IPCL. Incidentally, India is a good producer of alcohol, about 920 million litres (1990).

9.3.1 Natural Rubber Collection

Crude rubber from other plant sources is generally contaminated by an admixture of resins that must be removed before the rubber is suitable for use. Such crude rubbers include guttapercha and balata, which are products of various tropical trees in the sapodilla family, Sapotaceae. Other, nontropical sources of rubber, which were cultivated for economic reasons during World War II (1939-45), include two shrublike plants : guayule, Parthenium argentatum, native to Mexico, and the Russian dandelion, *Taraxacum kok-saghyz*, native to Western Turkistan.

To gather the latex from plantation trees, a diagonal cut angled downward is made through the bark; this cut extends one-third to one-half of the circumference of the trunk. The latex exudes from the cut and is collected in a small cup. The amount of latex obtained on each tapping is about 30 ml (about 1 fl oz). Thereafter, a thin strip of bark is shaved from the bottom of the original cut to retap the tree, usually every other day. When the cuttings reach the ground, the bark is permitted to renew itself before a new tapping panel is started. About 250 trees are planted per hectare (100/acre), and the annual yield for ordinary trees is about 450 kg/hectare (400 lb/acre) of dry crude rubber. In specially selected high-yield trees, the annual yield may range as high as 2225 kg/hectare (2000 lb/acre), and experimental trees that yield 3335 kg/hectare (3000 lb/acre) have been developed. The gathered latex is strained, diluted with water, and treated with acid to cause the suspended rubber particles within the latex to clump together. After being pressed between rollers to consolidate the rubber into 0.6 cm (0.25 in) slabs or thin crepe sheets, the rubber is air - or smoke-dried for shipment.

9.3.2 Properties of Rubber

Pure crude rubber is a white or colourless hydrocarbon. The simplest unit of rubber is *isoprene*, which has the chemical formula C_5H_8. At the temperature of liquid air, which is about - 195°C (about - 319°F), crude rubber is a hard, transparent solid; from 0° to 10°C (32° to 50°F) it is brittle and opaque, and above 20°C (68°F) it becomes soft, resilient, and translucent. When rubber is mechanically kneaded, or is heated above 50° C (122°F), it becomes plastic and sticky; above 200°C (392°F) it decomposes.

9.3.3 Compounding-Vulcanisation

For the majority of applications, the raw rubber is mixed with a variety of compounding ingredients to modify its characteristics.

Fillers that stiffen the rubber in the final product, but that do not materially increase its strength, include whiting, or calcium carbonate, and barite, or barium sulphate. Reinforcing fillers add materially to the strength of the finished product; they include carbon black, zinc oxide, magnesium carbonate, and various **clays**. Pigments include zinc oxide, lithopone, and a number of organic dyes. Softeners, which are necessary when the mix is too stiff for proper incorporation of the various ingredients like plasticisers usually consist of **petroleum** products, such as oils or waxes; pine tar.

The principal vulcanizing agent continues to be sulphur; selenium and tellurium are also used, but generally with large proportions of sulphur. In the hot process of vulcanizing, which is used for most rubber goods, the sulphur is ground and mixed with the rubber at the same time as the other dry ingredients. The proportion of sulphur to rubber varies from 1 : 40 in soft rubber goods to as much as 1:1 in hard rubber. Cold vulcanization, used chiefly for soft, thin rubber goods such as gloves and sheeting, is accomplished by exposing the uncured articles to the vapor of sulphur chloride (S_2Cl_2).

9.3.4 Masticating and Mixing

Before the compounding ingredients are mixed with the raw rubber, it is subjected to a mechanical grinding process called *mastication*, which makes the rubber soft, plastic, and sticky. Here mixing and blending takes place more easily and more thoroughly with the various fillers, pigments, vulcanizers, and other dry ingredients. Masticators are of two types. One type is the *rubber mill*, consisting of two power-driven steel rollers that rotate at different rates in a trough to shear and knead the rubber until it is broken down to a soft and pliable condition. After the 1920s the rubber mill was largely replaced by the *Gordon plasticator*, which consists of a power-driven variable-pitch screw operating in a jacketed cylinder. The churning action of the screw on the rubber generates temperatures of 182°C (360°F). The heat, rather than the mechanical action, breaks down the rubber. After the masticator, the next machine in the production line is the mixer. Mixers may resemble the rubber mill masticator in having two rollers, but in the mixer the rollers rotate in opposite directions, whereas in the rubber mill the rollers rotate in the same direction at different speeds. In most cases, however, the ingredients are dry-mixed for later calendering, extrusion, or other fabrication in the preparation for final vulcanization. Calenders are machines consisting of three to five rollers of equal diameter. The speed of rotation and spacing between the rollers is adjustable, depending on what kind of rubber product is desired. Calenders are used for sheeting, which produces sheets of rubber with or without impressed designs such as tire-tread markings; for frictioning, which squeezes rubber to give it the texture of fabrics or cords; or for coating, which covers frictioned or previously coated rubber with an additional layer of rubber.

9.3.5 Vulcanisation

Many articles are vulcanized in moulds that are compressed by hydraulic presses, but the high pressures necessary for effective vulcanization can also be achieved by subjecting the rubber to external or internal steam pressure during heating. Certain types of garden hose, for example, are coated with lead, and are vulcanized by passing high-pressure steam through the opening in the hose ; the rubber hose is compressed against the lead housing during the process. After the process is complete, the lead is stripped from the hose and melted down for reuse. Tin sheathing is used in the same way to process certain types of high-grade electrical insulation.

9.3.6 Foam Rubber Goods

Foam rubber is manufactured directly from latex by using emulsified compounding ingredients. The mixture is then whipped mechanically in a frothing machine into a foam containing millions of air bubbles. The foam is poured into moulds and vulcanized by heating to form such articles as mattresses and seat cushions. Latex can be shaped into such products as toys or gloves by dipping forms made of porcelain or plaster of paris into concentrated latex. A coating of latex adheres to the form and is stripped off after vulcanization. Compared to vulcanized rubber, uncured rubber has relatively few uses. It is used for cements; for adhesive, insulating, and friction tapes; and for crepe rubber used in insulating blankets and footwear. Vulcanized rubber, on the other hand, has numerous applications.

9.4 SYNTHETIC RUBBERS

The production of synthetic rubber principally involves two steps : (1) synthesis of monomer. (2) polymerisation of monomer. Synthesis of monomer depends upon the cheap availability of raw material. As an example, ethyl alcohol occupied a prominent position till fifties for butadiene (ethylene also). Later on tubestill crackers revolutionalised the production of olefins and with improved separation techniques, practically alcohol has gone into the shadow. Alcohol plays a vital role in countries where sugar industry is concentrated like ours. Rubbers are usually classified into two varieties as general purpose and special purpose. Butadine, SBR are treated as general type rubbers.

The most important monomer used in synthetic rubber is butadiene. Butadiene is produced conventionally from petroleum feedstocks. It is isolated from naphtha cracker gases. Also it can be produced by dehydrogenation or oxydehydrogenation[3] of butene, oxyde-hydrogenation is carried over antimony oxide at 400-450°C at 2 atmospheric pressure, and air to steam ratio is maintained at 1.8 : 1.0. Two methods of production of butadiene from alcohol are described below.

9.5 BUTADIENE FROM ALCOHOL

Ethyl alcohol, in the first step is converted into acetaldehyde, which on condensation gives rise to crotonaldehyde.

$$C_2H_5OH \xrightarrow{\text{Dehydrogenation}} CH_3CHO + H_2$$

$$2\,CH_3\,CHO \xrightarrow{\text{Condensation}} CH_3 \cdot CH{:}CH.\,CHO + H_2O$$

Crotonaldehyde reacts with alcohol giving crotonyl alcohol which on dehydration gives bútadiene
$CH_3 . CH : CH . CHO + C_2H_5 . OH \rightarrow CH_3 . CH : CH . CH_2OH + CH_3\,CHO.\ CH_3 .CH{:}CH . CH_2OH \rightarrow CH_2{:}CH.\ CH{:}CH_2 + H_2O$

9.6 ALTERNATE ROUTE[4]

Butadiene can be obtained from alcohol by another route. Alcohol and acetaldehyde vapours are passed over silica get impregnated with oxides of tantalum, zirconium at 320°-350°C. This results in butadiene.

$$C_2H_5\,OH + CH_3\,CHO \rightarrow C_4\,H_6 + 2\,H_2O.$$

Fig. 9.1 shows the process outlines. Fresh alcohol and process alcohol are vaporised and sent into Reactor (1), where copper packed in the tubes at 250°C, heated by dowtherm, acts as catalyst to dehydrogenate alcohol to acetaldehyde. The vapours are cooled, compressed and absorbed in water to drive off other gases accompanying like CO, CO_2 and H_2. The dilute phase contains alcohol and aldehyde. These are separated by distillation in columns 4 and 3 respectively. The aldehyde vapours from column (3) and alcohol vapour from column (4) are heat exchanged and sent into another packed tube reactor (5). Silica gel dehydrates the mixture to give butadiene. The vapours are cooled and compressed, and light gases are first driven out by distillation at high pressure and low temperature in column (6). The bottoms from column (6) are distilled to get the crude butadiene in the vapour phase. The overheads are now purified by extractive distillation with solvent chlorex in two columns (8) and (9). Pure butadiene is obtained as a over head product of column (9). Butadiene purification is essential because of the accompanying butylene and propylene (each 2%). It is now-a-days vastly produced from the petroleum feedstocks. It can be either isolated from cracker gases of C_4 fraction, or secondly it can be produced by dehydrogenation of butanes and butenes over usual catalyst (Cr_2O_3 – Al_2O_3) at 550°C. These were described in earlier chapters.

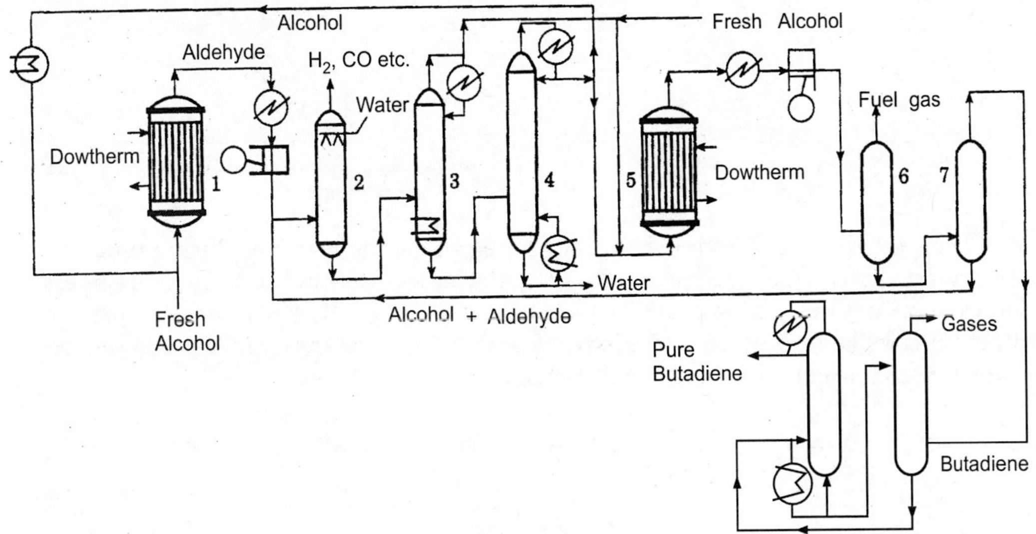

1. Reactor - Aldehyde
2. Scrubber
3. Aldehyde Column
4. Alcohol Column
5. Reactor - Butadiene
6. Light Gases Fractionator
7. Butadiene Separator

8, 9. Chlorex - Extractive Distillation system

Fig. 9.1: Production of Butadiene from Alcohol.

9.7 BUTADIENE RUBBER (BUNA) (PBR)

Rubber synthesis rests on monomer polymerisation. Polymerisation is attended by mass or emulsion techniques. Bulk and emulsion polymerisation gives branching and a wide distribution of molecular weight and molecular weight touches a million or more. Emulsion polymerisation is a free radical polymerisation.

9.8 RADICAL POLYMERISATION

In radical polymerisation the growth of chain is independent of initiating species, while in ionic polymerisation, counter ion has a major influence. Initiators like peroxides, oxygen, azo-compounds and others are actively useful under different characteristic conditions only. For example benzoyl peroxide decomposes to phenyl radicals at a rate of 10% per hour at 40-80°C, whereas oxygen requires 250°C to decompose to the same extent. Depending upon the method and catalyst, the structures of polymer varies. As an example when butadiene or a configurated diene like isoprene polymerises, different types of additions take place as shown below.

$$\begin{array}{c} CH_3 \\ | \\ CH_2-C- \\ | \\ CH \\ \| \\ CH_2 \end{array} \qquad \rightarrow \qquad \begin{array}{c} -CH_2-CH- \\ | \\ C-CH_3 \\ \| \\ CH_2 \end{array}$$

1, 2 addition 3, 4 addition

$$\begin{array}{ccc} -CH_2 & & CH_2- \\ & C=C & \\ CH_3 & & H \end{array} \qquad \begin{array}{ccc} -CH_2 & & H \\ & C=C & \\ CH_3 & & CH_2- \end{array}$$

cis 1, 4 Trans 1, 4

In case of butadiene 1, 2 and 3, 4 additions are not distinguishable. During butadiene (1, 3) polymerisation, 20% of 1, 2 structures, 80% of 1, 4 structures are noticed. As the temperature of polymerisation increases cis 1,4 polymer increases while 1, 2 polymer remains same. However, substituted diene like isoprene polymerises 1, 2 and 3, 4 cis and trans (1, 4) polymers result. But generally trans 1, 4 head to tail configurations results.

$$\begin{array}{ccc} -CH_2 & & H \\ & C=C & \\ CH_3 & & CH_2- \end{array}$$

With variation in temperature of polymerisation the different configurations likely to form during polymerisation of different dienes are as shown below[(5)].

(Temp of polymerisation) °C		*cis %*	*Trans %*	*1, 2%*	
For butadiene	–20	6	77	17	
	–10	9	74	17	
	5	15	68	17	
	100	22	58	20	
	233	43	39	18	
(Tg°C)		–107	–106	–15	
For Isoprene					3, 4 addition
	–20	4	90	5	1
	–5	7	82	6	5
	100	24	66	6	4
(Tg°C)		–73	–60		
Chloroprene					
	–45	5	94	1	—
	10	9	84	1-4	1-4
	100	14-22	70	4	4

At higher temperatures cisoid conformations increase. Conventional polymers prepared using butyl or ethyl lithium at temperature range 25-40°C are normally linear.

Polymers produced by Zeigler-Natta complexes show a great degree of stereospecificity. Aluminium alkyl/aluminium halide with transition metal like titanium, dichromium or molybdenum combinations are used for production of stereogregular butadiene.

Emulsion polymerisation is governed by radical formation. Except peroxides other like potassium persulphate generate less free radicals. Some ions like ferrous, accelerate the formation of radicals; thereby the system gets the name redoxi initiators. Bulk polymerisation using lithium and potassium catalysts show the above polymer(6) structures of different degrees as given below :

Bulk polymerisation	*Cis 1,4*	*% Trans 1, 4*	*% 1, 2 and 3, 4 addition*
Lithium cat.	35-36	52-56	8-13
Potassium cat.	15-16	37-40	52
Solution polymerisation	32-36	53-58	8-11
(lithium alkylcat)			
Emulsion Polymerisation	14-18	64-69	17-18

Butadiene polymerisation has been shown in Fig. 6.10. However, butadiene rubber is produced by mass polymerisation by using metallic sodium or lithium metal. Usually autoclaves of 2 to 3 cub. metres capacity (carbon steel make) with provisions of heating and cooling are used for this purpose. Sodium coated rods are dipped into the autoclave after charging the same with monomer. The reactor is kept at 40-60°C for six to eight hours under a pressure of 8-9 bars. The unconverted monomer is distilled off and the polymer is pushed off by nitrogen pressure. A yield of 95% is expected. Lower temperatures favour linear polymer with 1, 4 linking. Recently some modifiers are suggested to decrease branching; the modifiers listed are carbon tetrachloride hexa chloroethane, mercaptans, acetaldehyde, vinyl esters etc. During butadiene polymerisation, it is found that near sodium rods, the polymer becomes harder than in the bulk. To remove the disadvantage of such catalyst, Alfin catalysts have been widely used as suspension in monomer itself. Alfin catalysts are quite suitable for linking macro molecules in 1, 4 positions to yield regular structure and high molecular weight polymers. Improvement in strength and frost resistance is noticed with butadiene rubbers of Alfin catalysts. Butadiene rubber (PBR) is inferior in strength and other properties hence styrene butadiene copolymer rubber is widely produced.

9.9 STYRENE BUTADIENE RUBBER (SBR OR GR-S)

Styrene and butadiene can be copolymerised in any desired ratio. However, mixture of 70% butadiene and 30% styrene is copolymerised to yield SBR, which forms the major amount of all synthetic rubbers produced and constitute about 70%. Styrene may be replaced with (a)-methyl styrene. Styrene and a-methyl styrene are produced by dehydrogenation of ethyl and propyl benzenes over selected catalysts like iron or magnesium oxide,

$$C_6H_5.C_2H_5 \rightarrow C_6H_5CH:CH_2$$

$$C_6H_5.CH(CH_3)_2 \rightarrow C_6H_5.\overset{\overset{\displaystyle CH_3}{|}}{C} = CH_2 + H_2$$

Polymerisation of styrene, butadiene mixture is done at 5°C or 50°C; the product obtained under those conditions is known as cold or hot rubber respectively. SBR is known as general purpose rubber and is produced by free radically initiated emulsion polymerisation. Cold rubbers are produced under prolonged time of polymerisation using oxidation-reduction system catalyst.*

Fig. 9.2 shows the process-flow sheet on SBR production.

Essential steps are grouped as :

1. Emulsion preparation - monomers addition
2. Initiation-polymerisation
3. Separation of monomers
4. Coagulation of latex and separation of rubber.

Emulsion in aqueous phase (containing 5% ammonia) with Neicke is prepared in an autoclave. Potassium stearate, sodium sulphite are added to stabilise the emulsion. To this aqueous mix, hydrocarbon monomers are added in the proportion required alongwith initiator potassium sulphate and activator (isopropyl benzene hydroperoxide in water, and hydro-quinone and sodium sulphite solution). The mixture is heated by means of hot water circulation in the autoclave (1). Then the whole mixture is transferred into polymerisation reactors (autoclaves 10-12 in series, each fitted with a stirrer). Initiator and activator are added into the reactors and the temperature is kept at 5°C by circulation of brine. Regulators or modifiers, buffer salts are added into the fifth or sixth like benzoyl peroxide and ferric sulphate reactor so that the conversion proceeds upto 70%. Then the short stops (like quinols) are added in the last reactor (7). The product is later discharged into special type of still (3), where latex settles at the bottom and monomers are fractionated in individual columns (4) and (5) as shown in the figure. The monomers are useful for recycling. Anti oxidants are added to the latex and the latex from the still bottom is washed and coagulated by addition of electrolytes. After that the latex is washed, dried and sent for sheet making. Recovered butadiene and styrene vapours are purified and recycled. SBR vulcanisates generally resemble natural rubber in mechanical properties, and they possess better resistance to crack formation and abrasion. They resist oxidation better than natural rubber. They can be vulcanised to ebonite stage. Hot rubbers are low mole weight 30,000 to 100,000 rubbers with a wide spread. Solution polymerisation gives narrow molecular weight range.

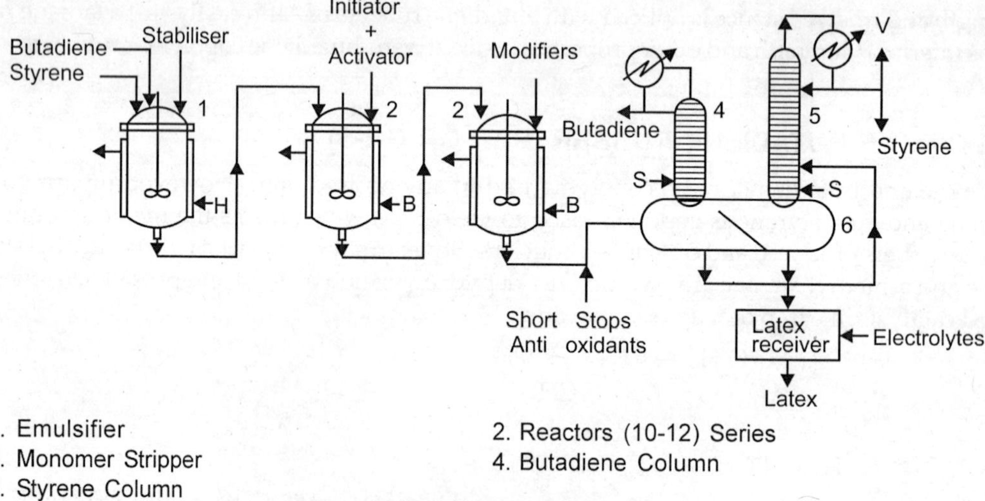

1. Emulsifier
2. Reactors (10-12) Series
3. Monomer Stripper
4. Butadiene Column
5. Styrene Column

B. Cold Brine
H. Hot Water
S. Steam
V. Vacuum

Fig. 9.2: Production of Styrene-Butadiene Rubber.

Various combinations like styrene/butadiene/vinyl pyridine (Buna tex VP) or α-methyl styrenebutadiene rubbers (SKMS) are produced by emulsion of polymerisations.

9.10 ACRYLONITRILE-BUTADIENE RUBBERS (NBR) (BUNA-N)

This is produced by copolymerising, various proportions of butadiene and acrylonitrile monomers in aqueous emulsions. Molecular weight of the product is controlled by adding suitable modifiers and inhibitors. NBR is quite resistant to hydrocarbon oils, hence coatings, tubings can be made out of this. It is also used for better finishing and paper coatings etc. Because of its versatility about 200 grades are produced.

9.11 EMULSION POLYMERISATION

Acrylonitrile production by Sohio process has been described. Required amount of acrylonitrile (25-50%) with butadiene in emulsion is polymerised by initiation with free radicals; this closely resembles SBR production. Usually cold rubbers are only produced (5-25°C). Redox system initiator with chain transfer agents like mercaptans are used in polymerisation. As the acrylonitrile content increases chemical resistance and strength increases while its resiliency naturally comes down with low molecular weight (3000-5000). NBR with terminal reactive groups like carboxyl mercaptan, vinyl moieties are available. These are useful in linking non rubbery polymers with rubbery polymers to give tough composites.

9.12 POLY ISOPRENE

Isoprene is produced by two stage dehydrogenation of isopentane :

$$CH_3-\underset{|}{\overset{CH_3}{CH}}\ CH.CH_3 \rightarrow CH_2-\overset{CH_3}{\overset{|}{C}}-CH_2-CH_3 \rightarrow CH_2=\overset{CH_3}{\overset{|}{C}}\,CH=CH_2$$

It can also be produced from amylenes like methyl butenes which are obtained from cat cracking units. IFP process produced by cracking 4, 4 dimethyl 1,3 dioxane, which is produced by the reaction of formaldehyde and i-butene, is as shown below.

$$\begin{matrix} CH_3 \\ CH_3 \end{matrix}\!\!>C=CH_2+2HCHO \rightarrow (CH_3)_2\!<\!\begin{matrix} C-CH_2-CH_2 \\ O-CH_2O \end{matrix} \longrightarrow$$

$$CH_2=\overset{CH_3}{\overset{|}{C}}-CH=CH_2+CH_2O+H_2O$$

Since isoprene molecule is not symmetrical all the additions as described earlier can take place. Polyisoprene is produced by polymerisation with lithium or lithium compounds or co-ordination catalyst; Zeigler type catalysts produce very high level of cis structure. Aluminium to titanium ratio dictates the cis content and rate of polymerisation. If the ratio of Al to Ti is less than one, then its trans structure predominates in the polymer.

9.13 CHLOROPRENE RUBBERS: (CR) (NEOPRENE)

Chloroprene rubbers, a special purpose rubbers are superior to natural or even isoprene rubbers in resisting heat, oils, ozone etc. Sulphur modified rubbers have high resistance, heat resistance and tear strength. The structure of chloroprene looks like isoprene, where chlorine appears instead of methyl group at position 2. Molecular weights vary from 20,000 to 200,000 depending upon sulphur or mercaptan modified polymer.

9.14 MONOMER PRODUCTION[8]

Chloroprene is produced by dimerising acetylene to monovinyl acetylene in presence of a catalyst, cuprous chloride at 80°C. This monovinylacetylene reacts with HCl to give chloroprene.

$$2\,CH\equiv CH \rightarrow CH\equiv C-CH=CH_2$$
$$(Cu_2Cl_2 + H_2O)\ \text{Monovinyl acetylene}$$
$$CH\equiv C-CH=CH_2 + HCl \rightarrow CH_2=C.Cl-CH=CH_2$$

Alternatively butadiene by chlorination, isomerisation and dehydrochlorination in sequence can give rise to chloroprene

$$CH_2:CH-CH:CH_2 + Cl_2 \longrightarrow CH_2\,Cl.\,CHCl-CH:CH_2$$
$$\text{Dichlorobutene (cis/trans)}$$
$$CH_2\,Cl.\,CH:CH.\,CH_2Cl \xrightarrow{\text{isomerisation}} CH_2Cl.\,CHCl-CH:CH_2$$
$$CH_2Cl.\,CHCl.\,CH:H_2C \xrightarrow[\text{chlorination}]{\text{dehydro}} CH_2:\underset{\displaystyle Cl}{\underset{|}{C}}-CH:CH_2$$

Chloroprene is unstable like isoprene and very easily polymerises. Polymers are obtained as aqueous emulsions with potassium sulphate catalysts which may be coagulated by addition of acetic acid. Sulphur modifies the polymers. Generally copolymer with acrylonitrile; neoprene rubbers; are used in hose pipes, tanks carrying oils and hydrocarbon fractions, alcohols etc; though concentrated mineral acids, chlorine etc. are likely to attack. The major uses of polychloroprene are in adhesives, gaskets, conveyor belts and cables. Because of its tensile, abrasive and impact strength and non-inflammable nature, it has much demand in industry.

9.15 BUTYL RUBBER[9]

Butyl rubber is obtained by polymerising i-butene with a small amount of isoprene. On vulcanisation it resembles natural rubber but possesses superior qualities in heat, stress and weather resistance. The cured rubber is highly impermeable to gases hence used for tubes, balloons etc. It is used for electrical insulation, gaskets, linings etc. Bromobutyl rubber is a recent addition to these family of rubbers. Butyl rubber is produced by cationic polymerisation using vanadium oxychloride as cocatalyst and ethyl or butyl trialkyl aluminium as catalyst of ziegler-Natta complexes. 2 to 3 per cent isoprene mixed with isobutylene dissolved in methylchloride and cooled to 9°C. Catalyst is also dissolved in methyl chloride. The polymer heat distills of methylchloride, maintains the reactor conditions. Polymer slurry is overflown into flash drum, which contains copious amounts of hot water to agitate the polymer and flash off the hydrocarbons, which are recovered and routed back. The polymer mass is washed and filtered to separate the polymer. Fig. 9.3 shows the outlines of butyl rubber production. Chloro, sulphonated rubbers under the name Hyplon are also another form of these chloroprenes. The importance of isoprene is to provide double bonds for vulcanisation. Chloro butyl rubber contains 1-2% combined chlorine, it shows outstanding resistance to heat and chemical attack. It is used for gaskets, tubes, and can be vulcanised with natural or SB rubbers. Halogenation provides polar groups hence adhesive properties and curing rates are improved.

9.16 TRANS POLYPENTAMER RUBBER (TPR)

TPR (trans 1, 5 polypentamer) is not only new because of its polymerisation mechanism but also of interesting properties it possesses (specific gravity 0.85) and this is the only solution-polymerised elastomer having maximum trans structure. It is amorphous at ordinary temperature having a glass transition temperature of 90°C; nearer as that of BR. Because of solution polymerisation a narrow molecular weight distribution is attainable (average mole weight 4×10^5). It can be processed without complications and can absorb large amount of fillers and plasticizers. As the basic unit being C_5 fewer double bonds exist unlike butadiene; hence high resistance to ozone, and aging. Compression and abrasion resistance should also be attractive.

9.16.1 Monomer Production

The source for the monomer is usually naphtha crackers; meant for ethylene production. Liquid hydrocarbon feeds shall produce good amount of C_5 during cracking. This contains isoprene, cyclopentadiene (CP) dicyclopentadiene (DCP) etc. upto an extent of 3-6%.

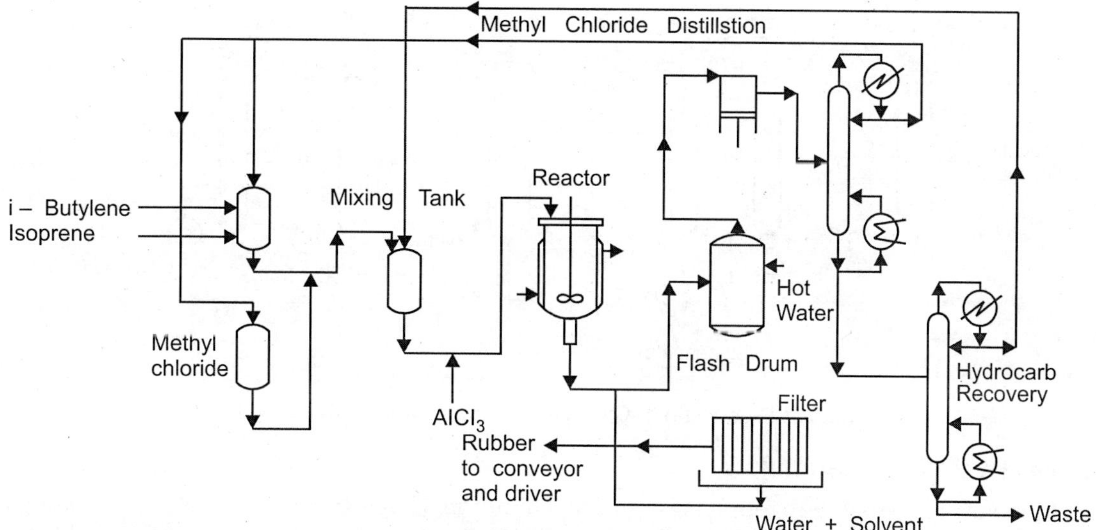

Fig. 9.3: Butyl Rubber Production.

9.16.2 Polymerisation

Polymerisation of cyclopentene by ring cleavage is an interesting reaction, where (cyclo olefins) C-C bonds are retained during cleavage polymerisation. Next stage is polymerisation of cyclopentene, propagation, depolymerisation and reconstruction of macro molecules take place simultaneously; polymerisation is conducted by tungsten hexachloride and aluminium alkyl catalyst along with co-catalysts to give a satisfactory degree of conversion with trans double bonds to maximum extent.

CH_2 CH_2 CH_2 CH_2

→ C

CH_2 CH_2 H CH_2 CH

The polymerisation speed and final degree of conversion depend to a small extent on the monomer concentration.

9.16.3 Isolation of DCP

The amount increases depending upon the molecular weight of the feed and cracking severity. The cracked gasoline fraction containing CP/DCP has to be isolated from the C_5 cut. IFP has developed a catalyst for selective hydrogenation of cyclopentadiene to polymer grade cyclopentene (CPD less than 50 ppm). The catalyst is soluble in the hydrocarbons and shows light activity resulting in product loss by dimerisation to less than 0.1 per cent. The yield is found to be 80-90% Palladium on alumina or Palladium titanium on Li-Al spinel was used as catalyst. The process for production of cyclopentene is shown in Fig. 9.4. C_5 cut from cracked gasolines is first isolated by careful distillation in column (1). The C_5 cut containing other olefins, cyclopentadiene and dicyclopentadiene is dimerised. Dimerisation is possible by storing the mixture at elevated temperature for a long time (2) alternatively it can be attended catalytically. The fraction now being DCP, is easily separated by distillation in two or more (3) columns. The remaining stream may be sent for isoprene extraction.

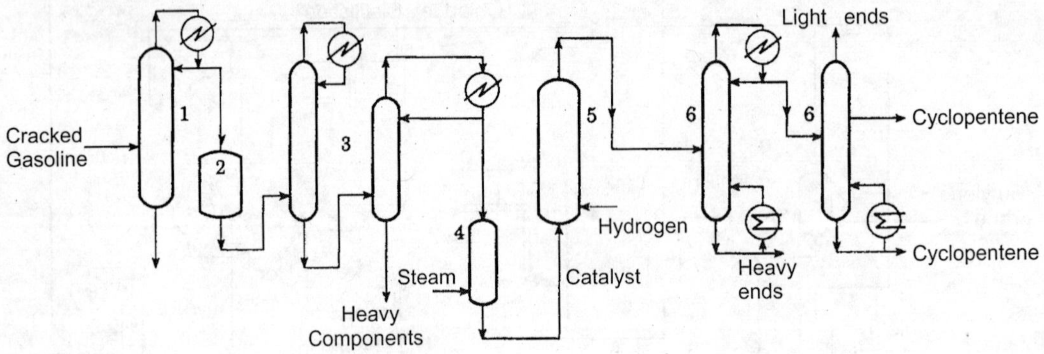

Fig. 9.4: Production of Cyclopentene from Cracked Fractions.

9.16.4 DCP Monomerisation

Monomerisation is done by applying high temperature for a few seconds. Steam, hydrogen, nitrogen act as diluent in this cracking reactor (4). The monomer cyclopentadience can be removed by distillation and submitted to hydrogenation (5) and further distillation to obtain cyclopentene in a battery of columns (6).

9.17 ETHYLENE-PROPYLENE RUBBERS (EPM)

Copolymerisation of ethylene-propylene has been subjected to discussion earlier. Addition of small amount of diene imparts suitable double bonds for vulcanisation. Ziegler-Natta catalysts are useful in bringing polymerisation in a hydrocarbon solvent. Vanadium compounds and aluminium alkyls are used as catalyst components. Though a wide range of ethylene to propylene ratios is chosen, conventional polymers show 50/50 to 75/25 variations of ethylene to propylene. The temperature of polymerisation is kept around – 20 to + 20°C. To control the exothermic reaction, some monomers are allowed to boil taking off the heat from the reactor. The slurry from the reactor is discharged (after mixing with antioxidant) continuously, into a stripper where monomer and solvent are stripped off. Water-polymer slurry is decanted to separate polymer. Ethylene-elastomers consist of randomly alternating blocks of 10-20 monomer units, as follows :

$$-(CH_2-CH_2-CH_2-\underset{\displaystyle CH_3}{\underset{|}{C}H}-)_n \quad \text{where } n = 10 \text{ to } 20$$

Molecular weight ranges between 1000,000 to 2,000,000. Elastomers soluble in many organic solvents and can be plasticised by mineral oils and phthalates. These have very low specific gravity of 0.85 to 0.87 and resemble natural rubber in permeability to gases; tensile strength and elongation. Glass transition temperature is around-58°C. It is better heat resisting than natural rubber, provided it is vulcanised by chemicals like zinc oxide and monobutyl phthalate. E-P rubbers possess excellent electrical properties and they are used like natural rubbers. Ethylene propylene-diene methylene (EPDM) rubbers are made by introducing certain amount of diene (1, 4 hexadiene) into polymer mix. This diene gives terminal double bond which maintains the activity for vulcanization etc. The polymer is made is a solvent like hexane with catalyst $VOCl_3$ and $(C_2H_5)_3\ AlCl_3$ complexes.

The structure looks as below:

```
    C
    |
C – C – C – C – C – C –
        |
        C
        |
        C
        ||
        C – C
```

9.18 POLY SULPHIDE RUBBERS

These are known as thiokols, elastothiomers or thioplasts. They are available in solid, liquid forms or as pre-vulcanised moulding powders. Though they have poor mechanical properties they have good oil resisting, gas and moisture impermeability. They have been in commercial scene for longer than any other synthetic rubber, since produced in 1924 by J.C. Patrick. These rubbers are usually linear, but copolymers can be made by using dihalides two or more. Ethylene dichloride[(12)] rubbers are usually called as Thiokol-A and others are designated as listed below.

Di - 2 - chloroethyl ether	Thiokol B
Di - 2 - chloroethyl formal	Thiokol ST
(i, 2, 3, trichloro propane for branching)	
Di - 2- chloroethyl formal ethylene dichloride	Thiokol FA

High molecular weight polymers as well as liquid polymers have been available. Wherever the end groups are hydroxyl or thiol different vulcanisation processes are used. Hydroxyl ends like thiokol FA is coupled with zinc oxides. Thiol terminated polymer are vulcanised by using lead oxide as shown below.

$$RSH + PbO_2 + HSR \longrightarrow RSSR - + PbO + H_2O$$

9.18.1 Production of Polysulphides

Aliphatic dihalide and sodium polysulphides are condensed together. The simplest Thiokol-A is obtained by reacting dichloroethane with sodium tetra sulphide.

$$ClCH_2CH_2Cl + Na_2S_4 \longrightarrow (CH_2-CH_2-\underset{\underset{S}{\|}}{S}-\underset{\underset{S}{\|}}{S}-CH_2-CH_2-) + 2NaCl$$

The reaction is conducted in aqueous medium at 70°C in presence of magnesium hydroxide to assist dispersion[11]. The latex obtained is later coagulated with acid and then dried. Several polysulphide rubbers can be produced by varying dihalides or using mixed halides and varying the type of sodium polysulphide. Suitable modifiers (Trichloropropane) may be added to control crosslinkage or chain length. The thiokol rubbers are costly and have specific gravity varying from 1.13 to 1.6 depending upon liquid state or solid state. These are not as good as natural rubbers. These rubbers do not possess good strength also, depending up on the intermediates and chain extending, the elastomers exhibit different properties. Thus the elastomers produced from different polyols are grouped as : (1) polyesters (2) polyethers and (3) natural products.

9.19 POLYURETHANES

Urethane is an engineering material that offers an endless applications list. Soft as marshmallow, harder than a bowling ball, urethane can be used in lieu of plastic, rubber, nylon, and many metals. Within the industry cast urethane parts are known for toughness. Toughness being resilient, abrasive resistant, cut and tear resistant, memory foam compression, and flexion. Urethane parts are not impacted by most harsh environments.

In the production of hot NBR, emulsifier (soap), 2-propenenitrile (acrylonitrile), various butadiene monomers (including 1,3-butadiene, 1,2-butadiene), radical generating activators, and a catalyst are added to polymerization vessels. Within the vessel, water seres as the reaction medium. The tanks are heated to 30° C - 40° C to facilitate the polymerization reaction and to promote branch formation in the polymer. Because several monomers capable of propagation of the reaction involved in the production of nitrile rubber, the composition of each polymer can vary (depending on the concentrations of each monomer added to the polymerization tank and the conditions within the tank). One repeating unit found throughout the entire polymer may not exist. For this reason, there is also no IUPAC name for the general polymer. The reaction for one possible portion of the polymer is shown below:

1, 3-butadiene + 1, 3-butadiene + 2-propenenitrile + 1, 3-butadiene + 1, 2-butadiene → acrylonitrile butadiene rubber

Monomers are usually permitted to react for 5 to 12 hours. Polymerization is allowed to proceed to ~70% conversion before a "shortstop" agent (such as dimethyldithioarbamate and diethyl hydroxylamine) is added to react with the remaining free radicals. Once the resultant latex has "shortstopped", the unreacted monomers are removed through a steam in a slurry stripper. Recovery of unreacted monomers is close to 100%. After monomer recovery, latex is sent through a series of filters to remove unwanted solids and then sent to the blending tanks where it is stabilized with an antioxidant. The yielded polymer latex is coagulated using calcium chloride, aluminum sulphate, and other coagulating agents in an aluminum tank. The coagulated substance is then washed and dried into crumb rubber.

Polyurethane elastomers have been in the market since Bayer introduced in 1950. The development of polyurethane rubbers has enabled alternative approaches to be devised, which differ from traditional techniques used in rubber industry. Some of the methods have been so perfectly developed to help polyurethane rubbers as one of the most important class of special rubbers. Polyurethanes can be viewed as mixed amide-esters of carbonic acid and thus have the properties intermediate between polyesters and polyamides. Polyurethanes usually are low melting polymers than polyamides and decompose at temperatures above 230°C to give free isocynate and alcohol. Phenolic urethanes decompose even at lower temperatures.

Polyurethanes have a wide range of applications in the fields of adhesives, textiles, flexible and rigid foams etc. The following information shows, how polyurethanes have occupied the different branches of engineering.

Furniture and upholstry	39%
Adhesives-coatings	7%
Automative	10%
Refrigerators	7%
Building and construction	13%
Shoes	4%
Textiles	6%
Others	8%

Presently the elastomer part of the resine only is dealt.

9.19.1 Polyurethanes-elastomers

Urethanes are produced by the reaction between alcohols and isocynates. Generally two isocynates *i.e.* 2, 4 and 2, 6 toluene diisocynates (TDI and 4, 4 methylene bis phenylisocynate, MDI) with functionality varying in between 2-3 are used, Alcohols of functionality up to eight are also used. Mixtures of 2, 4 and 2, 6 TDI are used in flexible foams. High resistance (HR) foams use TDI & MDI Aliphatic diisocynates (4, 4 dicyclo hexyl methylene diisocynate H_{12} MDI) are useful in coatings.

Creation of foams is attended either by mechanical frothing, volatile liquids, thermal decomposition of substances, known as blowing agents.

9.19.2 Polyester-elastomer

Polyester or polyurethane rubbers as they are known, are classed into two types containing urethane *i.e.* $-NH-\underset{\underset{O}{\|}}{C}-O-$ back bone in a linear fashion; the type is a perpolymer of linear polyester

or polyether that has been attended several fold in chain length by coupling through urethane linkages; giving rise to cross links. The molecular weight is of the order 10,000. These rubbers possess very good wear resistance with high elasticity and permeability. They can withstand hydrocarbon attack but are not suitable for low or high temperature operations polyurethanes are used to modify the surface of cotton fibres to impart stiffness and tear resistance. Polyurethanes are synthesised by reacting long chain diols with excess of diisocynates.

$$OCN{-}R{-}NCO + HO{-}R'{-}OH \longrightarrow -\overset{\overset{O}{\|}}{C}-NHRNH-\overset{\overset{O}{\|}}{C}O-R'O-$$

where *R* is benzene ring.

In practice polyurethanes contain linking groups other than urethane. These arise from the reactive isocynates with groups containing hydrogen in which reactions are of the following general nature :

$$NCO + HX- \longrightarrow -NHCOX-$$

With water

$$NCO + H_2O \longrightarrow NH.CO.OH$$

With hydroxyl group

$$-NCO + OH- \longrightarrow NHCOO-$$

With amine

$$-NCO + H_2N \longrightarrow -NH-\underset{\underset{O}{\|}}{C}-NH-$$

The reactivity of isocynates is due to resonance structure

$$R-N=C=0 \quad === \quad R=N-C^{+}=0 === RN=C^{+}-O$$

Thus as R increases from aliphatic amines to aromatic the rate of reaction is increased.

9.19.3 Synthesis of polyurethane (Polyesters)

Synthesis involves basically two stages. In the first stage polyester is produced by reacting adipic acid with glycol as shown below :

$$(CH_2)_4(COOH)_2 + (CH_2)_2(OH)_2 \longrightarrow -OOC(CH_2)_4COO(CH_2)_2OH$$

The second stage reaction depends upon the activity of isocynate (–NCO), which is produced by condensation of R amines with phosgene.

$$RNH_2 + COCl_2 \longrightarrow RNCO + 2HCl$$

The isocynate group readily reacts with hydroxyl or amino groups

$$NCO + HO \longrightarrow NHCO.O-$$

$$NCO + NH_2 \longrightarrow NHCONH-$$

Excess of isocynates react with the product to give cross links like

$$\begin{array}{l} -NCO.O- \\ \quad | \\ CONH- \end{array}$$

The best known polymer (perlon U) is produced by direct reaction of 1, 6 hexamethylene diisocynate with 1,4 butanediol in an inert solvent o-dichlorobenzene.

$$OCN\,(CH_2)_6\,NCO + (CH_2)_4\,(OH)_2 \rightarrow CONH\,(CH_2)_6\,NHCO.\,O\,(CH_2)_4\,OH$$

or reacting polyester with diisocynate

$$2\text{–}OC\,(CH_2)_4\,COO(CH_2)_2\,OH + R\,(NCO)_2 \rightarrow O\,(CH_2)_4\,COO(CH_2)_2\,OOCNHRHNCO + O\,(CH_2)_2\,COO(CH_2)_2\,COO$$

The production of di isocynates involves phosgene, a highly toxic compound. Some isocynates can be produced without using phosgene. An example being MDl using CO and ethyl alcohol and aniline.

$$C_6H_5NH_2 + CO_2 + C_5HOH + (O) \longrightarrow C_6H_5NH_2COOC_2H_5 + H_2O$$

Ethyl. Phenyl Carbonate

$$\text{EPC} + \text{HCHO} \xrightarrow{\text{Condensation}} \text{ONC}-C_6H_4-CH_2-C_6H_4-\text{NCO}$$

9.19.4 Polyethers-Urethanes

Polyethers when used instead of polyesters, the product cost decreases very much. However, it may effect certain properties. General reduction of molecular weight to increase the hardness and modulus by using polyethers is welcomed. Certain pendent groups in glycols (methyl group in propylene polyol) adversely affect the product. Natural polyols are still worse than these polyether in giving elastomers.

Similarly the established works show that aliphatic diisocynates give unsatisfactory products, while 2, 4 toluene diisocynate and 1, 5 naphthalene diisocynate are found to be yielding tough rubbers with high tearstrength.

By varying ratio of polyol to isocynate the physical properties of polyurethanes can be changed, *i.e.* tensile strength can be changed from 900 to 1500 bars and elongation 150 to 180 per cent. In fact polyurethanes can be made with a wide range of properties depending upon the polyol and cross linking triols chosen. These rubbers are available in three forms :

1. Millable rubbers are based on isocynates terminated prepolymers that can be chain extended by reaction with water, glycols and amines etc.
2. Liquid casting rubbers, usually prepared by long sequence of prepolymer, chain extension and cross-linking.
3. Thermoplastic rubbers suitable for cross-linking without any vulcanising additives.

9.19.5 Silicone Rubbers

Silicone rubbers possess very valuable properties like frost resistance upto 100°C and heat resistance upto 250°C. Silicone rubbers are used in special gaskets and insulations. These are harmless to human organism so extensively used for making surgical rubbers and other types to synthetic organs. They can be vulcanised easily.

It is produced from dimethyl dichlorosilanes which on hydrolysis yields monomer dimethyl silanediol which can condensate give rise a basic structure of

$$HO\left[-\underset{CH_3}{\overset{CH_3}{\underset{|}{\overset{|}{Si}}}}-O\right]_n-H$$

These silanes may be mono, di or tri functional

```
R                R            R
 \               |            |
O - Si - R    O - Si - O   O - Si - O
 /               |            |
R                R            O
Mono             Di        Tri functional
```

Mono shows that one oxygen atom is shared by two silicone atoms. Silicones are linear rubbers; cold resistant rubbers can contain benzyl instead of methyl radical. These rubbers can be easily cross-linked giving rise to mole weight of 10^5-10^6.

Various chlorosilanes, like the following are used

$(CH_3)_3\ SiCl$	trimethyl chlorosilane
$(CH_3)_2\ S1Cl_2$	dimethyl dichlorosilane
$CH_3\ SiCl_3$	methyl trichlorosilane
$(C_6H_5)_2\ SiCl_2$	diphenyl dichlorosilane
$(CH_3)_2\,(CH_2 : CH)\ SiCl_2$	methylvinyl dichlorosilane

Chloro silanes are produced by reacting methyl chloride with silicon at 250-280°C in presence of copper or silver as catalyst. Ethylene, acetylene can react with trichloro silanes ($H\,SiCl_3$) to give chloro silanes.

9.19.6 Fluoro Carbon Elastomers

Extreme thermal stability, chemical, solvent and oxidation resistance of Teflon, (tetrafluoro ethylene) polymers are largely attributed to C-F bonds rather than the conventional C-H bonds.

Copolymer of tera fluoroethylene and methyl vinyl ether containing 20-50% ether is obtained by free radical aqueous emulsion polymerisation of tetrafluoro ethylene and methyl vinyl ether. A latex containing 20-30% solid polymer is obtained. The latex is coagulated by adding acid or a salt. The copolymer is soluble in many fluorinated solvents but resists acids, alcohols, ketones etc. It has low permeability to many fluids and have good insulation properties. Polymer can be easily milled or calendered and handled as ordinary rubber. The small atomic radius of flourine, low polarisability and high linkage energy between flourine and carbon give a very high softening point 250°C to polymer having a specific gravity of 2.0-2.2 Tetra-flouroethylene-hexaflouro propylene copolymers have formulae

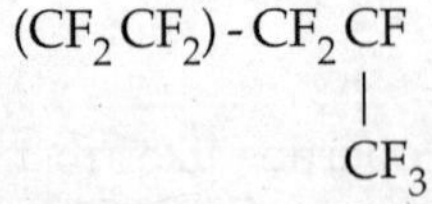

$$(CF_2\,CF_2) - CF_2\,\underset{\displaystyle CF_3}{\underset{|}{CF}}$$

They are generally having specific gravity 2.15 to 2.17 and they are useful as good insulation agents and can withstand fluctuating temperature ranging from -260 to 200°C or above. Poly chloro triflouro ethylene (fluoroethene) can almost approach in its qualities to PTFE. Vinyl flouride, vinylidine flouride are some of the halogenated monomers that can be easily polymerised by radical initiation.

REFERENCES

1. Rubber News, Polymer Publication, Bombay, July, 1985, 29-30.
2. Rubber Chem News, Elastomer Publication Co., May, 1985.
3. F. C. Newman, IEC, May, 1970, 43.

4. The Science of Petroleum Vol. 1, Part 11, 37 (Dunston, Oxford).
5. Malcolm. P. Stevens. Polymer Chemistry Addition Wesley Pub. 1975, Massachussets, 119.
6. a 11. W. J. Roff a J. R. Scott. Fibres, Films Plastics And Rubbers Batter Worthy, London, 1971, 374. 437.
7. P. Belov, Fundamentals of Petroleum Chemical Technology, Mir Publication Moscow, 191.
8. a 12. J. A Brydson, Rubber Chemistry Applied Science Publishers Ltd. London, 1978, 297, 394.
9. Maurice Morton. Rubber Technology, Van Nostrand, 1973
10. Wilhelm Graulich, Dieter Theisen a Wolfgang Swodenk, HCP Dieter Theisen a Wolfgang Swodenk, HCP, Dec. 72, 75.
11. Richard L. Rowton, Polyurethanes-looking Ahead to the Eighties Proceedings of SPI-25, Annual Urethane Division Tech. Conference Arizona, Oct. 29-30,1979, 76.
12. International Progress in Urethanes; Proce of Second Mexican Urethane Symposium. Sept. 319-75. Vol-I.
13. Polyurethanes. Proc of S PI - 6th International Tech/Marketing Conference. San Diego - Cali, 1983.

❑

CHAPTER 10

Plastics

10.1 CLASSIFICATION OF PLASTICS

In terms of tonnage, plastics are more important polymers than all other polymers like rubbers, adhesives, surface coatings combined together. Plastics are nothing but polymers that have been processed into rigid shapes due to high degree of crosslinking. In principle, thermosetting plastics would be expected to be harder and more brittle than thermoplastics, with over-lapping properties of mechanical, chemical and electrical. Exceptions are there, as it is not always true that all thermosets are very hard, the mechanical properties vary very much depending upon the structure and molecular weight of the plastic. Notably a material like poly-urethane cannot be specified into any group as it can be tailored to suit any type of application. Most usually plastics are classified on mechanical and electrical properties.

Thermosets are long chain molecules that give the polymer cross-links in three dimensional infusible structures. These are produced by condensation reaction. Setting is caused by unidirectional polymerisation under heat and pressure resulting in hard rigid masses; the way they differ from thermoplastics. Economically they are not regarded as important as there is no scrap value, hence they are not produced in such large tonnage as thermoplastics.

Cross-linking occurs, during curing or polymerisation or with cross- linking agents. The demands for strong and stiffest polymers for special uses cannot be over-looked, hence they have been produced on a limited scale in the category of engineering plastics. Reinforcement to improve the quality of thermosets is a common practice. The most widely used material for reinforcement is glass fibre; others include synthetic fibres, cotton, paper. Phenol formaldehyde resin production in the country stands at 52,000 tonnes while Bisphenol production amounts to only 1750 tons.

10.2 ENGINEERING PLASTICS

Phenol-formaldehyde, Urea-formaldehyde, Polycarbonates are among the older engineering plastics. These plastics are exceptionally strong, dimensionally stable, and resist abrasion, wear and impact; and work well over a wide temperature range.

Sheet moulding compounds (SMC) are made by sandwiching resin, reinforcement fillers etc. with the aid of thickners between two sheets of plastics. Thermosetting (plastics) resins are used in several ways, like moulding powders casting resins, impregnating resins or manufactured goods, sheets, rods and tubes (extruded shapes). Polysulphones, polyacetals because of the transparent nature, toughness, high heat distortion temperature have penetrated into medical appliances and devices. New combinations of resins like Styrene-maleic anhydride (SMA), impact modified Nylons-6; ABS/Nylons, ABS/PCs are also classified as new engineering polymers because of the high performance and rare properties they exhibit. And new developments to make them fire-proof and to replace metals are in progress round the world.

Plastics are divided into the following categories depending upon the prominence of use :
Resins For Moulding Purpose
1. Phenol-formaldehyde 2. Phenol-furfural 3. Urea-formaldehyde 4. Melamine formations.
Resins for Adhesives
1. Urea-formaldehyde 2. Melamine-formaldehyde 3. Phenol-formaldehyde.
Resins for coatings, varnishes etc.
1. Phthalic anhydride-glycerine 2. Maleic anhydride-glycerine 3. Phenol-formaldehyde
4. Urea-formaldehyde 5. Melamine-formaldehyde.

10.3 PHENOL-FORMALDEHYDE RESINS

These are the oldest type of resins, discovered by Backeland (1909) and go by the name 'Bakelites'. Since then, they have been widely used in electrical and automotive industry. Phenol-formaldehyde polymers can be produced in a wide variety of forms because of bi and tri functional groups.

The (condensation) reaction between phenol-formaldehyde results in two types of resins known as Novolacs and Resols. The condensation reaction proceeds in alkaline as well as in acid medium producing two distinct types of resins.

10.3.1 Chemistry of Reactions (Novolac)

When phenol and formaldehyde react in acidic medium with a slight excess of phenol or equimolal quantities of phenol and formaldehyde o and p-hydroxy benzyl alcohols result

OH (phenol ring) + CH_2O ⟶ OH (ring O) – CH_2OH ⟶ OH (ring O) with CH_2OH at para position

The reaction is an electrophilic substitution on ortho, para position in benzene ring. Polymerisation involves another substitution reaction with the formation of methylene bridges

OH (ring O) – CH_2OH + OH (ring O) ⟶ OH (ring O) – CH_2 – OH (ring O)

Further condensation gives a long straight chain compound by elimination of hydroxy methyl benzenes with methylol phenols.

OH (ring O) – CH_2 – OH (ring O) + HCHO ⟶ OH (ring O) – CH_2 – OH (ring O) – CH_2OH

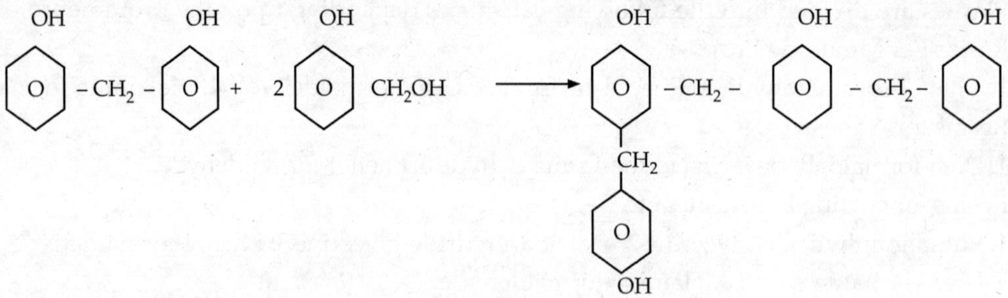

Acid catalysts with excess phenol produce linear resins.

These resins are soluble in many organic solvents and have a good demand in Paint and Varnish industry. To produce thermoset from this resin, it is heat cured by mixing with a small amount of hexamethylene tetramine to affect cross-linking. Instead of acidic medium if alkaline medium with excess of formaldehyde is used, the condensation reaction leads to di, or trimethylol phenols in stages. In the first stage phenoxide is formed.

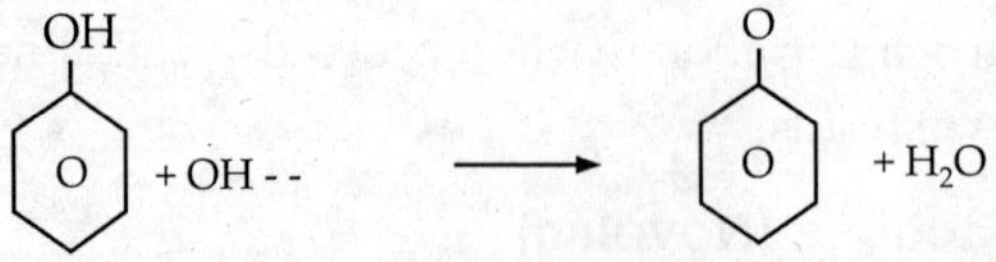

The phenoxide ion later adds to carbonyl carbon of formaldehyde, giving a substituted monomethylol phenols and dimethylol phenols.

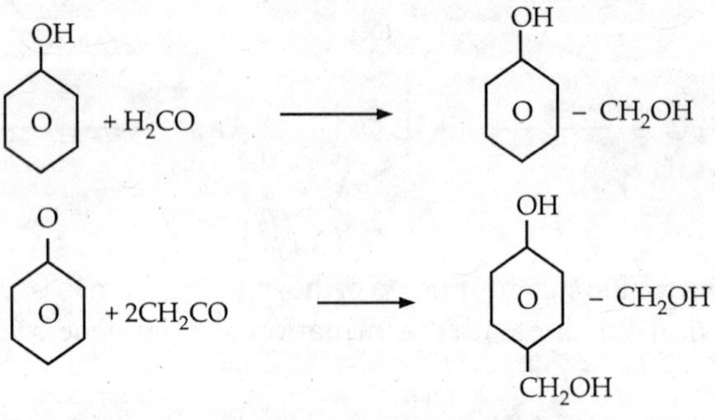

Methylol derivatives are stable in alkaline solution but in acid solution they enter readily into polycondensation reactions with one another to yield polymers

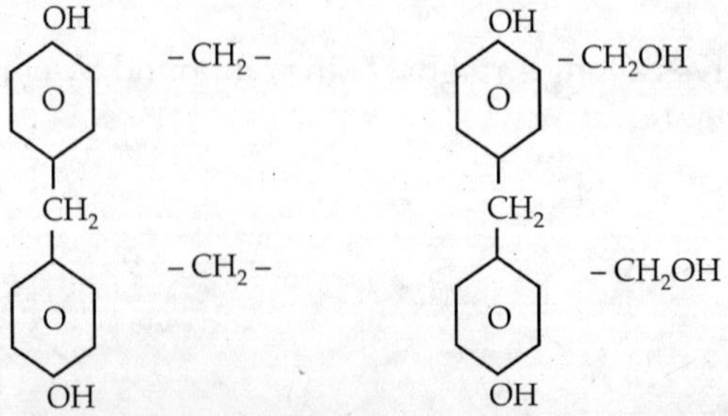

Hydroxy benzyl alcohols and mixtures of methylol phenols interact due to condensation, during which elimination of water and formation of fusible polymer appear. Two distinct types of bridges

methylene and ether $-CH_2O$ are noticed. While phenoxide, and methylol groups yield methylene bridges, ether bridges are formed due to condensation of two methylol groups.

In the manufacture of phenolic resins, it is possible to produce a resin fit for many different applications, like resins for moulding, heat resistance duties, or coatings, enamels, or adhesives, specially for plywood industry.

10.3.2 Manufacture of Phenol-formaldehyde Resins

Technically speaking resins of three stages A, B and C arc produced in industry[(2)]. 'A' stage resenoid is primary fusible end product ($H_2C - C_6H_4$ (OH)–O– C_6H_5). The resenoid can be heat hardened by using a catalyst. Resenoid of A, if produced from basic catalyst, can be very easily made into hardened 'C' stage. On the other hand, if 'A' resenoid is produced by acidic catalyst, 'C' stage resin will not be hard, hence suitable only for paints and varnishes. Commercial 'B' stage resins are those, which are sold to the moulder and contain A resin plus pigments, plasticizers, fillers, colourants, lubricants etc. and can directly go into moulds. After curing they become hardened. Fig. 10.1 shows the different routes by which these resenoids are differed for various designations. Phenol and formaldehyde alongwith the catalyst are charged into the reaction kettle equipped with heating coils and reflux condenser. Equal moles of phenol and formaldehyde with Na(OH) of 3% wt of phenol, make a charge for base catalysed resenoids. The reaction is allowed to proceed in a kettle (1) kept at 100°C; during the reaction water separates out and floats as a layer. If the base is much then neutralisation is required (usually lactic acid is used). The bottom layer from the kettle contains the resin in its plastic state. It can be mixed with alcohols or other solvents and used as paints. These resins of this type exhibit hydrophobic properties. The fusible 'A' resin with fillers and dyes is gently heated to harden for rigid plastics. The mixture after hardening is ground in a grinder to powder which is sold as moulding powder. In one step resenoid production always hydrophilic type resin is favoured. In this case, a high amount of formaldehyde to phenol is taken with Na(OH) catalyst in excess quantity. Some water remains in resin in colloidal form hence they are called hydrophilic resins.

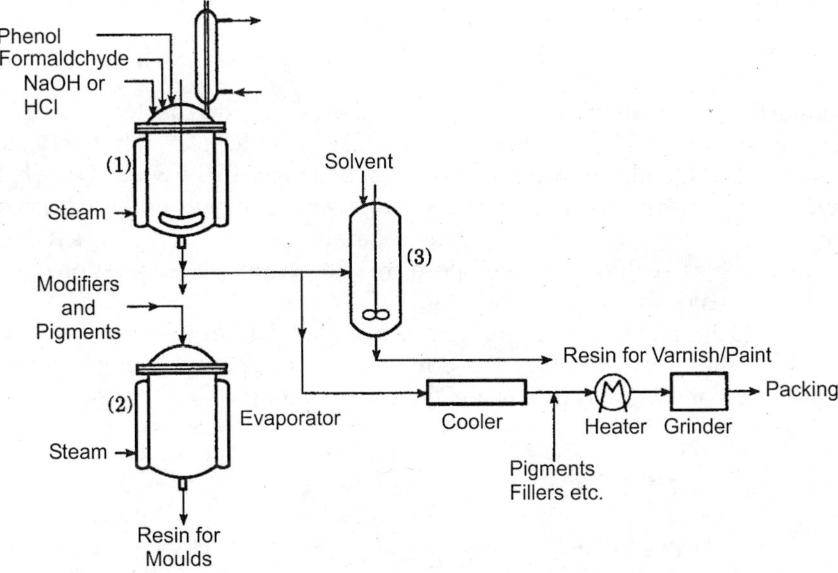

Fig. 10 1. Flow Diagram for Phenol Formaldehyde.

Neutralisation of excess alkali has to be done by using an acid. The resin produced is soluble in alcohols, and the resin can also be plasticised with other reagents. Further, mixture is fed into castings and kept at 80 – 90°C for several days so that heat curing may take place, then they become infusible. Alternatively resin from the kettle is run down into pans to evaporate (2) the water and the material is allowed to flow on to the floor to harden into thick mass. The resin is now in fusible state. Resin produced by acid catalysation also adopts the same technology except change in catalyst.

Acid catalysed reaction follows electrophilic substitution leading to novolac resins, while base catalyst gives a solid resin known as resol softening at 65 – 75°C. Resol when heated polycondensation sets-in resulting in a cross linked polymer, which now becomes a thermoset and resists heat up to 280°C. At higher temperatures degradation occurs. Novolacs can be transformed into resols by adding formaldehyde and resols can be converted into novolacs by adding phenol. Novolacs are mixed with hexamethylene diamine (initiator) before moulding powders are produced. Laminated plastics for fabricating very strong articles are obtained by impregnating layers of various fillers with resin solution. The solvent is removed and the sheets are compressed in press at 80 – 10 kg/cm^2 until they harden. Moulds are cured at a temperature of 140 – 160°C for general purpose as well as shock resistant materials. A good heat fusion and soluble polymer for varnishes and paints is produced by substituting cresol for phenol.

10.4 UREA-FORMALDEHYDE RESINS

These fall under aminoplastics. Urea and formaldehyde condensation products are as old as phenol formaldehydes and condensation reaction follows a number of steps. As in the above case the ratio of the reactants and the conditions of reaction alter the course of reaction. Nucleophilic addition of urea to formaldehyde results in monomethylol urea or dimethylol urea.

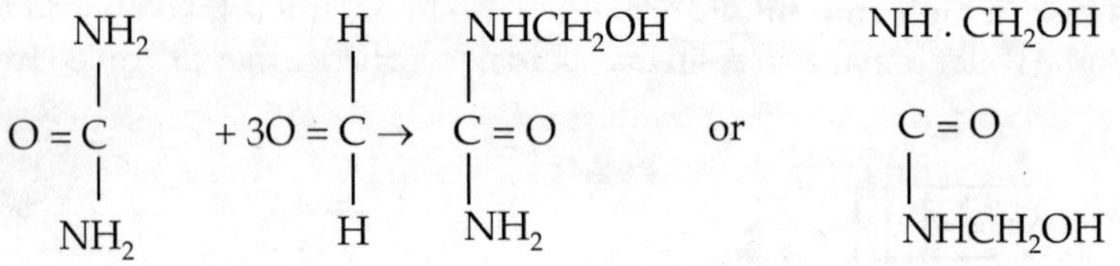

$$\begin{array}{c} NH_2 \\ | \\ O=C \\ | \\ NH_2 \end{array} + 3O = \begin{array}{c} H \\ | \\ C \\ | \\ H \end{array} \rightarrow \begin{array}{c} NHCH_2OH \\ | \\ C=O \\ | \\ NH_2 \end{array} \quad \text{or} \quad \begin{array}{c} NH.CH_2OH \\ \\ C=O \\ | \\ NHCH_2OH \end{array}$$

Condensation (Polymerisation) of methylol urea takes place when heated in the presence of acid or hexamethylene tetramine. The reaction proceeds slowly under carefully regulated conditions, temperature is kept at 40°C and P_H 7-8. Heating is continued to distill the water. Usually large amount of formaldehyde; 3 to 4 times of urea is used. After water removal the resinous mass becomes viscous. In case of phenol-formaldehyde resins and water separates out, but in this case it does not. Hence water repellent materials like thiourea or cellulose are added to aid the separation of water from the product. Highly viscous resin is allowed to set gradually after fillers are added. Acid catalysts like bromine, tartaric acid are added before the resin is kneaded. Under vacuum the acid catalysts are driven off, subsequently heat promotes the polymerisation. Thermosetting resin is mixed with ammonium chloride, which works as a hardner, before moulding.

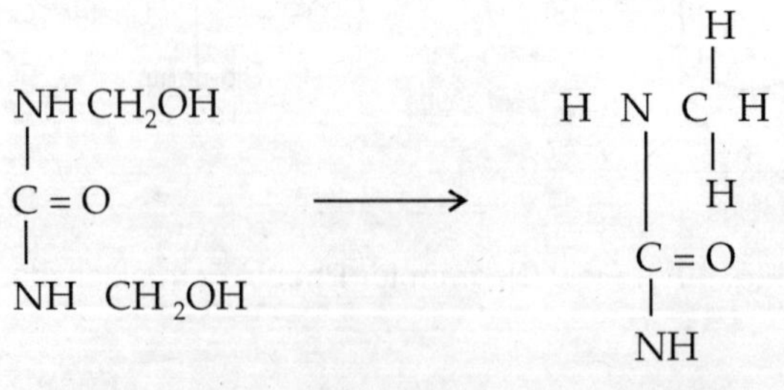

$$\begin{array}{l} NH\,CH_2OH \\ | \\ C=O \\ | \\ NH\;\;CH_2OH \end{array} \longrightarrow \begin{array}{c} \qquad\quad H \\ \qquad\quad | \\ H\;N\;C\;H \\ | \quad | \\ | \quad H \\ C=O \\ | \\ NH \end{array}$$

While it is only urea that can give a polymer with aldehyde[3], other substituted ureas cannot give, as methylene bridges with nitrogen are absent ($-N=CH_2$) *i.e.* NH_2 group acts as a primary amine and an amide. Hence the polymer takes the shape of

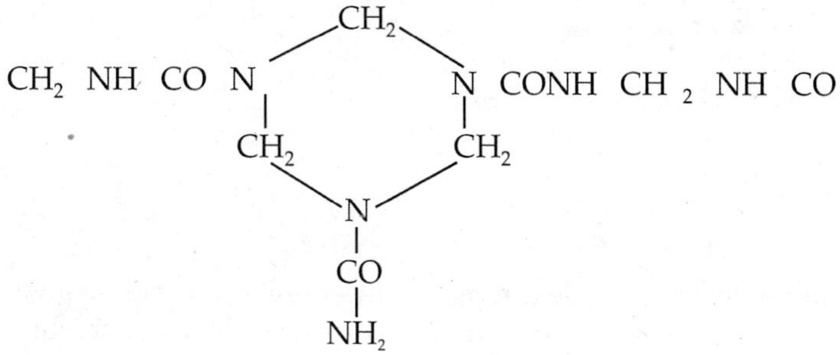

```
                          CH2
                       /      \
CH2 NH CO N                     N CONHCH2 NH CO
          |                     |
         CH2                   CH2
             \                /
                      N
                      |
                      CO
                      |
                      NH2
```

Another method of controlling the reaction is by adding hydrophilic colloid a-cellulose. It is essential to see that not much of formaldehyde is used, as it complicates the purification and leads to cracking of the moulds. Uncured urea moulding compounds are water soluble, so they are suitable for making plywoods. When they are used at 170–180°C they become hard infusible water resistant products. Though foams are also produced from UF resins they are not as popular as urethane foams. Density of foam is around 10–14 kg/m^3.

10.5 PHENOL-FURFURAL RESINS

Phenol-furfural resins do not have features of phenol-formaldehyde resins which exist in A, B and C stages. They can be kept in liquid state for a long period and brought into presses or moulds as heavy liquids. The resin in cold state is mixed with fillers and used as powder, as we see in phenol formaldehyde resin. The heat hardening temperature is approximately 180°C that is why they are resistant to heat. The fillers, lubricants accepted by phenol-formaldehyde resins are also accepted here. The reaction shows :

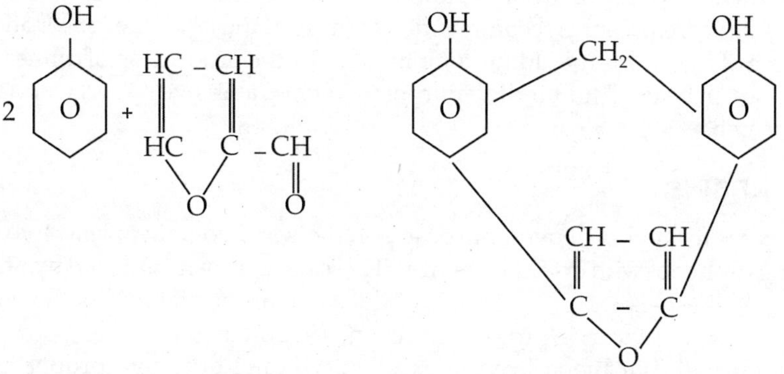

```
     OH                                 OH                  OH
     |                                  |        CH2        |
              HC - CH                  / \    /      \     / \
2  ( O )  +   ||    ||                ( O )             ( O )
              HC    C - CH             \ /                 \ /
                \  /    ||              \                  /
                 O      O                \   CH - CH     /
                                          \  ||    ||   /
                                            C  -  C
                                              \  /
                                               O
```

They are having superior moulding properties due to heat setting and good flow characteristics; like phenolic resins these are also translucent, and possess low coefficient of expansion and good chemical resistance.

10.6 UREA-FURFURAL RESINS

These resins have been known only for chemical interest and confined to laboratory only.

10.7 MELAMINE-FORMALDEHYDE RESIN

Melamine is produced from urea under high pressure in an atmosphere of ammonia.

$$6\ \mathrm{NH_2{-}C({=}O){-}NH_2} \longrightarrow \mathrm{C_3N_3(NH_2)_3} + 6NH_3 + 3CO_2$$

(Urea: NH_2, C = O, NH_2; melamine ring: C – N = C, N, C = N, with three NH_2 groups.)

Melamine reacts with formaldehyde through a nucleophilic addition to give methylolamine. A variety of methylols are possible with methylamine because of the availability of six replaceable hydrogen atoms.

$$\mathrm{H_2N{-}C_3N_3(NH_2)_2} + 3\ CH_2O \longrightarrow \mathrm{HOCH_2\,NH{-}C_3N_3(NH\,CH_2\,OH)_2}$$

(Reactant ring: C, N = C NH$_2$, H$_2$N, N, N C NH$_2$; product ring: N = C NH CH$_2$ OH, HOCH$_2$ NH C, N, N C NH CH$_2$ OH.)

Amino compounds, primarily the melamines, can be modified chemically or by use of fillers. Melamines are moulded at 150 – 175°C under a pressure of 150 – 350 bars. Thermo-setting moulding compounds are produced with a-cellulose as the filler.

Articles made from amine resins appear water-clear, hard, strong but can be broken. They are used like urea formaldehyde resins in laminations. These are good adhesives and can be safely used for moulds too. Textile treatment is recommended because of the less water absorption capacity. The single greatest use of melamine moulding compounds is in the production of dinnerware. Like other amino resins these can be used for binding purposes, such as abrasive paper, wood boards, flooring and furniture assembly.

10.8 EPOXY RESINS

Epoxy resins are termed as general purpose polymers and contain at least two epoxide groups made by reacting diphenol with epichlorohydrin. These have been introduced by Shell Chemicals in 1949; for coatings, adhesives etc. These can be produced in forms ranging from low viscosity liquids to high melting solids just by varying the repeating units in polymer molecule. Epoxy resins are cross-linked by reaction with hardners, leaving excellent wetting adhesive properties and can give composites with fillers like glass fibres, asbestos, paper fibres, cotton etc. Infact most important attribute of epoxies which they share with polyester resins is their ability to convert from liquids to thermosets without pressure and without any volatile by-products. Resins having melting point range of 64–76°C or 97–103°C are used for coatings. Commercially these resins are made by reacting Bisphenol-A and epichlorohydrin. Condensation takes place in basic medium at 60–70°C. It is a ring opening reaction; a nucleophilic substitution. The acid liberated is neutralised by base present in the medium[(4)].

$$(n+1)\ \underset{\diagdown\ \mid\ \ }{CH_2.\ CH}\ CH_2\ Cl + [OH - AR - \underset{R'}{\overset{R'}{\underset{|}{\overset{|}{C}}}} - AR - OH] \rightarrow$$
$$\qquad O$$

$$CH_2CH - CH_2\ [O - AR - \underset{R'}{\overset{R'}{\underset{|}{\overset{|}{C}}}} - AR - O\ CH_2 - CH\ (OH) -]\ O - AR - O - CH_2 - CH\ CH_2 + HCl$$
$$\diagdown O \diagup \qquad\qquad\qquad\qquad\qquad\qquad\qquad\qquad \diagdown O \diagup$$

where AR = benzene ring

R' = aliphatic radical

4% caustic acts as catalyst. Toluene is added to separate the salt from the crude product. Toluene can be later removed by distillation. Depending upon the molar ratio of chlorohydrin to phenol the softening point of the resin varies as shown below :

Ratio of reactants	*Softening point* °C
2.6	27
1.4	84
1.2	112

Most of these resins are soluble in organic solvents like acetone, toluene, MEK. These are having good mechanical properties, exhibit good electrical insulation properties even in the presence of water. High adhesive strength, good chemical and heat resistance have made them useful in composites with different fillers. Further delicate electrical circuits and instruments are encapsulated in epoxy castings. Obviously properties of these can be modified by chlorine blends, curing agents, modifiers etc[(5)]. Most of the glycidyl type epoxy resins are made of mono, bi or poly nuclear phenols. Epoxy resins are extensively used to bond concrete blocks and in other structural works, matrix materials in high performance composites, such materials are used in aggresive environments. Moisture affects badly the strength of resin. These resins can be made with aliphatic polyols or aromatic polyols.

Aliphatic polyols of ethylene, propylene and pentaerithrytol have been used in aliphatic resins. Epichlorohydrin reacts with polyols in presence of Lewis acids to form chlorohydrin ether which on hydrodechlorination in presence of sodium aluminate gives epoxy resins. Epoxy-phenolics, Epoxy modified silicones are also in service for better performance.

For Pressure-sensitive Adhesives: Michelman wax emulsions can be used for a variety of reasons in pressure-sensitive adhesives. Polyamide and paraffin wax emulsions offer adhesion control to provide required peelability and seal strength. Paraffin and microcrystalline wax emulsions can enhance water, moisture, vapor and ozone barrier properties. Carnauba and microcrystalline emulsions can be used as slip and anti-block additives to promote smooth machine operation without affecting sealability.

For Heat seal Adhesives: The low heatseal temperature of ethylene-acrylic acid (EAA) polymers, combined with excellent adhesion properties, make Michem® Prime dispersions excellent choices for a variety of adhesive applications. Common uses include laminating adhesives for paper, foil and film; heatseal adhesives for skin packaging and medical packaging; and binders for nonwoverns.

10.8.1 Epoxy Esters[6]

Epoxy resins when modified with fatty acids produce vehicles for surface coatings. Hydroxyl groups are esterified with two or three fatty acid groups by heating epoxy resin with acid at 150–200°C under an inert gas. Water produced in the reaction is removed by azeotropic distillation. Chain lengthening techniques are used to prepare light molecular weight epoxy ester directly from a low mole weight resin. Carboxylic anhydrides like pyromellitic dian-hydride (PMDA) curing agent for glycidyl ethers of bisphenol results in high heat resistant plastics.

10.8.2 Epoxy Acrylates

By using monobasic unsaturated acid and epoxy resin vinyl ester resins are formed. Carboxylic acids are used as curing agents. Primary[7] or secondary amines Lewis acids[8] (BF_3) and halogenated amines are also used as curing agents. These are cured by oxidative polymerisation catalysts, like cobalt naphthenate. Araldites available in market are such adhesives. Liquid resins from bisphenol, epichlorohydrins are polymerised by oxidation.

10.9 ABS PLASTICS

These are two phase systems consisting of acrylonitrile; butadiene-styrene copolymers and truly thermoelastic type. There are two basic types of ABS polymers[9]. One type is just a mechanical blend of styrene acrylonitrile copolymer and butadiene-acrylonitrile rubber. In the other type styrene acrylonitrile copolymer grafted on butadiene-acrylonitrile copolymer of varying amounts. Acrylonitrile gives the blend, the desired properties of chemical resistance and tensile and flexural strength styrene gives good thermoplastic characteristics. The heart of the process is grafting and follows :

Table 10.1: Production of A.B.S. (Tons)

		1985-86	1986-87	1987-88
June '87				
(Vadodra)	ABS Plastics	2,552	2,680	2,984
1980				
(Vadodra)	Polychem	1,328	1,820	1,947
1986				
(Ankaleswar)	Gujarat Binil			
	Chemicals	41	263	113
	Total	3,921	4,743	5,047

Table 10.2: A.B.S. Resins-Properties

Melt flow (G/Min)		2-45
Tensile Strength	(kg/cm^2)	350-540
Flexural Strength	(kg/cm^2)	520-800
Rockwel Hardness		94-110
Flexural Modulus	(kg/cm^2)	18,000-27,000
Impact Strength	(kg-cm/cm)	10-40
Specific Gravity		1.03-1.05

10.9.1 BEPL

BEPL is a multinational that manufactures ABS—a raw material used extensively across industries like automobiles, home appliances, telecommunications, luggage and other sectors. Bhansali Engineering Polymers Limited was incorporated in the year 1986 by B.M. Bhansali. BEPL was the first company in India to start manufacturing ABS resins using foreign technology. The company has a plant at Satnoor that has been setup in technical collaboration with Sumitomo Chemical Engineering Co. Limited/Sumitomo Nagautuck Limited, Japan at a capital cost of Rs. 36.7 crore. BEPL's first plant became operational in January 1990 and started manufacturing general purpose injection moulding, extrusion and speciality grades of ABS and its alloys. The company started manufacturing Merchant SAN with the acquisition of erstwhile Polychem Limted at Vadodara in 2000. Later on, in 2004, BEPL acquired Rajasthan Petrochemical Limited. BEPL has consolidated all operations at one location i.e. Abu Road by relocating the Bulk SAN manufacturing facility from Vadodara in order to mitigate uneconomic and unsustainable high cost operations at three different locations. The company is presently amidst diversifying into the manufacture of SBR latex and saleable SAN with an ultimate capacity of 40,000 tonne.

BEPL's Capacity at a Glance

Data in TPA

Product	*Abu Road Plant Capacity*	*Expansion Plans*
ABS	60,000	200,000

The company has also proposed to use the vacated plant at Satnoor to manufacture SBR and other allied latex.

Styrene and acrylonitrile monomers are added to a latex of polybutadiene and the mixture is agitated at 50°C. This allows the monomer to be absorbed by the polymer latex. Potassium persulphate is used as a catalyst in this system. The emulsion is coagulated after copolymerisation is over. The mix is washed and dried and compounded.

Production of copolymers can be done by usual methods like mass, suspension, emulsion and solution polymerisation. Free radical producing catalysts are common initiators. Emulsion polymerisation is used, where potassium persulphate acts as catalyst and sodium stearate acts as emulsifying agent. Modifiers are usually mercaptans. The reaction is carried out inabsence of oxygen at around 0°C. ABS resins may be fabricated by most methods used for thermoplastics, injection moulding gives housings for radios, telephones etc. Extrusion gives pipes and fittings and also sheets. Decorative pieces are added attractions; can be metallised, plated. Although these are flammable; are serviceable in a temperature range of –40°C to 107°C.

Fig. 10.2 shows the production flow sheet of ABS plastics. It contains two sections. First section gives butadiene-acrylonitrile polymer, while the second gives acrylonitrile styrene polymer. These two polymers are then blended to yield ABS polymer.

Reactor 1 contains the water-emulsion, modifiers, short stop and monomers. The polymerisation is carried out and the polymer is stripped off (2) to separate monomers for recirculation. From mixer (3) the acrylonitrile and styrene monomers in emulsion-water is carried to polymerisation reactor.

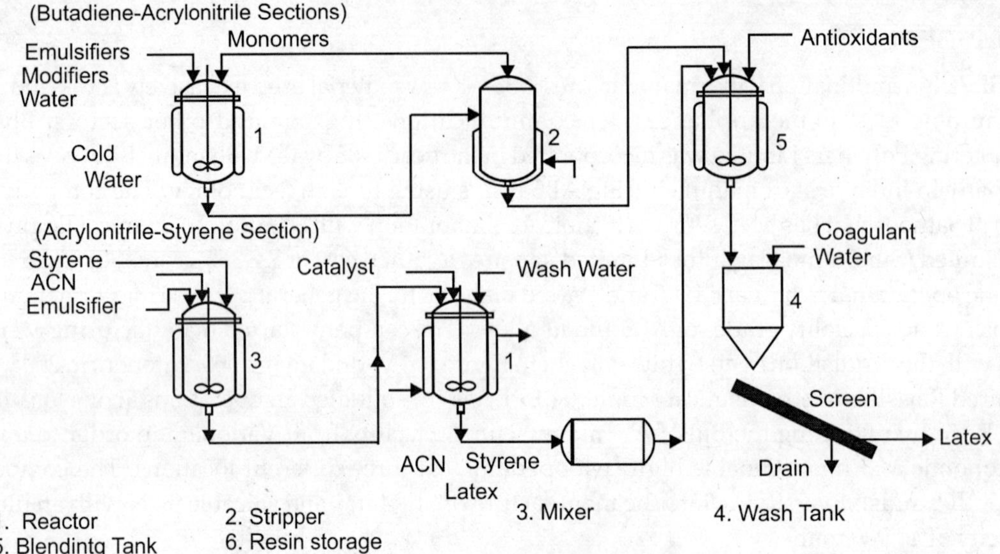

Fig. 10.2: Styrene Butadiene-Acrylonitrile Plastics.

The polymer material is separated in the form of latex. This latex and latex from the other section are sent proportionally into another graft-polymeriser (5) reactor where antioxidants and modifiers are added. The crumb is washed with water in a tank (4) to remove catalyst and chemicals and sent to drier and mill section.

10.10 POLY CARBONATES (PC)

These are relatively transparent and expensive thermoplastics which were developed by Bayer in 1957. These are esters of carbonic acid HOCOOH, and are produced by condensation polymerisation of sodium salt of bisphenol A and phosgene. They have excellent impact resistance with high heat distortion temperature. The credit of transparency and chemical resistance, are widely used in specialised applications. Mobays Merlon[10] Rx3-113 resin developed for medical appliances and devices has been successfully used in open heart surgery etc. Further, these are useful for housing various instruments and articles.

The reaction is similar to epoxy resin condensation. An aqueous solution of disodium salt of bispenol-A is brought into contact with phosgene in a solvent like methyl chloride. A phase transfer catalyst such as tetrabutyl ammonium chloride has to be added to initiate the reaction between the two immiscible phases. Sodium chloride is precipitated and the solvent assists this. Later the solvent is separated by distillation. The reaction is shown below.

$$\mathrm{NaO-Ar\overset{\overset{CH_3}{|}}{\underset{\underset{CH_3}{|}}{C}}-Ar-ONa+Cl-\overset{\overset{O}{\|}}{\underset{\underset{CH_3}{|}}{C}}-Cl\rightarrow[-O-\overset{\overset{CH_3}{|}}{Ar}-C-Ar-\overset{\overset{O}{\|}}{C}-]+NaCl}$$

Fig. 10.3: Shows the outlines of the process[11].

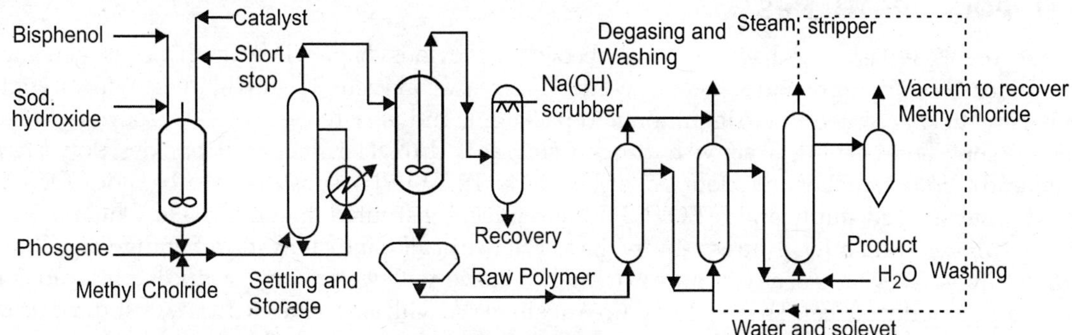

Fig. 10.4: Production of Polycarbonate.

Structurally these are polyesters derived from the reaction of carbonic acid. These are unaffected by water and many inorganic and organic solvents. Aromatic dihydroxy compounds when react with phosgene in an inert solvent give rise to interfacial polymerisation resulting in resins. Transesterification is also common. Catalysts like alkaline metal oxides, or alkalies are used during condensation reaction that takes place around 260 – 300°C. The reaction product shows a molecular weight of 100,000. These are partially crystalline and transparent with high impact strength. Because of transparency, window screens, pans, lenses etc. are made with these and can suitably be moulded at a high temperature 500–600°C. They are self-extinguishing type also. Depending upon the aromatic or aliphatic group the m.p. and glass transition temperatures vary as shown below.

$$OH-Ar-\underset{R}{\overset{R}{\overset{|}{\underset{|}{C}}}}-Ar-OH$$

	M P°C	T_G°C	
R = H	223 - 225	147	
= CH_3	220 - 230	149	
= C_3H_7	200	148	
= Ar	210	120	
OH – Ar – R – Ar – OH			
R = CH_2		240	167
R = C_6H_4		250	170

Polycarbonate-Properties

Density	1220 kg/m3
Young's modulus (*E*)	2000-2200 MPa
Tensile strength (s_1)	60-65 MPa
Elongation @ break notch test	80-150%
Glass temperature	150° C
melting point	—
heat transfer coefficient (l.)	0.21 W/m.K
linear expansion coefficient (a)	6.5 10^{-5}/K
Specific heat (c)	1.3 kJ/kg.K
Water absorption (ASTM)	0.16

Deformation temperature at 10kN needle load = 130°C

10.11 POLY URETHANES

Poly urethane elastomers have been described in the previous chapter. Poly urethanes are generally used for load bearing, abrasion resistance works. Their low conductivity and lightness with strength has been widely accepted in refrigeration and packaging industry respectively. Diisocynates react with any compound having an active hydrogen, such as hydroxyl terminated[12] epoxies, polyesters, furans and cellulose, rubber and groups like SH, NH, NHR, CO NH_2, substituted carbamide CONHR, SO_2NH_2, substituted sulphoamide SO_2 NHR, thionides, substituted thioamides etc. Only not very active hydrogen containing groups require the assistance of alkaline catalysts rather than metallic or acidic catalysts. Isocynates of all types are extremely reactive, with water* they produce carbon dioxide, with a large amount of heat. The produced urea again reacts with isocynates to form substituted ureas to form prepolymers. High molecular weight polyethenes, polypropylene glycols, polyesters are used in the soft foams. Rigid foams do not require high purification of reactants. Polymethylene poly phenyl isocynate is the most widely used isocynate in rigid foams.

Exothermic reaction heat of isocynates with alcohols usually is made to volatalise a volatile liquid like Freon[13] ($CFCl_3$) which gives rise to the foamy structure to the plastic. Isocynate when reacts with water, carbon dioxide accompanies the heat of reaction, this also provides rigid foams. The density of these[14] foams is supposed to be very less (32 kg/m^3). The construction industry is the largest consumer of rigid polyurethanes. Rigid polyurethanes, *i.e.* modified polycynurate (PIR) foams comprise of polymeric isocynates trimerised in presence of a surfactant catalyst and blowing agent; while polyols are added to help processing, *i.e.* to reduce internal friability and improve adhesion substrates.

Urethane paper based insulations with aluminium foils, produced by continuous laminations are used in buildings for insulation purpose. Glass fibre reinforced laminates yield fire-proof insulation material. The compressive strength of these goes upto 350 KPa with TS touching to 200 KPa. Majority of the laminates contain either TDI or MDI (Diphenyl methane diisocynate) and NDI (Naphthalene 1, 5 diisocynate)[15].

10.11.1 Cast Poly Urethanes

Castings are done by two short methods, *i.e.* first producing prepolymer and then polymerising prepolymer. Excess of diisocynate at 130°C with polyol gives liquid prepolymer. This liquid prepolymer, is heated to remove the gases. This polymer mixed with chain extenders, and curing agents is then cast into moulds. Curing takes place at 200 to 250°C for 15 to 20 hours.

Most types of polyurethane resin cross-link and become thermosetting plastics. However, some polyurethane resins have a linear molecular arrangement that does not cross-link, resulting in thermoplastics.

Thermosetting polyurethane molecules cross-link into a single giant molecule. Thermosetting polyurethane is widely used in various forms, including soft and hard foams. Soft, open-celled polyurethane foms are used to make seat cushions, mattresses and packaging. Hard polyurethane foams are used as insulation in refrigerators, freezers and homes.

Thermoplastic polyurethane molecules have linear, highly crystalline molecular structures that form an abrasion-resistant material. Thermoplastic polyurethanes are moulded into shoe soles, car fenders, door panels and other products.

---

*Bhopal Disaster which resulted in the death of more than 2500 people and crippled many thousands is due to this reaction methylisocynate with water and all isocynates are high toxic and special precautions must be taken while handling.

10.12 POLYIMIDES

Polyimides are heat resisting plastics and stable up to 500°C. Polyimides are produced by two stage condensation process. The initial condensation of pyromellitic anhydride with diaminodiphenyl ether in anhydrous dimethyl acetamide yields a soluble acid. This latter is heat cured to an insoluble infusible poly aromatic imide. These can be used for filled polymers and other industrial applications like piston rings, valve seats, bushings and glass sheets.

10.13 POLY ACETALS

Poly vinyl alcohol films and fibres are made water resistant by immersion in aqueous formaldehyde. The acetal reaction takes place between hydroxyl groups and aldehyde group resulting vinyl formal form. This form is soluble in many organic solvents.

These are suitable for coatings and laminating purposes. Some polyacetals derived from butyraldehyde can be plasticized and can be utilised in coating textiles.

10.13.1 Polyacetal

Also called polyoxymethylene (POM) or simply acetal, polyacetal has the simplest structure of all the polyethers. It is manufactured in a solution process by anionic or cationic chain-growth polymerization of formaldehyde ($H_2C = O$), a reaction analogous to vinyl polymerization. By itself, the polymer is unstable and reverts to monomer on heating to 120°C (250°F), for this reason the commercial product is reacted further with acetic anhydride to cap the ends of the chains (where depolymerization is initiated on heating) with acetate groups. The end-capped polymer is marketed by DuPont under the trademarked name of Delrin. It is a high-strength, highly crystalline engineering plastic that exhibits a low coefficient of friction and excellent resistance to oils, greases and solvents. Also marketed is a copolymer (trademarked as Celcon by Hoechst Celanese Corp.) prepared from trioxane (a trimer of formaldehyde) and small amounts of ethylene oxide to prevent the polymer from decomposing to formaldehyde on heating. Both polyacetal and the copolymer have been used as a replacement for metal in plumbing and automotive parts. Principal uses include appliance parts, electronics components, gears, bushings, bearings, plumbing fixtures, appliances, toys, toiletry and cosmetic articles, food-processing equipment, zippers and belt buckles.

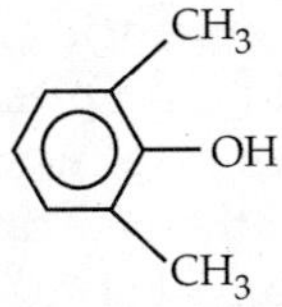

CH_3

OH

CH_3

10.13.2 Polyetherketone (PEK) and polyether ether ketone (PEEK)

PEK and PEEK are high-strength, radiation-resistant engineering plastics whose structures combine both ether and ketone groups. Both are thermally stable and highly resistant to chemicals. Principal uses are in machine parts, nuclear power-plant equipment, automobile parts, aerospace components, cable insulation, and pump parts.

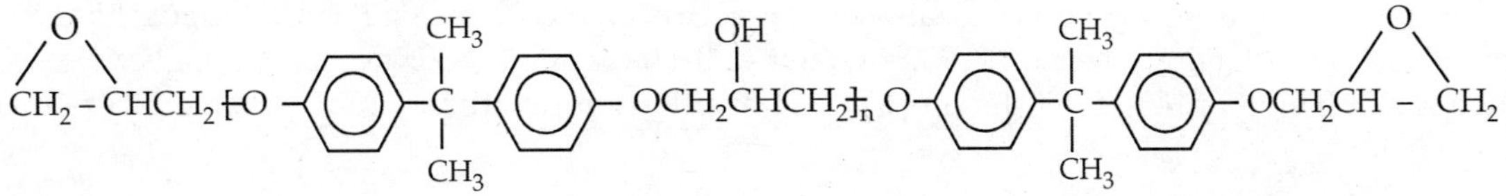

$$\overset{O}{CH_2 - CHCH_2}\!\!-\!\![O-C_6H_4-C(CH_3)_2-C_6H_4-OCH_2\overset{OH}{C}HCH_2]_n\,O-C_6H_4-C(CH_3)_2-C_6H_4-OCH_2\overset{O}{CH - CH_2}$$

In the above structure, n varies from about 2 to 25 repeating units, such low-molecular-weight prepolymers as these are called oligomers. Depending on their average chain length, the prepolymers vary from dense liquids to solids. In a typical epoxy reaction, the prepolymers are further polymerized through the opening of the terminal epoxide rings by amines or anhydrides. This process, called curing yields complex, thermosetting network polymers in which the repeating units are linked by linear ether groups. The highly polar network polymers characteristically exhibit excellent adhesive properties. In addition, because the curing reaction is easy to initiate and proceeds quite readily at room temperature, epoxy resins make very useful surface coatings. Most commonly a two-component system is used, in which one component is a low-molecular-weight polymer with amine end-groups and the other component is an epoxide-terminated polymer. The two components are mixed before application to the surface, where the polymer is allowed to cure. Epoxy resins are also made into structural parts such as laminated circuit boards, laminates and composites for aerospace applications and flooring. For these applications epoxies show high strength when reinforced with fibres of glass, aramid, or carbon.

10.14 POLYSTYRENE

Polystyrene is one of the most important resins and plastics. A wide range of formulations like crystal, impact and moulding resin are made out of this material. It is linear polymer having a density of 1050 kg/cu.m. It can be produced in the form of foam which is highly used in packing industry and a variety of disposable items are made out of this foam. Audio, video tapes and cabinets for different electronic goods, refrigerators etc. are also made out of the resin. It can also go into copolymers for the production of elastomers and resins. Formerly Poly Chem Ltd, Mumbai and Hindusthan Polymers (McDowell group), now LG Polymers, at Visakhapatnam are the main manufacturers for Polystyrene. The production of these plants is about 3600 tpa. Other manufacturers include Supreme Petrochemicals, Reliance which would add another 3600 tpa.

Expandable polystyrene would be available to 20,000 tpa. Being brittle and hard, a high performance product can be made by giving it some rubber properties by mixing with poly butadiene. Raw materials for polystyrene are benzene and ethylene. First sep involves in the production of styrene. This produced by dehydrogenation of ethylbenzene. Ethyl benzene is produced by the alkylation of benzene with ethylene using aluminium chloride or phosphoric acid or friendly catalysts. Benzene to ethylene ratio is 2.5 mol/mol and the reaction is carried out in glass lined reactor in presence of a catalyst at moderate temperature 150° and a pressure of 7-8 atm. After the reaction is completed, the reactants cooled and the organic mass which is lighter than the catalyst phase is separated by gravity settling. The catalyst phase is routed back to the reactor. Alkylate is distilled to separate benzene -Ethyl benzene is sent for dehydrogenation after washing. EB dehydrogenation is done wither by a. adiabatic b. isothermal, or c. oxidation process. First two methods work on the difference of heat in heat application. E dehydrogenation is carried out in presence of iron oxide catalyst promoted by potassium and chromium oxides. In isothermal method the mixture is heated in tube heater and then sent into a reactor kept at 800° C. As the reaction is endothermic heat has to be supplied. The mixture is then cooled to knock of hydrogen and then distilled under vacuum to separate styrene. Styrene cannot be stored for a long time due to polymerization in presence of oxygen, hence suitable inhibitors have to be added. Styrene is having a density of 930 kg/cu.m and a flash point of 30.6° C, having higher viscosity than water. It boils at 145° C, it is having a glass transition temperature of 95° C and melting point of 240° C.

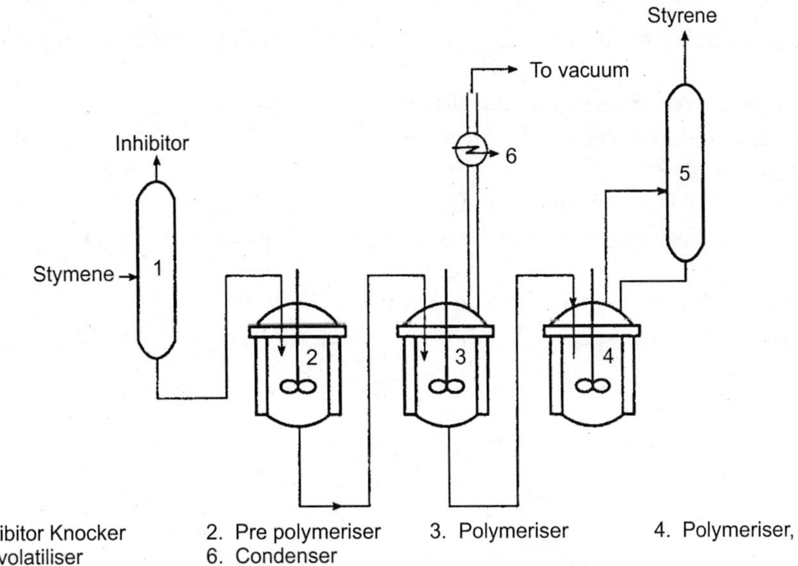

1. Inhibitor Knocker 2. Pre polymeriser 3. Polymeriser 4. Polymeriser, 5. Devolatiliser 6. Condenser

Fig.10.5: Polystyrene Production.

Styrene is polymerised in presence of free radical or by thermal method to produce polystyrene. Usual methods of production are (*a*) Suspension polymerisation (*b*) Solution polymerisation and (*c*) Mass polymerisation.

Suspension polymerisation is mostly used for general PS resins. As shown in Fig. 10.4 it begins with stripping off the inhibitor from styrene. Styrene is now directed into a reactor where water suspension is ready. Suspending agent can suspend the monomer in fine droplets. This reactor is fitted with condenser-reflux operating at reduced pressure. Prepolymer reactor is operated at 80°C. Usually peroxide catalysts initiate the reaction. 30% prepolymer is formed in this reactor. The material is transferred to next reactor where the final polymerisation is attended. The non polymer material and low polymers are separated by distillation. In thermal polymerisation usually a temperature of 130°C is maintained.

Polystyrene foams are produced by using blowing agents during polymerisation. Foams of density as low as 0.0015 gm/cc can be produced.

REFERENCE

1. Charles, E. Dryden; Outlines of Chemical Technology, Affiliated East-West Press, New Delhi, 1967, 467.
2. Raymond Nauth, The Chemistry And Technology of Plastics Reinhold Publishers, New York, 1947, 25.
3. Beat Meyer, Urea Formaldehyde Resins, Addison Wesley Pub. Co. 1979, 117.
4. H.P. Pyne, Co-organic Coatings Technology, Chapman & Hall Ltd., New York, 1954, 317.
5. Henry Lee, Kris Neville, Hand Book of Epoxy Resins P-2-5, McGraw Hill Book Company, N.Y.
6. Calvin, E. Schildknecht with Irving Skejst, Polymerisation Processes, A Wiley International Science Pub. N.Y. 609.
7. Jovan Mijovic, IEC Prod. Res. Rev., 1982, 21, 290-298.

8. Devendra Kumar & Gupta Alka, J. Macro Molecular Science-Chemistry, Vol. 22. No. 8,1985, 1103.
9. CH. Basdekis, ABS Plastics, Reinhold Pub. N.Y., 1964, 74.
10. International Modern Plastics, August 85, 45.
11. HCP Nov. 79, 216.
12. E.N. Doyle, The Development and Use of Polyethane Products, 240.
13. D.C. All port, Plastics and Rubber Processing and Applications, Elsveir Applied Science, Vol. 4, No. 2,1984.
14. Maurice Morton, Rubber Technology, Van Nostrand, 1973, 443.
15. William B. Seymour, Modern Plastics Technology, 227, Reston Pub. Co., Virginia, 1975.

❑

CHAPTER 11

Synthetic Detergents

11.1 A HISTORICAL SURVEY OF DETERGENTS

A detergent is defined as a cleaning agent. As such some are available in nature. Plant derived natural detergents, like soapnuts, shikakai etc. are widely used even today in our country. For washing clothes some types of soils possessing detergency action have been also used. Gradually soaps, the manufactured detergents, organic and inorganic mixtures for domestic and industrial applications appeared to rule for a long time till 1930. However, the soaps suffered from two main disadvantages: (*a*) the cost of manufacture of soaps (*b*) soaps work as detergents under certain conditions only. Indirectly shortage of edible oils has forced the soap manufacturers to go for animal oils, infact shortage of fats and oils has made the synthetic detergent industry to grow. Emergence of synthetic detergents, to satisfy the large scale and varied needs of industry was welcomed. To be precise synthetic detergents have become widely accepted since the Second World War period. The main industries that use synthetic detergents are textiles, leather, including domestic sources like laundry and washing utensils.

Synthetic detergent market is having excellent growth rate, approximately 2.2 million tons of non-toilet soaps and 3.50 million tons of detergents are being consumed yearly, with a growth rate of 5.2 %.

Turkey red oils were known as early synthetic detergents. The sulphation that gave Turkey red oils detergency action was equally available to other oils. This gave rise to sulphated fatty alcohols, *i.e.* a real break through for the production of modern detergents. Only difficulty was getting fatty alcohols, which were derived from fatty acids by catalytic hydrogenation under pressure. Pre-war detergents were mainly Igepons, Ultravons, Lamepons etc. Derivatives of fatty acids and amino acids, *i.e.* Iglevers were prepared by reacting a fatty acid chloride with an amino acid such as sodium sarcosinate.

$$C_{11}H_{23}COCl + NH(CH_3)CH_2COONa \text{ ® } C_{11}H_{23}CON(CH_3)COONa + HCl$$

This type of sarcosinate shows the properties of soaps, with improved behaviour in hard water. Short, chain alkyl-naphthalene sulphonates are regarded as first syndets. The advantages of synthetic detergents over soaps are grouped below :

1. High detergency in soft and hard water alike, since they do not react with calcium and magnesium salts present in water.
2. Small concentration of detergent is sufficient to give rise large detergency action.
3. Detergency action is reached even at low temperature.
4. Synthetic detergents do not hydrolyse hence almost neutral.

11.2 CLASSIFICATION OF DETERGENTS

Synthetic detergents are classified into three groups :

1. Anionic, 2. Cationic and 3. Non-ionic and amphoteric.

Detergency action is related to the molecular structure. When these detergents dissolve in water, the surface properties of the system are altered dislodging the dust particles that is clearly differing from the mechanical dislodging. That is the reason why these detergents are called surfactants.

11.2.1 Structure of Detergents

Detergents are made-up of hydrophilic and hydrophobic parts. Detergent when placed in water arranges in two constituents namely hydrophilic and hydrophobic parts. During washing of a fabric hydrophilic heads firmly remain in water carrying the dirt particles while the hydrophobic tails repelled by water project outwards. Thus these tails are attracted by the tails of another layer of hydrophobic tails and form chains. The rust or dirt particle which is surrounded by heads of hydrophilic molecules are easily removed by agitation.

Detergent molecules have ionisable inorganic groups such as SO_3^-, COO^-, CO^-, NH^+, NH_2^+, $C1^-$, which are soluble in water and the long chain hydrocarbons act as hydrophobic constituents. These will pierce the surface of water reducing the surface tension. The underlined portion of the detergents are water soluble, remaining portion is hydrophobic.

$$CH_3\,(CH_2)_{11}\,C_6\,H_4\,SO_3^-\,Na^+$$

$$CH_3\,(CH_2)_{11}\,C_6\,H_4\,CH_2\,N^+\,(CH_3)_3\,C1^-$$

All detergents should wet the fibres/materials so that dust can be dispersed. Hence they should possess good emulsifying, frothing and detergent properties.

The assessment of a detergent is very difficult, and only empirical methods are practised.

An important property in washing a fabric is removal of dust particle, and suspending it in the medium, *i.e.* not to allow it to redeposit on fabric concerned. In finishing textiles, properties like dying, the uniformity of wetting, acceptance and retention of dye are important. Similarly in painting, leather finishing etc. different properties or phenomena are given priority over the detergency action. Thus the physico-chemical properties of surfactants are selectively chosen for the action to be followed.

The ability of surfactant is judged by the formation and stabilisation of emulsion and froth. Emulsion is a suspension of fine droplets of a liquid in another liquid, thus, it may be possible to get oil-in-water or water-in-oil emulsion depending upon the nature of detergent. Fatty acids, alcohols etc. favour water-in-oil emulsion.

Dispersion, peptisation and suspension are based upon the surface phenomenon. Sometimes finely divided particles may form due to precipitation of a solid phase from a liquid medium. The surface activity(2) of detergent is due to the preferential adsorption at interfaces lowering the surface energy of the system. Water insoluble portions of detergent tend to be expelled soon the detergent is placed in water. This creates rapid wetting and foaming. Soil suspending power is the ability of a detergent to prevent redeposition of dust particles on fabrics. Soil suspending power increases with increase of chain length of molecule. Ethylene oxide based detergents seem to be the best in this respect. Active additives like carboxy methyl cellulose (CMC), some silicates are found to improve the soil suspending power of detergents.

Froth is essentially generated during detergency action, this froth is air cells separated by thin liquid films. Froth is formed when surface tension is lowered, hence all detergents must exhibit frothing capacity. Properties of detergents are governed by the balance between their hydrophilic and hydrophobic groups.

It may be mentioned that the detergency action starts by altering surface conditions, thus it differs from other washing methods conventionally followed. Being a sum of complex phenomena, qualitative assessment of a detergent becomes difficult. It is true that one detergent is suitable for one type of action, while it may fail in other cases. On the whole the detergency action depends upon the chemical nature of detergent and on the presence of active additives. Further, dislodging of particles is related to the nature, physico-chemical characteristics and geometry of dust particles, intensity of stirring, medium and temperature. Thus a useful definition arises out of all these things; the efficiency of a detergent is estimated by the quantity of dust present after and before washing. The amount of dust, may be in terms of whiteness, can be measured by photo techniques.

A formulation of a synthetic detergent is shown below:

Active deteregent	20%
Sodium tripolyphosphate: STPP	22%
Sodium silicate	10%
Soda ash	10%
SCMC	1 to 3%
Perborates	10%
Sodium sulphate	25%

11.3 ANIONIC DETERGENTS

Anionic detergents are the more versatile and cheap detergents ever since 1930. They are obtained in a number of ways. During fifties propylene tetramers played a vital role in detergent industry, even now the role continues. Carboxylic acid soaps, alkyl sulphates, alkyl sulphonates, alkyl aryl sulphonates are the principal detergents of this category. Leaving soaps, others may be treated as synthetic detergents. Sulphonates and sulphates of alkanes more or less resemble the early detergents. The present day detergent is principally composed of alkyl-aryl sulphonates. The most commonly used detergent of this type is sodium alkylbenzene sulphonate; R - $C_6H_4SO_3Na$, where R contains carbon atoms from 10 to 15. The sulphonates of simple aromatic hydrocarbons, benzene, toluene and naphthalene do not possess any detergent action, however, if a hydrogen atom is replaced by aliphatic chains, then the detergency properties are markedly improved.

Thus the properties of alky-aryl sulphates depend upon the kind of aromatic compound, the length of alkyl chain, the degree of branching, the position at which alkyl chain attached to benzene ring. Benzene can be easily alkylated either with an olefin or n-paraffin chloride.

It has been found that branched chain detergents dissolve better than the long chains. Compounds of 10 to 16 carbon atoms only are found to have detergent properties. The detergency action of the branched chain additions are found to be lower than that of straight chains and maximum detergency is noticed when third carbon atom is attached to benzene ring. The commercially important detergents of alkyl-aryl sulphonates are sodiumdodecylbenzene sulphonate, and sodium alkyl naphthalene sulphonate.

11.4 DODECYL BENZENE SULPHONATE

Sodium isododecylbenzene sulphonate consists of two distinct groups, alkylchains of dodecane ($C_{12}H_{26}$) and aromatic ring. Being used in laundry, these detergents are produced in large tonnage. The attachment of this straight chain compound to benzene, and then sulphonation of ring structure followed by neutralisation gives rise to this structure as shown below :

$$C_6H_6 + C_{12}H_{24} \longrightarrow C_{12}H_{25}C_6H_5$$

$$C_{12}H_{25}C_6H_5 + SO_3 + H_2O \longrightarrow C_{12}H_{25}C_6H_4SO_3H$$

$$C_{12}H_{25}C_6H_4SO_3H + Na + \longrightarrow C_{12}H_{25}C_6H_4SO_3Na$$

Another alkyl-aryl sulphonate is diisopropyl naphthalene sulphonate.

$$C_3 - C_{10}H_5(SO_3Na) - C_3$$

11.4.1 Dodecane

Dodecane can be obtained from propylene by tetramerisation :

$$4C_3H_6 \longrightarrow C_{12}H_{24}$$

Propylene tetramer is obtained by tetramerising propylene in presence of a Friedel-Crafts (Acid) Catalyst. Usually a mixture of propane and propene is tetramerised at a temperature of 180–200°C under a pressure of 5 MPa. Almost all propene is tetramerised giving an yield of 30% desired product. Being a heteropolymer the mixer has to be fractionated to separate the suitable cut, *i.e.* 180–250°C. Also the tetramerisation results in branching which on alkylation with benzene gives rise to a detergent of non-degradable nature. A wave of ecological consciousness and environmental pollution has affected major changes in detergent formulations. The biodegradable detergents like linear alkylbenzene-sulphonates (LABS) instead of tetra propylene benzene sulphonates (TPBS) have been widely inducted into industry. Biodegradability though partly assigned to the position of carbon atom of alkyl chain attached to benzene ring, truly benzene itself is nonbiodegradable. However, the addition of chain to benzene ring at any particular carbon atom cannot be controlled.

11.5 OXO, F - T, ALCOHOLS-SULPHATES

All these are branched chain primary alkyl sulphates, probably with a methyl branch at 2-carbon position. The alcohols are derived from a-olefins by catalytic substitution in which formaldehyde enters the position. On reduction with hydrogen, aldehyde becomes alcohol having terminal group CH_3CH_2 CH_2OH. These can be easily reacted with sulphuric acid. Usually impurities such as sulphones and sulphonates are formed. FT product of kerosene range can be chlorosulphonated in presence of ultra-violet light as catalyst. Chlorosulphonyl groups can enter anywhere along the chain. Sodium secondary alkyl sulphates (Teepols) are usually branched secondary alkyl sulphates which are derived from olefins by substituting in the double bond with sulphuric acid, and the double bond position varies too. This results in a product mixes of different chain lengths and sulphate isomers which thrusts high solubility to these products in water. Non-ionic sulphates are also obtained from alcohols. Alcohols can be ethoxylated and then sulphated. Ethanolamine alkylsulphates are also in service, these are derived from mono or triethanolamines. Amine groups impart good solubility-character hence they are used in shampoo formulations. Alkylphenol ethoxylate sulphates belong to another category of the detergents. In this aromatic molecule is sulphonated. Other sulphates include ethyleneglycol monoester, sodium or ammonium monoglyceride sulphates. Alkyl sulphates of branched chain type are formed mainly during chlorosulphonation of olefins. Hydrolysis will yield sulphonyl chlorides. Alkene sulphonates are similar to alkyl sulphonates; these are produced by the action of sulphur trioxide on olefins. Alongwith alkane sulphonates, impurities like alkene sulphonates, dihydroxy sulphonates are formed. If the fatty alcohol is a long chain one, the solubility is less, to increase the solubility, the alcohol is condensed with either ethylene oxide or propylene oxide. Other detergents sold in the form of Aerosols[3] are esters of sulphosuccinic acid; sulpho carboxylates have terrible wetting properties. These are derivatives of fatty acids sold as lgepon (RCONH $(CH_2)_4$ SO_3 Na).

Other types of alkyl aryl sulphonates are based on naphthalene or toluene. A typical example being benzimidazole sulphonates of the shown formula:

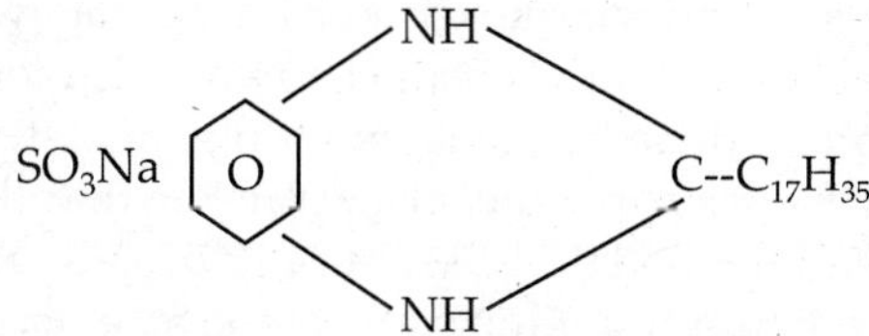

These types were used only for a short time. These were prepared by reacting o-Phynelene diamine with stearic acid chloride and subsequent sulphonation with sulphuric acid. Stearic acid derivatives are obtained by first reacting fatty acid chloride with benzene and then sulphonating the resulting ketone in the ring.

SO_3Na

O

$C_{17}H_{35}$ COCl

These resemble alkylbenzene sulphonates.

R — R

O — O $(CH_2CH_2)_{nl}$O CH O (CH_2CH_2O) n_2 — O

SO_3Na — SO_3Na

Formyl Di-alkylphenol ethoxylate sulphonate, has the formula shown above.

It may be obtained from vegetable or animal origin. Sulphation of polyglycol ethers of alkyl phenols which when reacted with ethylene oxide or formic acid give the above structure. These are having high foaming characteristics with effective detergency action, hence preferred in liquid formulations.

11.6 n-PARAFFLNS FOR PETROLEUM DETERGENTS

n-Paraffins suitable for detergent industry are in the range of C_{10} to C_{12} present in kerosene fraction. High temperature cracking also gives distillates of 150–300°C. This cut is suitable for detergent manufacture as it easily reacts with sulphuric acid giving random sulpho groups.

11.7 GENERAL MANUFACTURE OF SULPHONATES

Alkylation of benzene has been early discussed. Aluminium chloride acts as a catalyst for this reaction, either an olefin or chloride of a paraffin can be used. Fig. 11.1 shows the process flow sheet. After the alkylation is over the product is distilled to separate benzene, and is allowed to recycle. The heavy product is again distilled to separate light alkylate from heavy alkylate. A close fractionation gives rise to alkylate of suitable molecular weight for detergent manufacture. Heavy alkylate can be used for detergents of inferior quality manufacture. Oleum is used as a sulphonation agent and the reaction takes place in special type of agitated kettles. The mass is then pumped into a settler where acid is separated and sent for concentration. Excess acid is neutralised with caustic soda. Detergent slurry is now mixed with builders such as sodium triphosphate, silicates, sodium chloride etc. Carboxy methyl cellulose (CMC) is added at this stage. The liquid slurry is then concentrated and spray dried. Sulphur trioxide can also be used instead of oleum. There are three such processes namely Ballestra, Chemithon and Sulphan. Ballestra Process operates at atmospheric pressure, whereas Chemithon Process operates under a slight pressure. Sulphan process uses commercial grade sulphur trioxide stabilised with boron compounds. Alkyl benzene sulphonates are mainly used in heavy duty household detergents specially solid formulations. Long tensile chains are also soft towards skin. Sulphonation of kerosene range paraffins can be done by using either oleum (1-1.5 mol. of free SO_3 per mol of alkylate) or with sulphuric acid (4-10 mol per mol of alkylate) on the other hand to avoid the disposal of large acid sludge, a very useful method based on sulphur trioxide and air has been accepted. The process is shown in Fig. 11.2. A mixture of sulphur trioxide and air enters into four or five sulphonation reactors from the top into the reaction mixture which is kept in intense agitation. A proportioned mixture of sulphuric acid and alkane enter the first reactor and gradually leave into next reactor. The reactors are cooled by cold water circulation, otherwise local hot spots may be developed. Any liberated gas like sulphur dioxide and unconsumed air is collected, scrubbed with caustic before it is vented into the atmosphere. The sulphonated mixer leaving from the fifth reactor goes into a neutraliser, where washing with water and neutralisation are carried out. Builders, additives are added here. Sodium hypochlorite is also added to bleach and improve the colour of detergent. The slurry is then concentrated and dried.

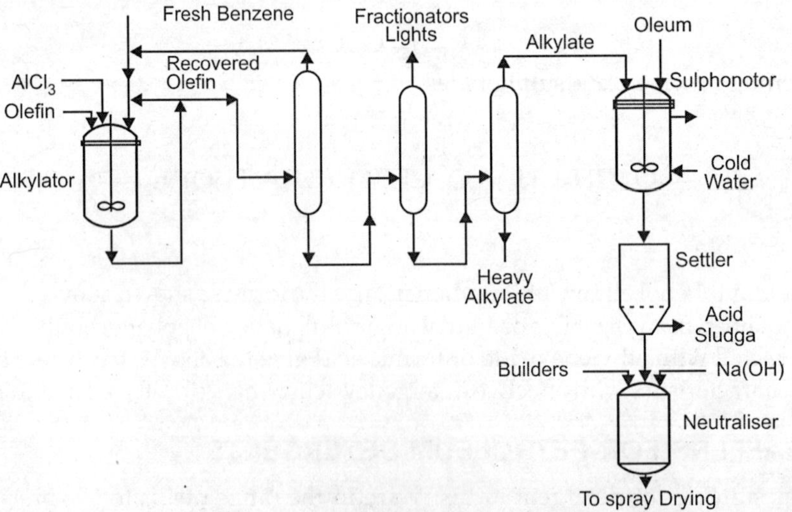

Fig. 11.1: Detergents Through Olefins.

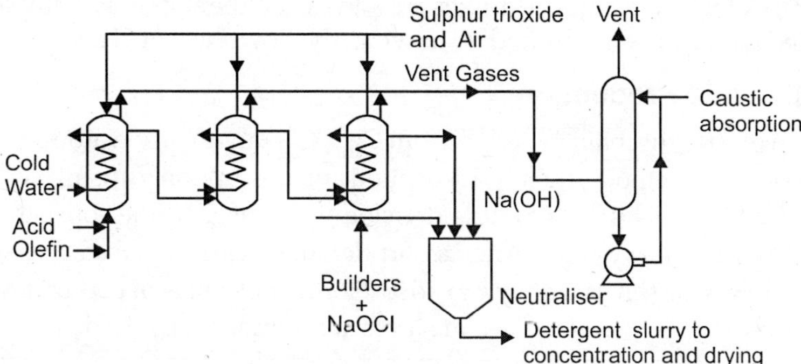

Fig. 11.2: Sulphonation of Olefins.

11.7.1 Keryl benzene Sulphonate (Surf)

India presently manufactures 5.2 x 10^5 tonnes of synthetic detergents. All household detergents in India are manufactured from paraffins found in kerosene range cut. These fractions are rich in aromatics too. These aromatics are removed by either adsorption, liquid extraction or by acid destruction. Kerosene range paraffins are chlorinated first. The ratio of paraffin to chlorine is varied between 1 : 3 to 1: 10 on mol basis. Liquid phase chlorination is conducted. Usually two reactors in series are used for chlorination. Excess of chlorine and hydrochloric vapour are stripped off for recycling. The stripped chloroparaffin contains some unconverted paraffin which can be separated by distillation and sent into the chlorinator. Heavy alkylate assists in separating chlorine and HCl from the chlorinated paraffin. It is dehydrated and then alkylated with benzene in presence of a catalyst, $AlCl_3$ (Friedel-Crafts). The alkylated hydrocarbon mixture is settled and washed to remove catalyst, then separated from the remaining hydrocarbons and impurities like HCl and Cl_2. The fraction is carefully sulphonated with oleum. After sulphonation the mixer is neutralised with sodium hydroxide.

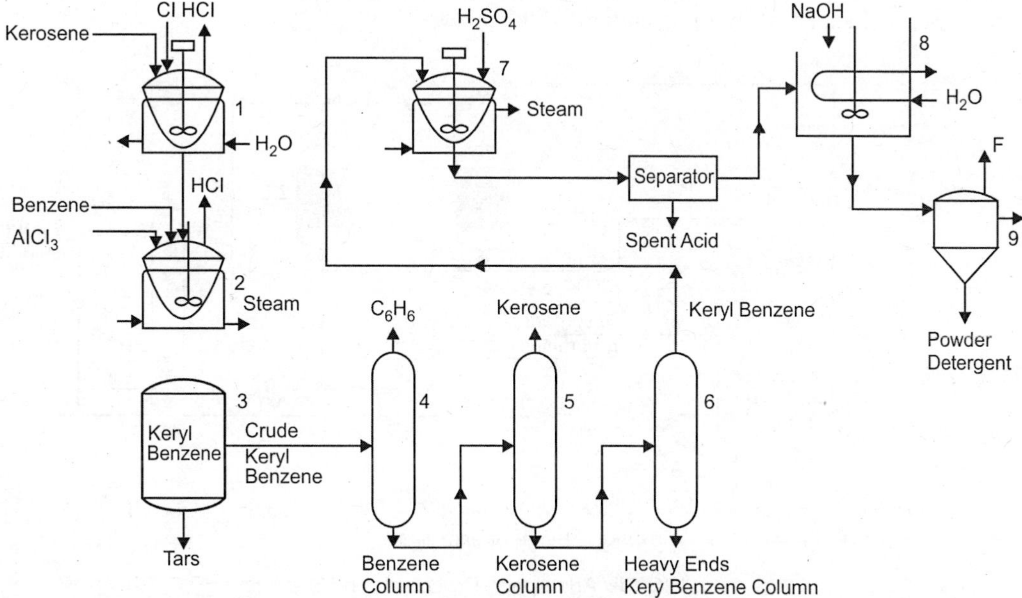

Fig. 11.3: Detergent from Kerosene.

Filler, Builders, Colours and other additives are added and the slurry is finally concentrated and spray dried. The technology of this method is shown in the flow sheet of Fig. 11.3.

11.7.2 n-Pararfins and Production

The production of straight chain paraffins from $C_{10}-C_{12}$ range carbon atoms from jet/kerosene range fuels is done by two methods, mostly adsorption and desorption by using synthetic zeolites or molecular sieves. Iso-Siv developed by Linde division of Union Carbide consists of vapour phase adsorption in one of the reactor systems. After adsorption the cycle is reversed to desorb this bed. The character of these molecular sieves is they can adsorb only molecules of certain pour size (5A) and reject the over[4,5] size. Three types of molecular sieves are commercially used.

Type		*Pore size*
4A	$Na_{12}\,(AlO_2)_{12}\,(SiO_2)_{12}\,27H_2O$	4A
5A	$Ca_6\,(AlO_2)_{12}\,(SiO_2)_{12}\,27H_2O$	5A
13X	$Na_{86}\,(AlO_2)_8\,(SiO_2)_{106}\,276H_2O$	10

Molex is a process developed by UOP and operates on this principle. Molex process differs from the previous one (iso-Siv) in that it is conducted in liquid.phase adsorption. Molex is a continuous process. Desorption in the case of Iso-Siv process is done by reduction in pressure or by light n-paraffin purging while Molex process resorts to purging by recycle of n-paraffins. Iso-SiV process does isomerisation of paraffins too. Fig. 11.4 shows the outlines of the process technology. Isomer produced can be separately distilled to utilise n-paraffin into purge stream. The paraffin column separates the paraffin into a light and heavy fraction. This heavy fraction that goes for detergent production. The lighter fraction goes for purging operation. Dehydrogenation of paraffins for getting a-olefins of linear structure is practised by Pacol Process[6]. Fixed bed catalyst reactor is used. A separator eliminates hydrogen and light gases from the fraction. Detergent range a-olefin is fractionated from the total product. Paraffin is recycled into the reactor.

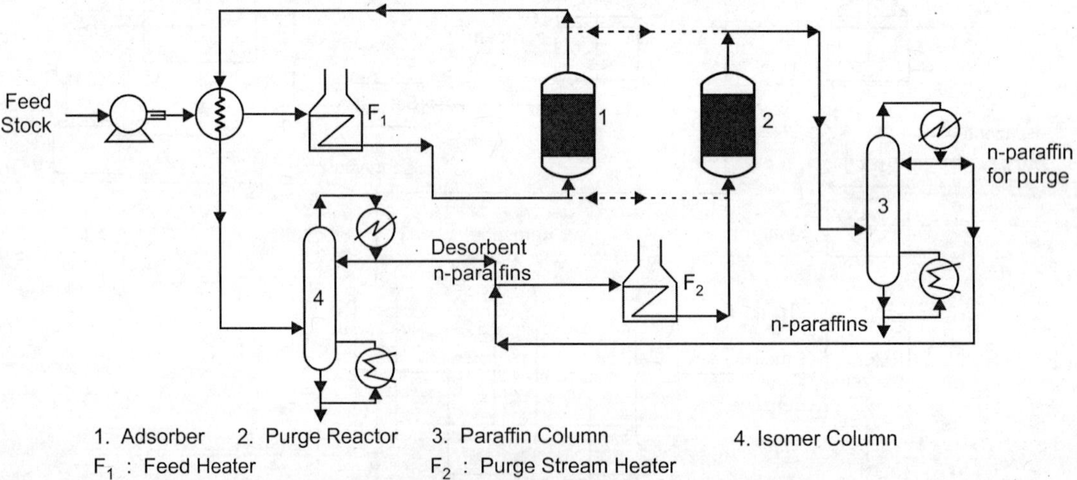

Fig. 11.4: ISO SIV Adsorption Process for n-Paraffins.

Another technique urea adduction is also useful for separation of n-paraffins Though it is in operation, it is not significant magnitude. Alkylation of phenols with olefins in presence of boron fluoride catalyst has been developed by Esso Research and Engineering.

Thus the main reactions sulphation, sulphonation, are critically dealt in industry. Sulphonation can be done by using SO_3, oleum, Air-SO_3, mixture SO_3 under vacuum in presence of a solvent or solvent flowing as a thin film and sulphamic acid (Chlorisulphonic acid for sulphation).

Sulphoxidation is a free radical reaction of paraffin with sulphur dioxide and oxygen. It can be initiated by light or water to give rise to RSO_3H.

$$RH + SO_2 + 1/2\,O_2 \rightarrow RSO_3H$$

Sulphoxidation is commercially useful in obtaining detergents from paraffin with chains from ten to twenty carbon atoms long. Sulphoxidation is used for saturated hydrocarbons only. Unsaturates and aromatics should be removed from the feed. In addition to long chain saturated hydrocarbons, alkylchlorides, nitrites, carboxylic acids, alcohols and ethers can also be sulphoxidised. The reaction can be initiated and carried out with ozone, peracids or peroxides. Commercially sulphoxidation is conducted in iron column lined with plastic and furnished with lamps to generate ultraviolet light. Sulphur dioxide and oxygen (2:1) are fed into the reactor where it comes across trickling paraffin. Water is fed continuously to remove the sulphonates formed. The circulating mass is externally cooled to keep the reactor temperature below 30°C. The sulphonated paraffin is allowed to separate in a separate vessel, whereby the sulphonated hydrocarbon is obtained as top layer and acid as bottom layer.

11.7.3 Alkyl Sulphates

Alkyl sulphates are obtained through the reaction between alcohol and sulphuric acid. The detergency of secondary alkyl sulphates obtained by sulphonation of olefins is less and then frothing power increases as the sulphate group shifted towards the middle of the chain. In alkyl sulphates when SO_3Na is in second position frothing power and detergency increases from C_{11} to C_{17}; further increase of chain length adversely affects the properties.

11.8 NON-IONIC DETERGENTS

Nonionic detergents do not dissociate into ions like alkyl sulphonates (anionics). The solubility of these compounds is basically assigned to hydrophilic nonionic ether or hydroxyl groups present. The excellent solubility is due to the formation of hydrates developing a hydrogen bond between oxygen atom and water molecules. As the temperature is raised the hydrogen bonds become weak, this causes the hydrates to decompose; resulting in non-soluble formations. Hydrophobic part may be a compound containing an active hydrogen atom condensable with ethylene oxide or propylene oxide. Different active atoms are derived from compounds of following nature.

$$HO\text{-}C_6H_4\text{-}R + nCH_2\text{--}CH_2 \text{ (O bridge)} \longrightarrow (OCH_2CH_2)_nOH\text{-}C_6H_4\text{-}R$$

$$ROH + nH_2C\text{--}CH_2 \text{ (O bridge)} \longrightarrow R(OCH_2CH_2)_nOH$$

$$RSH + nH_2C\text{--}CH_2 \text{ (O bridge)} \longrightarrow R.S.(OCH_2CH_2)_nH$$

$$R\overset{O}{\overset{\|}{C}}-NH_2 + 2n\,H_2C-CH_2\text{ (epoxide, }\backslash O/\text{)} \rightarrow R.\overset{O}{\overset{\|}{C}}N\begin{cases}(CH_2\,CH_2\,O)_n\,H\\(CH_2\,CH_2\,O)_n\,H\end{cases}$$

$$R\,.\,NH_2 + 2n\,H_2\,C-CH_2\text{ (epoxide, }\backslash O/\text{)} \rightarrow R.N\begin{cases}(CH_2\,CH_2\,O)_n\,H\\(CH_2\,CH_2\,O)_n\,H\end{cases}$$

Additionally fatty sugar esters, fatty amine and fatty amide ethoxylate ETO, PO copolymers are available. The alkyl phenols should contain 8-10 carbon atoms. In condensing with alcohols/acids, two moles of ethylene oxide must be added to every three carbon atoms of the acid or alcohol. As a rule non-ionic detergents have low frothing power and froth stability. Igepols (ethylene-oxide-phenols) were the first non-ionic detergents introduced. Water or ethylene glycol is used to produce long chains in alkyl phenols, classified as polyoxyethylene glycols. The distinguishing feature of non-ionic detergents is that they are capable of preventing redeposition of dust, hence they exhibit high detergency power. They are also used in mixtures with anionic detergents specially in liquid form Most of the non-ionic detergents serve in costly detergent mixes like shampoos, shaving creams, cosmetic preparations, further they are used as emulsifiers and de-emulsifiers in many industries.

11.9 PREPARATION OF ALKYLATED PHENOL DETERGENTS (OP)

In Fig. 11.5, generalised flow diagram for the production of OP type detergent is presented. Alkyl phenols are obtained by alkylation of phenols with isobutene or propylene dimers and trimers respectively. Alkylation of these chains with phenol can be done very easily in presence of acid catalyst. These alkyl phenols are condensed with ethylene oxide in presence of caustic soda or potash. Ethylene oxide and phenols are dehydrated before they enter the condensation reactor. The reactor and lines are blown with nitrogen, so that the reaction can take place in an inert atmosphere. Instead of ethylene oxide, propylene oxide is also used; or even ethylene oxide and propylene oxide copolymers can also be used.

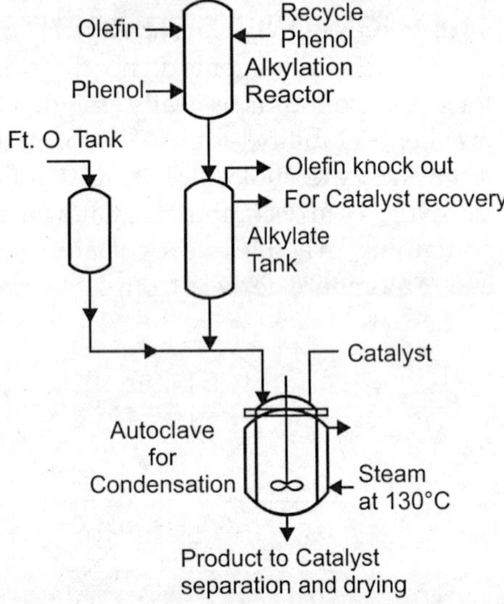

Fig. 11.5: O P Type Detergents Production.

11.9.1 Ampholytes and Amphoterics

These are also classified under non-ionic detergents. But they are capable of ionising depending upon the medium. Zwitterion structures exhibit this nature. These are generally compounds of quarternary nitrogen compounds. The long chain amines condensed with fatty acids exhibit ionic structures when the pH of the medium changes.

$$C_n\,H_{2n+1}\,NH_2\,(CH_2)_m\,COO$$

neutral structures

$$C_n\,H_{2n+1}\,NH\,(CH_2)_m\,COOH$$

The above structures are strictly neutral, but in acidic medium they behave like anionics, NH-$(COOH)^+$ and in alkali medium like cationic NH_2^+ (COO). Some structures of fatty acid-amine condensates having Zwitter ion structures exhibit, germicidal or bactericidal properties though they are mild to skin. Amphoterics based on quaternary nitrogen compounds are of either alkyl based or aromatic based as shown below :

$$\text{(O)} - CH_2 - \underset{CH_3}{\overset{CH_3}{\underset{|}{\overset{|}{N}}}} - (CH_2)\,N = SO_3H$$

$$(CH_2)_n - N - (CH_2) - SO_2(CH_2)_5\,SO_3H$$

Amphoterics are used as antistatic agents. These antistatic agents are useful in some plastics and polymers as well as in jet fuels.

11.10 CATIONIC DETERGENTS

Cationic detergents are the quaternary (ammonia) or[7,8] ternary nitrogen compounds of alkyl chains. Some compounds of fatty amine derivatives known as (Santomerse DT. of Monsanto Chemical Corporation). The structure of which looks

$$C_{12}H_{25}\,.\,NH\,(CH_2)\,CH_2\,NHCH_2\,CH_2\,.\,NHCl$$

Alkyl quaternary ammonium halides known as Cetavlon marketed by Rohm & Hars is of

$$C_{16}H_{33}\,N^+\,(CH_3)_3\,Br \text{ formula}$$

Cetyl benzyl dimethyl ammonium chloride is having a formula

$$C_{16}H_{33}\,N^+\,(C_6H_5-CH_2)\,(CH_3)_2\,Cl$$

In fact cationic surfactants are of little use in cleaning, as most of the surfaces carry a negative charge. The cations adsorb on the surface instead of solubilising the dirt. Of course they have their special applications like inhibiting the growth of mono-cellular organism, *i.e.* they have antiseptic and bactericidal properties. Propyl amines[9] etc. are used as corrosion inhibitors.

11.11 FINISHING OF DETERGENTS

11.11.1 Builders

A finished detergent customarily consists of two main components, *i.e.* surfactant and the builder. Function of surfactant is to wet the substrate that is to be cleaned. The builders aid in other supplementary functions like buffering PH foaming or suds forming, soil dispersion, soil anti-redeposition, solubilization of water insoluble materials, neutralisation etc. These builders are either organic or inorganic type. Metal carbonates, phosphates, borates, silicates etc. serve as inorganic builders[10]. Organic substances are mainly ammonium and substituted ammonium polycarboxylates. Although the widely accepted condensed inorganic polyphosphates are also having their shortcomings. Phosphates are propensive to hydrolyse into less condensed derivatives of less builder properties. Further, these are not subject to bacterial degradation; although they are recognised plant food; may not lead to any controversies regarding pollution; green algae may grow at an explosive rate in presence of phosphates. Though large scale operations to replace phosphate builders have been launched still success is eluding. Some non-phosphate builders like Maleic, Citric[11], Acrylic acid, succinic acid based ammonium derivatives in varying concentrations have been used in detergents. Builders that do not have disagreeable side effects are preferred. Trisodium nitrilo acetate is one such effective builder. Sodium carbonate builders include calcium sequestrant and calcium carbonate anti-deposition agent to inhibit the formation of inorganic deposits on washed fabrics. The per cent of such builders in detergents of all types vary from anything from 30 to 80%. The most important property, like soil suspending is assigned to many variety of chemicals like natural synthetic gums, cellulose derivatives, protein materials, synthetic polymers including polyvinyl alcohols. Widely accepted soil suspending agent is sodium carboxy methyl cellulose. Many such agents like polyvinyl pyrolidone retain the effectiveness of soil suspending even in the presence of relatively high concentrations of calcium such as would be prevalent in hard water. Usually these agents are added in moderate quantities only, *i.e.* 0.5 to 5%.

Optical brightners are lightly coloured organic compounds that are incorporated in washing soaps and detergents in small concentrations of less than 0.5%, these are adsorbed on fibres and can convert invisible ultra violet light into visible light thus creates an illusion of bright washing. The amount of adsorption is reflected by the brightness of the fibre. Many synthetic fibres fail to adsorb these brightners and the brightners for all fabric may not work in the same fashion. These optical bleaches are compounds of sulphonated triazinyl stilbene frequently referred as cynauric chloride (CC) and disodium salt of diaminostilbene (DAS).

11.11.2 Bleaches and Whiteners

Dirt that is not removed by ordinary methods can often be modified for easy removal by the use of bleach. Example being chlorine containing bleaches like sodium hypochlorite. Bleaches are supposed to decompose and give rise to hydrogen peroxide. Example ; sodium perborate decomposes into hydrogen peroxide. Active oxygen carriers can also be used.

11.11.3 Fillers and Processing Aids

They are required to decrease the bulk weight of detergent, *i.e.* to reduce the cost. Sometimes cheap processing aids also are added, they do also slight adjustment of PH and removal of moisture and prevention of caking during storage. They may be sodium sulphonates or sodium carbonates. Enzymes are also used to remove special stains made by proteins like milk, blood etc. Dyes and pigments are also added to detergents to give a slight blue tinge to fabrics. Further, the agents propagate mildness towards skin, are also incorporated in detergents. The effectiveness of these is not clearly understood. Some mild additives developed are compounds of taurine salt; esters of polymerised unsaturated fatty acids (C_{12}–C_{20}) and glycols.

11.11.4 Perfumes

Perfumes are important in marking undesirable offensive odours into a pleasant aroma. In presence of other compounds specially bleaches, these may become unstable and lose fragrance. Household detergents may not require these, but special formulations like shampoos, creams, cosmetics require. Citrus, lemon grass oils, rose oils, oil of peppermint etc. are favoured for such additions.

11.11.5 Corrosion Inhibitors

Specially when dish washing and metal washing are desirable inhibitors are added. Silicates, borax and buffering alkalies are some such agents.

11.11.6 Foams-Agents

Foam is an important factor for considering detergent performance, though it is found that excessive foam does not efficiently clean. However, in some cases excessive foaming is needed to satisfy the desire of consumer. Hence agents which produce enough foam are used. Foaming characteristics of detergents can be regulated by choosing carefully the tensile types and lengths. Besides sulphonates or sulphate types, nonionics of alkylol amide type are useful. Non-detergent foam inhibitors like n-alkylated amino-tri-azines (containing chlorine), propoxylated or butoxylated aminotriazines are used. Tertiary amine oxides are used to increase detergency action and modify the foaming properties.

11.11.7 Other Additives

Hydrotopes[14] are added to reduce the viscosity of the detergent slurry. One type of hydrotope is sodium toluene sulphate. These are used even to increase the solubility in water. Bacteriostats are added to prevent the growth of bacteria. These are mostly used in shampoos and cold creams.

11.12 BIO DEGRADABILITY AND PUBLIC CONCERN

The detergent industry sells a very high proportion of chemicals to public. The final outlet of all these detergents is into ponds, rivers and seas. The pollution problem is directly related to such wastes in large quantities. In India where detergent consumption is not more than 270×10^3 tonnes per year may not pose a big problem compared to industrial effluents; but western countries where the consumption is in an average 13-15 kg/head would be naturally a concern of pollution. Essentiality of bio-degradability of the detergents is not anticipated when such formulations were marketed. The biggest detergent on the nature of utility is aklylaryl sulphonates, which are not biodegradable due to benzene ring. Hence they cause unrelented harm to fish and aqua life. Biodegradable detergents must not contain any aromatic compositions. Hence long chain molecules derived from primary alcohols or olefins are useful as they confer to biodegradability. Ethylene oligomerisation based on Zeigler technology to long straight chains is a modern development[(13)]. Olefins are obtained by steam cracking of paraffin waxes. Oxo reactions give rise to straight chain alcohols.

Secondary alcohols are suitable for certain non-ionic surfactants and are obtained from mild oxidation of straight $C_{11} - C_{15}$ paraffins.

α-olefin sulphonates (AOS) of $C_{12} - C_{18}$ are sulphonated by sulphur trioxide. These produce mixture of compounds, like alkens sulphonates and hydroxy alkane sulphonates. Both are having similar detergency and foaming characteristics like LAS. Other group consists of mainly alcohol sulphates. Primary alcohols of $C_{12} - C_{16}$ are sulphonated and these are used in shampoos, tooth pastes etc. As secondary alkane sulphonates (SAS) ($C_{14} - C_{18}$ n–alkanes) are chlorosulphonated to produce these detergents.

Biomolecules or enzymes in tiny silica gel beads that could help to get clothes better washing these beads act as microscopic reservoirs of enzymes which can remove the stains.

Archer Dainer Middland Co. build commerical World's first biodegradable plastic plant PHAS. polyhydroxy alkanets from renewable sources like corn starch 50,000 tpa.

Alcohol ethoxy sulphates (AES) produced by condensation of 3 - 4 moles of ethylene oxide with linear primary alcohol, gives rise to ethoxylate which on sulphation becomes a detergent. Both sodium and ammonium salts are used commercially.

The above are the detergents which can be biodegradable very easily.

REFERENCES

1. C.F. Longman, The Analysis of Detergents and Detergent Products, John Wiley & Sons, 1975, 11.
2. Kenneth A. Kobe & John Mcketta Jr. Advance in Petroleum Chemistry and Refining Vol. 2, 511.
3. H.E. Garett, Surface Active Chemicals, Pergamon Press, 1975, 84.
4. Kenneth A. Kobe & John Mcketta Jr., Advances in Petroleum Chemistry and Refining, Vol. 4, 116.
5. H C P, Sept. - 72, 314.
6. B V. Vora, P.R. Pujado & J.B. Spiner, H C P, Nov. -84, 86.
7. A. Segales, H C P, March - 75, 71.
8. O.C. Kerfoot & H.R. Flammer, H.C.P. March 75, 74.
9. G.G. Leist. H C P March-75, 83.
10. P.F. Den, H C P, March - 75, 85.
11. R.C. Richmond, H C P. March - 75,88.
12. M. Sitting, Detergent Manufacture NDC, New Jersey, 1976, 225.
13. Harold A. Wittcoff & Brayan G. Reuban, Industrial Organic Chemicals in Perspective, Vol. II. 191.
14. K. Parasuram; Soaps and Detergents, Tata McGraw Hill: 159; 1995.

CHAPTER 12

Petroleum Coke-Carbon Black

12.1 PETROLEUM COKE

Petroleum coke and carbon black may not be exactly fitted into the category of calling them as petrochemicals, however, they can be called as the nostalgic derivatives of petroleum produced almost in the refinery itself. Prolific researches in either refining methods or in petrochemical production made these industries viable and competitive; on the contrary relatively less progress has been made on the better technologies for carbon black and coke.

Petroleum coke is produced in the refinery itself as an ultimate product of prolonged cracking. The resultant product a hard, set, porous mass is known as coke. Being a valuable material for electrodes it is considered to be of the highest significance to aluminium industry. It is also useful in other metallurgical applications such as furnace linings, coatings, crucibles etc. The production of petroleum coke in our country is about 1,52,000 tonnes[1] per annum and the systematic breakup[2] is shown in Table 12.1. The demands and utilisation are shown in Table 12.2. The growth of electrochemical industries does require atleast 20% more than that is present in Table 12.2 by the end of 7th plan.

12.1.1 Production —Theory

Cracking of hydrocarbons inevitably produces carbon called coke and the coke deposition progresses with increase of cracking temperature. All molecules crack at severe conditions, however, the big molecules are prone to crack easily at temperatures ranging from 450–550°C.

Table 12.1: Coking Plants in India

Raw Coke	
Assam Oil Company	16,000 tonnes/year
Barauni Refinery	78,000 ″
Gauhati Refinery	36,000 ″
Coke Calcining Plants Approved by Govt.	
Manali	50,000 tonnes/year
Visakhapatnam	50,000 ″
Marogo	50,000 ″
Haldia	50,000 ″
Bongaigaon	30,000 tonnes/year
India-Carbon Ltd. at Noonmati (1961)	65,000 (installed capacity)

(Raw coke obtained from Digboi. Gauhati (Anthracite coal also)	
Anthracite coal calcining plant Budge-Budge (1970)	12,000 tonnes/year
Actual production (1972-73)	58,701
IOC-Plant, Barauni, capacity	60,000
(1971) Production	40,000 tonnes/year
Goa Carbon Ltd.. Panjim (1967)	50,000 (Capacity)

Table 12.2. Demand Pattern In India for Raw Petroleum Coke (in tonnes)

Industries	*Requirement*	%
A. Electrothermal Industries		
1. (*a*) Primary secondary and alloy steel (steel plant)	1510	0.57
(*b*) Primary secondary and alloy steel Arc furnaces	37,800	14.15
2. Ferro-Alloys (FeSi, FeMn, FeCR)	13,410	5.02
3. Calcium Carbide, Silicon Carbide, elemental phosphorous	49,740	18.62
4. Phosphoric Acid	490	0.18
5. Midget electrode, cine arc pencils, carbon rush etc.	735	0.28
Total	103,685	38.82
B. Electrochemical Industries		
1. Aluminium or Anode paste	157,985	59.14
2. Caustic Soda Anodes	5,445	2.04
Total	163,430	61.18
Grand total	267,115	100%

Aromatics and naphthanes contribute to more carbon formation during thermal cracking. Infact the deposited carbon is not pure carbon, but a large molecule with a high ratio of carbon to hydrogen (more than 30). Further many impurities like metals, sulphur and ash concentrate in the product, as it is the residual product, affirmed as the "Bottom of the Barrel".

Thermal dissociation is always associated with the formation of carbon. The reactions leading to carbon formation are immense, however, they are likely to proceed assertively in a sequence *via* free radicals.

Any paraffin molecule cracks into a saturated and unsaturated molecules. The cracking goes on till the molecule dissociates into elements; and this all depends upon the conditions of cracking.

The pattern shows : $C_{30} \rightarrow C_{15} + C_{15}$ (unsaturated)

Both the molecules crack again $C_{15} \rightarrow C_7 + C_7$ (unsaturated)

Dehydrogenation is also common during cracking, leading to olefins formation:

$$C_7 \rightarrow C_7 + H_2$$

$$C_7 \rightarrow C_4 + C_3 \qquad \text{(unsaturated)}$$

$$C_4 \rightarrow C_4 + H_2/C_2 + C_2 \qquad \text{(unsaturated)}$$

$$CH_4 \rightarrow C + H_2$$

Olefins have the capacity to polymerise at suitable temperature may be under pressure but majority will go through diolefins formation due to dehydrogenation.

$$C_4H_8 \rightarrow C_4H_6 + H_2$$
$$C_4H_6 \rightarrow C_4H_4 + H_4 + H_2/C_2H_4 + H_2$$
$$C_4H_4 \rightarrow C{:}H + H_2C + H_2/C_3H_4 + C$$

(High carbon and small amount of hydrogen molecule)

However, aromatics crack in a different fashion; chain detachment, dehydrogenation in chain or scission in chain are the most usual happenings.

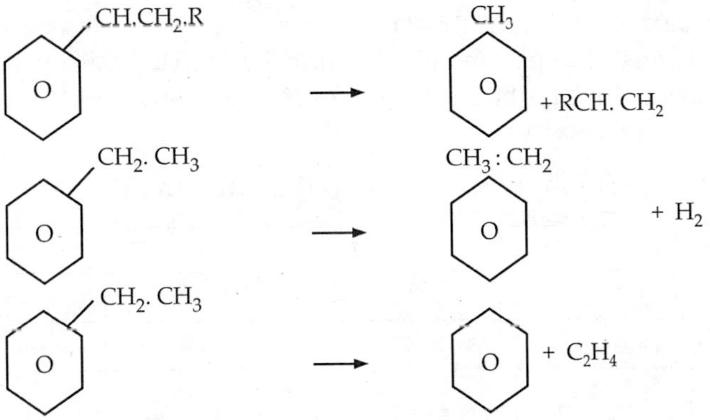

CH.CH$_2$.R (ring O) → CH$_3$ (ring O) + RCH. CH$_2$

CH$_2$. CH$_3$ (ring O) → CH$_3$: CH$_2$ (ring O) + H$_2$

CH$_2$. CH$_3$ (ring O) → (ring O) + C$_2$H$_4$

Saturated rings are easily dehydrogenated and the reaction sequence in chains is as usual.

$$R\,CH_2\,CH_3 \rightarrow RH + C_2H_6 + 3H_2$$

Ring ultimately opens under extreme conditions and the products are manifested into a large number and usually not explainable easily.

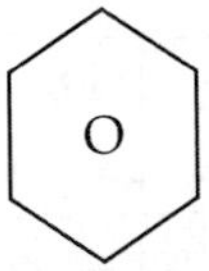

(ring O)
$$\rightarrow CH_2 : CH_2 + CH_2 : CH\,CH : CH_2 + H_2$$
$$\rightarrow 3C_2\,H_2/CH_2 : CH\,.\,CH : CH_2 + 2\,C$$
$$\rightarrow C_2H_4 + 4C + H_2$$

Usually residual oils for coke making must contain at least 10% asphaltene content. If the sulphur content is more than 1% then needle coke can not be produced. Best raw material for needle coke can be obtained by flash distillation of oil to form solid pitch. The ultimate decomposition leads to the formation of carbon and hydrogen. This is the source of carbon, and carbon is sold under different names as coke or carbon black depending on the method of production. High temperature coking is carried out at 800°C–980°C[(3)] while carbon black production requires a temperature above 1000°C.

Reactions in cracking (coking) are dependent upon the temperature, pressure, nature of the molecule. During coking, the amount of coke that is produced is more important. Hence empirical predictions have been formulated. Propensity of cracking is pronounced by Canradson Carbon Residue. Industrially a concise analogical term 'Conradson Decarbonising Efficiency (CDE)' is familiar. It advents more CDE means, more formation of coke. Prediction based on the parameters as well as feed materials are available in literature[(4)]. The coker yields are expressed by the following relations :

$$Wc = 2.0 + 1.66\ CR$$
$$wcg = 5.5 + 1.76\ CR$$

where Wc : Wt. of coke produced

Wcg : Wt. of gas + coke
CR : Conradson Carbon Residue.

The above expressions are useful for delayed coker only and similar correlations are many and available for other techniques too. The amount of coke and other products vary depending upon the process chosen. A comparison of delayed, contact and fluid coking are present in Table 12.3. Aromatics, residuums, asphaltenes are very good coking stocks as they have high CCR. The effect of sulphur seems to be positive during coking, however, the residual product, *i.e.* coke shall be highly contaminated with sulphur. Once it is trapped in the molecule of coke, in a cross linked (equivalent to vulcanisation) fashion it is impossible to remove it. The draw back of high sulphur crudes in coking process, even though they may be of naphthenic origin, must be always remembered. It seems IMP - LUMMUS have been successful in reducing the sulphur content in the coke from 4 to 0.5% using a retort type reactor at a temperature of 1300°C. This may be lauded in future depending upon the performance, specially where coke is made from high sulphur crudes.

Table 12.3: Comparison of Delayed Coking and Fluid Coking

Charge - for	*Delayed*	*Fluid*
Gravity° API	15	15
Conradson Carbon Residue	9	9
Sulphur Products	1.2	1.2
C_3 and Lighter	6.0	5.5
Coke %	22.0	11.0
CDE%	99.8	91.2
Charge for	Contact coking	Delayed coking
Gravity° API	18.9	18.9
Conradson Carobon	11.7	11.7
Sulphur	0.6	0.6
Products		
Lighter fractions%	14.9	7.5
Coke%	13.0	20.0
CDE%	99.3	99.3

Table 12.4. Ultimate Analysis and Physical Properties of Industrial Carbons

	Calcined petroleum coke	
	Delayed	*Fluid*
Carbon wt %	98.40	93.90
Hydrogen wt %	0.14	1.27
Nitrogen wt %	0.22	0.33
Oxygen wt %	0.02	0.09

Sulphur wt %	1.2	4.90
Ash wt %	0.35	0.20
Bulk density kg/lit	0.89	0.99
Real density gm/cc	2.05	1.94
Specific Resistivity ohm	0.089	0.115

Table 12.5: Typical Properties of Petroleum Coke Used In Carbon and Graphite Manufacture

Property	*Raw Coke*	*Calcined Coke*
Aluminium (ppm)	15-100	15-100
Ash %	0.1-1.0	0.2-1.5
Boron (ppm)	0.1-0.5	0.2 - 0.7
Calcium (ppm)	25 - 500	25-500
Fixed Carbon (%)	87-97	97-99
Hydrogen (%)	3.0-4.5	0.1
Iron (ppm)	50 - 2000	50 - 2000
Moisture (%)	0.5-2	Negligible
Nickel (ppm)	10.0 -100	10.0 -100
Nitrogen (%)	0.1-0.5	0.1
Real density (gm/cc)	1.6 - 1.8	2.08-2.13
Silicon (ppm)	0.2 - 2.5	0.2 - 2.5
Titanium (ppm)	2-60	2-60
Vanadium (ppm)	5-500	5-500
Volatile matter %	5-15	0.2
Manganese (ppm)	2-100	2-100
Magnesium (ppm)	10 - 250	10-250

12.1.2 Methods of Production

1. Hot oven method
2. Thermal cracking (Dubbs Two Coil Cracking)
3. Delayed coking
4. Fluid coking
5. Contact coking

Presently delayed coking enjoys the highest reputation in the industry while fluid and contact cokings are gaining position relatively being, recent innovations. Hot oven method and Dubbs cracking techniques were used only for a brief period in the initial stages only.

Afterwards these methods were completely obliviated in favour of the later efficient methods of production like Delayed Coking.

12.1.3 Delayed Coking

In India the refineries produce coke based on this technology only. Delayed coking, as the name conveys, the cracking is carried out in a furnace at a temperature of 500°C–550°C and the free radical reactions are allowed to go to completion leisurely in a separate drum at 450°C–500°C under a slight pressure of 4 - 5 bars. The advantages in such separation of furnace and coil are firstly heater tubes are not damaged and different stocks can be cogently handled. Fig. 12.1 shows the outlines of the technology. Tall huge, thick (3-4 cm) large steel stacks (5-7 metre dia and 15-20 m high) in a battery of 6-12 constitute the main coking drums where the cracking reaction continues to extermination. A hot charge stock first going through different heat exchanging systems enter the fractionator and the heavy oil from the fractionator after being heated in a furnace to the temperature of cracking is discharged into these drums one after another. A hard set mass results due to the coke forming reactions which are fostered due to free radicals. About 10-16 hours is required for setting the coke. During setting some volatiles are set free and from the bottom of the drums condensates are produced. All these products are sent into a fractionator. After coke is formed the dislodging of coke has to be begun; known as 'Decoking'. Decoking in olden days was practised by manual hammering of hard coke mass. Presently either hydraulic jets at a pressure of 150-300 atmospheres or chain pulling technique is employed. Chain pulling device is simple, as strong heavy chains are laid in the drum initially with ends connected to the top of the drum. After the molten mass sets, the chains are pulled, due to which crumbling of coke takes place.

Fig. 12.1 shows the delayed coking outlines. The process begins with the fresh feed entering the fractionator (1). The fractionator is useful as a separator of lights and volatiles from the cracked vapours obtained from coke drums (3). Fresh feed as quench is also sent into this fractionator, by this any volatiles available in the charge pass into the volatiles of the cracked products.

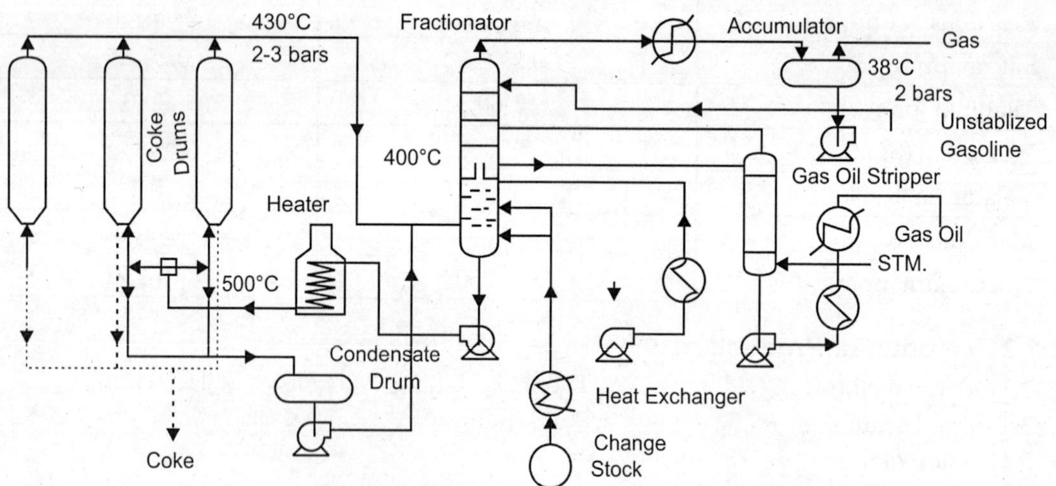

Fig. 12.1: Delayed Coking Unit

Uncoked heavy liquid, from the bottom of coke drums is also fed into fractionator. Hence the duty of fractionator is not only to separate the lights from heavy, but also acts as a good mixer for uncracked and cracked products. Side stream strippers are also connected to the fractionator. The bottom product being heavy forms the required charge. The charge is heated in a furnace (2) to 470°C–530°C and allowed to enter the coke drums (3) where the temperature is maintained around 470°C for a long time so that coke setting takes place under a pressure of 6-7 bars. Setting may take place 10-16 hours and decoking may last another 4 to 6 hours; charging into drums is monitored by cathode rays and is completed in 4-5 hours. Volatiles from the top and heavy product (unset) from the bottom of the coke drum are routed to fractionator. Large number of drums are required to make the process continuous. Each drum may yield 10 tonnes of coke per charge. Increase of pressure during coking has a spawning effect in formation of coke.

12.1.4 Fluid Coking

It was introduced by Standard Oil Development Corporation. This works on the principle of fluidisation. In this process there are three main components (1, 2, 3) reactor, burner and fractionator. In between the burner and the regenerator (2, 3) the red hot coke particles circulate. The heat for coking is supplied through burning of coke particles and oil matter over the particles in the heater. The incandescent particles are sent into reactor where the hot pitch/feed directly comes in contact with the coke particles and immediate cracking takes place. The cracker products (light) escape through the cyclones where fines are collected. The gases are allowed to go to sulphur removal unit. The fractionator which is on the top of the reactor separates volatiles. From the bottom of the reactor heavy coke particles which are formed on the seeds of coke particles due to deposition of cracked material settle down and reach the heater.

The heavy coke particles sink to the bottom Steam stripping dislodges the hydrocarbon from carbon, later these are conveyed into heater. Part of the coke-hydrocarbon mixture (on the surface of coke) is burnt in the heater for supplying heat, and the remaining coke is obtained as product. Reactor temperature is maintained at 530°C–560°C at atmospheric pressure, while the heater temperature goes upto 700°C. Fine coke particles escaping even through cyclone are caught in fractionator in the form of slurry in oil and the slurry is returned back to the reactor. Fluid coking gives more liquid distillates. Except these being continuous, other things basically remain same and match with delayed coking. Fig. 12.2 shows the outlines of this technology.

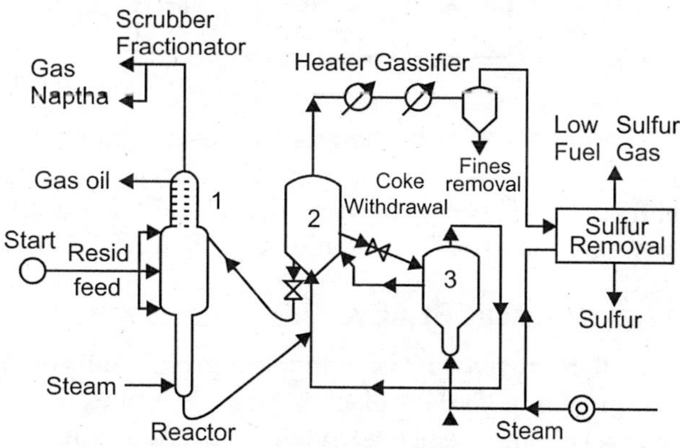

Fig. 12.2: Flexi Coking.

12.1.5 Contact coking

This is also a continuous process. Oil wetted circulating coke particles are allowed to travel downward in reactor in the form of a dense bed. This allows sufficient time for coking and drying. Fig. 12.3 shows the outlines. Vacuum residue is fed into a fractionator (1), where from the heavy (bottoms) oil is taken and fed through a heater. The hot feed enters the reactor (2), where the hot calcined coke particles obtained from reactor (3) travel concurrently. The oil drenched red hot coke particles decompose the oil to coke which is firmly deposited on the surface of these particles. The coke formed is quenched alongwith coke particle and steam stripped to separate hydrocarbons. Later the particles are transferred to a common auxiliary storage bin (4) situated at the top of reheater. Part of the coke is taken as product while the remaining is sent into reheater where air and fuel gas are injected to start combustion for providing required heat to the coke particles. The red hot coke particles from the lift tanks are transmitted into reactor. Part of the coke also goes into storage bin (5) where some times external seed coke are added to begin the process; otherwise it serves simply as a storage bin. Mostly 10-14 times coke circulation is prevalent for every unit of oil.

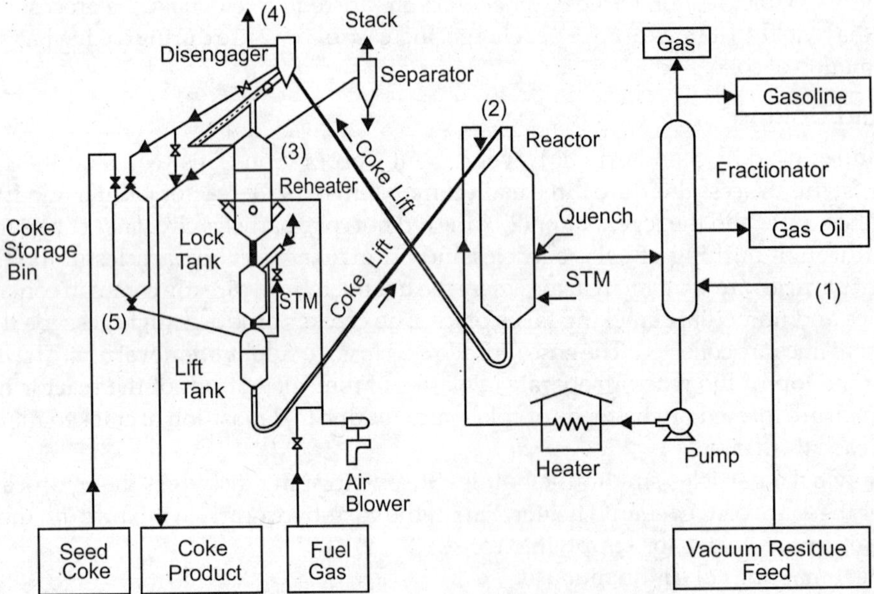

Fig. 12.3: Continuous Contact Coking.

In India 50% of the petroleum coke is almost met from imports as the total production is only 152,000 tonnes per annum. The analysis of calcined petroleum coke produced by Delayed Coking and Fluid Coking[9, 10] are shown in Table 12.4 for many applications raw coke is further calcined, and this raises carbon to hydrogen ratio more than 100. Table 12.5 shows the difference.

12.2 CARBON BLACK

There are five units making carbon black output to 58,000 tpa in 1987, 2,91,000 in 1998-99. Present production of Carbon Black is 4,82,000 tonnes (2007-08). These units use cheap aromatic sources (extracts from lube plants), however the new units are gas based.

It is called by different names like, Thermal Black (produced by decomposition of natural gas) Channel Black (produced by impingement of numerous regulated flames on cold, steel surfaces), Furnace Black (obtained by partial combustion of gas in a furnace and separated by cyclones or electrostatic precipitators), Lamp Black (a form of soot). Carbon Black is soft, fluffy particles formed by thermal decomposition of hydrocarbons. It has an immense use in rubber industry. Further as a pigment in paints, as an ingradient in carbon paper making etc. are some other uses. The carbon black is produced either using methane as feed stock, liquid petroleum fractions of highly aromatic; unconventional decomposition of acetylene is also sought depending upon the availability. The production on the basis of raw materials is briefly illustrated.

The Aditya Birla Group's Carbon Black business spans four companies in four countries. The Group's four Carbon Black companies are *Alexandria Carbon Black* in Egypt. *Hi Tech Carbon* in India, *Thai Carbon Black* in Thailand and *Liaoning Birla Carbon* in China. Egypt's company of capacity of 285,000 mtpa, is one of the fastest growing Carbon Black in the world. Thai Carbon Black has been listed among the world's best 200 small companies for three years from 2002 to 2005. With an annual capacity of 210,000 mt, with manufacturing plants at Renukoot in U.P and Gummidipoondi in Tamilnadu. Hi Tech Carbon is the second largest producer of Carbon Black in India. The company has a combined capacity of 240,000 mtpa and exports about one third of its production overseas.

Liaoning Birla Carbon, a pioneering venture in China, is a joint venture between Thai Carbon Black Co. Ltd. and Dashiqiao Ronghua Co. Ltd. With a capacity of 55,000 mtpa, the company uses the distributed control system (DCS) to ensure consistency of quality. With a total annual capacity of 790,000 mt, the Aditya Birla Group is the fourth largest producer of *Carbon Black* in the world. The four companies work in synergy to produce world-class *Carbon Black*. The varied locations give the companies access to most of the world's markets.

12.2.1 Methane Decomposition

Methane decomposes to form carbon and hydrogen at a high temperature, *i.e.* at 1200°C on the surface only. While in other gases the decomposition is throughout the mass and goes through free radical mechanism. Early works consisted of passing methane or purified natural gas through heated carbon particles in a retort at 1300°C, alternatively it was done by passing through aluminium tube at 1300°C–1700°C. Under this thermal gradient the gas decomposes into basic elements without any intermediate products. This gives a high quality carbon compared to other types of feed stocks. Usually 5% of the available carbon results as product. Industrial adoption of the process began with Rincker-Wolter process, in which two generators (coke packed) were used. The generators were heated to 1400°C, where a spray of tar is directed against this bed.

The decomposition of tars and pitches takes place over the red hot-coke particles. Thermatomic process was the first important process for the production of carbon black. It used refractory lined reactor in which refractory checker bricks were staked. Initially natural gas is burnt in those checker bricks by supplying air and gas through different ports. When the reactor becomes red hot, the exhaust valve was closed, natural gas was admitted from the top of the reactor. The gas decomposed and carbon was collected after cooling the reactor. This in due course, became the Wulffs' process.

12.2.2 Wulffs' Process

The process consists of decomposing feedstock in two reactor chambers fitted with checker bricks. These form the main equipment in this process. These two chambers have common refractory duct through which gases can be taken out or admitted. These two combinedly with common duct work on heat recuperative system. One reactor burns natural gas and then blue gas goes through second reactor and common duct. There the heating cycle is stopped, and the reaction cycle is started by reversing the gas feed. The hot gas from the reactor-2 goes and decomposes in reactor-1, finally gas goes through scrubbers and recovery system. So the alternative heating and reaction cycles produce carbon black. This carbon black carried away by blue gas is first showered with water to cool and then later collected in a bag filter or collected in cyclones, 3-28 kg of carbon black per 1000 m^3 of natural gas is obtained.

12.2.3 From Acetylene[9]

Acetylene being more unstable (3 to 28% acetylene with air is explosive) forms an easy feedstock for decomposition, also its carbon content is very high. Further it is more stable at very high temperature. It decomposes at a moderate temperature of 520°C under a slight pressure in presence of iron surface to carbon and hydrogen. It is found that catalysts like copper, silver assist decomposition even at 200°C. Even with or without air it can be decomposed. If electrically initiated it can be decomposed at a pressure of 2 bars. Commercially used method consists of decomposing acetylene at 800°C, the reaction is exothermic. It is usually done by passing acetylene through heated refractory tube. Heating is done by burning acetylene itself. The decomposed fluffy carbon blacks are collected at the base of the retort and conveyed to the presses.

12.2.4 From Hydrocarbon Oils

Aromatic rich oils (usually having a b.p. of 300°C–400°C) are used for decomposition. The incomplete combustion results in partial decomposition of oils into fine coke in the form of soot, which is collected by allowing it to impinge on the relatively cold surfaces.

12.3 PROCESSES - AVAILABLE

Different processes available for manufacturing carbon black are (*a*) Channel process (*b*) Furnace process (*c*) Regenerative cracking (*d*) Lamp Black.

12.3.1 Channel Process

It is mainly used for decomposing gases. The gas is burnt in burners in a row. A large system of channels covered with good amount of clearance, formed by beams of inverted channels. In the channels there are stationary burners in rows spaced at 1-1.5 metre distance and about 10 cm distance from the space of channel for burning the gas continuously. Air is admitted through separate ports. Due to combustion heat in the furnace, partial decomposition takes place. The decomposed gas deposits fine carbon particles on the surface of the channel beams which are given forward-backward (reciprocating) motions continuously. The deposited carbon is scraped by a blade when the reverse motion takes place. The collected carbon is carried away by a system of conveyors into ducts and collecting chutes. Depending upon the burner design different types of flames are produced, due to which the quantity and quality of carbon change.

12.3.2 Furnace Process

The bulk of carbon black is produced by this process. Partial combustion of highly aromatic oils produces this furnace black[10]. This is produced in a refractory lined steel furnace whose dimensions vary 1.5 to 4.5 m[11] long with 15 - 75 cm ID. Such furnaces are in a large number. For maximum efficiency, the furnace and burner are designed in such a manner that large separation is available to affect combustion and carbon formation reactions. Auxiliary burners are used for burning either refinery off gas or natural gas. The combustion liberates a heat flux of 7000 - 10,000 kcal/m^3. The zone of complete combustion is given a swirling motion and the make oil is preheated to 250°C and either vaporised or atomised and tangentially admitted into the centre of the zone. The hot combustion gases carry carbon particles and the light carbon black fluffy particles are carried over by blue gas. This blue gas enters series of cyclones and bag filters where the particles are collected in different sizes. Fig. 12.4

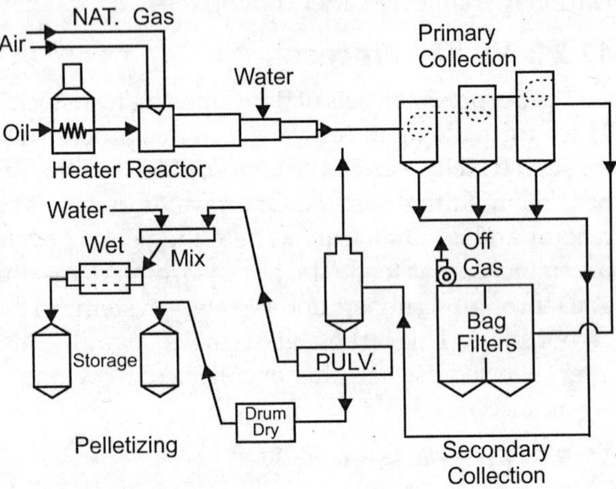

Fig. 12.4. Phillips Oil Furnace Process.

shows the process, in the processing plant grit and metallic bodies are separated from carbon black. Coke produced is characterised by the burner itself as general purpose type or abrasion type. Natural gas and air in the ratio of 1 : 12 to 1 : 18 are admitted through separate ports in specially designed burners into a fire brick lined furnace. The furnace is horizontal, either rectangular type, *i.e.* 1.2 × 3 × 4.3 metres or cylindrical (1.5 m dia 10 m long). There are about six burners in a furnace.

The furnace is filled with a rolling mass of luminous flame at a temperature of 1300– 1400°C where process oil is injected for decomposition. The blue gas carries the carbon black particles through a horizontal fire brick-lined flue duct of a diameter of 90 cm to the water sprays to cool down the temperature to 205°C. Finally cyclone separators and bag filters separate the fine carbon particles. Afterwards about 50% carbon deposited on the checker bricks is burnt away by admitting air and some more natural gas. The composition of thus produced carbon shows :

Carbon	91-99%
Oxygen	6-0.5%
Sulphur	0.01-0.12%
Hydrogen	0.8-1.2%
Fixed carbon	84-98%
Nitrogen	0.04-0.01%
Volatile matter	1.1-6%
Ash	0.05-0.25%
Carbon Hydrogen	18.1–24:1

Classification based on volatile matter is usually followed. Carbons which contain large amount of volatile matter are termed as long blacks and less amount of volatile matter are termed as short blacks.

Depending upon the size and purity of carbon particles uses are destined.

REFERENCES

1. Rao, B.K.B., Modern Petroleum Refining Processes, Oxford-IBH., 84; 343.
2. Indian Petroleum Handbook, 1973, 213.
3. W.C. Nelson, Petroleum Refinery Engineering, McGraw-Hill, Oil & Gas J.; 129.
4. James H. Gary & Glenn. E. Handwerk Petroleum Refining, Marcel Dekker, 58.
5. Fernendo Mazanilla, F.M. Sadilla, Oiliverio M.L.A. Moute, M.C. Size & W.V. Bawer, HCP, March, 79, 101.
6. Rao, B.K.B. & P. Sen. All India Seminar on Carbon Technology Korba (BALCO), March, 78, 33.
7. Chakraborty & Arun Kumar - (Ibid) -11.
8. Kenneth A. Kobe & John McKetta. Advances in Petroleum Chemistry and Refining, Vol. 2, Inter-science Pub. 1959, 371.
9. Charles L. Mantel; Carbon and Graphite Hand Book, Interscience, Pub., N.Y. (1968), 107.
10. Charleton Ellis, The chemistry of Petroleum Derivatives, The Chemical Co. (USA) 1934, Vol. 1,246.
11. Fredrick - A. Lowenhein & Magnetite K Morn Industrial Chemicals. A Wiley Publication (NY) 1975, 205.

CHAPTER 13

Pollution and Toxicity

13.1 POLLUTION AND TOXICITY

Man is born on the Earth, he dwells on Earth, he consumes the natural resources and for his comfortable way of living he synthesizes new materials, finally he becomes wicked, to fulfil his cupidity, he invented lethal destructive weapons. Finally he spoils the Earth. Hindus in their religious concept worshipped the planet along with its resources in the name of Pancha Bhuthas viz. Sun (Fire), Air, water, atmosphere (Akash), and finally the dwelling place the Earth. Today the concept has changed to scientific phenomena leading to the laws of prevent polluting these. The importance of clean environment may not be the ideal wish but nearing to that must be pleaded and a genuine effort from all directions should be made. Present day pollution is caused by all utilities and luxuries/comforts. Truly we cannot avoid these in the present day fast life. However care must be taken to minimize the pollution. Rapid industrialization, urbanization required enormous energy for transportation, maintaining industry and urbanity of the area. No doubt the air, water and land pollution begins and grow beyond dimensions in few years time The critical examination follows:

13.1.1 Air pollution

Practically the pollution is man made. The exhausts of automobiles, power houses, industrial units are the significant contributors of pollution, in the form of oxides of carbon, sulphur and nitrogen, besides toxic discharges of wastes into water and land. Hydrocarbons along with chloro flouro derivatives, make the Earth highly polluted. Besides these gases; particulate matter generated by mining operations, metallurgical industries, coke ovens, power houses and cement industries and construction activities, float in the atmosphere depending upon the size of the particles. The smaller size particles of less than 10 microns are more dangerous as they take long time to settle down entering respiratory system with out any difficulty. These minute particles form conglomerates with acid gases in presence of rain drops/snow or moisture. Principal acidic gas being carbon dioxide which is increasing at a rate of 1 ppm per year standing at present 365 ppm. Methane, another to increase Green House effect is present in atmosphere due to petroleum activity, wet lands, lakes etc. and paddy fields these contribute yearly 553 TG. Besides pollution due to these gases, radiation due to radioactive minerals and nuclear reactors cannot be ignored. Some pollutants are less severe, some are very toxic. Toxic pollutants are mostly man made chemicals.

13.1.2 Environmental Causes

India, the World's second largest populous country has seen its population explosion since 1947 when its population was only 300 millions to 1.37 billions (latest UN estimate 2019) as it stands today. Rapidly growing population has placed a strain on resources, needs and environmental security besides health and other social establishments. According to WHO New Delhi is one of the 10 top most polluted cities in the World. Many cities have pollution monitoring centers functioning under

Central Pollution Board and the facts are shown in Table 13.6. The pollution of atmosphere is mainly caused by two sources one is automobile emissions and other one is due to untreated industrial smoke. Coal is main source of energy in the country. The quality of Indian coal is also not good having an average 35% ash content or more. Increased coal consumption in the last five decades has led to unprecedented energy related carbon emission mainly due to power plants and cement industry. Further many industrial establishments donot perform energy audit thus energy intensity is not given any importance. Definitely the replacement of coal based plants to oil or gas plants do require heavy investment besides the question of availability of resources. Though many statistics of India on percapita basis are on very low index actual production is very high and they surpass many countries. The Government is harping on Renewable energy to reduce carbon emissions, the progress is moderate if not good. The major resources of renewable energy are Hydroelectrical and Wind followed by Biomass related programmes. Solar energy usage is also increasing.

Table 13.1: National Ambient Air Quality Standards (Central Pollution Control Board)

Pollutant	*Sulphur dioxide (SO_2)*		*Oxides of Nitrogen (NO_2)*		*Suspended particulate matter (SPM)*	
Time Weighted Average	Annual Average	24 hours average	Annual Average	24 hours Average	Annual Average	24 hours Average
Industrial Area	80 $\mu g/m^3$	120 $\mu g/m^3$	80 $\mu g/m^3$	120 $\mu g/m^8$	360 $\mu g/m^3$	500 $\mu g/m^3$
Residential Rural and other Area	60 $\mu g/m^3$	80 $\mu g/m^3$	60 $\mu g/m^3$	80 $\mu g/m^3$	140 $\mu g/m^3$	200 $\mu g/m^3$
Sensitive Area	15 $\mu g/m^3$	30 $\mu g/m^3$	15 $\mu g/m^3$	30 $\mu g/m^3$	70 $\mu g/m^3$	100 $\mu g/m^3$

13.1.3 Ozone Layer

Atmosphere contains many man made pollutants, out of these some can interact with the ozone layer. Chlorine and NO are the most important constituents which can deplete ozone through photo chemical reactions. Stratospheric Ozone layer which occupies 11-15% has been reduced by 1.3% during the last fifty years. In 1980 it was found to be 250 ppt in troposphere. Ozone is getting depleted at a rate of 2000 tons yearly, at 17 km height. This increases a vertical indent of UV by 6% on earth. Further air crafts release huge amount of particles and along with other particles, they reflect or absorb solar light causing Albedo Effect on Earth. Present level of CO is 0.1 ppm and it can stay for 4 months in atmosphere. Deforestation is increasing CO_2 at a rate of 1 ppm yearly.

$Cl + O_3 \quad ClO + O_2\text{:} \quad ClO + O \rightarrow Cl + O_2 : O_2 \rightarrow \dot{O} + \dot{O}$ Free radicles.

Similarly : $NO_2 + \dot{O} \rightarrow NO + O_2 \quad NO + O_3 \rightarrow NO_2 + O_2$

Thus chlorine and nitric oxide always regenerate eating away ozone. Freons, chloro hydrocarbons are the main sources of chlorine while high temperature sources along with IC engines are responsible for NO. Persistent organic pollutants undergo distillation in which they are vaporized into atmosphere in warmer regions. The role of active hydroxyl radicals in inducing smog cannot be ignored. It is true that pollution at one part of the globe either on land, water or atmosphere will affect the another part however apart they may be. Only constraint is time. For example the sulphur dioxide in atmosphere is increasing with the burning of about 4000 MMT of oil and similar equivalent of coal, which can generate on the average (even on the assumption of 1% sulphur in oil and 2% sulphur in coal) may give rise to half a million ton per day of sulphur oxides. This is likely to increase the SO_2 concentration in atmosphere by another 0.007 ppm. The atmospheric water vapor reacts with the oxides of sulphur to produce sulphuric acid and becomes a major component in rain water. It is expected that the life of SO_2 in atmosphere is about 4.5 hrs. Ozone is a dominant constituent in photochemical smog. Grit and dust particles of wide range 100 microns to less than 10 microns are available in atmosphere and these cause health problems. Dust particles are trapped in when inhaled by respiratory system and eliminated, these may not cause much severe damage. But particles less than the 10 micron size really cause much damage as they reach the lungs and block the tissues. Many of these materials are judged as not chemically active though silica dust, asbestos filaments etc. really lead to cancer after accumulation in lungs, and act as centers for smog formation with water vapor. Relatively less than 10 micron size particles hang on air for a long time. It can be calculated by Stokes law the time taken by these particles settle down to the Earth. Approximately 10 hr time is expected in still air.

$$U_t = \left[\frac{4D_s(\rho s - \rho)g}{3C\rho}\right]^{0.55}$$

Ut terminal velocity, ρ is the density of solids and p is the density of air, D_s is the diameter of the particle, *C* is a function of Reynolds number.

The settling of the particles depends upon the size, temperature, wind velocity, turbulence in air etc. One is aware that the earth temperature in the stratosphere falls very steeply, this creates downward movement of air giving rise to turbulence. The fallen dust finally adsorbed by earth or surface movement of the accumulated dust caused by wind, water or by live stock in course of time pollutes the ground water or carried away to lakes and rivers. Many countries and states have prescribed the standards of sewage and trade effluent discharged into streams and lands. Tables 13.7 and 13.8 show these standards.

13.1.4 Air Pollution in India

Industrialization and urbanization have resulted in a profound deterioration of India's air quality. Of the 3 million premature deaths in the world that occur each year due to outdoor and indoor air pollution, the highest number are assessed to occur in India. Surveys indicate that in New Delhi the incidence of respiratory diseases due to air pollution is about 12 times the national average.

According to another study, while India's gross domestic product has increased 2.5 times over the past two decades, vehicular pollution has increased eight times, while pollution from industries has quadrupled. Sources of air pollution, India's most severe environmental problem, come in several forms, including vehicular emissions and untreated industrial smoke. Apart from rapid industrialization, urbanization has resulted in the emergence of industrial centers without a corresponding growth in civic amenities and pollution control mechanisms.

Regulatory reforms aimed at improving the air pollution problem in cities such as New Delhi have been quite difficult to implement, however. For example, India's Supreme Court recently lifted a ruling that it imposed two years ago which required all public transport vehicles in New Delhi to switch to compressed natural gas (CNG) engines by April 1, 2001. This ruling, however, led to the disappearance of some 15,000 taxis and 10,000 buses from the city, creating public protests, riots, and widespread "commuter chaos. "The court was similarly unsuccessful in 2000, when it attempted to ban all public vehicles that were more than 15 years old and ordered the introduction of unleaded gasoline and CNG. India's high concentration of pollution is not due to a lack of effort in building a sound environmental legal regime, but rather to a lack of enforcement at the local level. Efforts are currently underway to change this as new specifications are being adopted for auto emissions, which currently account for approximately 70% of air pollution. In the absence of coordinated government efforts, including stricter enforcement, this figure is likely to rise in the coming years due to the sheer increase in vehicle ownership.

Most of the Indian rivers, especially the smaller ones are toxic drains. To name a few : Sabarmati, Bhadasr, Damodar, Betwa, Godavari and top of all these Ganga and Yamuna Ground water is essentially polluted due to various factors. Even the local governments are not bothered to do any systematic analysis and control the pollution. Recent soft drinks controversy is an example. Proper treatment, if envisaged, many attending diseases/epidemics can be controlled. The amount of expenditure spent on controlling the later effects is much more than the treatment of ground water for safe drinking. Air pollution is illustratively increasing leaps and bounds because of urbanisation and growing automobile industry. It is mentioned that Shillong is the only city in India with least air pollutants, particularly -particulates, the dreaded pollutant of Indian Urban area, round the year. In 69 other cities the pollution is moderate, high or critical. In some places the pollution was marginal some time but quickly became critical during certain times of the year. In about 33 cities the ari monitoring showed critical pollution and some times it was even to the threshold level of disastrous. As seen in Table 13.2.

13.1.4.1 Air Quality Index (AQI)

The AQI is an index for coverage of daily air quality. It shows the status of how clean or polluted the air is, and what associated health effects might be a concern to public. The AQI spotlights on health effects one may experience within a few hours or days after breathing polluted air. EPA calculates the AQI for five major air pollutants regulated by the Clean Air Act: ground-level ozone, particulate matter, carbon monoxide, sulphur dioxide, and nitrogen dioxide. Air quality standards for each of these pollutants has been clearly established with an interest to protect public health. Ground-level ozone and airborne particles are the two pollutants that pose the greatest threat to human health. AQI scale runs from 0 to 500. The higher the AQI value, the greater the level of air pollution and the greater the health concern. An AQI value of 50 represents good air quality with little potential to affect public health, while an AQI value over 300 represents hazardous air quality. An AQI value of 100 generally corresponds to the national air quality standard for the pollutant, which is the level EPA has set to protect public health and is believed to be satisfactory.

13.1.4.2 Good and bad ozone at different altitudes

At 50 kilometers above the ground, ozone is good: it filters dangerous ultraviolet radiation which harms life and materials at ground level. At 10 kilometers (6 miles), ozone is a greenhouse gas, as it keeps the planet warm, but bad if the warming continues to increase at too rapid a rate. At 5 kilometers, ozone is a source of the hydroxyl radical which cleanses the atmosphere of pollutants. But at ground level, ozone is a primary pollutant in smog, causing respiratory problems and damaging biosphere.

Table 13.2

Annual Mean Concentration Range (μg/m^3)				
Pollution Level	*Industrial*		*Residential*	
	SO_2 & NO_2	SPM	SO_2 & NO_2	SPM
Low (L)	0-40	0-180	0-30	0-70
Moderate (M)	40-80	180-360	30-60	70-140
High (H)	80-120	360-540	60-90	140-210
Critical (C)	>120	>540	>90	>210

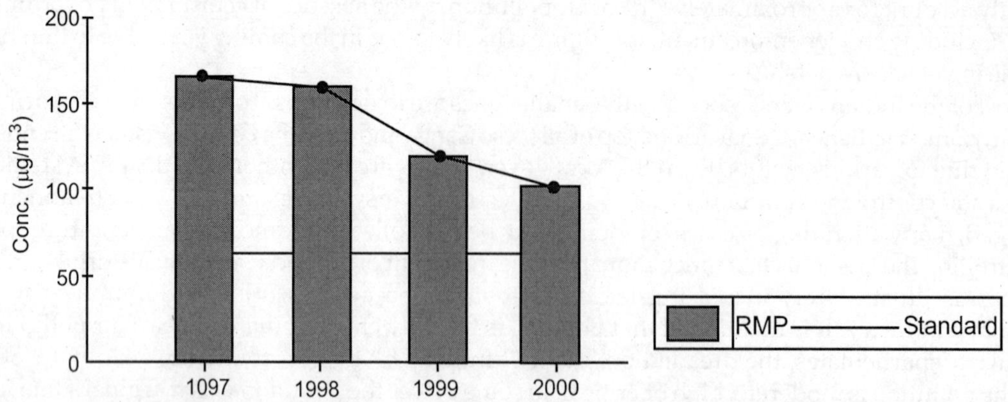

Fig. 13.1 Variation of RPM Concentration (Annual average) in Kolkata during 1997-2000

13.2 EARTH'S CLIMATE

Weather changes continuously, a thunderstorm can bring down the rapid drop in temperature in the atmosphere which otherwise, may be referred to long periods of time subtle and gradual way. Similarly volcanic eruptions, wild fires rapidly change the climate. Since the birth of the Earth, the climate has been continuously changing due to various factors. However the Industrial Revolution has changed the climate rapidly due to the release of different gaseous pollutants like CO_x, SO_x, CH_4, NO_x and fluoro-chloromethanes etc. Since these gases trap the heat in the atmosphere and cause abnormal changes in the Global Temperatures these gases are called Green House gases. Green House gases directly change the surface temperature triggering additional climate changes by evaporating more water, melting glaciers to increase Sea level. However some argue that the melting of glaciers produce more land and increase in the sea level is marginal because of the quantum sea water compared to glaciers and the water bodies on the Earth. However the increasing Green House gases in the atmosphere may bring about rapid changes in the climate, perhaps some effects spell disastrous. The power houses and automobile industry should adopt new techniques to minimize the release of these gases.

One of these climate changes is global warming, an increase in the Earth's average surface temperature. Global warming likely triggers additional climate changes, such as increased evaporation of water, that can change quantity of rain and snowfall.

These climate changes affect both people and the environment, rainstorms are becoming heavier, glaciers are melting, snow cover is decreasing, and sea level is rising. Because human-produced greenhouse gases continue building up in the atmosphere, many scientists project faster and more intense climate changes over the next century that will increase these and other effects. Some effects will he beneficial, while others will be harmful. Energy generation should emit least emissions. In other voluntary programs, EPA works with industries to reduce the emission of very strong greenhouse gases like methane, hydrofluorocarbons, perfluorocarbons and sulphur hexafluoride.

13.2.1 What are Green House Gases?

Some green house gases occur naturally in the atmosphere while others result from human activities. Water vapor, carbon oxides, nitrogen oxides, sulphur oxides, halogens (chlorine) HCs and many gases and particles are increased annually due to increased activity of industrialization Atmosphere reactions with each other or photochemical reactions or simple absorption in water vapor cause many stories. Perhaps carbondioxide, water vapor, methane may not affect the atmosphere much as they are available in nature too. However the quantity of carbon-dioxide generated may not return back to the earth as the photosynthesis by plants is photochemical and slow too. Thus accumulation increases. Oxides of nitrogen continue to return to the earth in the form of nitrates or acids. But this is conditional too, as the photo chemical activity with ozone is very high. Halogens specially chlorine is expected to react with ozone. All these gases, multiatomic gases, absorb heat eventhough not in the same intensity. Nitrous oxide absorbs 200 times more heat than carbon dioxide. The quantity of methane produced from biological depositions and natural formation is continuous as is evident from the following table. All these gases trap heat and cause green house effect.

13.2.2 Global Warming Potentials

Gases in the atmosphere can contribute to the greenhouse effect both directly and indirectly. Indirect radiative forcing occurs when chemical transformations of the original gas produce a gas or gases that are greenhouse gases, when a gas influences the atmospheric lifetimes of other gases, and/or when a gas affects other atmospheric processes that alter the radiative balance of the earth (*e.g.*, affect cloud formation or albedo). The concept of a Global Warming Potential (GWP) has been developed to compare the ability of each greenhouse gas to trap heat in the atmosphere relative to another gas. Carbon dioxide (CO_2) was chosen as the reference gas to be consistent with IPCC guidelines.

Table 13.3

Methane emissions	Tg/Y
Wetlands	150
Cattle	120
Paddy fields	95
Direct anthropogenic	40
Burning biomass	25
Ocean	13
Tundras	12
Lakes	10
Others	88
Total	553 (Tg = 10^{12} gms)

Global warming potentials were not provided for the criteria pollutants CO, NO_x, $NMVOC_s$ and SO_2 because there is no agreed upon method to estimate the contribution of gases that have only indirect effects on radiative forcing. All gases in the table 13.3 presented are in units of teragrams of carbon dioxide equivalents. The relationship between gigagrams of a gas and $TgC0_2$ can be expressed as follows :

$$TgCO_2 = (\text{Gg of a gas}) \times GWP \times (\text{Tg}/\ 1000\ \text{Gg})$$

The GWP of a green house gas is the ratio of global warming or radiative forcing from one unit mass of green house gas to carbon dioxide over a period. IPCC recommended 100 years as time period which was accepted by USA. GWP values are shown in Table 13.4.

Table 13.4: Global Warming Potentials (100 Year Time Horizon)

Gas	*GWP*
Carbon dioxide (CO_2)	1
Methane (CH_4)*	21
Nitrous oxide (N_2O)	310
HFC-23	11,700
HFC-125	2,800
HFC-134a	1,300
HFC-143a	3,800
HFC-152a	140
HFC-227ea	2,900
HFC-236fa	6,300
HFC-4310mee	1,300
CF4	6,500
C2F6	9,200
C4F10	7,000
C6F14	7,400
SF6	23,900

13.3 AMBIENT AIR QUALITY

Ambient Air Qualtiy Standards or criteria are concentrations of pllutants in the air, and typically refer to outdoor air. The criteria are specified for a variety of reasons including for the protection of human health, buildings, crops, vegetation, ecosystems, as well as for planning and other purposes.

The Current National Ambient Air Qualtiy Standards were notified on 18th November 2009 by Central Pollution Control Board (CPCB).

Table 13.5: Ambient air quality standards in India

Pollutant	Time Weighted Average	Concentration in Ambient Air	
		Industrial, Residential, Rural and Other Areas	Ecologically Sensitive Area (notified by Central Government)
Sulphur Dioxide (SO_2), $\mu g/m^3$	Annual *24 hours**	50 80	20 80
Nitrogen Dioxide (NO_2), $\mu g/m^3$	Annual *24 hours**	40 80	30 80
Particulate Matter (size less than 10 μm) or PM_{10} $\mu g/m^3$	Annual *24 hours**	60 100	60 100
Particulate Matter (size less than 2.5 μm) or $PM_{2.5}$ $\mu g/m^3$	Annual *24 hours**	40 60	40 60

Ozone (O_3) $\mu g/m^3$	8 hours* 1 hour**	100 180	100 180
Lead (Pb)$\mu g/m^3$	Annual* 24 hours**	0.50 1.0	0.50 1.0
Carbon Monoxide (CO) mg/m^3	8 hours* 1 hour**	02 04	02 04
Ammonia (NH_3) $\mu g/m^3$	Annual* 24 hours**	100 400	100 400
Benzene (C_6H_6) $\mu g/m^3$	Annual*	5	5
Benzo(a)Pyrene (BaP)- particulate phase only, ng/m^3	Annual*	1	1
Arsenic(As), ng/m^3	Annual*	6	60
Nickel (Ni), ng/m^3	Annual*	20	20

* Annual arithmetic mean of minimum 104 measurements in a year at a particular site taken twice a week 24 hourly at uniform intervals.

** 24 hourly or 8 hourly or 1 hourly monitored values, as applicable, shall be complied with 98% of the time, they may exceed the limits but not on two consecutive days of monitoring.Source: National Ambient Air Quality Standards, Central Pollution Control Board Notification in the Gazette of India, Extraordinary, New Delhi, 18th November, 2009

Table 13.6: Air quality in Bangalore city at different locations and other cities

City and location					
Component (mg/cu.m)	*Ozone*	*SO_2*	*NO_x*	*CO*	*SPM*
National standards		80	80	2000	100
Airport Road (O)	3	15	23	1.2	113
Indian Express (O)	4.3	18	67	3.6	152
Yeshwanthpur NH4 (O)	3.6	15	105	3.5	170
Town Hall (O)	8.7	7	47	3.4	207
Anand Rao Circle, Opp to Shastry hotel (O)	7.7	13	89	3	185
Delhi SSZ Marg		9	63		450
Kanpur, Vikas Nagar		4	35		140
Kolkata, JatinDas Park		4	30		92

13.3.1 State Pollution Control Boards (SPCB)

Various State Pollution Control Boards (SPCB) also monitor air quality and fix their own Ambient Air Qualtiy Standards.

13.3.2 Ambient Air Quality Standards-in West Bengal

Table 13.7

		Concentration in ambient air			
Pollutants	*Tine weighted average*	*Industrial Area*	*Residential Area*	*Sensitive & other*	*Method of measurement*
	2			5	6
Sulphur Dioxide (SO_2)	Annual Average 24 hours	80 μg/m³ 120 μg/m³	60 μg/m³ 80μg/m³	15μg/m³ 30 μg/m³	1. Improved West and Greak method
Oxides of N_2 as NO_2	24 hours	120 μg/m³	80μg/m³	30 μg/m³	Gas phase Chemi luminiscence
Suspended Particulate Matter (SPM)	Annual Average 24 hours	360 μg/m³ 500 μg/m³	140μg/m³ 200 μg/m³	70 μg/m³ 100 μg/m3	High volume sampling (Average flow rate not less than 1.1 m³ minute
Respirable particulate matter (size less 10um (RPM)	Annual 24 hours	120 μg/m³ 150 μg/m³	60 μg/m³ 100 μg/m³	50 μg/m3 75 μg/m³	Respirable particulate matter sampler
Lead (Pb)	Annual Average 24 hours	1.0 μg/m³ 1.5 μg/m³	0.75 μg/m³ 1.00 μg/m³	0.50 μg/m³ 0.75 μg/m³	AAS method after sampling using EPM2000 or equivalent Filter paper
Carbon Monoxide (CO)	8 hours 1 hour	5.0 μg/m³ 10.0 μg/m³	2.0 μg/m³ 4.0 μg/m³	1.0 μg/m³ 2.0 μg/m³	Non dispersive infrared spectroscopy

• Annual Arithmetic mean of minium 104 measurements in a year taken twice a week 24 hours.

13.3.3 Air quality in Tamil Nadu

The Tamil Nadu Pollution Control Board's Chennai Ambient Air Quality Monitoring Programme has shown that the particulate matter in T Nagar traffic intersection was highest at 598 micrograms/ cubic metre (mcg/cm) above the permissible limit of 200 mcg/cm followed by Vallalar Nagar with 344 mcg/cm and Kilpauk - 307 mcg/cm.

The Respirable Dust particle (RDP) level was high at T. Nagar - 113 mcg/cm, Kilpauk- 83 mcg/cm and vallalar Nagar- 78 mcg/cm.

13.3.4 Air Quality in Indian cities in 1997

Table 13.8: Air Quality in Indian cities in 1997 (based on measurements of Total Suspended Particulate Matter)

Level of Air Quality	*No. of towns with air quality round the year*	*Percentage of town where air quality is monitored*	*No. of towns with air quality on most polluted day*	*Percentage of towns where air quality is monitored*
Critical Pollution (above 1.5 times the standard)	33 (Vadodara, Mumbai Pune, Ludhiana, Lucknow, Jaipur, Haldia, *Indore, Ahmedabad, Ghaziabad, Bhopal Jodhpur, Gajraula, Chandigarh, Dehradun, Anpara, Gobindgarh, Guwahati, Vapi, Howrah, Alwar, Delhi, Ankleshwar, Agra, Kanpur, Faridabad, Jharia, Patna, Surat	47%	61 (Yamunanagar, Nagpur, Hyderabad, Thiruvananthapuram, Talcher, Kota, Chittor, Korba, Rourkela, Bhilai, Raipur, Bangalore,Satna, Nasik, Paonta Sahib, Dhanbad, Jalandhar, Pondicherry, Sindri, Sholapur, Visakhapatnam, Udaipur, vadodara, Mumbai, Pune, Ludhiana, Lucknow, Jaipur Haldia, Indore, Ahmedbad, Ghaziabad, Bhopal, Jodhpur, Gajraula, Chandigarh, Dehradun, Anpara, Gobindgarh, Guwahati, Vapi, Howrah, Alwar, Delhi, Ankleshwar, Agra, Kanpur, Faridabad, Jharia, Patna, Surat, Kochi, Tuticorin, Vasco, Damtal, Angul, Chennai, Coimbatore, Parwanoo, Nagda, Dombivali)	87%
High Pollution (be tween 1 and 1.5 times the standard)	18 (Yamunanagar, Nagpur, Hyderabad, Thiruvananthapuram, Talcher, Kota, Chittor, Korba, Rourkela, Bhilai, Raipur, Bangalore, Satna, Nasik, Paonta Sahib, Dhanbad, Jalandhar, Pondicherry, Sindri, Sholapur, Visakhapatnam, Udaipur)	26%	6 (Shimla, Rayagada, Mysore, Chanderpur, Jamshedpur, Ponda)	9%

Moderate Pollution (between 0.5 and 1 times the standard)	17 (Kozhikode, Jabalpur, Shimla, Kochi, Vasco, Rayagada, Damtal, Angul, Mysore, Chennai, Coimbatore, Chanderpur Jamshedpur, Ponda, Parwanoo, Nagda, Dombivali)	24%	2 (Kozhikode, Jabalpur)	3%
Clean Air (below 0.5 times the standard)	2 (Shillong, Tuticorin)	3%	1 (Shillong)	1%
TOTAL	70	100%	70	100%

Source : 1997 Air quality data from the Central Pollution Control Board.

13.4 WATER POLLUTION AND GROUND WATER SOURCES

Hazardous waste sites, septic tanks, municipal dumps, mining and metallurgical rejects, radio active ore residues, coal washery rejects, detergents, organic chemicals, oils, greases and every illegal dump would contaminate the ground over certain time if not immediately. Urbanistion has a fear of acute potable water supply. Advection is the movement of contaminants with the flowing ground water at the seepage velocity in the pores and is represented by Darcy's law :

$Vx = (K/\varepsilon)(dh/L)$ where K is hydraulic conductivity (m/s) dh pressure head, ε is porosity and L length, m.

Safe drinking water Act requires each State to submit a well head program of potable ground water. A[2] minimum and maximum set back zones are laid to restrict the land use near the well. Set back zone provides a buffer between well and potential contamination sources and routes. It gives clean up efforts of contaminated ground water to obtain an alternate water supply before the existing source becomes dirty. These pollutants are slowly carried away by seepage water. The Earth's adsorption capacity, filtration capacity are the only important parameters to determine how much contaminant goes into the well. Some chemical, biological reaction do proceed and they can alter the nature of contaminant structure as well as its toxicity. Minimum and maximum suspended particles, floating debris/solids, physical and biological residues, industrial dumps, underground storage tank seepages and radio active mineral, used catalysts dumps and many more are the real polluting components of ground water.

The distribution and concentration of contaminants in ground water is strongly influenced by interaction between the contaminants and physical, chemical, biological components of sub surface. The interaction of these contaminants with these processes ultimately change them to non hazardous compounds. This is known as Fate. Hydrophilic and hydrophobic materials will be partitioned in to aqua phase. In aqua phase the living organism interact by bio process known as abiotic, whereby hydrolysis, oxidation-reduction or elimination reactions can influences the fate[3]. In acute urbanizations sewage treatment plants and septic tanks are major problems. Advection is the phenomenon of movement of all the contaminants with the flowing water at the seepage velocity. The pore velocity Vx is more than the actual velocity found by Darcy's method. The discharge from a nearby septic tank to ground water well can be calculated by using the following equation.

$$Q = \frac{KH(P_e - P_w)}{\mu Ln(r_e / r_w)(1 - d^2 / r_e^2)}$$

where P_e–P_w is the pressure difference between the wells, μ is the viscosity of the fluid, d/r is the ratios of distance between the well and tank (drainage area) *re, rw* is the radius of the well.

13.4.1 Biological Oxygen Demand, BOD

This[5] is defined as the amount of oxygen required by bacteria while stabilizing decomposable organic matter under aerobic conditions. This test is one of the most important and pertinent test in stream-pollution-activities. However biological oxidation in nature may take infinite time *i.e.* to convert the organic matter to carbon dioxide, water and ammonia, which is definitely not possible in laboratory tests. Even 20 days time is also too long, though incubation for twenty days gives highest conversion. But for quick assessment five day incubation is performed which is equivalent to 70-80 % of BOD requirement of 20 days. The oxidation is due to dissolved oxygen in the water. At room temperature (25°C) when chloride ion concentration is zero, oxygen saturation in water is 8.4 mg/liter, in sea water it comes down to 6.7 mg/liter. In liquid water dissolved oxygen determines whether biological changes are brought about by aerobic or anaerobic organisms. The former bacteria consumes the organic material for its metabolic processes while the later in the absence of oxygen, carry out digestion of complex organic matter through h alcohols, acids which later convert into carbon dioxide, methane and water. Even the wastes may contain poisons like sulphur, lead and other inorganics usually the bacteria do not die but adjust to the environment. The advantage of such process is that the reaction can be conducted in depths without large surface area and less expensive too. This is very much noticed in the petroleum formation.

13.4.2 Chemical Oxygen Demand, COD

This is a measure of organic strength of domestic and industrial wastes. Measurement waste in terms of total quantity of oxygen required for oxidation of the matter to carbon dioxide and water. Chief limitation of COD is its inability to differentiate between biodegradable and biological inert organic matter. In addition it does not provide any evidence of the rate at which the biologically active material would be stabilized under conditions existing in nature. Usually 3-hr time is sufficient for estimation of biodegradability of the wastes.

13.4.3 Water pollution

Sweet and soft water as such is insufficient. In addition, the water sources are polluted at a very fast rate. The more the population and industrialization, more the pollution, industry and municipal corporations are mainly responsible. The sewage and industrial wastes without treatment is discharged into rivers and water bodies. The contaminants are plenty, right from organic to inorganic, heavy metals to chelates. Supplementing these big sources, live stock pollution cannot be forgotten. Though it has not attained a recognizable proportion, but likely in future. Dumping, wastes of live stocks, washing clothes, animals, machines etc though isolated instances when the population of the live stocks and meager water resources are taken into account it would be definitely reckoning.

In Water pollution by industry or by municipalities the bacteria and chemicals play vital role. Industrial wastes must be regarded as a subject itself as different industries send different effluents with different concentrations of chemicals. Many are poisionous, toxic and reactive too. They may spoil marine life, they may contaminate the soil and make unfit for cultivation.

Most of the sewage plants are based upon oxidation and reduction systems. These are carried out bacteria. When the electron activity is low eq. 1 follows, if the medium is oxidizing eq. 2 follows :

$$2H_2O + 2e \rightarrow H_2 \text{ gas} + 2\,OH \quad \text{... 1}$$

$$2\,H_2O \rightarrow O^2 + 4H^+ + 4e \quad \text{...(2)}$$

Chelating agents are potential water pollutants. These are found in sewage water and industrial wastes. Chelating agents have more than one atom that can be bonded to central metal ion at one time to form ring structure. EDTA, nitilo triacetate (NTA) are some examples. These are very stable compared to complexes. Coagulation is required to bind colloidal particles 0.001 to 1 urn. Autotrophic bacteria utilizes solar heat or chemical energy synthesizing organic matter from inorganics. Heterotrophic organism utilize organic substance produced by autotrophics for synthesis of their own biomass, algae, fungi are some of the examples.

13.5 LAND POLLUTION

Ever since the Earth was born many geological, environmental changes took place. The changes are continuing. Sudden changes in the topography due to volcanic eruptions, earth quakes, meteorite collisions, excessive floodings, do occur and these are natural, however some man made changes like, diversion of rivers, demolishing mountains, huge unnatural reservoirs, mining, activities, atomic explosions are some prominent reasons for changes in land and environments. In all these processes new generation of chemical atmosphere is born. Most of the chemicals in excessive doses become toxic. There are a number of ways in which man may be exposed to chemicals. Primary routes of exposure are dermal, oral and respiratory system. Highly volatile substances has high impact on respiratory system. Many hazardous things are taken through food. Occupational exposure involves inhaling vapors, dust and exposure to chemicals through skin. There is no special way description of land pollution than what is air and water pollution. Any material if soluble it is carried away by water. If the pollutant is biodegradable bacterial action predominates. The pollutant if adsorbed by soil and depending upon the adsorption coefficient and retention coefficient the material gradually goes into water streams, the accumulation in soil also dominates. Again physical, chemical, biological reactions singly or serially follow. The catalytic nature of earth cannot be forgotten. The pressure temperature and presence of water vapour all accelerate the decomposition. Only bio refractory materials like aromatics, chlorohydrocarbons, polyaromatics etc are refractory. These depending upon Henry's constant escape into atmosphere due to fractionation. Adsorption and/partition properties in non aqueous liquids also directly connect biotic and abiotic solids. The most pertinent process through which transportation, degradation, bio availability are resulted. Cometabolism is a term used to describe the partial transformation of an organic substrate by micro organism that cannot use the substrate for growth or energy.

Municipal dumps and industrial wastes are to be treated before they are let down into streams or lands. Every industrial effluent will be given a special method of tackling. Oil, leather, mining, paper, metallurgical industries adopt the suitable technologies to destroy the chemical activity of the substances. Some times toxic (extremely toxic 1 mg/kg weight oral administration, 1-50 mg/kg weight is highly toxic, moderately toxic upto 500 mg/kg weight) wastes are given special treatments.

Preliminary examination reveal the large amount of solids, wood, carcass etc are to be hand picked. Then filtration, clarification, adsorption with charcoal or other absobent, finally disinfection is required for sewage and municipal wastes. Pollution due to live stock cannot be ignored. The contact, bites, scratches are some ways of diseases spreading from animals and humans.

Deforestation ; soil erosion ; overgrazing ; desertification ; air pollution from industrial effluents and vehicle emissions ; water pollution from raw sewage and industrial wastes are some environmental effects on land, water and atmosphere. El Nilo effect and Albedo effect are direct result of these pollutions.

Sewage and trade effluents discharged into streams, Indian Standards Institution and many countries have formulated their own standards. These are shown in tables 13.7.

Similarly Air pollution standards are also prescribed depending upon the need of the country or locality. Such Tables are presented in 13.8.

Table 13.9: Standards for sewage and trade effluent discharged into streams

S.No.	*Characteristics*	*General tolerance limit*	*Relaxation for specific industries*
1.	Total suspended solids, mg/1, max	–	–
2.	Particle size of total suspended solids	Shall pass 850 micron I.S. sieve	–
3.	pH	5.5. to 9.0	5.5 to 9.0 of nitrogenous fer-tiliser industry
4.	Temperature	Shall not exceed 40°C in any section of the stream within 15 meters down stream from the effluent outlet	
5.	BOD for 5 days at 20°C, mg/l. max.	30	100 for existing distilleries and for existing Straw Board industry
6.	Oil and grease mg/l, max.	10	—
7.	Phenolic compounds (as (C_6H_5OH) mg/l. max.	1.0	—
8.	Cyanides (as CN) mg/l, max.	0.2	—
9.	Sulphides (as S) mg/l, max.	2.0	—
10.	Insecticides	absent	—
11.	Colour and odour	absent	absent as far as possible for paper industry
12.	Total Residual, chlorine mg/l, max.	1.0	—
13.	Flouride (as F) mg/l, max.	2.0	5 for phosphatic fertiliser industry
14.	Cadmium (as Cd) mg/l, max.	2.0	—

15.	Copper (as Cu) mg/l, max.	3.0	—
16.	Total Chromium (as Cr) mg/l, max.	0.5	—
17.	Haxavalent Chromium	0.1	5 for phosphatic fertiliser industry
18.	Lead (as pb) mg/l max.	0.1	—
19.	Mercury (as Hg) mg/l, max.	0.01	—
20.	Nickel (as Ni) mg/l, max.	3.0	—
21.	Selenium (as Sc) mg/l, max.	0.05	—
22.	Zinc (as Zn) mg/l, max.	5.0	5 for phosphatic fertiliser industry.
23.	Ammoniacal nitrogen (as N)	50	100 for Nitrogenous fertiliser industry
24.	Free ammonia (as NH_3)	5.0	—
25.	Total nitrogen (as N) mg/l. max.	100	150 for Nitrogenous fertiliser industry
26.	Chemical Oxygen demand, mg/l, max.	250	—
27.	Dissolved phosphate (as P) mg/l, max.	5.0	—
28.	Chlorides (as Cl) mg/l, max.	1000	—
29.	Radio active materials	10^{-7}	—
	Alpha emitters µc/*ml*, max.	10^{-8}	—
	Beta emitters µc/*ml*, max.	0.2	—
30.	Arsenic (as As) mg/l, max.		1.0 for existing plants of nitrogenous fertiliser industry

Table 13.10: Standard for the effluents to be discharged onland

1.	pH	5.5-9.0	—
2.	Total dissolved solids mg/l, max.	2100	—
3.	Sulphates (as SO_4) mg/l, max.	1000	—
4.	Chlorides (as Cl) mg/l, max.	600	—
5.	Per cent sodium, max.	60	—
6.	Biochemical Oxygen Demand for 5	200	500 for existing distillery up to 1984 only
7.	Oil and grease mg/l, max.	20	—

8.	Boron (as B) mg/l, max.	2	—
9.	Cyanides (as CN) mg/l, max.	0.5	—
10.	Total chromium (as Cr) mg/l, max.	2.5	—
11.	Zinc (as Zn) mg/l, max.	10	—
12.	Ammoniacal nitrogen (as N) mg/l, max.	100	—
13.	Copper (as Cu) mg/l, max.	3.0	—
14.	Nickel (as Ni) mg/l, max.	3.0	—
15.	Total mercury (as Hg) mg/l, max.	0.01	—
16.	Selenium (as Se) mg/l, max.	0.1	—
17.	Lead (as Pb) mg/l, max.	0.5	—
18.	Arsenic (as As) mg/l, max.	5.0	—
19.	Cadmium (as Cd) mg/l, max.	5.0	—
20.	Radioactive materials	10^{-9}	—
	Alpha emitters $\mu c/ml$, max.	10^{-8}	—
	Beta emitters $\mu c/ml$, max.		—

Table 13.11: Swiss standards for water and liquid effluents

		Tolerance limit		
S. No.	*Effluent Characteristics*	*Effluent discharged into*		
		River water & industrial/water	*Surface water*	*Open Sea water*
1.	TSS, mg/l, max.	—	20	—
2.	Temperature °C, max.	25	30	60
3.	pH	—	6.5-8.5	6.5-9.0
4.	BOD(5 days for 20°C)	4	20	*
5.	COD mg/l. max.	not worth	*	*
6.	Dissolved oxygen (as O_2) mg/l, min	6	*	*
7.	Oils, fats and greases mg/l, max.	not worth	20	—
8.	Ammoniacal nitrogen (as N) mg/l, max.	0.5	*	—
9.	Free ammonia (as N) mg/l, max.	0.1	—	—
10.	Nitrates as (NO_2) mg/l, max.	25	—	—
11.	Sulphates (as SO_4) mg/l, max.	100	*	300

12.	Sulphides (as S) mg/l, max.	—	0.1	1
13.	Fluorides (as F) mg/l, max.	1	10	10
14.	Chlorides (as Cl) mg/l, max.	100	*	1
15.	Cyanides (as CN) mg/l, max.	0.01	0.1	0.5
16.	Lead (as Pb) mg/l, max.	0.05	0.5	0.5
17.	Zinc (as Zn) mg/l, max.	0.2	2	2
18.	Copper (as Cu) mg/l, max.	0.01	0.5	1
19.	Nickel (as Ni) mg/l, max.	0.05	2	2
20.	Cadmium (as Cd) mg/l, max.	0.005	0.1	0.1
21.	Mercury (as Hg) mg/l, max.	0.001	0.01	0.01
22.	Arsenic (as As) mg/l, max.	0.01	0.1	0.5
23.	Iron (as Fe) mg/l, max.	1	2	20
24.	Phenolic compounds (as C_6H_5OH) mg/l, max.			
	(*a*) with water vapour	0.005	0.05	5
	(*b*) without water vapour	0.005	0.05	1
25.	Residual chlorine (as Cl) mg/l, max.	—	0.05	0.5-3
26.	Barium (as Ba) mg/l, max.	0.5	5	*
27.	Tetravalent chromium (as C_r^4) mg/l max.	0.01	0.1	0.5
28.	Trivalent chromium (as C_r^3) mg/l, max.	0.05	2	2
29.	Dissolved organic carbon substances (as C) mg/l, max.	2	10.15	*

* Decided case by case.

A major contributory factor to the present calmer and more thoughtful approach toward the problems of the environment has been the creation of the UN Environment Program (UNEP), as well as the involvement of specialised UN agencies and non-governmental organisations such as the International Council of Scientific Unions (ICSU) and its Scientific Committee on Problems of the Environment (SCOPE).

Table 13.12: Nitroarenes in the Environment

Nitroarene	Source	Cone. (*ppb*)[b]
1.3-Dinitropyrene	Diesel particles	600
	Heavy duty diesel	1600
	Gas heater emission	600
	Carbon black	6300

	Diesel particles	810
1.6-Dinitropyrene	Heavy duty diesel	1200
	Atmospheric particles. Michigan	310
	Home gas heater	1880
1.8-Dinitropyrene	Heavy duty diesel	3400
	Atmospheric particles, Michigan	460
	Photocopier toner	300
6-Nitroben -zo[a]pyrene	Diesel car exhaust	50,000
	Gasoline-powered car. leaded fuel	32,800
	Gasoline-powered car, unleaded fuel	17,300
	Home fireplace smoke	120[d]
2-Nitrofluorene	Atmospheric particles, Tokyo	21,800[d]
l-Nitronaphthalene	Diesel exhaust	700[d]
	Atmospheric particles, Philadelphia	3,500
1-Nitropyrene	Diesel car, 1979 model	2,030,000
	Gasoline-powered car	2500
	Used diesel oil	500[d]
	Home fire place smoke	110[d]
	Atmospheric particles, Bermuda	720[d]

[a]IARC, 1989.
[b]Concentration in solvent extract unless otherwise noted
[d]Made prior to 1980.
[d]Concentration in substrate rather than in extracts.

Table 13.13: Nitrosamines in the Environment[a]

Nitrosamine	Source	Maximum Cone, (*ppb*)
Nitrosodimethylamine	Fried bacon (Germany)	12
	Ham (Germany)	11
	Dried fish (Japan)	34
	Pilsner beer (Germany)	7
	Pale ale (England) (Kind of beer)	8
	Cured tobacco	16
	Cigarette smoke, U.S. mainstream	43
	Cigarette smoke, U.S. sidestream	1770
	Rubber toys (Germany)	25
Nitrosodimethylamine	Toy balloons (Germany)	150
	Air in a bar	0.2B
Nitrosopyrrolidine	Fried bacon (U.S.)	139
	Ham (Germany)	36
	Sausage (Canada)	105

Nitrosonornicotine	Cigarettes (U.S.) Small cigar (U.S.) Snuff (U.S.)	1.7 45 39
Nitrosodiethanolamine	Cosmetics (U.S.) Cigarettes (US.) Shampoo (US.)	48,000 194 260
Nitrosopiperidine	Nursing bottle nipples (Germany) Rubber gloves (Germany)	100 100

[a]Searle, 1984: IARC, 1978.
[b]Concentration in $\mu g/m^3$.

Table 13.14: Some Toxic Constituents of Hazardous Waste[a]

Constituent	*Regulatory level (mg/L)*	*Constituent*	*Regulatory level (mg/L)*
Arsenic	5.0	Hexachloro-1, 3-butadiene	0.5
Barium	100.0	Hexachloroethane	3.0
Benzene	0.5	Lead	5.0
Cadmium	1.0	Lindane	0.4
Carbon tetrachloride	0.5	Mercury	0.2
Chlordane	0.05	Methoxychlor	10.0
Chlorobenzene	100.0	Methyl ethyl ketone	200.0
Chloroform	6.0	Nitrobenzene	2.0
Chromium	5.0	Pentachlorophenol	100.0
o-Cresol	200.0	Pyridine	5.0
m-Cresol	200.0	Selenium	1.0
p-Cresol	200.0	Silver	5.0
2, 4-D	10.0	Tetrachloroethylene	0.7
1,4-Dichlorobenzene	7.5	Toxaphene	0.5
1,2-Dichloroethane	0.5	Trichloroethylene	0.5
1,1-Dichloroethylene	0.7	2,4,5-Trichlorophenol	400.0
2,4-Dinitrotoluene	0.13	2,4,6-Trichlorophenol	2.0
Endrin	0.02	2,4,5-TP (Silvex)	1.0
Heptachlor and exoxide	0.008	Vinyl chloride	0.2
Hexachlorobenzene	0.13		

13.6 PREPARATIONS AND MIXTURES

Pure substances can, by the systems set out above, be labelled with symbols and Risk and safety-phrases to indicate their toxicity in a relatively simple way. Preparations, *i.e.* substances which are mixtures prepared by the manufacturer from two or more 'pure' substances, present more difficult problems. Ideally, it would be desirable to evaluate the toxicity of each preparation by the methods set out in this book. This has in fact been applied in one specific area. *i.e.* pesticides, toxicity testing is always undertaken. Unfortunately this technique would prove impossibly expensive and time consuming if it were universally applied to the very large number of other preparations on the market, particularly in view of the frequency with which their composition is changed. A simpler first approach was therefore required, but a suitable system has proved very difficult to develop given possible additive and synergistic effects of the component parts.

The Commission of the European Communities and developed what can best be described as a compromise solution and which is perhaps an over-simplification of the problem. It ascribes classes of toxicity to each individual component and attempts to combine these by the use of a summation formula to give an overall indication of hazard by which the classification of the mixture is determined.

A substance other than an organic peroxide which, although not itself necessarily combustible, may by yielding oxygen or by a similar process cause or contribute to the combustion of other material.	OXIDIZING AGENT Yellow Background **Fig. 13.4:** Oxidizing substance
A substance which is— (*a*) an organic peroxide ; and (*b*) an unstable substance which may undergo exothermic self-accelerating decomposition.	ORGANIC PEROXIDE Yellow Background **Fig. 13.5:** Organic peroxide.
A substance known to be so toxic to man as to afford a hazard to health during conveyance or which, in the absence of adequate data on human toxicity, is presumed to be toxic to man.	TOXIC GAS White Background **Fig. 13.6:** Toxic substance.
A substance known to be toxic to man or, in the absence of adequate data on human toxicity, is presumed to be toxic to man but which is unlikely to afford a serious acute hazard to health during conveyance.	HARMFUL - Stom Away from Foodstuffs White Background **Fig. 13.7:** Harmful substance.

A substance which by chemical action will- (*a*) cause severe damage when in contact with living tissue ; (*b*) materially damage other freight or equipment if leakage occurs.	CORROSIVE White Background **Fig. 13.8:** Corrosive substance.
A substance which is listed in Part 1A of the approved list and which may create a risk to the health or safety of persons in the conditions encountered in conveyance by road, whether or not it has any of the character-properties set out above.	DANGEROUS SUBSTANCE White Background **Fig. 13.9:** Other dangerous substance
Packages containing two or more dangerous substances which have different characteristic properties.	DANGEROUS SUBSTANCE White Background **Fig. 13.10:** Mixed hazards.

Classification and symbols used for substances dangerous for conveyance

1	2	3
Characteristic properties of the substance	*Classification*	*Hazard Warning Sign.*
A substance which — (*a*) has a critical temperature below 50°C or which at 50°C has a vapour pressure more than 3 bar absolute; and (*b*) is conveyed by road at a pressure of more then 500 millibar above atmospheric pressure or in liquified from.	Non-flammable compressed gas	COMPRESSED GAS Green Background **Fig. 13.11**
Other than a toxic gas or a flammable gas. A substance which has a critical temperature below 50°C or which at 50°C has a vapour pressure of more than 3 bar absolute and which is toxic.	Toxic gas	TOXIC White Background **Fig. 13.12**

A substance which has a critical temperature below 50°C or which at 50°C has a vapour pressure or more than 3 bar absolute and is flammable.	Flammable gas	FLAMMABLE GAS Red Background **Fig. 13.13**
A liquid with a flash point of 55°C or below except a liquid which— (*a*) has a flash point equal to or more than 21°C and less than or equal to 55°C ; and (*b*) when tested at 55°C in the manner described in Schedule 2 to the Highly Flammable Liquids and Liquefied Petroleum Gases Regulations 1972 (*a*) does not support combustion.	Flammable liquid	FLAMMABLE LIQUID Red Background **Fig. 13.14**
A solid which is readily combustible under conditions encountered in conveyance by road or which may cause or contribute to fire through friction.	Flammable solid	FLAMMABLE SOLID Red Hatched Background **Fig. 13.15**
A substance which is liable to spontaneous heating under conditions encountered in conveyance by road or to heating in contact with air being then liable to catch fire.	Spontaneously combustible substance	SPONTANEOUSLY COMBUSTIBLE Red Lower Background **Fig. 13.16**
A substance which in contact with water is liable to become spontaneously combustible or to give off a flammable gas.	Substance which in contact with water emits flammable gas	SUBSTANCE FLAMMABLE Blue Background **Fig. 13.17**

Table 13.15: Air Pollution Facts & Toxic Chemicals

Administrator of the Environmental Protection Agency is empowered to bring suit on behalf of the United states in a Federal district court when air pollution presents an imminent and substantial endangerment to the health of persons

Note: A table showing the danger levels is attached.

Air Pollution Levels That Could Cause Significant Harm to the Health of Persons

Sulphur dioxide

2620 micrograms/cubic meter (1.0 part per million), 24-hour average

Particulate matter

1,000 micrograms/cubic meter or 8 COHs, 24-hour average

Sulphur dioxide and particulate matter combined. Product of sulphur dioxide in micrograms/cubic meter, 24-hour average, and particulate matter in micrograms/cubic meter, 24-hour average, equal to 490 x 10^3 or product of sulphur dioxide in parts per million, 24-hour average and COHs, 24-hour average, equal to 1.5.

Carbon monoxide

57.5 milligrams/cubic meter (50 parts per million). 8-hour average
86.3 milligrams/cubic meter (75 parts per million), 4-hour average
144 milligrams/cubic meter (125 parts per million), 1-hour average

Photochemical oxidants

800 micrograms/cubic meter (0.4 part per million), 4-hour average
1,200 micrograms/cubic meter (0.6 part per million), 2-hour average
1,400 micrograms/cubic meter (0.7 part per million), 1-hour average

Nitrogen dioxide

3,750 micrograms/cubic meter (2.0 parts per million), 1-hour average
938 micrograms/cubic meter 0.5 ppm - 24 hour

Table 13.16: Thresh hold Limit Values for some chemicals in the environment

Component	*TLV ppm*	*mg/m³*	*STEL≅*	*mg/m³*	*@ Short Time Exposure Limit 15 mts*
Ammonia	25	17.4	35	35	
Benzene	0.6	1.0	0.2	0.64	
CO	50	57.3			
Cl_2	0.5	1.5	1.0	2.9	
$CHCl_3$	10	49			
HCN	11 mg/m³	5	5	1600 ppm for 20 mts death/ 12800 ppm immediate death	

CO	35	25	200 ppm 50 days or 8 hr weighted average at 300 ppm		
NaCN	5	5	5		
CaCynamide		0.5	0.5		
Cynogen		20	0.25		

CCl_4 at 1 ppm will be equivalent to 5 billion kgs is troposphere

Trimethylbenzene

SYNONYM(s): 1, 2, 4-Trimethylbenzene; 1,2,3,-Trimethylbenzene; 1,3. 5-Trimethylbenzene; Hemimellitene; Pseudocumene; Mesitylene; Diesel Exhaust Component ; Flammable Liquid. MW: 120.20 Exposure Limits : OSHA CONSTRUCTION INDUSTRY PEL: 25 ppm, 120 mg/m^3 TWA : ACGIH TLV: 25 ppm, 123 mg/m^3 TWA

NIOSH REL : 25 ppm TWA

Health Effects Irritation-Eye, Nose, Throat, Skin-Marked (HE 14) Cumulative CNS effects (HE7): Anemia (HE12)

Chemistry of emissions. Several of the chemicals identified in the emissions of brands "A" and "C" have known toxic effects, anisaldehyde induces increased secretion of fluid and mucus in the respiratory tract of rabbits. The trimethylbenzene isomers are respiratory irritants, and styrene inhalation can damage the respiratory tract mucosa. Toluene, ethylbenzene, xylene, and dipentene have other documented toxic effects.

Toxic chemicals[#]

Dioxins are generated when 0-Chloro phenols are heated with alkali metal salts to high temperature

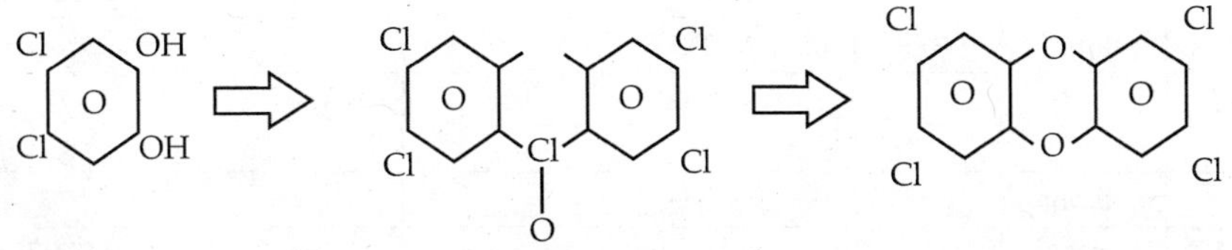

Chloro phenol Predoxin Dioxion TCDD

Fig. 13.18

There are 75 possible polychlorinated dibenzodioxin and 135 chlorinated dibenzofurans. 2, 3, 7, 8 Tetrachlorodibenzodioxin (TCDD) is the most toxic.

Table 13.17: ATSDR Minimal Risk Levels (MRLs)
December 2001

Factors	End-point	Final	Date	CAS	Route Number	Duration	MRL
ACENAPH THENE					Oral	Int. 0.6 mg/kg/day	
300	Hepatic	Final	08/95	000083-32-9			
ACETONE				Inh.	Acute	26 ppm	9
Neurol.	Final	5/94	000067-64-1		Int.	13 ppm.	100

Neurol.					Chr.	13 ppm	100
Neurol.				Oral	Int.	2 mg/kg/day	100
Hemato.							
ACROLEIN				Inh.	Acute	0.00005 ppm	100
Ocular	Final	12/90	000107-028				
					Int.	0.000009 ppm	1000 Resp.
				Oral	Chr.	0.0005 mg/kg/day	100
Hemato.							
ACRYLONITRILE				Inh.	Acute	0.1ppm	10
Neurol.	Final	12/90	000107-13-7				
				Oral	Acute	0.1 mg/kg/day	100
Develop.							
					Int	0.01 mg/kg/day	1000
Repro							
					Chr	0.04 mg/kg/day	100
Hemato.							
ANTHRACENE				Oral	Int	10mg/kg/day	100
Hepatic	Final	08/95	000120-12-7				
ARSENIC			Provisional	Oral	Acute	0.0005 mg/kg/day	10
Gastro.	Final	09/00	007440-38-2				
					Chr.	0.0003 mg/kg/day	3
Dermal Neurol							
BIS(2-CHLOROETHYL) ETHER				Inh	Int	0.02 ppm	1000 Body
Wt.	Final	12/89	000111-44-4				
BROMODICHLOROMETHANE				Oral	Acute	0.04 mg/kg/day	1000
Hepatic	Final	12/89	000075-27-4				
					Chr.	0.02 mg/kg/day	1000 Renal
BROMOFORM				Oral	Acute	0.6 mg/kg/day	100
Neurol.	Final	12/90	000075-25-2		Chr.	0.2 mg/kg/day	100

Hepatic							
BROMO-METHANE				Inh.	Acute	0.05 ppm	100
Neurol.	Final	09/92	000074-83-9				
					Int.	0.05 ppm	100
Neurol							
					Chr.	0.005ppm	100
Neurol							
				Oral	int.	0.003 mg/kg/day	100
Gastro							
CARBON Disulphide				Inh.	Chr.	0.3 ppm	30
Neurol	Final	08/96	000075-15-0	Oral	Acute	0.01 mg/kg/day	300 Hepatic
Carbon Tetra Chloride				Inh.	Acute	0.2 ppm	300
Hepatic	Final	05/94	000056-23-5		Int	0.05 ppm	100
Hepatic				Oral	Acute	0.02 mg/kg/day	300
					Int.	0.007 mg/kg/day	100
					Chr.	1mSv/yr	3 other
CHLOR OBENZENE				Oral	Int	0.04 mg/kg/day	100
Hepatic	Final	12/90	000108-90-7				
CHLORODIBROM OMETHANE				Oral	Acute	0.04 mg/kg/day	1000 Renal
	Final	12/90	000124-48-1				
Hepatic					Chr.	0.03mg/kg/day	1000
CHLOROETHANE				Inh	Acute	15 ppm	100
Develop.	Final	12/98	000075-00-3				
CHLORO-METHANE				Inh.	Acute	0.5 ppm	100
Neurol.	Final	12/98	000074-87-3		Int.	0.2 ppm	300
Hepatic							
Neurol					Chr.	0.05 ppm	1000
COBALT				Inh.	Chr.	0.0001 mg/m^3	10 Resp

Draft	*09/01*	*007440-48-4*		*Oral*	*Int.*	*0.01 mg/kg/day*	*100*
Hemato.				*Rad.*	*Acute*	*4mSv*	*3*
Develop.					*Chr.*	*1 mSv/yr*	*3 other*
CRESOL, META-				*Oral*	*Acute*	*0.05 mg/kg/day*	*100 Resp.*
Final	*07/92*	*000108-39-4*					
CRESOL, ORTHO-				*Oral*	*Acute*	*0.05 mg/kg/day*	*100*
Neurol.	*Final*	*07/92*	*000095-48-7*				
CRESOL, PARA-				*Oral*	*Acute*	*0.05 mg/kg/day*	*100*
Neurol.	*Final*	*07/92*	*000106-44-5*				
DI (2-ETHYL-HEXYL) PHTHA LATE			*Prov.*	*Oral*	*Int.*	*0.01 mg/kg/day*	*300*
Develop.	*Draft*	*09/00*	*000117-81-7*				
DI-N-BUTYL PHTHALATE				*Oral*	*Acute*	*0.5 mg/kg/day*	*100*
Develop.	*Final*	*09/01*	*000084-74-2*				
DI-N-OCTYL PHTHALATE				*Oral*	*Acute*	*3mg/kg/day*	*300*
Hepatic	*Final*	*09/97*	*000117-84-0*				

Table 13.18

Ethanol	1000	1885		
Formaldehyde	2	2.5		
Gasoline	300	890	500	1480
Methanol	200	262	250	328
Ozone	0.05	0.1	0.2	0.4
Trichloroethylene	50	269	100	53.6
Vinylchloride	5	13		

DDT 16 mg (oral dose) convulsions; Long term occupational exposures to 42 mg/day Technical DDT produce neurological problems. Because of the toxic nature many countries stopped using this chemical or allowed only restricted use. Since 1970 the human fat showed remarkable drop in DDT quantity from 10 ppm to 1 ppm in 1990. Similar is the case with BHC from 0.08 to 0.05 ppm in the same period.

Ethylene oxide : Colorless liquid or gas: bp 10.7°C, mp - 111.3°C

Miscible with water. Sweet odor detectable at 257 to 690 ppm (mean = 420 ppm) Vapor pressure; 1095 mm Hg at 20°C : Flash point : - 20°C : Autoignition Temperature 429°C

Toxicity Data

LD oral (rat) 72 mg/kg LC inhal (rat) 800 ppm(1600mg/m)
PEL (OSHA)1ppm (2mg/m)
TLV-TWA(ACGIH)1 ppm (2mg/m)

Major Hazards OSIIA "select carcinogen":

Toxicity : Ethylene oxide is a severe irritant to the eyes, skin, and respiratory tract and exhibits moderate acute toxicity by all routes of exposure Symptoms of overexposure by inhalation may be delayed and can include nausea, vomiting headache, drowsiness, and difficulty breathing.

13.7 NOISE POLLUTION

13.7.1 Introduction

Noise is unacceptable sound that creates annoyance due to distractions and diversions some times responsible for physical damages of health and property. Noise has long been recognized as one of the most prevalent workplace hazards. It was identified by NIOSH as one of the 10 leading occupational problems in its consensus conference, "Proposed National Strategies for the Prevention of Leading Work-Related Diseases and Injuries" **(NIOSH, 1988)** Hazardous levels of noise are defined here as time-weighted average levels of 85 dB(*A*) and above, although it is well known that some more susceptible workers will incur hearing losses at levels below 85 dB (A). According to OSHA'S 1981 estimates, occupational hearing impairments greater than 25 dB at the averaged audiometric frequencies of 1000, 2000 and 3000 Hz, is hearing handicap definition.

Noise[7] exposure does not kill people, at least directly. There is, that high levels of noise and the resulting hearing losses contribute to industrial accidents, may actually impair work safety under certain conditions. In addition, there is growing evidence that noise adversely affects general health, and the cardiovascular system in particular, **(Ising and Krupt 1993 Peterson et al., 1978,1981 and 1983; Rehm, 1983; Zhao, et al. 1993),** which directly affects mortality. Standards may have to be adjusted to prevent such damages. Human endurance to these sounds is given in Table 13.9. Exposure to 120 dB, greater than 5000 Hz causes hearing loss. Low pitches noise produce most serious hearing loss. 3 octaves contribute to the intelligibility of speech.

13.7.2 Pollution-Concerns

Like pollutions of other categories noise pollution was not made a contentious issue hence no legislations were laid. Now the importance of noise pollution has been highlighted and has become the point of discussion, following upon legislations have been formulated, changes incorporated and vigilance has also been stepped up for the implementation of the legislation. The legislations are based on many considerations. The identification of sounds, sources and time of duration are all overdue considerations for any legislation to be enacted.

Table 13.19: Typical Sound Levels[8]

Sound source or environment	*Decibels (dB)*	*Listener's Perception*	*Sound pressure N/M²*	*Energy in Watts*
Jet aircraft at take- off	140-120	Thresh hold of pain	200-20	100
Steam engines Boiler factory	110-120	Deafening		
Trains Noisy factor, loud street	100-90	Very loud	2	
Free way traffic, Noisy office, average	80-70	Loud	2×10^{-1}	
Market places Average office, noisy home	60-50	Moderate	2×10^{-2}	
Private office, quiet conversation	40-30	Faint	2×10^{-3}	
Whisper	10	Very faint	2×10^{-5}	

13.7.3 Refinery Out Fits

In a refinery noise control problem has to be addressed based upon the capacity and units to be installed, duration of working in a day, considerations for future expansions or retro-fittings, the repair cycles, addition or removal/resting of equipment temporarily,

A refinery having 2 stage distillation will have the following major units.

1. Distillation units : Crude distillation units *a.* ADU, *b.* VDU c. Stabilisers d Prefractionators (some cases only) and e. Process fractionators of different sizes.*f.* C_2, C_3, C_4 splitters. Xylene separators.
2. Furnaces : *a.* For distillation unit *b.* Steam raisers etc *c.* Treatment operations like solvent, clay, bitumen blowing etc. and Secondary processes like thermal cracking, catalytic cracking, reforming, hydrogen processes etc.
3. Compressors: *a.* Air supply to burners, regeneration units, pneumatic controls, *b.* Liquid Air production for cryogenic operations.
4. Pumps : Transportation of *a.* Crude to storage *b.* To different units like furnace, distillation units, blenders, treatment processes c. Products, fractions from unit to unit (U2U), units to storage (U2S) *d.* For distribution.
5. Steam equipment and auxiliaries : *a.* Super heaters, ejectors, steam traps, *b.* Steam for driving *c.* Process equipment like distillation units, FCCS etc.
6. Firefighting equipment

All these units will produce noise of different decibles. Usually at 3 feet distance, the safe limits of workmen for sound tolerance is given below.

Table 13.20: Noise Limits[10]

Equipment	*Octave Band sound Pressure Levels. DBA re 0.0002 microbar*								
	31	63	125	250	500	1000	2000	4000	8000
Furnaces, Heat Exchangers compressors, pipe, line pumps etc.	100	96	92	88	85	82	80	80	80
Motors	100	96	92	86	83	83	93	83	83
Turbines, vents, flares	100	96	92	88	85	85	85	85	85

REFERENCES

1. Arthur C. Stern; Vol VI Air pollution, 3rd ed Academic Press Orlando, 1989,33
2. Philip B. Bediant, Hanadi S. Rifai, Charles J Newell: Ground Water Contamination Prentice Hall 202.
3. H. Bank, Environmental Engineering Calculations, McGraw Hill, 1.255.
4. B. K. B. RAO, Fundamentals of Reservoir Engineering Oxford IBH.
5. Clair N. Sawyer Perry L.McCarty and Gene F. Parkins : Chemistry For Environmental Engineering; 4th Ed McGraw Hill, 537.
6. Department of Environmental Central Directorate on Environmental pollution London Pollution paper no. 15,1979.
7. OSHA Standards Planning Committee Novermber 28,1994
8. Source : Harris, David A, ed. Noise Control Manual New York : Van Nostrand Reinhold, 1991.
9. B.K.B. Rao, PTQ,, 4 Q. 2007. 109
10. B.K.B. Rao, HCP, Nov., 2007

Table Some Industries

Table 1: Demand-gap projections (Performance Plastics)

	Demand	*Gap ('000 tonnes)*
Poly Ethylene (HDPE/LDPE LLDPE)	860	462
PVC	485	273
ME Glycol	165	34
Poly propylene	286	150
Acrylo nitrile	140	66
Styrene	147	81
Butadine	156	88

Table 2: Thermosetting Plastics-Demands (Provisional estimate)

TONNES

CONSUMPTION		PROJECTED DEMAND	
	1976-77	*1985*	*2000*
Urea Formaldehyde	2100	3500	24000
Phenol Formaldehyde	4042	6000	89000
Melamine formaldehyde	150	700	2700

Performance Plastics (or Engineering Plastics)

	1989-90	*1994-95*
ABS	12000	24000
Polyamides	8000	16000
Polyacetates	1300	4500
Polycarbonate	1500	5000

Table 3: Summary of Demand Projection for Petrochemicals other than Polymers and Synthetic Fibres

SI. No.	*Product*	*Projections*	*Demand*	*000MT*
		1989-90	*1994-95*	*1999-2000*
1.	Linear Alkyl Benzene	145	287	497
2.	Alpha Olefins	NIL	60	96
3.	Petroleum Resin (Aliphatic)	6	9	16
4.	P-Xylene	147	255	400
5.	O - Xylene	71	106	151
6.	Benzene	353	578	858
7.	Toulene	51	62	76
8.	EO (Ethylene Oxide)	19	24	30
9.	EG (Ethylene Glycols)	91	165	264
10.	Formaldehyde	131	174	212
11.	Propylene Oxide	19	NA	NA
12.	Plasticizer Alcohol	37	57	84
13.	DMT/PTA	226	393	615
14.	Acrylonitrile	73	140	215
15.	C aprolactum	116	191	302
16.	Methanol	200	245	286
17.	Phenol	49	68	92
18.	n-Butanol	18	23	33

Table 4: Some more petro chemical industries

Concern	*Product*	*Capacity*	*Location*	*Commencement*
Aakarsha	Polycarbonate resins	2000 tpa	Alul. Guj.	1990
Engg.	Maleic Anhydride	10,000 tpa		–
Plastics.	Propylene rimer	5,000 tpa	Cochin	
Balmer Lawrie & Co. Ltd	Poly butenes	5,000 tpa		
Baansali Engg. Polymers Ltd.	ABS	7,500 tpa	Satnoor M.P.	
Bihar Chemical and Fertiliser Co.	Caprolactum	50,000 tpa	Bokaro	1992
Bombay Dyeing & MFG. Co. Ltd.	DMT	150,000 tpa	Patalganga	–
Birla Jute & Industries Ltd.	Trichloro-Ethylene	5,000 tpa	WB.	–
Chemicals & Fertiliser Corp. Ltd.	Caprolactum	50,000 tpa	Dhanbad	1992

Dalmia Aromatics Ltd.	Naphthalene, 15.000 tpa	Bongaigaon	1990	
	BTX	15,000 tpa		
Ganesh Anhydrides	Maleic Anhydride	7,000 tpa	Thane	1990
GAIL	Propylene P.P.	10,000 tpa	Guna	1992
		75,000 tpa		
Gujarat Apas Polymers	SBR/NBR	5,000 tpa	Ankaleswar	1991
Gujarat Godrej Innovate	Alpha Olefins	35,000 tpa	Bharuch	1991
Gujarat Narmada Valley Fertiliser Corporation Co. Ltd.	Methanol	100,000 tpa	Bharuch	1990
	Formaldehyde	30,000 tpa		1991
	Melamine	15,000 tpa		1991
	Caprolactum	100,000 tpa		1991
Deepak Fertilisers & Petrochemicals Ltd.	Methanol	100,000 tpa	Maharashtra	1991
	Poly acetal resins	10,000 tpa		1992
	Aromaticamines	3,000 tpa	Baroda	1992
Diamines & Chemicals Ltd.	Ethylene diamine	3,000 tpa	Baroda	1992
Dunlop India Ltd.	Carbon Black		W.B.	–
Haldia	Ethylene (Linde)	A.G, WG		–
	EG Sohio	U.S.A.		
	HDPE Chemtex			
	Styrene Lummus			
	ACN Phillips Petro			
	Butadiene Scientific			
	Design & I F P.			
Haryana State Indl. Dev. Corp.	Carbon Black	50,000 tpa	Panipat	–
Herdillia Polymers Ltd	EP Copolymers	10,000 tpa	Thane	1990
	Poly butene	5,000 tpa	Nagothane	1990
Hindustan Lever	LAB	80,000 tpa	Punjab	–
Hindustan Organic Chemicals Ltd.	Polyurethanes	30,000 tpa	Rasayani. Maharashtra	–
	Propylene trimers			
I.G. Petro Chemical Ltd.	Phthalic anhydride	45,000 tpa	Taloja Maharashtra	1992
Indian Rayon & Industries Ltd.	Carbon black	50,000 tpa	U.P.	1992
Indo Nippon Chemicals & Co. Ltd.	ABS Resins	5,000 tpa	Nagothane	1990

Indo Nissan Oxochemicals	OXO-alcohols	25,000 tpa	Baroda	–
Industrial Organics Ltd.	VAM	3,300 tpa	Barnala Punjab	1990
Modi Enterprises	Caprolactum	50,000 tpa		1993
Modi Pan Ltd.	Expansion of Polyster	3,500 tpa	(Planning)	
	Filament yarn			
	Nylon Yarn	150 tpa	(Planning)	–
Gujarat Ambuja Proteins Ltd. & IPCL	Polyols	10,000 tpa		
	PO	6,000 tpa	Baroda	1992
J.K. Petro Chemicals	Benzene	30,000 tpa	Salempur	1993
	Paraxylene	30,000 tpa		
	Ortho Xylene	30,000 tpa		
	PTA	1,50,000 tpa		
Karnataka State Industrial Development Corp. Ltd.	Carbon black	50,000 tpa	(Planning)	–
Kothari Sugars & Chemicals	Polybutenes	5,000 tpa	Manali	1990
Ltd.	Maleic Anhydride	10,000 tpa	Pondicherry	1991
Lubrizol India Ltd	Poly Olefins	10,000 tpa	Hazira	1991
	Poly butene Polymers	7,500 tpa		
Maharashtra Petro Chemicals Corp. Ltd.	Epoxy resins	2,500 tpa	Ratnagiri	–
	SBR	60,000 tpa		
	MTBE	70,000 tpa		
	Butane-1	1,200 tpa		
	Petro Chem. Res.	5,000 tpa		
	Iso Butyl benzene			
Manali Petro Chemicals	PO	12,000 tpa	Man ali	1990
	PG	6,200 tpa		
	Mono Chloro Acetic	1,000 tpa		
Mc Dowell & Co. Ltd.	Styrene	80,000 tpa	Vizag	1991
	PS	55,000 tpa		
Modi Rubber Ltd.	Butyl rubber	25,000 tpa	Nanded	1992
Modi Pun Ltd.	Bisphenol	20,000 tpa	Alwar	1993
National Aroma tics & Petro Chemicals Corp. Ltd	O-Xylene	30,000 tpa	Manali	1992
	P-Xylene	1,05,000 tpa		
	Benzene	30,000 tpa		
	PTA	1,50,000 tpa		

National Organic Chemical Industries Lt.d	Ethylene	3,00,000 tpa	Thane	–
	PP	50,000 tpa		
	SBR	67,000 tpa		
Niraj Petro Chemicals Ltd.	Maleic Anhydride	22,000 tpa	Visakhapat-nam	–
Oswal Agro Mills Ltd.	EPDM	10,000 tpa	Ratnagiri	–
	Polyethylene Oxide	12,000 tpa		
	PS	30,000 tpa		
	Ethylene	1,00,000 tpa	Bombay	
	HDPE/LDPE	50,000 tpa		
	VAM		Punjab	
Petro Polyols Ltd.	Propylene Glycol	6,000 tpa	–	–
Phillips Carbon Black Ltd.	Carbon Black	40,000 tpa	Visakhapat-nam	1992
Polychem Ltd.	ABS	5,000 tpa	Baroda (Nimbut)	
	PVA	2,000 tpa		1990
	VAM	10,000 tpa		
Polyolefins Industries Ltd.	Ethylene, Propylene,	10,000 tpa	Thane	1990
	EPDM			
	Chlorinated Polyethy lene	6,000 tpa		
	HDPE	1,50,000 tpa		
	Polyacetals	5,000 tpa	Panoli	
	Aromatic Amines	20,000 tpa	Gujarat	

APPENDIX – I

INDIAN PETROLEUM INDUSTRY

Ap 1.1 Indian Oil Industry

Except the first oil refinery at Digboi started in 1932, all other activities of petroleum like discovery of onshore and offshore fields, establishment of new refineries and petrochemical units are post independence scene. ONGC, Oil India Ltd, IOC, Reliance are the crude producers and ONGC the biggest producer of oil and gas in India. The production in the country is about 32.2 MMT and an improvement was noticed in 1995-96 when the production touched 37.45 MMT. After commissioning of Neelam and Panna complexes the production is likely to touch 38 MMT far from a modest goal of 48 MMT as envisaged by Union Ministry of Petroleum and Natural Gas, but in vain.

Cat reformers are being introduced into Mathura. Digboi Gujarat and Barauni refineries. Introduction of Propylene recovery unit at Barauni will be another factor for meeting the demands of propylene. Gujarat refinery capacity will be increased to 15 MMTPA. To meet the requirements of middle distillates various modifications and new units are being added. One million ton visbreaker is being added to Gujarat refinery. One million tons soaker mode coil is being added to Mathura refinery, this can work at higher temperatures and residence time. BRPL's coker unit is 0.5 MMT. Numaligarh refinery will have a delayed coker. Barauni refinery a needle coke plant based upon the phenol extracts is planned. Panipat refinery will have a visbreaker of soaker type.

Many pipeline net works for transportation of field and imported gas and oil. Haldia-Barauni pipeline for crudes, Khandla-Bhatinda-Delhi diversion of HBJ line. Pipe line to Vijayawada-Nellore from KG basin. Madras-Bangalore-Mangalore line etc are some of the important transportation net works going to come. Efforts are made to lay sub sea pipelines from Oman and Iran to India to overcome shortage in gas. On Eastern side also efforts are going to link Burma-Bangladesh and India via pipe lines. For importing more terminals are commissioned or the existing ones improved.

Production and consumption of crude (In million tons):

Production	*1994-95*	*95-96*	*96-97*	*97-98*
crude production	33.2	35.2	38.87	34
crude imports	27.35	27.3	33.9	34.49
products	14.5*	20.3	20.3	19.53
* Including 6 lakh tons of M.S				
Natural gas		22.639	23.254	24.7 (Billion cubic meters)
Consumption of petroleum products	65.5	74.7	79.3*	

*Light distillates; Middle distillates; Heavy distillates

LPG	4.4	ATF	2.1	FO/LSHS	10.7
Naphtha	5.1	SKO	10.7	Bituman	2.3
Gasoline	7.0	HSD	30.5	Lubes	1.2
Others	1.6	LDO	1.7	Others	1.3
		Others	0.7		
Total	18.1		45.7		15.5

LPG availability :

Uran	0.45	Hazira	0.5	Bijapur	0.4
OIL, Duliajan	0.4	MRL	0.12	Ankaleswar	0.4
Koyali	0.14	Haldia	0.2	Cochin	0.12
Bharat	0.14	Vizag	0.08	Barauni	0.04
Gauhati	0.01	Mathura	0.2	HPCL, Bombay	0.14
				Total 3.15	MMTPA

Facilities available at various refineries

Atmospheric distillation : All refineries

Vacuum distillation: Not having : HPCL, Gauhati. Gujarat, BRPL Mathura, CRL

Cat Reformer: AOC, Barauni, Mathura, Haldia,, CRL, Gujarat BRPL, Panipat, Mangalore (MRPL), Chennai Refineries Ltd.

Haldia, Mangalore, Mathura, Koyali, Bongaigon (BRPL),

Visbreaker : MRPL, Barauni, Gujarat, Mathura, CRL, HPCL, BPC

FCC Vizag, Koyali, Cochin, MRL, BRL, Mathura, MRPL, HPCL

Hydrocracker Numaligarh*, MRPL, Koyali, Panipat*

Thermal cracker; MRPL

Coker: AOC, Barauni, MRL, Numaligarh, BRPL, Haldia

Bitumen :MRPL, Haldia, AOC. BPL, HPCL, CRL, MRL, Barauni, Mathura,

Lubricant: AOC, Barauni, Haldia, Cochin, HPCL (Mumbai)

HDS: Haldia, Mathura, Panipat, Gujarat. AOC-wax, HPCL, Barauni, MRL

Merox; MRL, Haldia

Amine Treaters : All refineries

Wax MRL. AOC,

Phenol Treatment: Barauni, BPL,

Furfural Treatment: MRL, Haldia

Sulphur dioxide extractor; CRL, AOC

Essar, Reliance, Mangalore Grass root refineries. .

The booming automobile industry and race in refrigeration and batting telecommunication industry along with esssential demands for meeting the population requirements, the petroleum industry must be able to supply atleast 120 MMTPA products by 2001. No healthy deposits could be discovered in the last decade. The exisiting prospective fields are likely to supply about 800 MMT. Hence a long term contingency plan must be designed.

Ap.[2] Petrochemicals

The consumption of petrochemicals from 7.5 lakh tons in 1979 to over 3.5 million tons in 1995-96 shows the growing needs of petrochemicals and the growth rate is likely to maintain the trend till the end of the century. IPCL expansion of its naphtha cracker to 3 MMTPA is a partial story to meet the demands, besides Nagothane unit based on natural gas and Gandhar and Aurya gas crackers will shortly join the string of petrochemical units of the country which would supplement the demands of various petrochemicals, another gas cracker plant in Tripura would be soon operational. Polypropylene production at IPCL was 30,000 tons during 1978 but now it has gone upto 1.8 lakh tons. Reliance plant at Hazira (Surat) is producing 3.5 lakh ton of polypropylene.

The second plant at Jamnagar will produce the same amount by 1988—1989. The other units are NOCIL, Square-D and Krishna Petrochemicals and Haldia petrochemicals which are at various stages of planning. Thus by the end of the century the capacity of PP would be 7 lakh tonnes per annum. When Nagothane project begins to use natural gas, the cost of production of petrochemicals would come down.

The production of PSF may outstrip the demand by the end of this century as the capacity is likely to touch 20 lakh tonnes.

Petrochemical Industries-Planned

Alkyl Amines Chemicals (I) Ltd, Kurukshetra, Amides 3,000 tpa
Andhra Petrochemicals Vizag: 2-ethyl hexanol 30000 tpa
ATV Chemicals Ltd Sakhar ; Terephthalic Acid 120,000 tpa,
Babrala (Tata) Ammonia 1350 TPD (Three units), urea 1300 TPD
Balmer Lawrie Propylene trimer and propylene tetramer
Bharat Petroleum Cor. Mahul Bombay. MTBE 30.6 MT
Bombay Dyeing Patalaganga (biggest producer) DMT 1.5 lakh ton/y
Century Enka, Pune, PET 1200 t/y
Cibatul at Atul Bis phenol - A 5,000 t/y
Finolex Plastics : Ratnagiri PVC 80,000 t/y
Vinyl monomer, 10,000 t/y
GAIL ; Bijapur PP : 100,000 t/y
Gujarat Narmada Valley Fertilisers, Bharuch, TDI 10,000 t/y
Gujarat State Fertilisers, Vadodra Melamine 10,000 t/y,
KRIBHCO, Hazira Ammonia 1350 × 2 tpd,
urea 1100 × 4 tpd
CAN 1000 tpd
Herdilia Unimers, Thane EPD Polymers, 10,000 t/y
Hindustan Organics Ltd. TDI 10,000 t/y
Nagarjuna Urea 46, 000 (Gas based),
Neyvelli Lignite Corporation; Urea 325 TPD (lignite),
Poly Olefins, Thane ACN : 25,000 t/y
Sun Carbon Cor, Haldia; Carbon Black 5,000 t/y,
Supreme Petrochemical Ltd. Nagothane. Ethylbenzene 106,000 t/y
Polystyrene 60,000 t/y
Styrene 100,000 t/y
Thirumalai Chemicals, Ranipet Maleic acid 7,500 t/y,
UB Elastomers Visakhapatnam Butyl Elastomer 25,000 t/y
Uniglobe Glycols Ltd. Sandila Ethylene glycol 25,000 t/y
Zuria (Birla group) and Chambal Fertilisers Urea, 1 MMTPA

Concern	*Product*	*Capacity*	*Location*	*Commencement*
Poly Propylene India Ltd.	P.P.	30,000 tpa	Mathura	1991
Rashtriya Chemicals & Fertilisers Ltd.	DMF	25,000 tpa	Thal Vaishet	1990
	Methyl Amine	9,000 tpa		
RPG Petro Chem. Ltd.	Ethylene	4,50,000 tpa	Chingleput T.N.	
	Propylene	2,20,000 tpa		
	Benzene	1,00,000 tpa		
	Butadiene	80,000 tpa		
	Carbon black	60,000 tpa		
Reliance Industries Ltd.	P-Xylene	2,70,000 tpa	Surat	—
	PTA	2,70,000 tpa		
Sip Resins Ltd.	Epoxy Resins	1,200 tpa	Chingleput T.N	1990
S.M. Dyechem Ltd	Methanol	3,00,000 tpa	Machilipatnam,	
	Ethylene Oxide	10,000 tpa	AP	
	MEG	50,000 tpa		
	DEG	2,500 tpa	Bihar	1991
Sanghi Leathers	Methanol	6,00,000 tpa	Narsapur	—
South Indian Viscose Ltd	Acrylamide	5,000 tpa	—	—
Plus Gujarat IIC Ltd.	Monomer	3,000 tpa	—	—
	Polyacryl amide			
South Indian Viscose	VAM	10,000 tpa	S. Arcot	—
Ltd.				
Somiaya Organic Ltd.	Butanol	15,000 tpa	Bara Banki	1991
Standard Alkali (Industries)	Maleic Anhydride	10,000 tpa	Jaiged	1991
Straw Product Ltd.	LAB	60,000 tpa	Mathura	—
Super Byteron	Phthalic Anhydride	1,50,000 tpa	Salem pur	1991
Systems Ltd.	Maleic Anhydride	3,000 tpa		
Supreme Petro Chem. Ltd.	PS	60,000 tpa	Maharashtra	1992
Synthetics & Chemicals Ltd.	Styrene SBR	80,000 tpa	Bareilly	1991
	Nitrile Butadiene Rubber	80,000 tpa		

Tamil Nadu Petro Products Ltd.	LAB	75,000 tpa	Manali	—
Thirumali Chemical Ltd.	Maleic Anhydride	10,000 tpa	T.N.	1991
Tripura Petro Chemical Ltd.	Methanol	1,00,000 tpa	Tripura	1991
U.B Elastomers	Ethylene	3,00,000 tpa	Visakhapatnam	1991
	Butyl rubber	25,000 tpa		
	PO	12,000 tpa	Manali	1990
	Propylene Glycol	7,000 tpa		
	Polyether polyols	12,000 tpa		
U.P. Carbide & Chemicals Ltd	Trichloro Ethylene & Tetra Chloro ethane	5,000 tpa	Saharanpur	—
U.P. State IDC Ltd.	MEK	5,000 tpa	Mathura	1991
Vam Organics Ltd.	DMF	5,000tpa	Muradabad	1991
	Vinyl Pyride Latex	5,000 tpa		
	LAB	80,000 tpa		
Vam Petro Chemi-	Ethylene	3,00,000 tpa	Dhuliajan	—
cals Ltd.	Propylene	60,000 tpa		
	MEG	60,000 tpa		
	HDPE/LLDPE	2,00,000 tpa		
	Alpha Olefins	45,000 tpa		
	Butadiene	22,000 tpa		
Vinyl Chemicals Ltd.	VAM	10,000 tpa	Raigad	1990

Expansion of IPCL

Butadiene recovery unit (after expansion)	35,000 tonnes
PBR (after expansion)	50,000 tonnes
Grass root polyethylene plant	75,000 tonnes
Carbon fiber	12,000 tonnes
Poly carbonates	15,000 tonnes
Poly pheneylene oxide	15,000 tonnes
Orthoxylene	31,600 tonnes
Paraxylene	31,600 tonnes 45,000

Maharashtra Gas Cracker Complex (MGCC)

Installed at a cost of 1450 crores, under the IPCL banner, is put on trials. Feed stock for this plant will consist of 4.75 lakh tonnes of ethane and propane from Uran plant of ONGC and 51.000 tonnes of propane and propylene from Bharat Petroleum Refinery. The gas on cracking would furnish 66,000 tonnes of propylene and 3 lakh tonnes of ethylene.

Products

LLDPE/LDPE	1,35,000 tonnes
HDPE	80,000 tonnes
Ethylene Oxide	5,000 tonnes
Poly propylene	60,000 tonnes (Himout technology)

Gandhar Petro Chemical Complex

This is also commissioned by IPCL at a cost of 2290 cores of Rupees. The gas feed stock would be available from ONGC fields of Gandhar. The feed stock consists of 4.5 lakh tons of ethane, 3 lakh tons of propane to produce 3 lakh tonnes of ethylene, 20,000 tons of propene and 10,000 tons of butadiene. This mix would be utilised for the production of 2.4 lakh tonnes of ethylene dichloride, 1.5 lakh tons of vinyl chloride, 1.5 lakh tons PVC, 20,000 tons of ethylene oxide, 1.2 lakh tons of alcohol ethoxylate. This project is expected to be on stream by 1995.

Reliance Petro Chemicals

First phase of project costs 1050 crores of rupees and nearing completion. Based on the available NGLS from Hazira, it would be able to produce crack mix of Ethylene, propylene and Butadiene, to yield chemical and polymers as shown below :

LLDPE	1,00,000 tonnes
HDPE	1,00,000 tonnes
PVC	1,00,000 tonnes (Imported ethylene)
Ethylene Oxide (EO)	5,000 tonnes
Ethylene Glycol (MEG)	80,000 tonnes
Acrylonitrile (AN)	70,000 tonnes
PP	60,000 tonnes
SBR	80,000 tonnes

Auraiya Gas Cracker Complex

It may be the first gas cracker plant on HBJ pipe line of GAIL. Primary olefins like ethylene, Propylene, and butadiene are expected from cracking operations. This plant is likely to go on stream by 1999.

National Aromatics Petro Chemical Complex

To supplement the growing demands of aromatics, this plant when commissioned in 1993 would yield benzene 30,000 tons, Orthro Xylene 30,000 tons, paraxylene 1,40,000 tons and PTA 200,000 tons. This is based on aromatics extracted from Bombay High Crude.

Manali Chemical Complex

Based on the site of Madras Refineries it would commence production shortly to yield 6,250 propylene glycol, 600 tonns polyol 12,000 tonnes of propylene oxides.

PRODUCTION OF MAJOR PETROCHEMICALS DURING 2002-03 TO 2007-08

	Production				*(Figure in 000'MT)*	
Group / Product *Basic Petrochemicals*	2002-2003	2003-2004	2004-2005	2005-2006	2006-2007	2007-2008(P)
1. SYNTHETIC FIBRES						
Acrylic fibre	105.1	117.1	127.6	113.8	107.3	83.6
Polyester staple fibrefill	29.5	38.8	39.8	47.4	47.1	47.4
Nylon filament yarn	29.8	31.0	36.8	39.6	32.4	26.2
Nylon industrial yarn/tyre cord	51.4	55.9	48.0	55.4	71.6	83.7
Polyester filament yarn	946.4	1003.1	970.3	1015.0	1193.6	1334.4
Polyester staple fibre	574.7	603.7	638.7	623.1	784.6	891.6
Polyprodylene filament yarn	15.1	15.3	11.3	8.8	9.7	10.2
Polypropylene staple fibre	2.5	2.8	3.0	3.1	3.6	3.5
TOTAL	**1754.6**	**1867.7**	**1875.4**	**1906.2**	**2249.9**	**2480.6**
II. POLYMERS						
Low density polyethylene	197.7	183.8	204.8	200.6	194.9	200.3
High density polyethylene	977.8	958.1	1035.1	1034.7	958.4	981.6
Polyestyrene	224.4	274.0	275.2	310.5	284.9	280.6
Polypropylene (inc. co-polymer)	1430.2	1567.2	1690.2	1540.9	2000.9	1913.3
Expandable polyestyrene	29.3	31.4	35.7	39.2	45.8	48.2
Polyvinyl chloride	821.6	878.4	884.8	953.4	926.4	945.4
Liner low density polythylene	500.1	606.2	650.0	688.7	771.5	804.4
TOTAL	**4175.1**	**4499.2**	**4775.7**	**4768.1**	**5182.9**	**5173.6**
III. SYNTHETIC						
Styrene butadiene rubber	16.4	18.9	16.1	15.5	13.1	17.7
Poly butadiene rubber	53.6	55.8	64.1	67.3	71.7	73.4
Ethly propylene dimers	5.0	5.5	3.5	4.1	3.5	0.7
Ethyl vinyl acetate	0.0	0.0	4.0	14.2	3.0	0.0
Nutrile butadiene rubber	6.5	7.3	9.0	9.1	10.1	11.6
TOTAL	**81.5**	**87.4**	**96.8**	**110.2**	**101.4**	**103.5**
IV. SYNTHETIC DETERGENT INTERMEDIATES						
Linear alkyl benzene	375.2	382.0	408.6	468.0	459.9	470.3
Ethylene oxide	72.2	71.4	79.3	87.5	96.4	112.5
TOTAL	**447.4**	**453.4**	**487.9**	**556.0**	**556.3**	**582.9**

V. PERFORMANCE PLASTICS						
ABS Resin	49.5	48.4	62.2	75.8	74.2	79.6
Nylon-6 & 6.6	9.0	11.0	13.0	13.0	14.0	14.0
Polymethyl methacrylate	2.6	2.7	2.2	2.5	2.7	2.7
Styrene acrylonitrile	33.6	37.0	35.7	35.5	41.4	62.1
TOTAL	**94.6**	**99.1**	**113.1**	**126.9**	**132.4**	**158.4**
Major Petrochemicals (I to V)	6553.2	7006.8	7348.9	7467.4	8222.9	8499.0
B: INTERMEDIATES						
FIBRE INTERMEDIATES						
Acrylonitrile	33.1	36.4	38.8	33.2	37.0	40.2
Caprolactum	100.4	110.3	122.1	116.8	121.0	88.3
Dimethylterephthalate	199.1	216.8	238.6	197.4	27.5	7.1
Monoethylene glycol	611.2	651.9	713.9	881.4	872.5	926.1
Purified terephthalic acid	1704.7	1675.1	1738.0	1734.2	2379.2	2472.8
TOTAL	**2648.5**	**2691.1**	**2851.4**	**2963.0**	**3437.3**	**3534.5**
OLEFINS						
Butadiene	68.5	114.1	131.4	206.9	222.7	237.6
Ethylene	2304.6	2421.2	2644.6	2719.2	2683.4	2774.1
Propylene	1588.5	1746.3	1891.9	1744.6	2089.3	2084.4
TOTAL	**3961.7**	**4281.7**	**4667.9**	**4670.7**	**4995.3**	**5096.0**
AROMATICS						
Benzene	595.0	608.3	640.1	685.8	885.5	905.1
Mixed xylene	52.3	55.6	63.1	55.7	57.7	71.4
Ortho-xylene	191.7	206.8	145.6	242.4	431.2	277.8
Toluene	162.7	165.5	177.2	158.9	147.2	153.9
Paraxylene	1424.4	1389.0	1425.3	1393.7	1925.5	2248.4
Total	**2426.3**	**2425.0**	**2451.3**	**2536.5**	**3447.1**	**3656.5**

SOURCE : Chemical News. December 2008

APPENDIX – II

INDIAN PETROCHEMICAL INDUSTRY

The petrochemical industry can be broadly divided into two divisions. One is producer of building blocks and intermediates: second one being utilising these as raw materials and process for the finished materials. The giant stride taken by the government to utilise the feed blocks available from Indian Refineries/oil fields has resulted in establishing prodigious Indian Petrochemicals Corporation Ltd. Indian Petrochemicals Corporation Limited (IPCL.), registered on March, 22, 1969 as a company producing polymers, synthetic fiber, fiber intermediates, solvents, surfactants, industrial chemicals, of late catalysts and adsorbents. (IPCL's total production for FY2005 was 5. 772 K TA and the company clocked an operating margin of 1 8.82%), IPCL has built two other complexes based upon gas in Gandhar and Nagothane. Besides the complex at Vadodara that runs on naphtha, feed. The complex was based on associated gas from Mumbai High Fields as a feed stock at an estimated cost of Rs 1.167 crore during the Sixth Five Year Plan. Consequently, the production of Linear Low Density Polyethylene (LLDPL) was undertaken for the first time in the country. The plant has now been retrofitted and expanded to 220.000 M TA. Simultaneously, IPCL was also entrusted to set up a third petrochemical complex based on gas from the gas fields at Gandhar in Gujarat

The Indian polymers industry is small by international standards, accounting insignificant of 3% of the global production. However polymer consumption is increasing compared to China and other key Asian nations. The Indian polymer industry has progressed largely due to the protective policy of the Government. Although the plant sizes have increased over the years, they cannot be compared with the TNCS. The part played by private players cannot be ignored in the development of the industry, depending upon the resources available to them example being 750,000 tpa of ethylene by RIL, Oswal Agro Mills Limited '(former Union Carbide) owns a very small naphtha cracker of 22,000 tpa of ethylene.

The Indian polymer industry has witnessed a 13% growth from 1981-2002.that increased the size of the industry from 300,000 tonnes in 1981 through to 4.8 MMTPA in 2005-06. The part played by the reprocessing units in small scale sector in manufacturing article of utility cannot be underestimated They get the raw polymers from the major corporations and manufacture the needful articles and some times reprocess the old polymeries and rubbers. The Indian polymer industry started with NOCIL and Union Carbide followed by IPCL. IPCL Is the major producers of commodity plastic raw materials namely Polypropylene (PP), Polyethylene (PE) and Poly Vinyl Chloride (PVC). Total production of cracker products viz. Ethylene was 946 KT, Propylene was 358 KT, Butadiene was 47 KT and Benzene was 51 KT. IPCLs Polyethylene (PE) production was 598 KTA . IPCL produced 266 KT of PVC from both Gandhar and Vadodara plants, which was 8% higher than in 2004-05. Poly Butadiene Rubber (PBR) production was 67 KTA, an increase of 3 KT when compared to the previous year. The company's chemical business consists of downstream products like Benzene Acrylonitrile, Acrylates, LAB and Caustic Soda. The production of Benzene during the year 2005-06 was 51 KT against 53 KT during 2004-05. During the year, Caustic Soda production increased to 164 KT as compared to 159 KT in the previous year.

INDIAN OIL CORPORATION : IOC was formed in 1954 through merger of Indian Oil company Limited (Estd. 1959) and Indian Refineries Limited (Estd. 1958). The Indian Oil Group of companies owns and operates 10 of India's 18 refineries, including two refineries of subsidiary Chennai Petroleum Corporation Limited (CPCL) and one of Bongaigaon Refinery and Petrochemicals Limited (BRPL), with a combined refining capacity of 54.20 million tonnes per annum (1.1 million barrels per day). IOC and its subsidiaries account for 47% petroleum products and 41% of the national refining capacity. The year 2004-05 marked Indian Oil's entry into petrochemicals with the beginning, high-value projects such as Methyl Tertiary Butyl Ether (MTBE) and Butene-1 at Gujarat Refinery, followed by commissioning of the country's largest Kerosene-to-LAB unit at Gujarat Refinery. The plant produces LAB for manufacturing environment-friendly biodegradable detergents, using state-of-the-art Detal technology from M/s UOP, USA. The raw materials for the plant are Kerosene and Benzene produced at Koyali Refinery. Other activities in different places show the feed stocks of the industry. Naphtha will be sourced from Indian Oil's Panipat and Mathura refineries. Pxylene is to produce Purified Terephthalic Acid. In natural gas business, Indian Oil is already marketing 1.43 million tonnes of gas per annum. A Naphtha Cracker complex with downstream polymer units is coming up at Panipat and a refinery-cum-petrochemicals complex is proposed at Paradip on the east coast of India. An integrated plant at Panipat Refinery has commenced production of PX/PTA. The Paraxylene plant is designed to process Naphtha to produce Paraxylene.

IOCL'S CONTRIBUTION

Data in MTPA

Plant Capacity	*LAB Unit, Gujarat*	*PX/PTA Unit, Panipat*	*NaphthaCracker, Panipat*
IAB	120000	NA	NA
Naphtha	NA	500000	NA
Paraxylene	NA	360000	NA
Purified Terephthalic Acid	NA	553000	NA
Ethylene	NA	NA	800000
Propylene	NA	NA	600000
LDPE	NA	NA	NA
HDPE	NA	NA	NA
PP	NA	NA	NA
MEG	NA	NA	NA

The key surfactants LAB and EO increased by 6% and 8% respectively in 2005. Demand growth for LAB has remained same and is expected to be at 6% in 2006 and 2007 as shown in the following table. Demand growth for EO has decreased to 8% in 2005' from the previous year of 12%. It is expected that demand will continue to grow at 8% in 2006 and 2007. Since, there is no further capacity addition expected in 2006 & 2007 the EO capacity has remained unaltered at 107 KT since 2003. LAB capacity increased by 120 KT in 2004 to reach 460 KT.

PATA COMPLEX

Gail has set up gas cracker plant at Pata (UP), capacity of which is 400,000 TPA of ethylene feed being natural gas, this complex recovers ethane-propane (C2/C3) from natural gas going through HBJ pipeline PATA complex has the capacity to the following.

	Capacity (TPA)	
Gas Cracker	Ethylene	3,00,000
	Propylene	11,000
Products	Linear Low Density	2,10,000
	Polyethylene/High Density	
	Polyethylene	
	High Density Polyethylene	1,00,000
	Butene-1	10,000

ASSAM GAS CRACKER PROJECT

Brahmaputra Cracker and Polymer Limited (BCPL) has started a new petrochemical project in Lepetkata Assam, to be completed by 2011. The Feedstock for the Petrochemical Complex is 6.0 MMSCMD gas from Oil India Limited (OIL) Duliajan and 1.35 MMSCMD gas from-ONGCL. The Petrochemical Complex will also utilize 160,000 tons per annum (TPA) of petrochemical grade Naphtha from Numaligarh Refinery Limited. The complex has been configured to process

Feedstock		*Gas, Naphtha*
		Capacity (TPA)
Gas Cracker	Ethylene	220,000
	Propylene	60,000
Products	HDPE/LDPE	220,000
	PP	60,000
	Raw Pyrolysis Gasoline	55,000
	Fuel oil	12,500

BUILDING BLOCKS OF PETROCHEMICAL INDUSTRY

The major building blocks for the petrochemical industry are shown below:

Major Polymer Building Block of *(TPA)*

Company	*Capacity*
Reliance Industries Limited	27,000,000 Total
	750,000 Ethylene*
	400,000 Propylene*
IPCL Baroda	300,000
IPCL Nagothane	300,000
IPCL Gandhar	400,000

Haldia Petrochemicals	420,000 Ethylene
	200,000 Propylene
GAIL	300,000 Ethylene
NOCIL	150,000

FEED STOCKS

The basic feed stocks for petrochemical industry are natural gas and naphtha. The production and consumption of these are shown below:

Olefins and Aromatics are the main building blocks in the production of various petrochemicals. At refineries, olefins are produced mainly from hydrocarbons by chemical cracking such as steam cracking and by catalytic reforming of naphtha. Aromatic hydrocarbons are also produced in refineries by catalytic reforming or other such processes.

Production of Gas by Different Companies

Company	*2000-01*	*2001-02*	*2002-03*	*2003-04*	*2004-05*	*2005-06*
	MMSCMD	MMSCMD	MMSCMD	MMSCMD	MMSCMD	MMSCMD
ONGC	24.02	24.042	24.244	23.584	22.975	21.41
OIL	1.861	1.615	1.743	1.887	2.01	1.96
Pvt/JV	3.596	4.054	5.407	6.491	6.782	7.321
PLL-RLNG	NA	NA	NA	NA	NA	6.12
TOTAL	29.477	29.711	31.394	31.96	31.767	36.811

Industry-wise Off-takes of Natural Gas in India

Industry	*1990-91*	*2000-01*	*2004-05*
1. Power Generation	3634	8801	12099
2. Industrial Fuel	827	2870	3569
3. Tea Plantation	89	151	142
4. Domestic Fuel	50	335	343
5. Captive use/LPG Shrinkage	1775	5004	4944
6. Others	0	38	231
Total (*a*)	6375	17199	21328
1. Fertilizer Industry	5612	8480	8173
2. Petro-Chemicals	409	779	1236
3. CNG	0	0	0
4. Others	370	1402	37
Total (*b*)	6391	10661	9446
Grand Total (*a* + *b*)	12766	27860	30774
%age to Grand Total			
Energy Purposes	49.9	61.7	69.3
Non-Energy Purposes	50.1	38.3	30.7

Availability of Gas from different sectors: (Statistics: Ministry of Petroleum and Natural Gas)

Company	*2007-08*	*2008-09*	*2009-10*
	MMSCMD	MMSCMD	MMSCMD
ONGC	46.12	42.12	35.47
OIL	6.79	3.99	3.99
ONGC Additional (GI, C22 series, KG Basin)		2.28	4.18
BHEEMA (Surat-Niko On shore	0.042	3.034	3.009
NORTH SURAT-Niko On shore	0.133	0.093	3.071
GAURI	1.040	3.960	3.480
LAKSHMI	2.980	0.880	
HAZIRA-Niko/GSPC On shore	3.830	2.660	1.870
PANNA-MUKTA	3.410	3.270	3.240
TAPTI	12.000	12.000	12.000
RAVVA	1.760	4.120	3.950
PY-I	1.300	1.300	1.300
GSPC			
RIL		40.00	80.00
AMGURI CONSORTIUM	0.012	0.012	0.012
PLL-RLNG (7.5 MTPA BY 2009-10)	17.58	17.58	26.28
TOTAL	97.000	132.022	171.675

KG basin (5.37 MMSCMD from the KG-D6 field). Is the most important center gas activity and availability. Many power plants outside Andhra take off gas from KG Basin amounting to 17.94 MMSCMD. The state sector projects, with a capacity of 2,372 MW, require 7.69 MMSCMD of gas to operate at a 60.5% PLF. While 7.03 MMSCMD has already been tied up, the KG-D6 field would supply an additional 1.30 MMSCMD. The 464 MW Gautami and 445 MW Konaseema power plants, which have not yet been commissioned due to the paucity of gas, have been allocated their entire fuel requirements from the field. These projects would require 3.4 MMSCMD of gas to operate at 70% PLF. The private sector plants benefited from the additional gas supplies are the Vatwa, Trombay, GPEC Puguthan, GIPL Stage-II and Essar IMP, with a combined capacity of 1,395 MW. The government intends to supply 2.38 MMSCMD of gas to such projects from Reliance's field. In addition, the government has also allocated 2.74 MMSCMD of gas to Sugent Power's 1128 MW upcoming project in Gujarat.

Naphtha Position

Next important raw material is naphtha .The production is from refineries, the availability is good as the refining activity is nearly 150MMTPA. The availability and consumption are shown below:

Naphtha Production-Import-Sales and Export MTPA

Year	*Total Supply Production*	*Import*	*Total*
1999-2000	8170	1917	10087
2000-01	9908	3165	13073
2001-02	9180	3308	12488
2002-03	9650	2293	11943
2003-04	11317	2371	13688

Category wise, Company wise Naphtha Upliftments

Category	*April-June*		*Variation*
	2006-07	2005-06	
Fertiliser	302	327	–25
Petrochemicals	1594	1673	–80
Processors	2	4	–3
Power/Steel	301	376	–75
Other Misc.	1	D	1
Grand Total	2499	2681	–182
IOC (M)	517	628	–11
BPC	264	283	–19
HPC	287	246	41
IBP	D	0	0
IOC (AOD)	0	0	0
CPCL	44	56	–13
MRPL	18	5	13

AROMATICS

Almost all aromatics come from crude oil, although small quantities are made from coal Aromatics are the major ingredients of all polymers and fibers, rubbers etc. Major producers are refineries Catalytic reformers. Thus Reliance and IPCL form the major sources of aromatics. Reliance Aromatics Division is one of the world's largest manufacturers of para anisic aldehyde, para cresol and para anisic alcohol, supplying.

Its products to diverse industries, including cosmetics, flavours and fragrances, bulk drugs, dye intermediates; and plant and animal micro-nutrients. The Division's manufacturing site is located at Ankleshwar in Gujarat, about 350 km north of Mumbai. It also produces many Bulk Chemicals and Intermediates Division ' [' The volume of Benzene imported into India in 2004 was 5 KT and it increased by 273% to 19 KT in 2005. Imports of Ortho-Xylene declined significantly from 61 KT in 2004 to 39 KT in 2005 and imports of Toluene went up from 27 KT in 2004 to 70 KT in 2005. 2005 saw a negative growth in demand for all aromatics except Toluene while 2006 is expected to witness a positive growth. The year 2004 witnessed a 2% contraction in demand of benzene while the year 2005 witnessed a reduction in demand for all aromatics except Toluene, which registered a growth of

10%. The production of Benzene also decreased from 654 KT in 2004 to 623 KT in 2005 but Benzene exports rose by 46% from 128 KT in 2004 to 188 KT in 2005. Benzene exports constituted 93% of the total exports of other petrochemicals which stood at 201 KT.

Production of Elastomers Product wise Installed Capacity of Elastomers (in TMT)

Product	*2001-02*	*2002-03*	*2003-04*	*2004-05*	*2005-06*
Styrene Butadiene Rubber (SBR)	62	62	62	62	62
Poly Butadiene Rubber (PBR)	50	50	50	50	50
Nitrile Butadiene Rubber (NBR),	9	9	12	12	13
Ethyl Propylene Dimer (EPDM)	10	10	10	10	10
Ethyl Vinyl Acetate (EVA)	13	13	13	13	13
Total	144	144	147	147	147

Recent Developments in Indian Petroleum And Petrochemical Industry

Over View of the Industry : India is the sixth largest manufacturing country and may be fifth strong economy in the world India's economy is growing consistently (7.2 percent in 2017–18) Perhaps nominal GDP may project India to be 2nd economy by 2050 if the progress in oil field continues. The petrochemicals industry is growing evenly at a compounded annual growth rate about 14 percent, However the scene is not so much when the per capita plastic consumption visioned at 7.0 kgs as compared to 46 kgs in China and 65 Kgs in Europe. The break through in production of more plastics to increase percapita consumption may not be essential because of 1.3 billion population, considerable amount of population without any regard to environments and interest in conserving natural resources can lead to top most position in plastic pollution. Te petrochemical industry is impacted by the price of the crude, the price fluctuations n crude since first oil shock, made the industry to adjust to the price fluctuations, like revamping, modernization, expansion of capacity to additional chemicals at the same plant, to cope up with the economy. Similarly the GHGs emission, and chemical pollutions ,plastic disposales etc are some Global issues, which when strict laws are enforced, industry might suffer more than the price hike in crudes. Technological modifications with reference to adoption of new technologies, better catalysts , manipulating co-products etc are vital to the progress of the industry. Though this is a demand based industry, the price hiking cannot result continuous demand as this is not an essential product . Recent introduction of fracking, horizonal drilling of plays to produce shale gas and oil have also have considerable influence in choosing raw materias.

Indian refining capacity in the last twenty years ie. Since 1998 at a capacity of just 65MMTPa shot up to over 230MMTPA..With its 23 refineries likely to add one or two in near future ,18 under public sector,3 under purely private sector remaining as joint ventures, thus the resent refining capacity with over 247 MMTPA , the third largest refiner in the world and is expected to reach over 400MMTPA by 2030, in which Reliance contribution will be 40% expansion.. The World's largest refinery with 32 MMT in Reliance sector with 70MMTPA at single location, the complex has number of facilities both in refining and petrochemical sections , with a gas cracker facility to produce 8MMTPA polymers. It has provision for more capacity additions in refining sector as well chemical section in coming times,. IOC will be expanding from 80.7 to 150MMTPA during this period to match the requirements of 350MMTPA. Plans to raise capacity of its Panipat refinery in Haryana to 25 MTPA from current 15 MMTPA while Koyali refinery in Gujarat would be expanded to 18 MMTPA from 13.7 MMTPA. 15

MMTPA Paradip refinery in Odisha will see a capacity addition of 5 MMT while 3 MMT will be added in IOC's Barauni refinery , with an addition of 1.2-MMTPA Mathura refinery capacity will touch 9.2 MTPA. 15 MMTPA. BPCL at its Kochi complex achieves 36.5 MMTPA due to addition of 5 MT to its Kochi unit and Bina refinery with addition of 9 MT to 15 MTPA. Numaligarh capacity will go up to 9 MTPA from 3 at present . Hindustan Petroleum Corp Ltd (HPCL) also plans to expand its Mumbai refinery to 9.5 MTPA from 7.5 MTPA and that of ong pending Vizag unit to 15 MTPA from 8.3 now, and a proposal for a new refinery of 9 MMTPA at Barmer in Rajasthan is on board. 15 MMTPA Paradip refinery in Odisha will see a capacity addition of 5 MMT while 3 MMT will be added in IOC's Barauni refinery , with an addition of 1.2-MMTPA Mathura refinery capacity will touch 9.2 MTPA. IOC is ma add a 9 MTPA refinery at its subsidiary Chennai Petroleum (CPCL). State refiners will also share the capacity additions to meet rising demand. In that spree Bharat Petroleum Corp (BPCL) would reach al capacity of 56 MTPA, from 36.5 MMTPA currently, by adding 5 MT to its Kochi unit and Bina refinery with addition of 9 MT to 15 MTPA. Numaligarh capacity will go up to 9 MTPA from 3 at present . Hindustan Petroleum Corp Ltd (HPCL) also plans to expand its Mumbai refinery to 9.5 MTPA from 7.5 MTPA and that of long pending Vizag unit to 15 MTPA from 8.3 now, and a proposal for a new refinery of 9 MMTPA at Barmer in Rajasthan is on board. IOC, BPCL and HPCL are together putting up a new 60 MPTA refinery on the west coast in Maharastra The refineries and their capacities are shown in Table.1

Polymers and Plastics

Indian polymer industry is dominated by polyolefins., about 80% of all synthetics and the industry is growing at a rate of over 12% . Primary olefin in this industry is Ethylene a source for for intermediates and polymers, the capacity of which in the World (2018) has reached 175MMTPA (USA about 40MMTPA) slightly over demand (demand 157MMTPA, USA 40MMTPA). In India ethylene capacity has crossed 3.1 MMT in 2011 since its inception and is expected to reach 7.2 MMTPA by 2018, in which Reliance share will be 3.2 MMTPA. Ethylene main feed stocks are naphtha about 63%, propane and ethane form 33% remaining is based on other feed stocks. Next important olefin is Propylene, its World capacity reached more than 133MMTPA during 2018 and is expe,cted to touch 160 MMTPA by 2025 while the demand is far less than production.(production in 2016 is 120 while demand was 103MMT) Indian capacity is 4.6MMT in 2015-16 and is expected to be 5.0 MMT in 2018. The petrochemical scene in India is rapidly changing with average 5% yearly growth rate in polymer. Maximum growth rate and demands ae noticed for ethylene ,propylene polymers, ie. PE ,PP. VCM and EDC are consumed by PVC producers, as these are not produced to the required demands, about 600KTPA of EDC and 500 kTPA of VCM are imported and would continue to increase as the PVC market is growing .PVC demand is about 3300KTPA while the indigenous production is almost half. Next important resin is PET, Reliance is the major producer followed by Dhunseri Petrochemical Ltdand JBF industries.. Reliance industry manufactures bottle grade PET has a capacity of 640KTPA.., presently the capacity of PET is about 1915KTPA while the demand is about 1000KTPA.

Table 2.1: All polymers demand during last few years ; (000't)

Projected demand for all polymers in KTPA				
Capacity	9049	9084	9558	12017
Production	8161	7907	9181	10847
Import	2673	3330	3611	3555
Exports	1061	797	887	1536
Demand	9356	10004	11378	12621

Olefin (monomers) demand is shown

Table 2.2: Ethylene capacity additions (KTPA)

Location	2014	2015	2016	2017	2018
BCPL, Assam	55	220	220	220	
GAIL, Aurya, UP	500	500+100	500+450	500+450	500+450
Haldia, W.Bengal	670	670	670	670	670
India Glycols, Kashipur UP.	93	93	93	93	93
IOC, Panipat	857	857	857	857	857
Opal, Dahej		275	1000	1000	1000
Reliance, Dahej.	420	420	420	420	420
Hazira	860	860	860	860	860
Jamnagar	675	1350	1350		
Vadodra	500	500	500	500	500
Nagothane (Maharastra)	180	180	180	180	180
Total	4080	4650	6525	7200	7200

Table 2.3: Propylene availability in India (000tons)

Year	2014 – 15	2015 – 16	2016 - 17	2017 – 18
Capacity	4230	4600	4773	4842
Production	4020	4590	4770	4840
Imports	0	0	0	0
Exports	10	10	15	0
Net availability	4020	4590	4770	4840

All polymers are showing robust growth rate of 6-7.3% annually.

Table 2.4: Polymers of olefns

Year	2014-15	2015-16	2016-17	2017-18
LLDPE				
Capacity	960	960	1640	2190
Production	744	942	1156	2050
Imports	600	613	634	466
LDPE				
Capacity	206	206	206	550
Production	144	196	200	550
Imprts	328	465	500	445
HDPE				
Capacity	1830	1830	2810	2810
PP				
Capacity	4130	4520	5120	5120
Production	4176	4335	5007	5460
Imports	511	659	517	580
Demand	4192	4389	4396	4765

OPAL (ONGC POLY OLEFINS PLANT) – The import of all polymers during 2015-16 was about 32% of the consumption or about 3330KT while the same dependency may come down to negative or exports increasing because of new ventures . OPAL unit of Gujarat. Produces Ethylene 1.1MMTPA, Propylene 0.4MMTPA, PP 0.34 MMTPA, HDPE/ LLDPE 0.36MMTPA, HDPE (34MTPA,), benzene (1.35MTPA), butadiene (95MTPA), pyrolysis gasoline (165MTPA), CBFS (70,MTPA). On commissioning of new installed capacity , ethylene increased from 446 KTA to 900 KTA and Propylene from 14.8 KTA to 49 KTA after a new Gas Cracker of 450 KTA capacity has been commissioned in January 2016 , capacity of Polyethylene was enhanced from 410 KTA to 810 KTA after a new LLDPE/HDPE Swing Unit of 400 KTA capacity has been commissioned in January 2016 at GAIL, Pata, District-Auraiya , GAIL is als o producing 20,000TPA i-butene by dimerising ethylene. I-butene is in good demand for the production LLDPE.

Mangalore refinery & Petrochemical Complex . a subsidiary of ONGC with its full operation of PP plant the production is envisaged to touch 440KTPA by 2018

Bharat Petroleum Corporation Limited – Kochi Refinery after expanding the refining capacity (15.5MMTPA) .A Fluid Catalytic Cracker Unit (FCCU) is being set up by the BPCL that will produce 500 KTA of Propylene. Propylene production is expected to commence by November, 2016.

IOCL Refinery at Paradip, Odhisa has been commissioned on 8th February 2016 with capacity at 15 MMTPA. IOCL is building a Fluid Catalytic Cracker (FCC Ind Max Technology), which is integrated with Paradip Refinery. This Propylene unit will produce 680 KTPA(700KTPA of PP to be commissioned by June 2018) and ethylene based MEG and PVC plants will be added-be 200KTA . Propylene plant is expected to be completed and made operational by the end of 2017. Refinery will produce both ethylene & propylene streams, which will be further processed in the upcoming petrochemical units to produce Mono Ethylene Glycol & Polypropylene. The refinery also features a 24,000bpd paraxylene production unit, a 10,000bpd alkalination unit, a naphtha hydrotreater, and other secondary processing facilities like hydrogen plant.

Assam Gas Cracker Project (AGCP), approved in 2006 and implemented by Brahmaputra Cracker and Polymer Limited (BCPL), was commissioned on 2nd January 2016. BCPL has built a dual feed Cracker Complex to produce 220 KTA of ethylene, 60 KTA of propylene, 226 KTA of LLDPE-HDPE and 60 KTA of PP.

ONGC Petro Additions Ltd. (OPAL), a JV of ONGC, GAIL, and Gujarat State Petroleum Corporation : is building a grass roots petrochemical complex at Dahej The complex's' ethylene plant will be a dual-feed cracker with capacity of 1.1 million /year of ethylene and 400,000 m.t./year of propylene . In the downstream units two 360,000- m.t./year Swing PE units capable of producing 340,000m.t./ year HOPE plant and a 340,000-m.t/year PP unit will e put into operation.

Butadiene : the demand for butadiene is gradual with the commissioning OPAL 115KT and IOC L added 40KT during 2016-17 with these additions butadiene is likely to go up to 550 KTPA while the production in 2018 including 150KT exports.

Reliance petroleum is the major petroleum complex spreading its wings on refining and petrochemicals. It has manufacturing divisions at Patalganga,Vadodra, Gandhar, Dahej, Hazira, and Jamnagar. It has capacity chamicals ,like aromatics (BTX), PTA,LAB units.,besides polymeers and fibers unts.

Aromatcs 3.6 MMTPA ,Para Xylene the mostt important product is having a caacity of 2.MMTPA and PTA 2nd largest producer in the world 4.2MMTPA,o-xyene production is about 562KTPA,benzene s about 1.2MMTPA, LAB 180KTPA (Patalaganga plant120 VadOdra plant 60KTPA). Also TDI is produced upto100JTPA and cyclohexane 50KTPA.

IOC will expand its refining capacity to 116.55 million tonnes per annum (MTPA) by 2030 from the current 80.7 MTPA .India's current refining capacity of 247. MTPA exceeds consumption but with demand growing at 2-3% annually ,refining capacity is likely to reach over 350MMTPA by 2030. .. IOC plans to raise capacity of its Panipat refinery in Haryana to 25 MTPA from current 15 MMTPA while Koyali refinery in Gujarat would be expanded to18 MTPA from 13.7 MMTPA.

Global demand of propylene is 102MMTPA (2016) while the capacity is over 125MMT. Indian capacity is likely to touch 4.8 MMT by 2017 -18.

Commissioning of new installed Capacity

Capacity of Ethylene increased from 446 KTA to 900 KTA and Propylene from 14.8 KTA to 49 KTA after a new Gas Cracker of 450 KTA capacity has been commissioned in January 2016 , capacity of Polyethylene was enhanced from 410 KTA to 810 KTA after a new LLDPE/HDPE Swing Unit of 400 KTA capacity has been commissioned in January 2016 at GAIL, Pata, District-Auraiya , .

Bharat Petroleum Corporation Limited – Kochi Refinery after expanding the refining capacity (15.5MMTPA) .A Fluid Catalytic Cracker Unit (FCCU) is being set up by the BPCL that will produce 500 KTA of Propylene. Propylene production is expected to commence by November, 2016.

IOCL Refinery at Paradip, Odhisa has been commissioned on 8th February 2016 with capacity at 15 MMTPA. IOCL is building a Fluid Catalytic Cracker (FCC Ind Max Technology), which is integrated with Paradip Refinery. This Propylene unit will produce 680 KTA and etHindustan Petroleum Corp Ltd (HPCL) also plans to expand its Mumbai refinery to 9.5 MTPA from 7.5 MTPA and that of ong pending Vizag unit to 15 MTPA from 8.3 now, and a proposal for a new refinery of 9 MMTPA at BPCL Kochi will commence production of Acrylic acid, (47 KTPA) ,92KTPA oxo-alcohols,,190KTPA of acrylates using 250KTPA propylene.

Brahma Putra Cracker and Polymer Ltd (BCPL) At Assam (70% OWNED BY GAIL) has a capacity to produce ethylene 220KTPA, 80KTPA propylene, to produce pp of 60KTPA and 226KTPA of LLDPE will be joining the stream of petrochemicals by 2017.

Table 2.5: Refineries in India Capacity by 2018

S.No.	Refinery	Authority	Location	Capacity (mmtpa)
1	Barauni Refinery	Indian Oil Corporation	Barauni	6.00
2	Gujarat Refinery	—do—	Koyali	13.70
3	Haldia Refinery	—do—	Haldia	7.50
4	Mathura Refinery	—do—	Mathura	8.00
5	Panipat Refinery	—do—	Panipat	15.00
6	Digboi Refinery	—do—	Digboi	0.65
7	Bongaigaon Refinery	—do—	Bongaigaon	2.35
8	Guwahati Refinery	—do—	Guwahati	1.00
9	Paradip Refinery	—do—	Paradip	15.00
10	Mumbai Refinery	Hindusthan petroleum	Mumbai	7.50
11	Visakhapatnam Refinery	—do—	Visakhapatnam	15.30
12	Guru Gobind Singh Refinery	Hindusthan Petroleum and Hindustan Mittal Energy Limited (HMEL)	Bathinda	11.3
13	Mumbai Refinery Mahaul	Bharat Petroleum Corporation Limited	Mumbai	12.00
14	Kochi Refineries	—do—	Kochi	15.5
15	Bina Refinery	Bharat Oman Refineries Limited	Bina	6.00
16	Manali Refinery	Chennai Petroleum Corporation	Chennai	10.50
17	Nagapattnam Refinery	—do—	Nagapattinam	1.00
18	Numaligarh Refinery	Bharat Petroleum, Oil India and Government of Assam	Numaligarh	3.00
19	Tatipaka Refinery A.P	Oil and Natural Gas corp.	Tatipaka	0.07
20	Mangalore Refinery and Petrochemicals Limited	Mangalore Refinery and Petrochemicals	Mangalore	15.00
21	Jamnagar Refinery	Reliance Industries Limited	Jamnagar (DTA)	27.00
22	Jamnagar Refinery	—do—	Jamnagar (SEZ)	33.00
23	Essar Refinery	Essar Oil Limited	Vadinar	20.00
	TOTAL			345.37

Ethylene, the most important olefin capacity in the World has reached 169MMTPA, slightly over demand. In India ethylene capacity reached 3.1 MMT in 2011 and is expected to reach 7.2 MMTPA by 2018 ,in which Reliance share will be 3.2 MMTPA. Ethylene main feed stocks are naphtha about 63%, propane and ethane form 33% remainng is based on other feed stocks. Propylene capacity reached 4.6MMT in 2015-16 and is expected to be 5.0 MMT by 2018

Table. 2.6: Ethylene capacity additions(KTPA)

Location	2014	2015	2016	2017	2018
BCPL, Assam		55	220	220	220
GAIL, Aurya, UP	500	500+100	500+450	500+450	500+450
Halia, W.Bengal	670	670	670	670	670
India Glycols, Kashipur UP.	93	93	93	93	93
IOC, Panipat	857	857	857	857	857
Opal , Dahej		275	1000	1000	1000
Reliance, Dahej.	420	420	420	420	420
Hazira	860	860	860	860	860
Jamnagar			675	1350	1350
Vadodra	500	500	500	500	500
Nagoth (maharashtra)	180	180	180	180	180
Total	4080	4650	6525	7200	7200

Tabe 2.7: Propylene availability in 000'tons

Year	2014 – 15	2015 - 16	2016 - 17	2017 - 18
Capacity	4230	4600	4773	4842
production	4020	4590	4770	4840
Imports	0	0	0	0
Exports	10	10	15	0
Net availaility	4020	4590	4770	4840

The World's largest grass root refinery with 32 MMTPA single unit capacity , total 70MMTPA is located in Jamnagar at one location in India. It has a number of gas cracker complexes with downstream facility producing 8 MTPA plastics and polymers

OPAL (ONGC) Projected demand for all polymers in KTPA				
Capacity	9049	9084	9558	12017
Production	8161	7907	9181	10847
Import	2673	3330	3611	3555
Exports	1061	797	887	1536
Demand	9356	10004	11378	12621

Synthetic Fibers

The main products of the India synthetic fiber industry include man-made fiber and yarn. These man-made fiber and yarn mainly constitutes textile of cellulose and non-cellulose origin. The cellulose fiber or yarn industry is under the administrative control of the Ministry of Textiles while non-cellulose industry productionis the global pioneer in polyester yarn and fiber production. In petrochemicals. , Reliance is the 4th largest producer of paraxylene, 5th largest producer of purified terepthalic acid, and 7th largest producer of polypropylene. Reliance's manufacturing operations are located at Jamnagar, Patalganga, Naroda, and Hazira. :

Reliance Industries are classified into Petrochemicals, Refining sector and Marketing wing and crude & gas Exploration activity .This RIL industry accounts for about 2.09 billion kg i.e. 44%, of man-made fiber or yarn industry. Industrial output of India , during 2004-05 amounted to 774.65 million kg, products include polyester staple fiber, acrylic staple fiber and polypropylene staple fiber. The import of India synthetic fiber industry showed a mixed trend over the last five years and its imports doubled to 143.54 million kg during 2004-05. Further, the export of India synthetic fiber industry had grown by 50% during the same period. The synthetic fiber industry is under the control of Ministry of Chemicals and Fertilizers (Department of Chemicals and Petro Chemicals). The Indian synthetic fiber industry plays a vital role in the Indian Textile industry. . PTA AND MEG are the primary ntermediates for synthetic fibers occupying 40% and 60% spectively of the total 1730KTPA imported. However with Reliance full capacity running PTA imports would decrease Polyester is the mst important synthetic fibers. . The total global polyester market, including all fibres, PET resin, film and other resins, is close to reaching 65 million tonnes in 2014, growing at between 5.5 per cent to 6 per cent as polyester polymer. Total fibres production, including recyclates, was around 46 million tonnes . PET resin industry is close to 20 million tonnes and film around 3.5 million tonnes..

The following table shows the production and capacity of all the synthetic fibers and the production is likely to exceed 7 lakh tons by the end of 2018.

Table 2.8: Demand Supply Balance of Synthetic Fibre in 000TONS

YEAR	2014-15	2015-16	2016-17	2017-18 E
PSF (polyester staple fiber)				
Capacity	1260	1260	1260	1404
Production	953	974	979	1123
Imports	46	59	59	30
ASF (Acrylic staple fiber)				
Capacity	98	98	98	98
Production	95	95	95	95
Imports	34	34	34	34
PPSF (poly phenylene sulfide)				
Capacity	13	13	13	13
Production	4	4	4	4
PFY (polyester filament yarn)				
Capacity	4623	4731	4666	4746
Production	2766	2747	2905	3062
VSF(viscose staple fiber)				
Capacity	490	490	490	490
Production	364	354	364	364

Nylons are also known as polyamides discovered by Wallace n 1935 at Du Pont and occupy 20% of the manufactured fibers of the World .approximately 8billion pounds. Nylon 6uses caprolactam as the raw material

To meet the requirements of synthetic fibers ,fiber intermediates like ACN , Caprolactam,, MEG are important and rtouced about 6000KTPA of which glycols and PTA constituted about 90% GSFC is the major producer of Caprolactam with a capacity of 70.000KTPA , only 20000 KTPA is consumed internally remaining exported. ACN and caprolactam occupy the remaining share. However the share can is decreasing since 2015 and almost all the requirements, about 180KTPA are met through imports, while the scene is different with respect to PTA production exceeds the demand which is around 5.8 lakh tons while production is over 6.8 lakh tons . Caprolactam on the other hand meets nly half the demands of 180KTPA

ABS and SAN are part of engineering plastics. ABS demand is around 220 KTPA while the capacity amounts to 190 ktpa

15-MMTPA Paradip refinery in Odisha will see a capacity addition of 5 MMT while 3 MMT will be added in IOC's Barauni refinery , with an addition of 1.2-MMTPA Mathura refinery capacity will touch 9.2 MTPA. IOC is mayt add a 9 MTPA refinery at its subsidiary Chennai Petroleum (CPCL). State refiners will also share the capacity additions to meet rising demand. In that spree Bharat Petroleum Corp (BPCL) would reach al capacity of 56 MTPA, from 36.5 MMTPA currently, by adding 5 MT to its Kochi unit and Bina refinery with addition of 9 MT to 15 MTPA. Numaligarh capacity will go up to 9 MTPA from 3 at present . HiSAN Demand Supply (KT)

	2010-11	2013-2014	2014-15	2015-16	2016-17
Capacity	80	130	130	150	150
Production	73	82	84	113	154
Demand	114	138			

Synthetic Rubbers; Global demand of propylene is 102MMTPA (2016) while the capacity is over 125MMT. Indian capacity is likely to touch 4.8 MMT by 2017 -18.

Polyurethane is thermosetting plastic having variety of applications. Right from flexible foam to rigid rubbers or varnish to coatings and condoms. Total elastomers demand is shown to be around 625KTPA, Principally SBR,and PBR Play

A vital role role in elastomer industry. While most of the EPDM is imported. Reliance is the sole producer of PBR in india ,At Hazira plant it produces PBR 40,000t/y At Vadodra it has two plants producing 40,000t/y each.. Various grades with CO.; NI based resins are useful flexible cables,rollers, tyres, conveyor belts automotive components etc. SBR plant capacity in Hazira location is producing about 140KTPA. Extended SBR product is useful for variety applications, like, belts, shoes, ,conveyor belts,,gaskets etc.GAIL is also planning to set up 110,000tpa capacity plant in Gujarat in the vicinity of OPAL where butadiene nwill be available.GAIL and RIL together may produce about 230,000t/y of PBR BY 2018 . Styrene Butadiene Rubber(SBR) accounts for about 41% of the synthetic rubbers., whose growth is dependent mostly on automobile industry . SBR plant capacity in Hazira location is producing about 140KTPA. Extended SBR product is useful for variety applications, like, belts, shoes, ,conveyor belts,,gaskets etc.GAIL is also planning to set up 110,000tpa capacity plant in Gujarat in the vicinity of OPAL where butadiene nwill be available.

Table 2.9: Production of different Rubbers (000tons)

Year	2014-15	2015-16	2016-17	2017-18
PBR Capacity	114	124	124	124

Commissioning of new installed Capacity

Capacity of Ethylene increased from 446 KTA to 900 KTA and Propylene from 14.8 KTA to 49 KTA after a new Gas Cracker of 450 KTA capacity has been commissioned in January 2016 , capacity of Polyethylene was enhanced from 410 KTA to 810 KTA after a new LLDPE/HDPE Swing Unit of 400 KTA capacity has been commissioned in January 2016 at GAIL, Pata, District-Auraiya , .

Bharat Petroleum Corporation Limited – Kochi Refinery after expanding the refining capacity (15.5MMTPA) .A Fluid Catalytic Cracker Unit (FCCU) is being set up by the BPCL that will produce 500 KTA of Propylene. Propylene production is expected to commence by November, 2016.

Year	2014-15	2015-16	2016-17	2017-18
Production Demand	104 170	112	112 180	120 193
SBR Capacity Production Demand	 140 70 290	 290 143 300	 290 213 313	 290 230 330
NBR Capacity Production Imports	 20 17 27	 20 17	 20 18	 20 18 38
EPDM Capacity Demand	 10 30	 10 30	 10 30	 10 49

Table 2.10: Butadiene production : (000tons)

Year	2014-15	15-16	16-17	17-18
Capacity	435	435	650	650
Production	229	318	416	550
Imports	46	79	96	164

Table 2.11: Styrene Acrylonitrile Demand - Supply (KT)

Year	2010-11	2012-13	2013-14	204-15	2015-16	2016-17 E
Capacity	80	90	130	130	150	150
Production	73	78	82	84	113	154
Polystyrene Capacity	472				472	472
Production	270				308	378

Entire demand of 600 – 700 KTPA of Styrene requirements is met from imports.

Table 2.12: Next important polymer is PVC, with 10% demand growth is shown below:

Year	2014 -15	14 – 15	15 - 16	17 - 18
Capacity	1390	1470	1480	1482
Production	1250	1362	1388	1423
Imports	1172	1330	1600	1565

Intermediates like EDC and VCM are in great demand , almost all the production is consumed and looking for imports EDC production is stagnant at 205 KTPA since last four years requiring imports of 500KTP during 2017-18 . Similarly VCM capacity is marginally improved from 900 to 960KTPA while the demand increasedbfrom1200 to 1500KTPA during last four years.(2014 to18).

Aromatics

Benzene capacity was 1200KTPA in 2014 while the demand was only 475 KTPA this over capacity is continuing while the demand is marginally grown up. the production is exrt oriented. negative growth. However the growth was positive by 2017 ending in production of 1870 KTPA while exports registered a growth of over 1300KTPA., however the toluene demand maintained steady rate of over 500 KTPA while production was 140KTPA against installed capacity of 270 thus imports demand increased. Similarly the demand for O-xylene is steadily maintained around 280 to 290KTPA against a production of 450KTPA . Thus India's overall demand for petrochemicals shot over 40MMTPA in 2017-18 and is likely to maintain steady growth rate of 6.3% yearly in coming years.

The demand for engineering plastics is growing against a requirement of 220KTPA ,ABS resin production is only 190KTPA. Similarly SAN production is about 120KTPA against a demand of 135KTPA.

PET is another resin which exceeds demand requirement of 3900KTPA, while the production is 5000KTPA/(Capaciy 5600)

Polystyrene demand is almost constant around 300KTPA. However the capacity is around 500KTPA. LG Polymers formerly, Hindustan Polymers is the main producer of styrene India.

INDEX

D

E

F

G

N

O

P

Q

R

BULK CHEMICALS FROM PETROLEUM

DR. BHASKAR RAO

FEEDBACK FORM

1. Which topics from your examination point of view are inadequately or not covered in this book?

____

____

2. Is there any misprints/ mistakes/ factual inaccuracy in this book? If so, please specify.

____

____

3. What is your assessment of this book as regards the presentation of the subject-matter, expression, precision and price in relation to the other recommended books available on this subject?

____

____

4. Which competing books you regard as better than this? Please mention their author and publishers.

____

____

5. Any further suggestions/comments you would like to make for the improvement of the book?

____

____

PS: If needed, please attach a separate sheet.

Name : ____________________

College : ____________________

Address : ____________________

Email : ____________________

Mobile : ____________________

Please mail the feedback form at:

KHANNA BOOK PUBLISHING CO. (P) LTD.

4C/4344, Ansari Road, Daryaganj, New Delhi-110002

You can also email this feedback form at contact@khannabooks.com or Whatsapp at +91 – 99109 09320

BEST FEEDBACK WILL BE REWARDED SUITABLY